Springer Collected Works in Mathematics

For further volumes:
http://www.springer.com/series/11104

Photo by Mrs. Margo Weinstein, 1978.

Shiing-Shen Chern

Selected Papers I

Reprint of the 1978 Edition

Springer

Shiing-Shen Chern
(1911 Jiaxing, China –
2004 Tianjin, China)
The University of Chicago
Chicago, IL
USA

ISSN 2194-9875
ISBN 978-1-4614-4333-9 (Softcover)
978-0-387-90339-2 (Hardcover)
DOI 10.1007/978-1-4614-4334-6
Springer New York Heidelberg Dordrecht London

Library of Congress Control Number: 2012954381

Printed on acid-free paper

Springer is part of Springer Science+Business Media (www.springer.com)

Table of Contents

Foreword vii
S. S. Chern as Geometer and Friend, by André Weil. ix
Some Reflections on the Mathematical Contributions of S. S. Chern, by Phillip A. Griffiths. xiii
A Summary of My Scientific Life and Works, by Shiing-shen Chern. xxi
[4]* Abzählungen für Gewebe. 1
[5] Eine Invariantentheorie der Dreigewebe aus n-dimensionalen Mannigfaltigkeiten in $2n$-dimensionalen Räumen. 9
[7] Sur la possibilité de plonger un espace à connexion projective donné dans un espace projectif. 35
[13] The geometry of a differential equation $y''' = F(x, y, y', y'')$. 45
[18] On integral geometry in Klein spaces. 60
[19] On the invariants of contact of curves in a projective space of n dimensions and their geometrical interpretation. 72
[23] A generalization of the projective geometry of linear spaces. 76
[25] A simple intrinsic proof of the Gauss–Bonnet formula for closed Riemannian manifolds. 83
[30] On the curvatura integra in a Riemannian manifold. 90
[33] Characteristic classes of Hermitian manifolds. 101
[35] Sur une classe remarquable de variétés dans l'espace projectif à n dimensions. 138
[39] On the multiplication in the characteristic ring of a sphere bundle. . . 148
[47] A theorem on orientable surfaces in four-dimensional space (with E. H. Spanier). 159
[51] On the characteristic classes of complex sphere bundles and algebraic varieties. 165
[54] Pseudo-groupes continus infinis. 199
[55] On isothermic coordinates (with P. Hartman and A. Wintner). 217
[61] On a generalization of Kähler geometry. 227
[62] On the total curvature of immersed manifolds (with R. Lashof). . . . 246
[63] On the index of a fibered manifold (with F. Hirzebruch and J-P. Serre). 259
[68] Integral formulas for hypersurfaces in Euclidean space and their applications to uniqueness theorems. 269

*Numbers in brackets refer to the Bibliography on pp. 469–476.

[69] A uniqueness theorem on closed convex hypersurfaces in Euclidean space (with J. Hano and C. C. Hsiung) . . . 279
[70] Complex analytic mappings of Riemann surfaces I . . . 283
[71] The integrated form of the first main theorem for complex analytic mappings in several complex variables . . . 298
[75] Holomorphic mappings of complex manifolds . . . 315
[78] Pseudo-riemannian geometry and the Gauss–Bonnet formula . . . 325
[79] Minimal surfaces in an Euclidean space of N dimensions . . . 335
[88] On holomorphic mappings of Hermitian manifolds of the same dimension . . . 347
[89] Simple proofs of two theorems on minimal surfaces . . . 361
[90] Intrinsic norms on a complex manifold (with H. Levine, L. Nirenberg) . . . 371
[91] Minimal submanifolds on the sphere with second fundamental form of constant length (with M. do Carmo, S. Kobayashi) . . . 393
[92] Some formulas related to complex transgression (with R. Bott) . . . 410
[94] On minimal spheres in the four sphere . . . 421
[101] Meromorphic vector fields and characteristic numbers . . . 435
[103] Characteristic forms and geometrical invariants (with James Simons) . 444

Bibliography of the publications of S. S. Chern . . . 467

List of Ph.D. theses written under the supervision of S. S. Chern . . . 475

Foreword

In recognition of Professor Shiing-shen Chern's long and distinguished service to mathematics and to the University of California, the geometers at Berkeley have planned an International Symposium in Global Analysis and Global Geometry in his honor, to be held in June 1979 at Berkeley. The publication of the present volume is an outgrowth of the Symposium.

The papers in this Selecta volume comprise approximately a third of Professor Chern's total output to date. In making the selections, Professor Chern has given preference to shorter and less accessible papers. It should also be mentioned that the list of Ph.D. theses written under Professor Chern's direction, which is appended at the end of this volume, refers only to those in the United States. During the two years 1946–1948 when he was in charge of the Mathematics Institute of the Academia Sinica, he personally supervised the graduate training of a group of students which included, among others, S. C. Chang, K. T. Chen, Wu Wen-Tsün and C. T. Yang. In addition, H. C. Wang was an earlier student in Kunming.

The undersigned, in his capacity as coordinator of the publication of this Selecta volume, would like to express his appreciation to the following individuals: Professors André Weil and Phillip A. Griffiths for being so gracious and thoughtful in undertaking the writing of the introductory essays; to the staff of Springer-Verlag for the utmost courtesy and considerateness in seeing this project through from beginning to end; the editors of all the journals and books documented in this volume for their cooperation in resolving the technical problem of copyright; last but not least, Professor Chern himself, for his active and tireless participation.

Berkeley, June 1978 — H. Wu

S. S. Chern as Geometer and Friend

by André Weil

The friends and colleagues of S. S. Chern who have planned this volume in his honor have asked me for a contribution. Such an invitation is also an honor and could not easily be declined. At the same time, while I have no doubt that future historians of differential geometry will rank Chern as the worthy successor of Elie Cartan in that field, I do not feel competent to give an assessment of his work, nor called upon to do so, since the best part of it, or at least a very representative selection from it, is reproduced in this volume and speaks for itself. All I can do on this occasion is to evoke memories from a friendship of long standing—a friendship which has been among the most valuable ones, personally and scientifically, that I have been privileged to experience.

I must confess that even Chern's name was unknown to me in 1942 when I was asked to review his *Annals of Mathematics* paper on integral geometry ([18] of his bibliography). As I found out later, I had met him briefly in Paris during the year that he spent there in 1936–1937; I was then on the Faculty in Strasbourg and came to Paris regularly to attend the Julia seminar, organized by my friends and myself, which met there every other week. The topic of the seminar for that year was Elie Cartan's work; of course it was of special interest to Chern. However, I was invited to spend the second term (January to April 1937) at the Institute for Advanced Study in Princeton, and did not come back until the fall. Thus I did see him in the fall of 1936, but formed no acquaintance with him; to me he was just an anonymous young man from China, soon lost sight of and forgotten.

Five years later Chern was hardly a beginner any more, but somehow none of his published papers had attracted my attention; in part they had appeared in journals which were not even accessible to me at the time. Having left France early in 1941, I was that year at Haverford and had just written, in collaboration with Carl Allendoerfer, a paper on the Gauss–Bonnet formula. My work on the Haar measure and invariant measures in homogeneous spaces, and the interest I was taking in de Rham's work, had brought me close to "integral geometry," which had been a favorite subject for Blaschke and his students in the thirties. This made it natural for *Mathematical Reviews*, then in its infancy, to send me Chern's article [18] for review.

As I duly mentioned, the paper had some weak points. Nevertheless, it lifted the whole subject at one stroke to a higher plane than where Blaschke's school had left it, and I was impressed by the unusual talent and depth of understanding that shone through it. I tried to indicate this in my review, and also pointed it out to Hermann Weyl. As it happened, Veblen was well aware of Chern's work on projective differential geometry, and he and Weyl were considering an invitation to Chern to come to the Institute in Princeton, in spite of the enormous practical problems which this involved; since Pearl Harbor, war was everywhere; a trip from China to America was more than adventurous, it was risky; merely to obtain the necessary visas and priorities on airplanes required setting the whole diplomacy of the USA in motion. Needless to say, none of this fell to my share; I was a helpless refugee myself, officially classified as "enemy alien." All I could do was to express to Hermann Weyl my warm approval of the whole plan, and it is a matter of no little self-satisfaction to me to think that thus, in a small way, I may have contributed to Chern's coming to Princeton in 1943.

When he reached America, I was still not too far from Princeton, and he soon came to visit me. As we found out at once, we had many interests in common. Both of us had been deeply impressed by Elie Cartan's work and by the masterly presentation that Kähler had given of part of that work in his *Einführung in die Theorie der Systeme von Differentialgleichungen*; both of us had known Kähler in Hamburg. We were both interested in the Gauss–Bonnet formula. We were both beginning to realize the major role which fibre-bundles were playing, still mostly behind the scenes, in all kinds of geometrical problems. Better still, we seemed to share a common attitude towards such subjects, or rather towards mathematics in general; we were both striving to strike at the root of each question while freeing our minds from preconceived notions about what others might have regarded as the right or the wrong way of dealing with it.

Chern and I had been particularly intrigued by the little which was then known about characteristic classes (for which no name had been devised yet). Some mystery seemed to hide behind the fact that some Stiefel–Whitney classes were only defined modulo 2. I was able to tell Chern about the "canonical classes" in algebraic geometry, as introduced in the work of Todd and Eger. Their resemblance with the Stiefel–Whitney classes was apparent, while they were free from the defect (if it was one) of being defined modulo 2; their status, however, was somewhat uncertain, since that work had been done in the spirit of Italian geometry and still rested on some unproved assumptions. As to Pontrjagin classes, they had not yet been heard of at that time.

Such were the topics which came up during Chern's first visit and on subsequent occasions, which we sought to renew as often as we could. Very soon, as every geometer knows, they were completely transformed at the hands of Chern, first with his proof of the Gauss–Bonnet formula and then with his fundamental discovery of the role played by complex or quasicomplex structures in global differential geometry. Any comment would be superfluous; I will merely point out what can now be realized in retrospect about Chern's proof for the Gauss–Bonnet formula, as compared with the one Allendoerfer and I had given in 1942, following in the footsteps of H. Weyl and other writers. The latter proof, resting on the consideration of "tubes," did depend (although this was not apparent at the time) on the construction of a sphere-bundle, but of a non-intrinsic one, viz., the transversal bundle for a given immersion in Euclidean space; Chern's proof operated explicitly for the first time with an intrinsic bundle, the bundle of tangent vectors of length 1, thus clarifying the whole subject once and for all.

Chern and I had then to part for a while; I left for Brazil at the end of 1944, while he had to wait until 1946 before being able to go home to his family which he had been compelled to leave behind when first coming to America. There was not much communication between us during those years. My own ideas about fibre-bundles in algebraic geometry were maturing slowly, under the influence of Chern's work on complex manifolds. I knew that he was organizing an Institute of Mathematics in Nanking; I was also watching the political and military developments in China, with increasing anxiety for his fate. In 1947 I came to Chicago, where Marshall Stone had thoroughly reorganized the department of mathematics; he wrote to Chern, offering a visiting appointment. In the Fall of 1948, the civil war was coming closer and closer to Nanking; Veblen and Weyl, obviously feeling as I did about Chern, sent him an invitation to the Institute, backed up, as he later told me, by a friendly cable from Oppenheimer. Chern realized that he had to act quickly; he sent two cables, one to me and one to Princeton, telling us that he was coming to the United States.

I have criticized, sometimes severely, the American system of higher education; but I have often quoted the episode of Chern's second coming to America as an example of the flexibility which is perhaps its best feature. When Chern's cable reached me, Stone was travelling in South America. A single exchange of cables with him was enough; on his suggestion and mine, the professors in the department voted to ask for Chern's immediate appointment as a full professor. In the following months, there was some difficulty with the administration; obviously they thought that Chern, as a refugee, could be acquired more cheaply; I knew this attitude, which I had personally experienced during the war. It took Stone's coming back, a threat of resignation from him, and a personal appeal to Robert Hutchins to clinch the matter. Hutchins was in bed with the flu; also, he was technically on leave from the University. But the appointment went through, to take effect in the summer of 1949. In the meanwhile, the Institute in Princeton had provided the funds for him to come over with his family and stay in Princeton until his appointment would start in Chicago.

So it came about, in January 1949, that I could welcome Chern in Chicago at the railway station (not the airport; one still had that option) when he stopped there on his way to Princeton. On that day I met his wife and children for the first time, and remember the occasion vividly. Chern, in his fur cap, looked very much the Manchurian general. But to me the most unforgettable sight was his daughter May, a small girl not yet two years old, all wrapped up in white furs; nothing more lovely could have been imagined.

Thus we became colleagues in Chicago, and remained so for the better part of the decade that followed. We were also close neighbors, housed in the same Faculty building; the University had just built it across the Midway. These were fruitful years scientifically, for him and for me. Fiber-bundles, complex manifolds, homogeneous spaces were prominent among our interests; we discussed them in our offices in Eckhart Hall, or at home, or, better still, on long or short walks through the neighborhood parks, where it was still possible to take a walk and come out alive. Relations with colleagues and graduate students were cordial;

visitors, American or foreign, succeeded one another in a steady flow, for longer or shorter stays. With Ed Spanier's appointment a true topologist was added to the team. A quick look through Chern's list of publications in those years, or through mine, will bear witness to the stimulating influence which this scientific atmosphere had on our work.

The time came when circumstances persuaded both Chern and myself to seek elsewhere, among other things, a better climate and more pleasant physical surroundings. As we had sometimes jokingly predicted, he found them by moving closer to China, and I to France. This did not slacken the bonds of friendship, but it is only natural that from then on we followed each other's work less closely, even though we did arrange to get together at not too distant intervals. It is entirely to him and to the ties he had kept up with his colleagues in China that I owed my invitation there in the fall of 1976–an unusual experience which left on me a deep impression. But, rather than commenting upon such personal matters, or upon Chern's work of the last fifteen years (which others would be more competent to discuss, and whose value is recognized by all), it is perhaps appropriate to conclude with a few words about the place of geometry in mathematics—the mathematics of today and presumably also the mathematics of tomorrow.

Obviously everything in differential geometry can be translated into the language of analysis, just as everything in algebraic geometry can be expressed in the language of algebra. Sometimes mathematicians, following their personal inclination or perhaps misled by a false sense of rigor, have turned their mind wholly to the translation and lost sight of the original text. It cannot be denied that this has led occasionally to work of great value; nevertheless, further progress has invariably involved going back to geometric concepts. The same has happened in our times with topology. Whether one considers analytic geometry at the hands of Lagrange, tensor calculus at those of Ricci, or more modern examples, it is always clear that a purely formal treatment of geometric topics would invariably have killed the subject if it had not been rescued by true geometers, Monge in one instance, Levi-Civita and above all Elie Cartan in another.

The psychological aspects of true geometric intuition will perhaps never be cleared up. At one time it implied primarily the power of visualization in three-dimensional space. Now that higher-dimensional spaces have mostly driven out the more elementary problems, visualization can at best be partial or symbolic. Some degree of tactile imagination seems also to be involved. Whatever the truth of the matter, mathematics in our century would not have made such impressive progress without the geometric sense of Elie Cartan, Heinz Hopf, Chern, and a very few more. It seems safe to predict that such men will always be needed if mathematics is to go on as before.

Some Reflections on the Mathematical Contributions of S. S. Chern

by Phillip A. Griffiths

It is a special honor to have been asked to offer some comments and observations on Chern's mathematical work. Because of his own mathematical summaries and Weil's discussion of Chern's life and contributions provided from the perspective of a contemporary, perhaps this would be a good opportunity to explore how a few of Chern's papers and his view of mathematics appear to me, coming as I did a generation following the publication of his earliest work.

As a graduate student in Princeton in the early 1960s I was interested in algebraic and differential geometry and several complex variables, and so Chern's name was among those most frequently encountered. My initial direct exposure came when I studied the paper [33]* introducing Chern classes, one whose reading even now provides many of the essentials in an education in differential geometry and complex manifolds. Here is a brief description of how it appears to me looking back.

From classical geometry it is well known (and rediscovered in alternate generations) that the structure of the complex Grassmann manifold $G(n, N)$ of n-planes in $\mathbb{C}^N$ provides the key to enumerative problems, the simplest of which is determining the number of lines in P^3 meeting each of 4 skew lines. From a technical standpoint one needs to know the intersection relations among the various Schubert cycles, which are those n-planes failing by a specified amount to be in general position with respect to a flag $(0) = V_0 \subset V_1 \subset \cdots \subset V_{N-1} \subset V_N = \mathbb{C}^N$. In his thesis written under Elie Cartan, Ehresmann proved that the homology of $G(n, N)$ has as free basis the classes of Schubert cycles, thus providing a topological description of the basic enumerative relations such as Pieri's formula.

On the other hand the Grassmannian appears naturally in considering the Gauss mapping on an n-dimensional algebraic or complex manifold in N-space. Now the usual Gauss mapping for real manifolds in $\mathbb{R}^N$ and the intrinsic topological invariants to which it might lead seem to have been very much in the air in the early 1940s with the Allendoerfer–Weil discovery of the general Gauss–Bonnet theorem and Chern's own intrinsic proof [25] of it. A synthesis of

*Numbers in brackets refer to the bibliography.

enumerative and differential geometry occurs in [33] where it was found that the cohomology ring of the Grassmannian is generated by the Chern classes, which on the one hand are Poincaré dual to the basic Schubert cycles, and which on the other hand by using the newly emerging theory of fibre bundles led to intrinsic invariants of complex vector bundles. The differential geometry enters in that the cohomology ring of $G(n, N)$ is isomorphic to the algebra of invariant differential forms, and the differential forms representing the Chern classes turn out to be the elementary symmetric functions in the curvature matrix of the universal n-plane bundle.

One might say that the basic relations in geometry are discovered extrinsically but it is by finding combinations which have intrinsic meaning that fundamental new insights occur. An example of this is Max Noether's proof of his famous formula $\chi(\mathcal{O}_S) = \frac{1}{12}(c_1^2 + c_2)$ for the arithmetic genus of an algebraic surface, in which he took the plethora of Plücker formulas for S in P^N and found that combination which was birationally invariant. Another is Heinz Hopf's theorem that the Euler characteristic of a compact oriented hypersurface in Euclidean space is the degree of its Gauss map. The Chern–Weil theory of characteristic classes, classes in de Rham cohomology constructed from the curvature and leading to the Gauss–Bonnet theorem and its higher codimensional analogues, is again in this same spirit. The marvelous thing about the complex case, of course, is that here algebraic geometry and topology came together, and the formulas relating these two areas are most frequently expressed in terms of Chern classes.

We now turn to a discussion centered around one of Chern's papers in integral geometry. My own introduction to the subject came in the spring of 1965. I was a beginning instructor at Berkeley and sat in on a course which Chern taught leading up to the material in [84], which was just then being written. This was just at the beginning of turbulent times in Berkeley, and among Chern's courses which always gave an excellent intuitive and overall view of a topic this one was especially well done, covering the whole subject in depth and with clarity in a single semester. Looking back it is probably the class which most influenced my own view of mathematics.

Integral geometry is a pleasantly diffuse subject, one blessed with many elegant formulas and their application to elementary questions—see Chern's review (*Bull. A.M.S.*, vol. **83**(1977), pp. 1289–1290) of the recent treatise by Santaló. The simplest of these is Crofton's formula

$$\int n(L \cap C)\,dL = 2l(C) \tag{1}$$

expressing the length $l(C)$ of an arc in the plane as the average of the number $n(L \cap C)$ of intersections of a variable line L with the arc. By average we mean with respect to the unique normalized density on the space of lines which is invariant under the Euclidean group. The taking of averages with respect to invariant measures over homogeneous spaces of Lie groups is characteristic of formulas in integral geometry. One of the most famous of these is Blaschke's kinematic formula

$$\int \chi(D_1 \cap gD_2)\,dg = 8\pi^2\left(4\pi\chi_1 V_2 + 4\pi\chi_2 V_1 + M_1^{(1)}M_2^{(2)} + M_1^{(2)}M_2^{(1)}\right);$$

here, dg is the bi-invariant measure on the Euclidean group in $\mathbb{R}^3$, and D_1 and D_2 are domains with smooth boundaries of which V_i, $M_1^{(i)}$, $M_2^{(i)}$, χ_i are respectively the volume, area of the boundary, integral of the mean curvature of the boundary, and Euler characteristic.

Chern's general kinematic formula [84] is in a spirit similar to Blaschke's, and together with the foundational paper [18] serves to establish much of the general pattern of such results. To explain it we must first recall Hermann Weyl's formula for the volume of the tube T_ρ of radius ρ around a compact k-dimensional manifold M in $\mathbb{R}^n$. Setting $m = n - k$ it is

$$\operatorname{vol} T_\rho = \sum_{\substack{0 \leqslant e \leqslant k \\ e \equiv 0\,(2)}} c_e \mu_e(M) \rho^{m+e},$$

where c_e are positive constants and

$$\mu_e(M) = \int_M I_e(R)$$

with $I_e(R)$ being an orthogonally invariant polynomial of degree $e/2$ in the curvature tensor, and therefore a quantity intrinsic to the induced Riemann metric on M. Chern's kinematic formula, which was independently discovered by Federer, is

$$\int \mu_e(M^k \cap gM^l)\,dg = \sum_{\substack{0 \leqslant i \leqslant e \\ i \equiv 0(2)}} c_i \mu_i(M^k) \mu_{e-i}(M^l), \tag{2}$$

where the c_i are positive constants, ones incidentally whose determination led to several interesting questions.

Chern's proof of (2) exhibits a number of characteristic features. Of course, one is the use of moving frames, which in this case is particularly apt because on the one hand the invariant measures are given as constant expressions in the differential coefficients of a moving frame, while on the other hand they offer the usual advantage of calculating in a setting adapted to the geometric problem. Another is that the proof proceeds by direct computation rather than by establishing an elaborate conceptual framework; in fact, upon closer inspection there is such a conceptual framework as described in [18], however, the philosophical basis is not isolated but is left to the reader to understand by seeing how it operates in a non-trivial problem.

Now we come to the subject of value distribution theory, or holomorphic mappings between complex manifolds. This may be viewed as a generalization of the Picard–Borel theory of entire meromorphic functions, which was a major focal point of function theory in the generation following the publication of Picard's theorem in 1879. The subsequent quantitative refinements of the Picard–Borel theory first by R. Nevanlinna and then by H. and J. Weyl and Ahlfors constitute one of the mathematical highlights of the first half of this century. Chern wrote, or was a joint author of, several papers on the subject which stimulated considerable interest and which established much of the formalism and conceptual framework underlying recent developments in the area. I shall briefly sketch some of my interpretation of his viewpoint on the subject.

Suppose we consider a holomorphic mapping $f : \mathbb{C} \to P^n$, to be called a holomorphic curve. Denoting by $\Delta_r = \{z : |z| < r\}$ the disc of radius r and by $n(r, H)$ the number of points of intersection of $f(\Delta_r)$ with a hyperplane H in P^N, a variant of Crofton's formula (7) is

$$\int_{P^{N*}} n(r, H)dH = \int_{\Delta_r} f^*\omega, \tag{3}$$

where ω is the standard Kähler form on P^n. Actually, the exact analogue of Crofton's formula would have the area of $f(\Delta_r)$ on the right, but by the Wirtinger theorem this is equal to $\int_{\Delta_r} f^*\omega$. We are adopting here the viewpoint of [75], which is being paraphrased in the present discussion. Since intersection numbers in complex-analytic geometry are always positive, the left-hand side of (3) has topological significance. Indeed, the closed form ω represents the Poincaré dual of the cycle carried by the hyperplane H, and so we may write $\omega = dd^c\mu$ for a potential function $\mu \geqslant 0$ having singularities on H. Applying Stokes's theorem to the right-hand side of (3) gives the first main theorem of value distribution theory

$$N(r, H) + m(r, H) = T(r) + O(1), \tag{4}$$

where

$$N(r, H) = \int_0^r n(t, H)dt/t,$$

$$T(r) = \int_0^r \left\{ \int_{\Delta_t} f^*\omega \right\} dt/t$$

is the order function, and

$$m(r, H) = \frac{1}{2\pi} \int_{\partial\Delta_r} f^*\mu d\theta \geqslant 0$$

is the proximity form. Chern frequently spoke of (4) and the subsequent Nevanlinna inequality

$$N(r, H) \leqslant T(r) + O(1) \tag{5}$$

as relations in non-compact algebraic geometry; we may also view them as formulas in non-compact Chern classes. Comparison of (5) with the averaged form

$$\int N(r, H)dH = T(r)$$

of (3) suggests the deeper aspects of the theory. For example, it follows immediately that if $T(r) \to \infty$ then the image curve cannot miss an open set of hyperplanes, which is the Casorati–Weierstrass theorem.

In order to arrive at the main results of Nevanlinna theory, at least in the classical case $n = 1$, the same procedure as in the derivation of (4) is used only with the potential function being $\log v$ where $f^*\omega = (\sqrt{-1}/2)v \cdot dz \wedge d\bar{z}$. The resulting formula is called the second main theorem; it is just the (integrated) Gauss–Bonnet theorem for the region $\Omega_r = \Delta_r -$ (ramification divisor of f) with the metric $f^*\omega$ having constant Gaussian curvature $+2$ (due to $\chi(P^1)=2$). Comparing the first and second main theorems one may use a calculus trick to eliminate the boundary integrals, arriving at an inequality of the sort

$$\sum_{i=1}^{3} N(r, H_i) \leqslant 2T_f(r) + O(\log T_f(r)), \tag{6}$$

where the H_i are distinct points on P^1 and the explanation of the number 2 is the same as that given above. (6) is basically the famous defect relation of R. Nevanlinna; the present mode of derivation taken from [70] suggests how to look for generalizations.

It will be noticed that even though the subject is complex analysis the method is differential-geometric. The use of differential geometry to study complex manifolds and holomorphic mappings began in the 1930s and 1940s with the work of Hodge, Ahlfors, Bochner, and others. Almost from the beginning Hermitian differential geometry and the especially nice features peculiar to the complex case have been a main theme in Chern's work. In general he has always emphasized the geometric aspect of analysis, a viewpoint carried on by many of his students. While in Berkeley I recall several seminars in which he spoke on problems in analysis arising from geometry; often these covered difficult papers which at that time were not so well-known but were of fundamental importance, such as Calabi's work on the affine hyperspheres.

Another recurrent theme in Chern's work is the emphasis on the actual differential forms representing cohomology classes. Already implicit in his proof of the general Gauss–Bonnet theorem, this principle arises again in Chern's approach to Nevanlinna theory in which the selection of a potential function plays a crucial role. Together with Bott in [80] this is explored in generality when they proved that not only is there a canonical connection associated to an Hermitian metric h in a holomorphic vector bundle $\mathbb{C}^r \to E \to M$ over a complex manifold, but to each holomorphic section s there is canonically associated an $(r-1, r-1)$ form μ_s such that the rth Chern form is given by

$$c_r(E, h) = dd^c\mu_s$$

in $M-$ (zeroes of s). Moreover, $\mu_s \geqslant 0$ in case the curvature of E is positive in a suitable sense.

This theme is again formalized in the joint work [103] of Chern and Simons. Associated to a Riemannian metric ds^2 on a manifold M are the Pontrjagin forms $P_k(ds^2)$, together with the dual forms $P_j^{\perp}(ds^2)$ defined by $\left(\sum_{k>0} P_k\right)\left(\sum_{j>0} P_j^{\perp}\right)=1$. Up on the principal frame bundle P are canonically defined transgression forms $T_k(ds^2)$ satisfying $\pi^*P_k(ds^2)=dT_k(ds^2)$, where $\pi : P \to M$ is the projection. In cases where $P_k(ds^2)$ on $P_j^{\perp}(ds^2)$ are identically zero for symmetry or dimension reasons, the transgression forms acquire cohomological meaning, and the following theorem illustrates the flavor of the results: Let M be a compact Riemannian manifold of dimension n. Necessary conditions for its conformal immersion in $\mathbb{R}^{n+h}$ are

$$P_j^{\perp}(ds^2)=0, \qquad [h/2]+1 \leqslant j$$

$$\frac{1}{2}T_j(ds^2) \in H^{4j-1}(P, \mathbb{Z}), \qquad [h/2]+1 \leqslant j \leqslant \left[\frac{n-1}{2}\right].$$

Just as algebraic geometers seem always to periodically turn to questions on algebraic curves, throughout his work Chern exhibits a continuing interest in classical differential geometry, especially curves and surfaces in Euclidean space. Here, a particularly fruitful area is the study of minimal surfaces about which Chern, frequently with collaborators, has written several papers. Among these, to me, an especially suggestive one is [96], which was motivated by work of Calabi on minimal 2-spheres in an n-sphere S^n. First the general formalism of osculating spaces to a submanifold is developed, and then it is observed that for minimal surfaces in a Riemannian manifold of constant sectional curvature this osculating sequence has a particularly simple description. Now for surfaces in $\mathbb{R}^{m+2}$, being minimal is equivalent to the Gauss mapping being an anti-holomorphic mapping to the complex quadric in P^{m+1} [79], and so one may apply methods of complex differential geometry, especially the Frenet frame attached to a holomorphic curve in projective space and global function theory to study these surfaces. Here the function theory involves the theorems of Picard and Borel and therefore ties in with the value distribution theory. In the present study of minimal 2-spheres in an S^n, there are analogues of both of these; the osculating sequence has been mentioned, but for the function theory one essential difference is that the domain S^2 is compact. After some computation the conclusion is reached that the highest osculating space leads to a compact algebraic curve in the complex hyperquadric, a result dating to Boruvka (1933) and which now assumes a conceptually clear form.

Another manifestation of Chern's interest in classical differential geometry is the Math. 140–240 series at Berkeley, which together with its predecessor at Chicago and the widely circulated lecture notes [1], [2] more or less spawned a whole generation of differential geometers. While a student I well remember struggling over the infamous $dp = \Sigma\omega_i \cdot e_i$ in [1], as well as one particularly notorious computation in which some quantity with a large number of indices suddenly disappeared from the calculation. When I asked Don Spencer, my advisor, he looked at it and then shrugged and said that whatever it was it must not be a tensor, since Chern's definition of the latter is a quantity which could *not* be annihilated by a special choice of coordinates. Actually, this works quite well, at least if you know what you are doing—one is reminded here of the last paragraph of the review by Hermann Weyl (*Bull. A.M.S.*, 1938), pp. 598–601) of Cartan's book on moving frames.

The joint paper [105] with Moser treats the equivalence problem for a non-degenerate real hypersurface M in $\mathbb{C}^{n+1}$ $(n \geqslant 1)$. Recall from classical differential geometry that a hypersurface in $\mathbb{R}^{n+1}(n \geqslant 2)$ is determined up to rigid motion by its first and second fundamental form. Moreover, the first and second fundamental forms may be prescribed subject to certain differential conditions, the Gauss–Codazzi equations; and finally there is a difference between classical surface theory $(n = 2)$ and the case $n \geqslant 3$, where in general the first fundamental form alone suffices to determine the hypersurface. The Chern–Moser theory may be viewed as an analogue where the group of Euclidean motions is replaced by the infinite pseudo-group Γ of local biholomorphic transformations, and where there is corresponding distinction between the cases $n = 1$ (due to E. Cartan) and $n \geqslant 2$.

The basic observation is that if M is locally given by $r(z)=0$ for a real C^∞ function r, then the E. E. Levi form

$$\mathcal{L}(\xi)=\sum_{i,j}\frac{\partial^2 r}{\partial z_i\partial\bar{z}_j}(z)\xi_i\bar{\xi}_j,\qquad r(z)=0,\tag{7}$$

defined on complex tangent vectors $\xi=(\xi_i)$ satisfying

$$\sum_i \frac{\partial r}{\partial z_i}(z)\xi_i=0,\qquad r(z)=0\tag{8}$$

is, up to a conformal factor, invariant under the pseudo-group Γ. If the Levi form (7) is non-degenerate on the complex tangent spaces (8) to M, then there is in this bundle an intrinsic conformal Hermitian structure, which at least suggests that Cartan's equivalence procedure should terminate after two prolongations. This turns out to be the case, and in fact the paper is a good place to learn the equivalence method by seeing how it goes in a significant special case.

This paper is also characteristic of much of Chern's work. For example, geometric structures other than the Riemannian and Hermitian, which have been the main preoccupation of differential geometers in recent years, come into play. We recall that Klein's Erlangen program deals with properties of a space X invariant under the transitive action of a Lie group or pseudo-group. With Elie Cartan the outlook was broadened considerably in that one need not be given the group but should be given on X a structure such as Riemannian, conformal, contact, etc. The structure is generally non-homogenous, but if this is so then we are in Klein's situation. Cartan's equivalence method attaches, by the method of successive prolongations, to each point $x\in X$ a sequence of groups leading to a complete set of local invariants for the structure. The general theory has in recent years been examined in detail but, as with much of Cartan's legacy, the intricacy and particular features of a carefully chosen example have a special appeal. One may say that, in the work of both Cartan and Chern, of almost equal importance to what is being proved is just how the argument is carried out, with particular emphasis on selection of notation which most efficiently isolates the essential point. Another common feature is that the main technical step frequently appears in an algebraic manipulation, one which in the final form appears deceptively simple.

It is a pleasure to thank Hung-hsi Wu for extremely helpful suggestions in preparing this commentary.

April 11, 1978
Cambridge, Massachusetts

Photo by Dr. Harry Wang, 1950.

A Summary of My Scientific Life and Works

By Shiing-shen Chern

I was born on October 26, 1911 in Kashing, Chekiang Province, China. My high school mathematics texts were the then popular books *Algebra* and *Higher Algebra* by Hall and Knight, and *Geometry* and *Trigonometry* by Wentworth and Smith, all in English. Training was strict and I did a large number of the exercises in the books. In 1926 I enrolled as a freshman in Nankai University, Tientsin, China. It was clear that I should study science, but my disinclination with experiments dictated that I should major in mathematics. The Mathematics Department at Nankai was a one-man department whose Professor, Dr. Li-Fu Chiang, received his Ph.D. from Harvard with Julian Coolidge. Mathematics was at a primitive state in China in the late 1920s. Although there were universities in the modern sense, few offered a course on complex function theory and linear algebra was virtually unknown. I was fortunate to be in a strong class of students and such courses were made available to me, as well as courses on non-euclidean geometry and circle and sphere geometry, using books by Coolidge.

The period around 1930, when I graduated from Nankai University, saw great progress in Chinese science. Many students of science returned from studies abroad. At the center of this development was Tsing Hua University of Peking (then called Peiping), founded through the return of the Boxer's Indemnity by the U.S. I was an assistant at Tsing Hua in 1930–1931 and was a graduate student from 1931–1934. My teacher was Professor Dan Sun, a former student of E. P. Lane at Chicago. Therefore, I began my mathematical career by writing papers on projective differential geometry.

In 1934 I was awarded a fellowship to study abroad. I went to Hamburg, Germany, because Professor W. Blaschke lectured in Peking in 1933 on the geometry of webs and I was attracted by the subject. I arrived at Hamburg in the fall of 1934 when Kähler's book *Einführung in die Theorie der Systeme von Differentialgleichungen* was published and he gave a seminar based on it. In a less than two-year stay in Hamburg I worked in more depth on the Cartan–Kähler theory than any other topic. I received my D.Sc. in February 1936.

The completion of the degree fulfilled my obligation to the fellowship. It was natural to look forward to a carefree postdoctoral year in Paris with the master himself, Elie Cartan. It turned out to be a year of hard work. In 1936–1937 in Paris I learned moving frames, the method of equivalence, more Cartan–Kähler theory, and, most importantly, the mathematical language and the way of thinking of Cartan. Even now I frequently find Cartan easier to follow than some of his expositors.

I returned to China in the summer of 1937 to become Professor of Mathematics at Tsing Hua University. I crossed the Atlantic on the S.S. Queen Elizabeth and, after a month long tour of the United States, I crossed the Pacific on the S.S. Empress of Canada. The Sino-Japenese war broke out while I was on board and I never reached Peking.

During the war Tsing Hua University moved to Kunming in Southwest China and became a part of Southwest Associated University. Mathematically it was a period of isolation. I taught courses on advanced topics (such as conformal differential geometry, Lie groups, etc.) and had good students.

In 1943 I became a member of the Institute for Advanced Study; both Veblen and Weyl were aware of my work. During the period 1943–1945 I learned algebraic topology and fiber bundles and did my work on characteristic classes, among other things. The war ended in 1945 and I decided to return to China. Postwar transportation difficulties delayed my trip so that I did not arrive in Shanghai until March 1946. I was called to organize a new Institute of Mathematics of the Academic Sinica in Nanking. The work lasted only for about two years. On December 31, 1948 I left Shanghai for the United States, again on an invitation of the Institute for Advanced Study. (See Weil's article above. Before leaving China I was offered a position at the Tata Institute in Bombay, then at a planning stage, which I was not able to accept. The offer must have come on the initiative of D. D. Kosambi, the first professor of mathematics at Tata, who knew well my work on path geometry.) I spent the winter term of 1949 at the Institute. During 1949–1960 I was a Professor at the University of Chicago. I moved to Berkeley in 1960 and have remained in Berkeley ever since.

In the following I will try to give a summary of my mathematical works.

1. Projective Differential Geometry

Einstein's general relativity provided the great impetus to the study of riemannian geometry and its generalizations. Before that geometry was dominated by Felix Klein's Erlangen Program announced in 1871, which assigns to a space a group of transformations which is to play the fundamental role. Thus the euclidean space has the group of rigid motions and the projective space has the group of projective collineations, etc. Along the lines of classical curve and surface theory in the tradition of Serret–Frenet, Euler, Monge, and Gauss, projective differential geometry was founded by E. J. Wilczynski and G. Fubini and E. Cech. Its main problem is to find a complete system of local invariants of a submanifold under the projective group and interpret them geometrically through osculation by simpler geometrical figures. The main difficulty lies in that the projective group is relatively large and invariants can only be reached through a high order of osculation. Moreover, the group of isotropy is non-compact, a fact which excludes many beautiful geometrical properties.

In my first papers [1], [2] I avoided the first difficulty by studying more complicated figures. The papers are nothing more than exercises, but the philosophy behind them found an echo in the recent works of P. A. Griffiths on webs, Abel's theorem, and their applications to algebraic geometry. For example, instead of studying an algebraic curve of degree d in the plane, one can study the configuration consisting of d points on each line of the plane, its points of intersection with the curve. One gets in this way d arcs in correspondence. Paper [1] studies two arcs in correspondence.

My next paper [3] concerns projective line geometry, now a forgotten subject. A line complex is, in modern terminology, a hypersurface in the Plücker–Grassmann manifold of all lines in the three-dimensional projective space. While the consideration of tangent spheres of a surface leads to the fundamental notions of lines of curvature and principal curvatures and that of the tangent quadrics of a projective surface leads to the quadrics of Darboux and Lie, the use of quadratic line complexes in the study of general line complexes was initiated in this paper.

Several years later I returned to projective differential geometry by introducing new invariants of contact of a pair of curves in a projective space of n dimensions, and also of surfaces [17], [19]. They include as a special case, the invariant of Mehmke–Smith, which plays a role in some questions on singularities in several complex variables. Generally speaking, the study of diffeomorphism invariants of a jet at a singularity has recently attracted wide attention (H. Whitney, R. Thom). The projective invariants, studied extensively by Italian differential geometers, should enter into the more refined questions.

The Laplace transforms of a conjugate net was a favorite topic in the theory of transformations of surfaces. It is a beautiful geometric construction which leads to a transformation of linear homogeneous hyperbolic partial differential equations of the second order in two variables. In [24], [35] a generalization was given to a class of submanifolds of any dimension. This generalization could be related to the recent search of high-dimensional solitons and their Bäcklund transformations.

From projective spaces it is natural to pass to spaces with paths where the straight lines are replaced by the integral curves of a system of ordinary differential equations of the second order, an idea which could be traced back to Hermann Weyl. Such spaces are said to be projectively connected or to have a projective connection. Projective relativity (O. Veblen, J. A. Schouten) aims at singling out the projectively connected spaces whose paths are to be identified with the trajectories of a free particle in a unified field theory. They are defined by a system of "field equations." A new system of field equations was proposed in [112].

From the mathematical viewpoint projectively connected spaces are of intrinsic interest. Relating projective spaces and general projectively connected spaces is the imbedding problem. Given a submanifold M in a projective space, an induced projective connection can be defined on M by taking a field of linear subspaces transversal to the tangent spaces of M and projecting neighboring tangent spaces from them. In [7] I proved an analogue of the Schläfli–Janet–Cartan imbedding theorem for riemannian spaces of which the following is a special case: A real analytic normal (in the sense of Cartan) projective connection on a space of dimension n can be locally induced by an imbedding in a projective space of dimension $n(n+1)/2+[n/2]$. The dimension needed is thus generally higher than in Schläfi's case.

The fundamental theorem on projective connections is the theorem associating a unique normal projective connection to a system of paths. I announced in [23] that the same is true when there is in a space of dimension n a family of k-dimensional submanifolds depending on $(k+1)(n-k)$ parameters and satisfying a completely integrable system of differential equations. The case $k=1$ is classical and the case $k=n-1$ was the main conclusion of M. Hachtroudi's Paris thesis. My derivation was long and was never published. A geometrical treatment was later given by C. T. Yen (*Annali di Matematica* 1953).

In the Princeton approach to non-riemannian geometry led by Veblen and T. Y. Thomas, a main tool is the use of normal coordinates relative to which the normal extensions of tensors are defined. Normal coordinates in the projective geometry of paths can be given different definitions; their existence is generally not easy to establish. In [8] I showed that Thomas's normal coordinates are in general different from the normal coordinates defined naturally from Cartan's concept of a projective connection.

In my recent joint works with Griffiths on webs [113] we came across a theorem characterizing a flat normal projective connection as one with ∞^2 totally geodesic hypersurfaces suitably distributed; the classical theorem needs ∞^n totally geodesic hypersurfaces, n being the dimension of the space.

In concluding this section, I wish to say that I believe that projective differential geometry will be of increasing importance. In several complex variables and in the transcendental theory of algebraic varieties the importance of the Kähler metric cannot be over-emphasized. On the other hand, projective properties are in the holomorphic category. They will appear when the problems involve, directly or indirectly, the linear subspaces or their generalizations.

2. *Euclidean Differential Geometry*

Before the nineteen forties a mathematical student was usually introduced to differential geometry through a course on curves and surfaces in euclidean space, known in European universities as "applications of the infinitesimal calculus to geometry." I was particularly fascinated by Blaschke's book for its emphasis on global problems. I was, however, able to do some work only after I began to treat surface theory by moving frames. In [29] I observed that Hilbert's proof of the rigidity of the sphere gives the more general theorem that a closed strictly convex surface in E^3 (= three-dimensional euclidean space) is a sphere if one principal curvature is a monotone decreasing function of the other. My study of closed surfaces of constant mean curvature was less successful; it is still not known whether there is a closed immersed surface of constant mean curvature in E^3, whose genus is positive.

More generally, a natural area of investigation in euclidean differential geometry is concerned with the W-hypersurfaces, where there is a functional relation between the principal curvatures. If a hypersurface is closed and strictly convex, its Gauss map into the unit hypersphere is one-to-one and we can identify functions on the hypersurface with those on the unit hypersphere. Let σ_r, $1 \leqslant r \leqslant$

n, be the rth elementary symmetric function of the reciprocals of the principal curvatures of a convex hypersurface in E^{n+1}. In [68] I proved that if, for a certain r, the σ_r functions of two closed strictly convex hypersurfaces Σ, Σ^* in E^{n+1} agree as functions on the unit hypersphere, then Σ and Σ^* differ by a translation. The condition means geometrically that σ_r are the same at points of Σ, Σ^* at which the normals are parallel. In [69], Hano, Hsiung, and I proved a similar uniqueness theorem by replacing the conditions by $\sigma_r \leqslant \sigma_r^*$, $\sigma_{r+1} \geqslant \sigma_{r+1}^*$ for a certain r. The proofs depend on the establishment of some integral formulas.

In [81] I considered hypersurfaces with boundary in the euclidean space and found upper bounds on their size if certain curvature conditions are satisfied. This generalized some work of E. Heinz and S. Bernstein for surfaces in E^3.

Again using integral formulas Hsiung and I proved in [77] that a volume-preserving diffeomorphism of two k-dimensional compact submanifolds in E^n is an isometry if a certain additional condition is satisfied.

In [62] and [66] Lashof and I studied the total curvature of a compact immersed submanifold in E^n. The total curvature is defined as the measure of the image of the unit normal bundle on the unit hypersphere of E^n under the Gauss map. (Observe that independent of the dimension of a submanifold the unit normal bundle has dimension $n-1$, which is the dimension of the unit hypersphere of E^n.) The total curvature was considered by J. Milnor following his work on that of a knot. Generalizing the classical theorems of Fenchel for the total curvature of a closed space curve, Lashof and I proved that the total curvature of a compact immersed submanifold in E^n, when properly normalized, has a universal lower bound and that it is reached when and only when the submanifold is a convex hypersurface. As a corollary it is proved that a closed surface of non-negative Gaussian curvature in E^3 is convex, generalizing a classical theorem of Hadamard. In this work a lemma on the local behavior of a hypersurface with degenerate second fundamental form plays a fundamental role. Total curvature and tight immersion have received many interesting developments in recent years (Kuiper, Banchoff, Pohl, and Chen).

Given an oriented (two-dimensional) surface in E^4, its Gauss map has as image the Grassmann manifold of all oriented planes through a point. The latter is homeomorphic to $S^2 \times S^2$. As a result the map defines a pair of integers. Spanier and I [47] proved that if the surface is imbedded, these two integers are equal when the spaces are properly oriented.

In [50]Kuiper and I introduced two integers to an immersed manifold in E^n: the indices of nullity and of relative nullity. Inequalities are established between them and the dimension and codimension of a compact submanifold in E^n. The origin of this work was a theorem of Tompkins that there is no closed surface in E^3 whose Gaussian curvature is identically zero.

The smoothness requirements of various theorems in surface theory have been thoroughly investigated by P. Hartman and A. Wintner in a long series of papers. In [55] we studied the critical case for the isothermic coordinates, namely, the minimum conditions so that the metric in the isothermic coordinates has the same smoothness.

Finally I wish to mention a result on complex space-forms. In his thesis Brian Smyth determined the complete Einstein hypersurfaces in a Kählerian manifold of constant holomorphic sectional curvature, by using the classification of symmetric Hermitian spaces. The result turns out to be a local one. The problem leads to an over-determined differential system and I showed in [87] that the theorem follows from a careful study of the integrability conditions. The hypersurfaces in question are either totally geodesic or are hyperspheres.

Euclidean differential geometry is comparable to elementary number theory in its beauty of simplicity. Unlike the latter more remains to be discovered.

3. *Geometrical Structures and Their Intrinsic Connections*

A riemannian structure is governed by its Levi-Civita connection, and a path structure by its normal projective connection. A fundamental problem of local differential geometry is to associate to a structure a connection which describes all the properties. An effective way of doing this is by Elie Cartan's method of equivalence. In the years 1937–1943 when I was isolated in the interior of China I carried out the program in many cases:

The geometry of the equation of the second order

$$y'' = F(x, y, y'), \qquad y' = dy/dx, \qquad y'' = d^2y/dx^2$$

in the (x, y)-plane was studied by A. Tresse. Tresse's results were formulated in terms of the Lie theory; it would be more geometrical to say that a normal projective connection can be defined in the space of line elements (x, y, y'). I studied the equation of the third order

$$y''' = F(x, y, y', y'')$$

under the group of contact transformations in the plane and showed that in an important case a conformal connection can be defined intrinsically [6], [13]. I also defined affine connections from structures arising from webs [9] (cf. §8).

Local differential geometrical structures are defined either by differential systems or by metrics, the two typical cases being projective geometry and euclidean geometry. When the paths are the integral curves of a system of ordinary differential equations, the allowable parameter change has an important bearing on the resulting geometry. D. D. Kosambi considered a system of differential equations of the second order with an allowable affine transformation of parameters and attached to the structure an affine connection. I proved in [10] the result by the method of equivalence and went on in [11] to solve the corresponding problem when the paths are defined by a system of differential equations of higher order.

Geometrically it is more natural that a family of submanifolds is given with unrestricted parametrization, i.e., the parameters are allowed arbitrary (smooth) changes. Generalizing Tresses's problem to n dimensions, the given data should be $\infty^{2(n-1)}$ curves satisfying a differential system such that through any point and tangent to any direction at the point there is exactly one such curve. With these curves taking the place of the straight lines, a generalized projective geometry, i.e., a normal projective connection, can be defined. As mentioned in §1, I extended this result to the case when there is given $\infty^{(k+1)(n-k)}$ k-dimensional submanifolds satisfying a differential system. In the same vein I defined in [20] a Weyl connection, giving ∞^2 surfaces in $\mathbb{R}^3$ as "isotropic surfaces." This was extended to

n dimensions in [21], but the details of the n-dimensional case were never published.

In [22], [42] I studied the connections to be attached to a Finsler metric and showed that there is more than one natural choice.

In 1972 Moser found a local normal form of a non-degenerate real hypersurface in C_2 and asked me to identify his invariants with those of Elie Cartan. Years before I had extended Cartan's work to a real hypersurface in C_{n+1}. I have not published the results, partly because a paper of Tanaka on the same subject appeared in the meantime, although Tanaka made an assumption on the hypersurface (which he removed in a later paper). In [105] Moser and I gave both the normal form of a non-degenerate real hypersurface in C_{n+1} and its intrinsic connection as a CR-manifold and identified the two sets of invariants. When the hypersurface is real analytic, I defined in [107] a projective connection. The latter does not give all the invariants, but has the advantage that its invariants are in the holomorphic category.

All these are special cases of a G-structure. Some G-structures, such as the complex structures, admit an infinite pseudo-group of transformations. In [54] I gave an introduction to G-structures, including the notion of a torsion form and an exposition of Cartan's theory of infinite continuous pseudo-groups. A more complete account of G-structures was given in [83].

In [61] I observed that the Hodge harmonic theory is valid for a torsionless G-structure, with $G \subset O(n)$; the Hodge decomposition can then be generalized to the decomposition of a harmonic form into irreducible summands under the action of G. This viewpoint also gives a better understanding of Hodge's results.

Among mathematical disciplines the area of geometry is not so well defined. Perhaps the notion of a G-structure is of sufficient scope to fulfill the current requirements for the mainstream of geometry.

4. Integral Geometry

I went to Hamburg in 1934 when Blaschke, in his usual style, started a series of papers entitled "Integral geometry." Although I have a keen interest in the subject, my works on it have been scattered.

I observed that integral geometry in the tradition of Crofton deals with two homogeneous spaces with the same group. Call the group G. If the homogeneous spaces are realized as coset spaces G/H and G/K, H and K being subgroups of G, two cosets aH and bK, $a, b \in G$, are called incident if they have an element in common. With this notion of incidence, Crofton's formula was established in a very general context [14], [16], [18]. This notion of incidence was appreciated by Weil and found useful in later works of Helgason and Tits.

My other work on integral geometry concerns the kinematic density of Poincaré. With Chih-Ta Yen I gave a proof of the fundamental kinematic formula in E^n [15], [48].

In his formula for the volume of a tube Weyl introduced a number of scalar invariants of an imbedded manifold in E^n, half of which depend only on the induced metric. If M^p and M^q are closed imbedded manifolds of E^n, with M^p fixed and M^q moving, I proved in [84] a simple formula expressing the integral of an invariant of the intersection $M^p \cap M^q$ over the kinematic measure. This complements the fundamental kinematic formula, which deals with hypersurfaces.

5. Characteristic Classes

My introduction to characteristic classes was through the Gauss–Bonnet formula, known to every student of surface theory. Long before 1943, when I gave an intrinsic proof of the *n*-dimensional Gauss–Bonnet formula [25, 30], I knew, by using orthonormal frames in surface theory, that the classical Gauss–Bonnet is but a global consequence of the Gauss formula which expresses the "theorema egregium." The algebraic aspect of the proof in [25] is the first instance of a construction later known as transgression, which is destined to play a fundamental role in the homology theory of fiber bundles, and in other problems.

The Gauss–Bonnet formula is concerned with the Euler–Poincaré characteristic. It was natural to look at corresponding results for the general Stiefel–Whitney characteristic classes, then newly introduced. I soon realized that the latter are essentially defined only mod two and relating them with curvature forms would be artificial. Technically its cause lies in the complicated homology structure of the orthogonal group, such as the presence of torsion. The Grassmann manifold and the Stiefel manifold over the complex numbers have no torsion, and the same is true of the unitary group. In [33] I introduced the characteristic classes of complex vector bundles and related them via the de Rham theorem, with the curvature forms of an hermitian structure in the bundle. Actually this paper contains, through the explicit construction of differential forms, the essence of the homology structure of a principal bundle with the unitary group as structure group: transgression, characteristic classes, universal bundle, etc. These characteristic classes are defined for algebraic manifolds, but their definition, whether via an hermitian structure or via the universal bundle, is not algebraic. In [51] I showed that by considering an associated bundle with the flag manifold as fibers the characteristic classes can be defined in terms of those of line-bundles. As a consequence the dual homology class of a characteristic class of an algebraic manifold contains a representative algebraic cycle.

The study of the homology structure of a fiber bundle through the use of a connection merges local properties into global properties and combines differential geometry with differential topology. The general case of a principal bundle with an arbitrary Lie group as structure group, of which my work above is a special case concerning the unitary group, was carried out by Weil in 1949 in an unpublished manuscript. Part of Weil's results was presented in [I,1]. The main conclusion is the so-called Weil homomorphism which identifies the characteristic classes (through the curvature forms) with the invariant polynomials under the action of the adjoint group. This identification, whose importance should be immediately recognizable, has recently been found crucial in the heat equation proof of the Atiyah–Singer index theorem and in Bott's theorem on foliations.

Actually the characteristic forms themselves, which represent the characteristic classes via the de Rham theorem, contain more information. The vanishing of the characteristic forms, not just their classes, (which only means that the forms are exact), leads to the secondary characteristic classes. These were studied with James Simons in [98, 103]. The secondary characteristic classes depend on the choice of the connection, but enjoy strong invariance properties under a change of the connection. They have been found to play a role in various problems, such as conformal immersions and the η-invariant defined by the spectrum of a compact riemannian manifold. A duality theorem for characteristic forms was given in a joint paper with White [108].

In [39] I determined the mod two cohomology ring of the real Grassmann manifold. As a consequence it follows that the Stiefel–Whitney classes generate the mod 2 characteristic ring of a sphere bundle. The result plays a role in the estimation of the number of closed geodesics on a compact riemannian manifold.

When the base manifold has a complex structure, its ring of complex-valued exterior differential forms has also a more refined structure. Forms have a bidegree and there are two exterior differentiations, one with respect to the complex structure and the other to its conjugate complex structure, denoted usually by ∂ $\bar{\partial}$ respectively. In [80, 92] Bott and I studied the forms of a holomorphic hermitian vector bundle relative to the operator $i\partial\bar{\partial}$. This has applications to complex geometry, and in particular to the study of the zeroes of holomorphic sections, which contains as a particular case the classical theory of value distributions of meromorphic functions.

In [101] I gave an elementary proof (without sheaf cohomology) of Bott's theorem on characteristic numbers and the residues of a meromorphic vector field on a compact complex manifold. The proof is in the spirit of a transgression.

On a manifold it is necessary to use covariant differentiation; curvature measures its non-commutativity. Its combination as a characteristic form measures the non-triviality of the underlying bundle. This train of ideas is so simple and natural that its importance can hardly be exaggerated.

6. *Holomorphic Mappings*

The simplest case of a holomorphic mapping is $\mathbb{C} \to P_1$, where $\mathbb{C}$ is the complex line and P_1 is the complex projective line. In usual terminology $\mathbb{C}$ is called the Gaussian plane and P_1 the Riemann sphere; the mapping is known as a meromorphic function. The geometrical basis of the classical value distribution theory consists of two theorems, known as the first and second main theorems, which are but the Gauss–Bonnet theorem applied to the Hopf bundle and the canonical bundle of P_1 respectively. From these the Nevanlinna defect relation follows by calculus-type inequalities.

In [70] these viewpoints were made precise by the study of holomorphic mappings of a non-compact Riemann surface into a compact one. As a differential geometer I have naturally been interested in the theory of a family of meromorphic functions interpreted as a holomorphic curve in P_n, the complex projective space of n dimensions, as developed by Henri Cartan, H. and J. Weyl, and Ahlfors. A geometrical treatment was given in [99] for P_2; the corresponding results for P_n were worked out by H. Yamaguchi in an unpublished manuscript. An essential ingredient for the good distributional behavior of a non-compact holomorphic curve lies in the validity of Frenet-type formulas. Cowan, Vitter, and I considered holomorphic curves in any complex manifold M and showed that Frenet formulas will be valid only when M has very special properties, which are close to being of constant holomorphic sectional curvature [104].

It is natural to consider holomorphic mappings in higher dimensions, a broad subject of which much remains to be understood. In [75] I gave some general observations. Following some work of H. Levine, done with my supervision, I studied in [71] a holomorphic mapping $f: C_n \to P_n$ and proved that under some growth conditions the set $P_n - f(C_n)$ is of measure zero.

In [80] Bott and I reformulated the value distribution problem as one on the distribution of zeroes of the holomorphic sections of a holomorphic vector bundle. A preparatory algebraic problem consists of the study of complex transgression, i.e., transgression relative to the operator $i\partial\bar{\partial}$ [80], [92]. Characteristic classes are defined in a refined sense, which is of importance in applications to problems pertaining to the holomorphic category.

In [88] I proved a Schwarz lemma in high dimensions as a volume-decreasing property. With S. I. Goldberg [106] an analogous theorem was proved for a class of harmonic mappings of riemannian manifolds.

In [90] I introduced with H. Levine and L. Nirenberg intrinsic pseudo-norms in the real cohomology vector spaces of a complex manifold M. The definition utilizes the pluri-subharmonic functions. The pseudo-norm becomes a norm when there are enough pluri-subharmonic functions in M. Under a holomorphic mapping the pseudo-norm is a non-increasing function.

Geometry occupies an important position in complex function theory. Its role in several complex variables will be even greater in the future.

7. *Minimal Submanifolds*

The Grassmann manifold $\tilde{G}_{2,n}$ of all the oriented planes through a point in E^n has a complex structure invariant under the action of $SO(n)$. On the other hand, an oriented surface in E^n has a complex structure through its induced riemannian metric. The surface is minimal if and only if the Gauss map is anti-holomorphic [79]. This theorem was proved by Pinl for $n = 4$ and is clearly the starting point in relating minimal surfaces with complex function theory. One of the fundamental theorems on minimal surfaces is the Bernstein uniqueness theorem which says that a minimal surface $z = f(x,y)$ in E^3 defined for all x, y must be a plane. As generalized by Osserman, Bernstein's theorem is a consequence of the theorem that if a complete minimal surface is not a plane, its image on the unit sphere under the Gauss map is dense. In [79] this was generalized to a density theorem on the image under the Gauss map of a complete minimal surface in E^n, which is not a plane. More refined density theorems were established in [86], a joint paper with Osserman.

The geometry of minimal surfaces on a sphere S^n takes a different course, because one naturally starts by studying the compact ones. As the codimension is arbitrary, it is necessary to consider higher osculating spaces. This was carried out in [96], in which it is shown that their dimensions are successive even integers. In fact, this is true for a minimal surface in any space form. According to E. Calabi, minimal two-spheres in S^n can be enumerated, because of the simple conformal behavior of the two-sphere. In [94] I gave a simple construction of all the minimal two-spheres in S^4. This work was continued by Lucas Barbosa for minimal 2-spheres in S^n.

In [91], following the celebrated work of Simons on closed minimal submanifolds (of dimension k) on the unit sphere S^n, do Carmo, Kobayashi, and I considered the square of the length of the second fundamental form, to be denoted by $\|h\|^2$. It follows from Simons' inequality that if $\|h\|^2 \leqslant k(n-k)/(2n-2k-1)$, then $\|h\|^2 = 0$ or $k(n-k)/(2n-2k-1)$. We determined all the closed minimal submanifolds of S^n with $\|h\|^2 = k(n-k)/(2n-2k-1)$.

In [115] I studied affine minimal hypersurfaces, where the integral to be minimized involves second partial derivatives. I computed the first variation and showed that its vanishing is characterized by the vanishing of the affine mean curvature.

Affine minimal surfaces play a role in Bäcklund-type transformations, which remains to be better understood. In fact, Terng and I proved in [116] that if the focal surfaces of a W-congruence have parallel affine normals at corresponding points, then both are affine minimal surfaces.

One of the objectives of the geometry of minimal submanifolds is to provide examples. This is of importance, because general theorems are frequently derived through comparison with them.

8. *Webs*

The papers [4], [5] constitute my thesis in Hamburg. In [4] I gave a sharp upper bound for the rank of a d-web of hypersurfaces in $\mathbb{R}^n$. Paper [5] was a natural outgrowth in a web-geometry atmosphere of my efforts to understand the Cartan–Kähler theory. I defined an affine connection, and from which a complete system of invariants, of a three web of r-dimensional submanifolds in $\mathbb{R}^{2r}$, and gave geometrical interpretations of some of these invariants.

It was an irony of fate that I should return to the subject forty years later. When Phillip Griffiths visited Berkeley in 1975–1976, he was interested in web geometry as a generalization of the geometry of projective varieties. For instance, an algebraic curve of degree d in P_n is met by a generic hyperplane in d points; by duality this gives a d-web of hyperplanes in the dual space P_n^*. We were able to clarify and extend what is perhaps the deepest result of the web geometry of Blaschke and Bol. In [110], [113], we proved the linearization theorem that a d-web of hypersurfaces in $\mathbb{R}^n$ of maximum rank, $n \geqslant 3$, $d \geqslant 2n$, is linearizable, i.e., there exists a local coordinate system relative to which all web hypersurfaces are hyperplanes.

In [114] we gave a sharp upper bound for the k-rank of a d-web of codimension k in $\mathbb{R}^{kn}$.

9.

I would not conclude this account without mentioning my wife's role in my life and work. Through war and peace and through bad and good times we have shared a life for forty years, which is both simple and rich. If there is credit for my mathematical works, it will be hers as well as mine.

Abzählungen für Gewebe.

Von SHIING-SHEN CHERN in Hamburg.

In einer früheren Arbeit[1]) hat BLASCHKE einige Sätze bewiesen, die sich auf Abzählungen für Kurvengewebe der Ebene und Flächengewebe des Raumes beziehen. In der vorliegenden Arbeit wollen wir diese Sätze auf höhere Dimensionen verallgemeinern. Es wird also der „Höchstrang" für alle Hyperflächengewebe eines N-dimensionalen Raumes bestimmt.

§ 1. Der allgemeine Fall.

Es sei R_N ein N-dimensionaler Euklidischer Raum mit den Koordinaten $x_1, x_2, \cdots, x_N$. Wir sprechen von einem „n-Gewebe von Hyperflächen" in einem zusammenhängenden Gebiet G des Raumes R_N, wenn n Hyperflächenscharen

(1) $$t_i(x_1, x_2, \cdots, x_N) = \text{konst.} \qquad (i = 1, 2, \cdots, n)$$

sich dort so darstellen lassen, daß für alle ungleichen $i_1, i_2, \cdots, i_N$

(2) $$\frac{\partial (t_{i_1}, t_{i_2}, \cdots, t_{i_N})}{\partial (x_1, x_2, \cdots, x_N)} \neq 0$$

in G gilt. Um den trivialen Fall zu vermeiden, nehmen wir $n > N$ an. Wenn es nun m und nicht mehr linear unabhängige Identitäten in $x_1, x_2, \cdots, x_N$ von der Gestalt

(3) $$\sum_{i=1}^{n} f_i^{(k)}(t_i) = 0 \qquad (k = 1, 2, \cdots, m)$$

gibt (dabei bedeutet k einen Index), so sagen wir: das Hyperflächengewebe hat den „Rang" m. Linearkombination ist dabei mit festen Koeffizienten gemeint, und die vorkommenden Funktionen sollen regulär und analytisch in G sein.

Wir denken uns die Identitäten (3) abgeleitet und finden zwischen den „Pfaffschen Formen" dt_i die linearen Abhängigkeiten

(4) $$\sum_{i=1}^{n} \frac{df_i^{(k)}}{dt_i} dt_i = 0 \qquad (k = 1, 2, \cdots, m).$$

---

[1]) W. BLASCHKE, T40, Hamb. Abhandl. 9 (1933), S. 299–312.

Wir deuten die $df_i^{(k)} : dt_i$ als homogene Koordinaten eines Punktes $\mathfrak{p}_i$ im linearen Raume R_{m-1}. Jeder solche Punkt

$$(5) \qquad (t_i) = \left\{ \frac{df_i^{(k)}}{dt_i} \right\}$$

beschreibt eine Kurve $\mathfrak{P}_i$ in diesem R_{m-1}. Dann können wir die Abhängigkeiten vektoriell so schreiben

$$(6) \qquad \sum_{i=1}^{n} \mathfrak{p}_i(t_i)\, dt_i = 0.$$

Führen wir einen Differentiator Δ im Raum R_N

$$(7) \qquad \Delta = \alpha_1 \frac{\partial}{\partial x_1} + \alpha_2 \frac{\partial}{\partial x_2} + \cdots + \alpha_N \frac{\partial}{\partial x_N}$$

so ein, daß

$$(8_1) \qquad \Delta t_1 = \Delta t_2 = \cdots = \Delta t_{N-1} = 0$$

und

$$(8_2) \qquad \Delta t_i \neq 0, \qquad (i \geq N)$$

ist, so folgt aus (6)

$$(9) \qquad \sum_{i=N}^{n} \mathfrak{p}_i(t_i)\, \Delta t_i = 0.$$

Wir nennen die Punkte $\mathfrak{p}_i(t_i)$ „zusammengehörig", wenn sie denselben Werten $(x_1, x_2, \cdots, x_N)$ entsprechen. Dann folgt aus (9), daß irgend $n - N + 1$ „zusammengehörige" Punkte $\mathfrak{p}_i$ linear abhängig sind.

Wir setzen im allgemeinen voraus, daß $n - N$ von den Punkten $\mathfrak{p}_1, \mathfrak{p}_2, \cdots, \mathfrak{p}_n$, z.B. $\mathfrak{p}_1, \mathfrak{p}_2, \cdots, \mathfrak{p}_{n-N}$, linear unabhängig sind. Dann liegen $\mathfrak{p}_{n-N+1}, \mathfrak{p}_{n-N+2}, \cdots, \mathfrak{p}_n$ in dem durch $\mathfrak{p}_1, \mathfrak{p}_2, \cdots, \mathfrak{p}_{n-N}$ bestimmten linearen Raume R_{n-N-1}. Somit ist gezeigt:

S_1: *Je n zusammengehörige Punkte $\mathfrak{p}_i$ der n Kurven $\mathfrak{P}_i$ liegen in einem R_{n-N-1}.*

Wir bezeichnen im folgenden den linearen Raum, welcher von einigen Punkten oder Räumen aufgespannt wird, als eine Summe davon, und zwar mit einem besonderen Summenzeichen $\dot{+}$. Nennen wir den „Summenraum" von

$$(10) \qquad \mathfrak{p}_i \dot{+} \mathfrak{p}_i' \dot{+} \cdots \dot{+} \mathfrak{p}_i^{(p)} = \mathfrak{S}_{i,p}$$

den „p-ten Schmiegraum" von $\mathfrak{P}_i$ in $\mathfrak{p}_i$ und den Summenraum zusammengehöriger

$$(11) \qquad \mathfrak{S}_{1,p} \dot{+} \mathfrak{S}_{2,p} \dot{+} \cdots \dot{+} \mathfrak{S}_{n,p} = R_{k(p)},$$

dann ist nach S_1 die erste Dimensionszahl

(12) $$k(0) = n - N - 1.$$

Wir haben im allgemeinen

(13) $$R_{n-N-1} = \mathfrak{p}_1 \dotplus \mathfrak{p}_2 \dotplus \cdots \dotplus \mathfrak{p}_{n-N}.$$

Wenden wir Δ auf diese Punkte an, so folgt, daß

$$R_{2n-3N} = R_{n-N-1} \dotplus \Delta\mathfrak{p}_N \dotplus \Delta\mathfrak{p}_{N+1} \dotplus \cdots \dotplus \Delta\mathfrak{p}_{n-N}$$

oder

(14) $$R_{2n-3N} = R_{n-N-1} \dotplus \mathfrak{p}'_N \dotplus \mathfrak{p}'_{N+1} \dotplus \cdots \dotplus \mathfrak{p}'_{n-N}$$

ist, wenn $n - 2N + 1 \geqq 0$ ist. Da $\mathfrak{p}_{n-N+1}, \mathfrak{p}_{n-N+2}, \cdots, \mathfrak{p}_n$ lineare Kombinationen von $\mathfrak{p}_1, \mathfrak{p}_2, \cdots, \mathfrak{p}_{n-N}$ sind, liegen $\Delta\mathfrak{p}_{n-N+1}, \Delta\mathfrak{p}_{n-N+2}, \cdots, \Delta\mathfrak{p}_n$ und somit $\mathfrak{p}'_{n-N+1}, \mathfrak{p}'_{n-N+2}, \cdots, \mathfrak{p}'_n$ auch im R_{2n-3N}. Daher können wir schreiben

(15) $$R_{2n-3N} = R_{n-N-1} \dotplus \mathfrak{p}'_N \dotplus \mathfrak{p}'_{N+1} \dotplus \cdots \dotplus \mathfrak{p}'_n.$$

Wir nehmen als Allgemeinfall an, daß unser R_{2n-3N} wirklich von der Dimension $2n - 3N$ ist. Dieser R_{2n-3N} ist im allgemeinen durch R_{n-N-1} und irgend $n - 2N + 1$ der Punkte $\mathfrak{p}'_N, \mathfrak{p}'_{N+1}, \cdots, \mathfrak{p}'_n$, z. B. $\mathfrak{p}'_{2N}, \mathfrak{p}'_{2N+1}, \cdots, \mathfrak{p}'_n$ bestimmt. Somit bleibt R_{2n-3N} fest, wenn wir die Marken $1, 2, \cdots, N-1$ der Reihe nach mit N vertauschen. Dies zeigt, daß die Punkte $\mathfrak{p}'_1, \mathfrak{p}'_2, \cdots, \mathfrak{p}'_{N-1}$ ebenfalls in diesem R_{2n-3N} liegen. Daher ergibt sich

(16) $$R_{2n-3N} = R_{n-N-1} \dotplus \mathfrak{p}'_1 \dotplus \mathfrak{p}'_2 \dotplus \cdots \dotplus \mathfrak{p}'_n,$$

und wir können den Satz aussprechen:

S_2: *Je n zusammengehörige Tangenten $\mathfrak{p}_i \dotplus \mathfrak{p}'_i$ an die $\mathfrak{P}_i$ liegen in einem R_{2n-3N}, d. h.*

(17) $$k(1) = 2n - 3N.$$

Nehmen wir allgemein an

(18) $$R_{k(p)} = R_{k(p-1)} \dotplus \mathfrak{p}_1^{(p)} \dotplus \mathfrak{p}_2^{(p)} \dotplus \cdots \dotplus \mathfrak{p}_q^{(p)}$$

mit

(19) $$q = k(p) - k(p-1).$$

Dann ist nach der obigen Schlußweise

(20) $$R_{k(p+1)} = R_{k(p)} \dotplus \mathfrak{p}_N^{(p+1)} \dotplus \mathfrak{p}_{N+1}^{(p+1)} \dotplus \cdots \dotplus \mathfrak{p}_q^{(p+1)}$$

also

(21) $$k(p+1) - k(p) = q - N + 1.$$

Nach (19) und (21) erhalten wir durch Subtraktion

$$(22)\qquad \{k(p+1)-k(p)\}-\{k(p)-k(p-1)\} = -N+1.$$

Aus (22) folgt wegen (12), (17)

$$(23)\qquad \begin{aligned} &\{k(1)-k(0)\} &&= n-2N+1,\\ &\{k(2)-k(1)\}-\{k(1)-k(0)\} &&= -N+1,\\ &\ldots\ldots\ldots\ldots\ldots\ldots\ldots\ldots &&\\ &\{k(p)-k(p-1)\}-\{k(p-1)-k(p-2)\} &&= -N+1 \end{aligned}$$

und durch Addition

$$(24)\qquad k(p)-k(p-1) = n-(p+1)N+p.$$

Daraus folgt wiederum

$$\begin{aligned} k(0) &= n-N-1\\ k(1)-k(0) &= n-2N+1,\\ &\ldots\ldots\ldots\ldots\\ k(p)-k(p-1) &= n-(p+1)N+p. \end{aligned}$$

$$(25)\qquad k(p) = (p+1)n-\frac{1}{2}(p+1)(p+2)N+\frac{1}{2}(p-1)(p+2).$$

Nach (24) wird $R_{k(p_1)}$ fest sein, wenn

$$n-(p_1+1)N+p_1 \leqq N-1$$

oder

$$(26)\qquad n-(p_1+1)N+p_1 = N-1-s,$$

wo $0 \leqq s \leqq N-2$ durch die Bedingung definiert wird, daß p_1 eine ganze Zahl sein soll. Die Bedingung wird dann und nur dann erfüllt, wenn

$$(27)\qquad \begin{aligned} s &\equiv -n+1 \mod N-1,\\ 0 &\leqq s \leqq N-2. \end{aligned}$$

So erhalten wir

$$(28)\qquad p_1 = \frac{1}{N-1}(n-2N+s+1)$$

und

$$(29)\qquad k(p_1) = \frac{1}{2(N-1)}\{(n-1)(n-N)+s(N-s-1)\}-1.$$

Somit ist der folgende Satz bewiesen:

S_8: *Die Kurven $\mathfrak{P}_i$ liegen notwendig in demselben $R_{k(p_1)}$.*

Ich behaupte: der Rang $m \leqq k(p_1)+1$. Wäre nämlich $m > k(p_1)+1$, so gäbe es eine lineare Gleichung zwischen den Koordinaten der $\mathfrak{p}_i$ mit festen Koeffizienten:

$$(30)\qquad \sum_{k=1}^{m} c_k \frac{d f_i^{(k)}}{d t_i} = 0 \qquad (i = 1, 2, \cdots, n).$$

Hieraus folgt durch Integration bei geeigneter Wahl der Anfangswerte

$$(31)\qquad \sum_{k=1}^{m} c_k f_i^{(k)} = 0 \qquad (i = 1, 2, \cdots, n).$$

Somit ist gezeigt:

S_4: *Der Höchstwert M für den Rang eines n-Hyperflächengewebes im linearen Raum R_N wird durch*

$$(32)\qquad M \leqq \frac{1}{2(N-1)} \{(n-1)(n-N) + s(N-s-1)\}$$

gegeben.

§ 2. Die Ausnahmefälle.

Im vorhergehenden Abschnitt haben wir nur den allgemeinen Fall betrachtet, d. h. wir haben immer angenommen, daß die Räume $R_{k(p)}$ für alle p zwischen Null und p_1 wirklich von der Höchstdimension sind. Jetzt wollen wir uns überlegen, wie die Diskussion sich ändert, wenn die Dimension von $R_{k(p_0)}$, wo p_0 eine ganze Zahl zwischen Null und p_1, niedriger als die durch (25) gegebene wird.

Es sei p_0 eine ganze Zahl zwischen Null und p_1, so daß

$$(33)\qquad k(p_0) = (p_0+1)n - \frac{1}{2}(p_0+1)(p_0+2)N + \frac{1}{2}(p_0-1)(p_0+2) - r, \qquad r \geqq 0.$$

Wir haben dann

$$(34)\qquad k(p) - k(p-1) = n - (p+1)N + p - r, \qquad p \geqq p_0.$$

Daraus folgt

$$k(p_0) = (p_0+1)n - \frac{1}{2}(p_0+1)(p_0+2)N + \frac{1}{2}(p_0-1)(p_0+2) - r,$$

$$k(p_0+1) - k(p_0) = n - (p_0+2)N + (p_0+1) - r,$$

$$\cdots\cdots\cdots\cdots$$

$$k(p) - k(p-1) = n - (p+1)N + p - r.$$

$$(35)\qquad k(p) = (p+1)n - \frac{1}{2}(p+1)(p+2)N + \frac{1}{2}(p-1)(p+2) - r(p-p_0+1), \qquad p \geqq p_0.$$

Nach (34) wird $R_{k(p_1)}$ fest sein, wenn

$$n-(p_1+1)N+p_1-r \leqq N-1.$$

Daraus

$$p_1=\frac{1}{N-1}(n-2N-r+s+1), \tag{36}$$

wo s durch

$$\begin{aligned} &s \equiv -n+r+1, \qquad \text{mod. } N-1, \\ &0 \leqq s \leqq N-2 \end{aligned} \tag{37}$$

definiert wird. Durch Einsetzung des Wertes p_1 von (36) in (35) haben wir

$$\begin{aligned} k(p_1)=\frac{1}{2(N-1)}&\{(n-r-1)(n-r-N)+s(N-s-1)\} \\ &+rp_0-1. \end{aligned} \tag{38}$$

Somit ergibt sich der folgende Satz:

S_5: *Wenn die Dimensionszahl des Raumes* $R_{k(p_0)}$ *durch* (33) *gegeben wird, dann wird der Höchstwert M für den Rang des n-Hyperflächengewebes durch die Formel*

$$M \leqq \frac{1}{2(N-1)}\{(n-r-1)(n-r-N)+s(N-s-1)\}+rp_0 \tag{39}$$

gegeben.

§ 3. Beispiele für Gewebe höchsten Ranges.

In diesem Abschnitt werden wir nach Castelnuovo ein Beispiel eines Gewebes geben, dessen Rang den Höchstwert in (32) wirklich erreicht.

Wir betrachten eine algebraische Kurve C von der Klasse n im linearen Raume R_N, die in keinem R_{N-1} liegt. Wir setzen voraus, daß C vom höchstmöglichen Geschlecht ist. Dann ist das Geschlecht von C gleich[2])

$$\alpha=\chi\left\{n-\frac{N+1}{2}-\chi\,\frac{N-1}{2}\right\}, \tag{40}$$

wo χ die kleinste ganze Zahl, die nicht kleiner als $\frac{n-N}{N-1}$ ist, bedeutet. Es ist in der Tat

$$\chi=\frac{n-N}{N-1}+\frac{s}{N-1}, \tag{41}$$

wo s durch (27) gegeben wird. Durch Einsetzung des Wertes χ in (40) erhalten wir für α gerade den Ausdruck auf der rechten Seite von (32).

[2]) G. Castelnuovo, Atti Torino **24** (1889), S. 368.

Wir beschränken uns auf das Gebiet, in dem die n Schmieghyperebenen der Kurve C durch jeden Punkt reell sind*). Die n Scharen der

*) Um zu zeigen, daß es unter diesen algebraischen Geweben reelle gibt, diene folgendes einfache Beispiel. Wir betrachten im R_N eine normale rationale Regelfläche der Ordnung $N-1$:

$$\text{(A)} \qquad x_0 = 1, \quad x_1 = \lambda, \ \ldots, \ x_m = \lambda^m, \quad x_{m+1} = \mu, \ \ldots, \ x_N = \mu \lambda^{N-m-1},$$

und eine Kurve γ der Ordnung n auf dieser Fläche, die jede Erzeugende in $\chi+1$ Punkten schneidet. Diese Kurve sei durch die folgende Gleichung gegeben:

$$\text{(B)} \qquad f(\lambda, \mu) = 0,$$

wo $f(\lambda, \mu)$ ein Polynom in λ, μ von den Graden $\eta, \chi+1$ bezüglich λ, μ ist. Der Einfachheit halber werden wir annehmen, daß das Glied $\lambda^\eta \mu^{\chi+1}$ in $f(\lambda, \mu)$ wirklich auftritt. Wenn N ungerade ist, nehmen wir $m = \frac{N-1}{2}$. Da γ von der Ordnung n sein soll, muß sie mit einer allgemeinen Hyperebene n Schnittpunkte haben, d. h. die Gleichung (B) hat mit

$$\text{(C)} \qquad c_0 + c_1 \lambda + \cdots + c_m \lambda^m + \mu (c_{m+1} + \cdots + c_N \lambda^{N-m-1}) = 0,$$

wo $c_0, c_1, \cdots, c_N$ irgendwelche Konstanten sind, n Paare gemeinsamer Lösungen in λ, μ. Dann ist nach einem bekannten Theorem

$$\frac{N-1}{2}(\chi+1) + \eta = n.$$

Falls die Kurve (B) in der λ, μ-Ebene keine singulären Punkte hat, ist ihr Geschlecht

$$p = \chi(\eta - 1) = \chi\left[n - \frac{N+1}{2} - \chi \frac{N-1}{2}\right].$$

Jetzt sei N gerade. Wir setzen $m = \frac{N}{2} - 1$ und nehmen ein Polynom $f(\lambda, \mu)$ von der folgenden Form

$$\text{(B')} \qquad \begin{aligned} f(\lambda, \mu) = {} & (a_{00} \lambda^\eta + \cdots + a_{0\eta}) \mu^{\chi+1} \\ & + (a_{11} \lambda^{\eta-1} + \cdots + a_{1\eta}) \mu^\chi + \cdots + (a_{\chi+1, \chi+1} \lambda^{\eta - \chi - 1} + \cdots + a_{\chi+1, \eta}), \end{aligned}$$

wo $a_{00} \lambda^\eta + \cdots + a_{0\eta} = 0$ und $a_{11} \lambda^{\eta-1} + \cdots + a_{1\eta} = 0$ keine gemeinsamen Wurzeln haben. Dann ist

$$\frac{N}{2}(\chi+1) + \eta = n + (\chi+1),$$

da die Gleichungen (B') und (C) jetzt $\chi+1$ feste gemeinsame Lösungen haben. Die Kurve (B') in der λ, μ-Ebene hat einen $(\chi+1)$-fachen Punkt $\lambda = \infty$, $\mu = 0$, und zwar mit lauter verschiedenen Tangenten, vorausgesetzt, daß die Gleichung

$$a_{00} + a_{11} y + \cdots + a_{\chi+1, \chi+1} y^{\chi+1} = 0$$

lauter verschiedene Wurzeln in y besitzt. Wenn (B) keine weiteren singulären Punkte hat, was wir voraussetzen können, ist ihr Geschlecht

$$p = \chi(\eta - 1) - \frac{\chi(\chi+1)}{2} = \chi\left[n - \frac{N+1}{2} - \chi \frac{N-1}{2}\right].$$

In beiden Fällen hat die Kurve γ dasselbe Geschlecht wie die Kurven (B) und (B'). Es läßt sich leicht einrichten, daß γ in keinem niederen Raum liegt und daß ihre Schnitt-

Schmieghyperebenen der Kurve C bilden ein n-Hyperflächengewebe im betrachteten Gebiet. Auf der Kurve C gibt es genau α linear unabhängige Abelsche Integrale erster Gattung $f^{(k)}$. Sind t_i die Parameterwerte der Berührungspunkte von n durch einen Punkt gehenden Schmieghyperebenen, so ist nach dem Abelschen Theorem[3]) bei geeigneter Normierung

$$\sum_{i=1}^{n} f^{(k)}(t_i) = 0. \tag{42}$$

Daher ist unser Gewebe vom Rang α, und es gilt der folgende Satz:

S_6: *Der Höchstrang eines n-Hyperflächengewebes im linearen Raume R_N ist*

$$M = \frac{1}{2(N-1)}\{(n-1)(n-N)+s(N-s-1)\}. \tag{43}$$

punkte mit einer passenden Hyperebene reell und verschieden sind. Gehen wir von γ zur dualen Figur über, so bekommen wir eine Kurve mit den gewünschten Eigenschaften.

[3]) Vgl. etwa Enzyklopädie II B 2, W. Wirtinger, Nr. 42, S. 160.

Hamburg, den 4. Dezember 1934.

Eine Invariantentheorie der Dreigewebe aus r-dimensionalen Mannigfaltigkeiten im R_{2r}.

Von SHIING-SHEN CHERN*) in Hamburg und Peiping.

Übersicht.

In der vorliegenden Arbeit beschäftigen wir uns mit einer Invariantentheorie von Dreigeweben von r-dimensionalen Mannigfaltigkeiten in einem $2r$-dimensionalen Raum R_{2r}[1]).

Es sei R_{2r} ein affiner Raum mit den Koordinaten $x_1, x_2, \cdots, x_{2r}$. Unter einer Schar von r-dimensionalen Mannigfaltigkeiten versteht man das topologische Bild einer Schar von parallelen r-dimensionalen linearen Räumen in einer Hyperkugel im R_{2r}. Aus der Definition folgt, daß durch jeden Punkt des betrachteten Gebietes G eine und nur eine Mannigfaltigkeit aus der Schar geht. Ein Dreigewebe besteht aus drei solchen Scharen, wobei als Regularitätsbedingung angenommen wird, daß sich Mannigfaltigkeiten aus verschiedenen Scharen weder in Kurven noch in höher-dimensionalen Mannigfaltigkeiten schneiden. Ferner verlangen wir, daß es in dem Gebiet G eine Umgebung U gibt, so daß zwei Mannigfaltigkeiten in U aus verschiedenen Scharen genau einen Punkt in G gemein haben. In den folgenden Betrachtungen werden wir uns ausschließlich auf die Umgebung U beschränken.

Das Gewebe und somit die auftretenden Funktionen seien analytisch. Dann kann man jede Schar durch ein vollständig integrierbares Pfaffsches System definieren. Auf Grund dieser Überlegung werden wir in § 1 eine Invariantentheorie entwickeln, welche das Äquivalenzproblem löst. In § 2 bestimmen wir die Bedingungen, deren Erfülltsein mit dem Bestehen gewisser Konfigurationen, die als Axiome in einem abstrakten Aufbau der Gewebetheorie zuerst von THOMSEN[2]) zugrunde gelegt worden sind, äquivalent ist. Diese Bedingungen — im Gegensatz zum Falle

---

*) "Research Fellow" der staatlichen Tsing Hua Universität, Peiping.

[1]) Eine entsprechende Invariantentheorie von Flächendreigeweben im R_4 hat G. BOL von einem anderen Standpunkt aus entwickelt. Man vergleiche G. BOL, Über Dreigewebe im vierdimensionalen Raum, Mathematische Annalen, Bd. 110 (1934), S. 431—463.

[2]) G. THOMSEN, T_{12}, Schnittpunktssätze in ebenen Geweben, Hamburger Abhandlungen, Bd. 7 (1930), S. 99—106.

des Kurvendreigewebes in der Ebene — haben für $r > 1$ verschiedene Bedeutungen, wie die in § 3 angeführten Beispiele zeigen.

Natürlich enthält diese Invariantentheorie die von Kurvendreigeweben in der Ebene als einen speziellen Fall.

Als Vorkenntnis setzen wir das Rechnen mit alternierenden Differentialformen und die Theorie der Systeme von Differentialgleichungen voraus, wie sie in dem Buch von KÄHLER[3]) behandelt sind. Im übrigen ist die Arbeit unabhängig von den anderen dieser Reihe lesbar.

Für Anregung zu dieser Arbeit und für viele wertvolle Ratschläge bin ich den Herren BLASCHKE und KÄHLER zu Dank verpflichtet.

§ 1. Invariantensystem.

Ein Dreigewebe von r-dimensionalen Mannigfaltigkeiten werden wir durch die folgenden drei vollständig integrierbaren Pfaffschen Systeme definieren:

(1) $$\omega_{1k} = 0; \quad \omega_{2k} = 0; \quad \omega_{3k} = 0$$ [4]),

die die drei Scharen bestimmen. In diesen Gleichungen bedeuten die $\omega_{\varrho k}$ Pfaffsche Formen der Gestalt:

(2) $$\omega_{\varrho k} = a_{\varrho k i}\, dx_i + a_{\varrho k r+i}\, dx_{r+i},$$

wobei die $a_{\varrho k i}$, $a_{\varrho k r+i}$ Funktionen der x sind. Die vollständige Integrierbarkeit der betreffenden Systeme drücken wir so aus:

(3) $$d\omega_{\varrho k} \equiv 0 \qquad (\mathrm{mod}\, \omega_{\varrho 1}, \cdots, \omega_{\varrho r}),$$

d. h. $d\omega_{\varrho k}$ ist eine Linearkombination der $\omega_{\varrho k}$.

Da die Mannigfaltigkeiten aus verschiedenen Scharen sich nicht in Kurven schneiden, sind insbesondere die ω_{1k}, ω_{2k} in bezug auf die dx linear unabhängig. Deshalb kann man mit ihrer Hilfe die ω_{3k} linear davon ausdrücken und man bekommt die Gleichungen:

(4) $$\omega_{3k} = z_{ki}(x)\, \omega_{1i} + t_{ki}(x)\, \omega_{2i}.$$

Man erhält dieselben Scharen, wenn man auf die $\omega_{\varrho k}$ mit festem ϱ irgendeine lineare Transformation ausübt. Da

$$|z_{ki}(x)| \neq 0, \qquad |t_{ki}(x)| \neq 0$$

---

[3]) E. KÄHLER, Einführung in die Theorie der Systeme von Differentialgleichungen, Hamburger Mathematische Einzelschriften, Heft 16 (1934).

[4]) Im folgenden werden wir verabreden, daß jeder freie lateinische Index von 1 bis r läuft, und daß ein griechischer Index von 1 bis 3 läuft. Außerdem benutzen wir die Tensorschreibweise, so daß zweimal derselbe Index eine Summation von 1 bis r bedeutet.

gilt, können wir

$$\omega_{1k}^* = -z_{ki}(x)\,\omega_{1i},$$
$$\omega_{2k}^* = -t_{ki}(x)\,\omega_{2i},$$

setzen und erhalten so:

$$\omega_{1k}^* + \omega_{2k}^* + \omega_{3k} = 0.$$

Nun denken wir uns diese Transformation schon ausgeführt, so daß also zwischen den $\omega_{\varrho k}$ Beziehungen der Form

(5) $$\omega_{1k} + \omega_{2k} + \omega_{3k} = 0$$

bestehen. Jedes System $\omega_{\varrho k}$ läßt eine lineare Transformation mit nichtverschwindender Determinante zu, ohne die entsprechende Schar zu ändern. Die allgemeinste lineare Transformation der $\omega_{\varrho k}$, die die Beziehungen (5) erhält, ist von der Form

(6) $$\Omega_{\varrho k} = u_{ik}\,\omega_{\varrho i},$$

wo die u_{ik} Funktionen von x_i, x_{r+i} mit $|u_{ik}| \neq 0$ sind.

Wir betrachten die u_{ik} als r^2 neue Variablen. Dann ist die Invariantentheorie des Gewebes mit der des Formensystems $\Omega_{\varrho k}$ gleichwertig.

Genauer soll das so heißen:

Hat man ein anderes, in dem Raum $(\bar{x}_1, \cdots, \bar{x}_{2r})$ *gelegenes Gewebe, das durch die Gleichungen*

(7) $$\bar{\omega}_{1k} = 0,\quad \bar{\omega}_{2k} = 0,\quad \bar{\omega}_{3k} = 0,$$
$$\bar{\omega}_{1k} + \bar{\omega}_{2k} + \bar{\omega}_{3k} = 0$$

beschrieben sei, und bildet man mit r^2 *weiteren Variablen* $\bar{u}_{ik}$ *wie oben die Pfaffschen Formen*

(8) $$\bar{\Omega}_{\varrho k} = \bar{u}_{ik}\,\bar{\omega}_{\varrho i},$$

so ist für die topologische (genauer biholomorphe) Äquivalenz der beiden Gewebe notwendig und hinreichend, daß es eine biholomorphe Transformation

(9) $$\begin{aligned} x_i &= x_i\,(\bar{x}_1, \cdots, \bar{x}_{2r}, \bar{u}_{11}, \cdots, \bar{u}_{rr}),\\ x_{r+i} &= x_{r+i}\,(\bar{x}_1, \cdots, \bar{x}_{2r}, \bar{u}_{11}, \cdots, \bar{u}_{rr}),\\ u_{ik} &= u_{ik}\,(\bar{x}_1, \cdots, \bar{x}_{2r}, \bar{u}_{11}, \cdots, \bar{u}_{rr}) \end{aligned}$$

gibt, die das Formensystem $\bar{\Omega}_{1k}$, $\bar{\Omega}_{2k}$ *in* Ω_{1k}, Ω_{2k} *überführt.*

Die Notwendigkeit ist trivial. Um zu beweisen, daß die Bedingungen auch hinreichend sind, nehmen wir an, daß die Gleichungen

$$\Omega_{1k} = \bar{\Omega}_{1k},\quad \Omega_{2k} = \bar{\Omega}_{2k}$$

24*

identisch in $\bar{x}_i, \bar{x}_{r+i}, \bar{u}_{ik}$ erfüllt seien, wenn man die Funktionen (9) in Ω_{1k}, Ω_{2k} einsetzt. Dann folgt

$$\frac{\partial x_i}{\partial \bar{u}_{kj}} = \frac{\partial x_{r+i}}{\partial \bar{u}_{kj}} = 0.$$

Die ersten beiden Zeilen in (9) stellen also eine topologische Abbildung

$$(10) \qquad \begin{aligned} x_i &= x_i(\bar{x}_1, \cdots, \bar{x}_{2r}), \\ x_{r+i} &= x_{r+i}(\bar{x}_1, \cdots, \bar{x}_{2r}) \end{aligned}$$

des Raumes $(\bar{x}_1, \cdots, \bar{x}_{2r})$ auf den Raum $(x_1, \cdots, x_{2r})$ dar, welche die durch

$$\bar{\Omega}_{1k} = 0, \qquad \bar{\Omega}_{2k} = 0, \qquad \bar{\Omega}_{1k} + \bar{\Omega}_{2k} = 0$$

definierten Scharen in die durch

$$\Omega_{1k} = 0, \qquad \Omega_{2k} = 0, \qquad \Omega_{1k} + \Omega_{2k} = 0$$

definierten überführen.

Um die allgemeine Theorie der Äquivalenz von Systemen Pfaffscher Formen[5]) auf den vorliegenden Fall anwenden zu können, müssen wir versuchen, uns zu den $2r$ unabhängigen Formen Ω_{1k}, Ω_{2k} r^2 weitere unabhängige zu verschaffen, die in eindeutiger und invarianter Weise mit den Ω_{1k}, Ω_{2k} verbunden sind. Dies gelingt durch Aufstellung der Ableitungsgleichungen.

Jedes System $\omega_{\varrho k} = 0$ $(k = 1, 2, \cdots, r)$ ist vollständig integrierbar. Man kann also setzen:

$$(11) \qquad \begin{aligned} d\omega_{1k} &= \alpha_{kij}\,\omega_{1j}\,\omega_{1i} + \beta_{kij}\,\omega_{2j}\,\omega_{1i}, \\ d\omega_{2k} &= \gamma_{kij}\,\omega_{1j}\,\omega_{2i} + \delta_{kij}\,\omega_{2j}\,\omega_{2i}, \end{aligned}$$

wo $\alpha_{kij}, \beta_{kij}, \gamma_{kij}, \delta_{kij}$ Funktionen von x_i, x_{r+i} sind mit den Zusatzforderungen

$$(12) \qquad \begin{aligned} \alpha_{kij} + \alpha_{kji} &= 0, \\ \delta_{kij} + \delta_{kji} &= 0. \end{aligned}$$

Aus der vollständigen Integrierbarkeit des Systems $\omega_{1k} + \omega_{2k} = 0$ folgt weiter

$$(13) \qquad 2(\alpha_{kij} + \delta_{kij}) - \beta_{kij} + \beta_{kji} - \gamma_{kij} + \gamma_{kji} = 0.$$

Durch Ableiten der Gleichung (6) erhält man

$$\begin{aligned} d\Omega_{1k} &= -\omega_{1i}\,du_{ik} + u_{ik}\,d\omega_{1i}, \\ d\Omega_{2k} &= -\omega_{2i}\,du_{ik} + u_{ik}\,d\omega_{2i}. \end{aligned}$$

Führt man die Größen v_{ik} ein, die durch

$$(14) \qquad u_{ik}\,v_{kl} = \delta_{il}, \qquad v_{ik}\,u_{kl} = \delta_{il}$$

---

[5]) Man vergleiche E. Cartan, Les sousgroupes des groupes continus de transformations, Chap. I, § 1–3. Annales de l'École Normale Supérieure, série 3, t. 25 (1908).

definiert sein sollen, so kann man schreiben

$$(15)\qquad \begin{aligned} d\,\Omega_{1k} &= \Omega_{1i}\,\vartheta_{ik}^{(1)},\\ d\,\Omega_{2k} &= \Omega_{2i}\,\vartheta_{ik}^{(2)}, \end{aligned}$$

wo die allgemeinste Form von $\vartheta_{ik}^{(1)}$, $\vartheta_{ik}^{(2)}$ die folgende ist:

$$(16)\qquad \begin{aligned} \vartheta_{ik}^{(1)} &= -v_{ij}\,du_{jk} - u_{nk}\,v_{ij}\,v_{ml}\,\alpha_{njl}\,\Omega_{1m}\\ &\quad - u_{nk}\,v_{ij}\,v_{ml}\,\beta_{njl}\,\Omega_{2m} + \xi_{ikm}\,\Omega_{1m},\\ \vartheta_{ik}^{(2)} &= -v_{ij}\,du_{jk} - u_{nk}\,v_{ij}\,v_{ml}\,\gamma_{njl}\,\Omega_{1m}\\ &\quad - u_{nk}\,v_{ij}\,v_{ml}\,\delta_{njl}\,\Omega_{2m} + \eta_{ikm}\,\Omega_{2m}. \end{aligned}$$

Dabei bleiben die ξ_{ikm}, η_{ikm} außer den Einschränkungen

$$(17)\qquad \xi_{ikm} = \xi_{mki}, \qquad \eta_{ikm} = \eta_{mki}$$

noch willkürlich.

Man bildet

$$\begin{aligned} \vartheta_{ik}^{(2)} - \vartheta_{ik}^{(1)} &= u_{nk}\,v_{ij}\,v_{ml}\,(\alpha_{njl}\,\Omega_{1m} - \gamma_{njl}\,\Omega_{1m} + \beta_{njl}\,\Omega_{2m} - \delta_{njl}\,\Omega_{2m})\\ &\quad - \xi_{ikm}\,\Omega_{1m} + \eta_{ikm}\,\Omega_{2m}. \end{aligned}$$

Es ist also $\vartheta_{ik}^{(2)} - \vartheta_{ik}^{(1)}$ von der Form

$$(18)\qquad \vartheta_{ik}^{(2)} - \vartheta_{ik}^{(1)} = c_{ik}^{(m)}\,\Omega_{1m} + d_{ik}^{(m)}\,\Omega_{2m},$$

wo

$$(19)\qquad \begin{aligned} c_{ik}^{(m)} &= u_{nk}\,v_{ij}\,v_{ml}\,(\alpha_{njl} - \gamma_{njl}) - \xi_{ikm},\\ d_{ik}^{(m)} &= u_{nk}\,v_{ij}\,v_{ml}\,(\beta_{njl} - \delta_{njl}) + \eta_{ikm}. \end{aligned}$$

Fordert man die Beziehungen

$$(20)\qquad \begin{aligned} c_{ik}^{(m)} + c_{mk}^{(i)} &= 0,\\ d_{ik}^{(m)} + d_{mk}^{(i)} &= 0, \end{aligned}$$

so hat man

$$(21)\qquad \begin{aligned} u_{nk}\,v_{ij}\,v_{ml}\,(\alpha_{njl} - \gamma_{njl} + \alpha_{nlj} - \gamma_{nlj}) - 2\,\xi_{ikm} &= 0,\\ u_{nk}\,v_{ij}\,v_{ml}\,(\beta_{njl} - \delta_{njl} + \beta_{nlj} - \delta_{nlj}) + 2\,\eta_{ikm} &= 0. \end{aligned}$$

Da diese Gleichungen auch in den Indizes i und m symmetrisch sind, sind die ξ_{ikm}, η_{ikm} dadurch eindeutig bestimmt. Mit den durch (21) bestimmten Werten von ξ_{ikm} und η_{ikm} bekommt man wegen (13)

$$(22)\qquad c_{ik}^{(m)} = d_{ik}^{(m)}.$$

Wir können also die Ableitungsgleichungen so schreiben:

$$(23)\qquad \begin{aligned} d\,\Omega_{1k} &= \Omega_{1i}\,\vartheta_{ik} - c_{ik}^{(m)}\,\Omega_{1i}\,\Omega_{1m},\\ d\,\Omega_{2k} &= \Omega_{2i}\,\vartheta_{ik} + c_{ik}^{(m)}\,\Omega_{2i}\,\Omega_{2m}, \end{aligned}$$

wo die ϑ_{ik} r^2 unabhängige Pfaffsche Formen bedeuten, die eindeutig durch die folgenden Ausdrücke bestimmt sind:

$$(24) \qquad \vartheta_{ik} = -v_{ij}\, du_{jk} - u_{nk}\, v_{ij}\, v_{ml}\, \gamma_{njl}\, \Omega_{1m} - u_{nk}\, v_{ij}\, v_{ml}\, \beta_{njl}\, \Omega_{2m}.$$

Außerdem ist

$$(25) \qquad c_{ik}^{(m)} = \tfrac{1}{2} u_{nk}\, v_{ij}\, v_{ml} (2\alpha_{njl} - \gamma_{njl} + \gamma_{nlj}),$$

also

$$(26) \qquad c_{ik}^{(m)} + c_{mk}^{(i)} = 0.$$

Durch den Ansatz (23) für die Ableitungsgleichungen werden so auf eindeutige Weise r^2 untereinander und auch von den Ω_{1k}, Ω_{2k} linear unabhängige Pfaffsche Formen ϑ_{ik} erklärt.

Da

$$d(d\,\Omega_{1k}) = d(d\,\Omega_{2k}) = 0$$

ist, folgen durch Ableiten der Gleichungen (23) Beziehungen der Form

$$\Omega_{1i}(d\,\vartheta_{ik} - \vartheta_{ij}\,\vartheta_{jk}) = P_{kij}\,\Omega_{1i}\,\Omega_{1j},$$
$$\Omega_{2i}(d\,\vartheta_{ik} - \vartheta_{ij}\,\vartheta_{jk}) = Q_{kij}\,\Omega_{2i}\,\Omega_{2j},$$

wo P_{kij}, Q_{kij} gewisse Pfaffsche Formen bedeuten. Aus diesen Relationen ist zu erkennen, daß $d\vartheta_{ik}$ von folgender Gestalt sein muß:

$$(27) \qquad d\,\vartheta_{ik} = \vartheta_{ij}\,\vartheta_{jk} + f_{ilm}^{(k)}\,\Omega_{1l}\,\Omega_{2m}.$$

Zum späteren Gebrauch rechnen wir die $f_{ilm}^{(k)}$ als Funktionen der Koeffizienten in (11). Man braucht dazu bloß die Gleichung (24) abzuleiten und (23) zu berücksichtigen. Definiert man

$$(28) \qquad \begin{aligned} d\beta_{kij} &= \beta_{kij,m}^{(1)}\,\omega_{1m} + \beta_{kij,m}^{(2)}\,\omega_{2m}, \\ d\gamma_{kij} &= \gamma_{kij,m}^{(1)}\,\omega_{1m} + \gamma_{kij,m}^{(2)}\,\omega_{2m}, \end{aligned}$$

so bekommt man das Resultat:

$$(29) \qquad f_{ilm}^{(k)} = u_{nk}\, v_{ip}\, v_{lq}\, v_{ms}\, A_{npqs},$$

wobei

$$(30) \qquad A_{npqs} = \gamma_{npq,s}^{(2)} - \beta_{nps,q}^{(1)} + \beta_{tqs}\,\gamma_{npt} - \beta_{npt}\,\gamma_{tsq} + \beta_{tps}\,\gamma_{ntq} - \beta_{nts}\,\gamma_{tpq}.$$

Wir machen die Ansätze:

$$(31) \qquad \begin{aligned} dc_{ik}^{(m)} &= c_{ik,j}^{(m,1)}\,\Omega_{1j} + c_{ik,j}^{(m,2)}\,\Omega_{2j} + c_{ik,jn}^{(m)}\,\vartheta_{jn}, \\ df_{ilm}^{(k)} &= f_{ilm,j}^{(k,1)}\,\Omega_{1j} + f_{ilm,j}^{(k,2)}\,\Omega_{2j} + f_{ilm,jn}^{(k)}\,\vartheta_{jn}, \end{aligned}$$

und leiten die Gleichungen (23), (27) ab. Die $c_{ik,j}^{(m,1)}$ usw. nennt man bekanntlich die kovarianten Ableitungen von $c_{ik}^{(m)}$, $f_{ilm}^{(k)}$ in bezug auf Ω_{1k}, Ω_{2k}, ϑ_{ik}. Da die Ω_{1k}, Ω_{2k}, ϑ_{ik} linear unabhängig sind, erhalten wir die Gleichungen

$$
(32)\qquad
\begin{aligned}
c_{ik,l}^{(m,1)} &= \tfrac{1}{2}\left(f_{mli}^{(k)} - f_{ilm}^{(k)}\right),\\
c_{ik,l}^{(m,2)} &= \tfrac{1}{2}\left(f_{mil}^{(k)} - f_{iml}^{(k)}\right),\\
c_{ik,jn}^{(m)} &= -\delta_{nk}\, c_{ij}^{(m)} + \delta_{ij}\, c_{nk}^{(m)} + \delta_{jm}\, c_{ik}^{(n)},\\
& f_{mjl}^{(k)} + f_{jlm}^{(k)} + f_{lmj}^{(k)} - f_{ljm}^{(k)} - f_{mlj}^{(k)} - f_{jml}^{(k)}\\
& - 4\left(c_{ik}^{(m)} c_{ji}^{(l)} + c_{ik}^{(j)} c_{li}^{(m)} + c_{ik}^{(l)} c_{mi}^{(j)}\right) = 0,\\
& f_{ilm,j}^{(k,1)} - f_{ijm,l}^{(k,1)} - 2\, c_{jn}^{(l)} f_{inm}^{(k)} = 0,\\
& f_{ilm,j}^{(k,2)} - f_{ilj,m}^{(k,2)} + 2\, c_{jn}^{(m)} f_{iln}^{(k)} = 0,\\
f_{ilm,jn}^{(k)} &= -\delta_{nk} f_{ilm}^{(j)} + \delta_{ij} f_{nlm}^{(k)} + \delta_{jl} f_{inm}^{(k)} + \delta_{mj} f_{iln}^{(k)}.
\end{aligned}
$$

Also haben wir den Satz:

Einem Dreigewebe von r-dimensionalen Mannigfaltigkeiten im R_{2r} *kann man durch Einführung von* r^2 *neuen Variablen* r^2 *linear unabhängige Pfaffsche Formen* ϑ_{ik} *auf eindeutige und invariante Weise zuordnen, wobei die Ableitungsgleichungen* (23), (27) *und die Integrierbarkeitsbedingungen* (32) *gelten.*

Es gilt auch die Umkehrung:

Zu einem System von $r^2 + 2r$ *linear unabhängigen Pfaffschen Formen* Ω_{1k}, Ω_{2k}, ϑ_{ik} *in* $r^2 + 2r$ *Variablen, etwa* $y_1, \cdots, y_{r^2+2r}$, *deren Ableitungsgleichungen die Gestalt* (23) (27) *haben, und deren Koeffizienten mit ihren kovarianten Ableitungen die Beziehungen* (32) *erfüllen, gehört in einem gewissen 2r-dimensionalen Raum ein Dreigewebe, das durch die Gleichungen*

$$\Omega_{1k} = 0, \quad \Omega_{2k} = 0, \quad \Omega_{1k} + \Omega_{2k} = 0$$

beschrieben wird.

Um diese Umkehrung zu beweisen, beachten wir zunächst, daß wegen (23) jedes System

$$\Omega_{1k} = 0, \quad \Omega_{2k} = 0, \quad \Omega_{1k} + \Omega_{2k} = 0$$

in dem Raume $(y_1, \cdots, y_{r^2+2r})$ vollständig integrierbar ist. Bezeichnen wir mit $x_1(y), \cdots, x_r(y)$ (Abkürzungen von $x_1(y_1, \cdots, y_{r^2+2r})$, usw.) die Integrale des ersten Systems und mit $x_{r+1}(y), \cdots, x_{2r}(y)$ die des zweiten Systems, so haben wir

$$
(33)\qquad
\begin{aligned}
\Omega_{1k} &= A_{ki}(y)\, dx_i,\\
\Omega_{2k} &= B_{ki}(y)\, dx_{r+i}.
\end{aligned}
$$

Da die Ω_{1k}, Ω_{2k} linear unabhängig sind, sind die x_i, x_{r+i} $2r$ analytisch unabhängige Funktionen der y. In dem Raume $(x_1, \cdots, x_{2r})$ wird dann und nur dann ein Dreigewebe verwirklicht werden, wenn es $3r^2$ Funktionen $L_{ki}(y)$, $M_{ki}(y)$, $N_{ki}(y)$ mit

$$|L_{ki}(y)| \neq 0, \quad |M_{ki}(y)| \neq 0, \quad |N_{ki}(y)| \neq 0$$

gibt, so daß die Linearkombinationen

$$L_{ki}\,\Omega_{1i}, \qquad M_{ki}\,\Omega_{2i}, \qquad N_{ki}(\Omega_{1i}+\Omega_{2i})$$

nur von den Variablen $x_1, \cdots, x_{2r}$ abhängen. Dazu ist notwendig und hinreichend, daß die Differentialformen

$$d(L_{ki}\,\Omega_{1i}), \quad d(M_{ki}\,\Omega_{2i}), \qquad d(N_{ki}\,\Omega_{1i}+N_{ki}\,\Omega_{2i})$$

sich aus den Ω_{1k}, Ω_{2k} allein zusammensetzen lassen. Die gesuchten Funktionen L_{ki}, M_{ki}, N_{ki} genügen dann, für festes k, dem folgenden Differentialgleichungssystem

$$dL_i = L_{ij}^{\cdot 1}\,\Omega_{1j} + L_{ij}^{\cdot 2}\,\Omega_{2j} + L_k\,\vartheta_{ik}. \tag{34}$$

Für die Umkehrung braucht man also nur den allgemeinen Existenzsatz für Differentialgleichungen auf den vorliegenden Fall anzuwenden. Unser Differentialgleichungssystem besteht in Wirklichkeit aus den Gleichungen (23), (27), (34). Die Integralelemente genügen außer diesen Gleichungen noch den aus ihnen durch Ableiten entstehenden Gleichungen. Wegen (32) geben die Ableitungen der Gleichungen (23), (27) bloß Identitäten, und man bekommt alle weiteren Gleichungen, indem man (34) ableitet:

$$dL_{ij}^{\cdot 1}\,\Omega_{1j} + dL_{ij}^{\cdot 2}\,\Omega_{2j} + dL_k\,\vartheta_{ik} + L_{ij}^{\cdot 1}\,d\Omega_{1j} + L_{ij}^{\cdot 2}\,d\Omega_{2j} + L_k\,d\vartheta_{ik} = 0, \tag{35}$$

welche mit den folgenden äquivalent sind:

$$\begin{aligned} & dL_{ij}^{\cdot 1}\,\Omega_{1j} + dL_{ij}^{\cdot 2}\,\Omega_{2j} + L_{kj}^{\cdot 1}\,\Omega_{1j}\,\vartheta_{ik} + L_{ik}^{\cdot 1}\,\Omega_{1j}\,\vartheta_{jk} \\ & + L_{kj}^{\cdot 2}\,\Omega_{2j}\,\vartheta_{ik} + L_{ik}^{\cdot 2}\,\Omega_{2j}\,\vartheta_{jk} \\ & - L_{il}^{\cdot 1}\,c_{kl}^{(j)}\,\Omega_{1k}\,\Omega_{1j} + L_{il}^{\cdot 2}\,c_{kl}^{(j)}\,\Omega_{2k}\,\Omega_{2j} + L_l\,f_{ikj}^{(l)}\,\Omega_{1k}\,\Omega_{2j} = 0. \end{aligned} \tag{35a}$$

Nach Eintragen des weiteren Ansatzes

$$\begin{aligned} dL_{ij}^{\cdot 1} &= L_{ijk}^{\cdot 11}\,\Omega_{1k} + L_{ijk}^{\cdot 12}\,\Omega_{2k} + L_{ij,kl}^{\cdot 1}\,\vartheta_{kl}, \\ dL_{ij}^{\cdot 2} &= L_{ijk}^{\cdot 21}\,\Omega_{1k} + L_{ijk}^{\cdot 22}\,\Omega_{2k} + L_{ij,kl}^{\cdot 2}\,\vartheta_{kl} \end{aligned} \tag{36}$$

in (35) folgen für die Bestimmung der Größen $L_{ijk}^{\cdot 11}, \cdots, L_{ij,kl}^{\cdot 2}$ die Gleichungen

$$\begin{aligned} L_{ijk}^{\cdot 11} - L_{ikj}^{\cdot 11} &= 2\,L_{il}^{\cdot 1}\,c_{kl}^{(j)}, \\ L_{ijk}^{\cdot 21} - L_{ikj}^{\cdot 12} &= -L_l\,f_{ikj}^{(l)}, \\ L_{ijk}^{\cdot 22} - L_{ikj}^{\cdot 22} &= -2\,L_{il}^{\cdot 2}\,c_{kl}^{(j)}, \\ L_{ij,kl}^{\cdot 1} &= \delta_{ik}\,L_{lj}^{\cdot 1} + \delta_{jk}\,L_{il}^{\cdot 1}, \\ L_{ij,kl}^{\cdot 2} &= \delta_{ik}\,L_{lj}^{\cdot 2} + \delta_{jk}\,L_{il}^{\cdot 2}. \end{aligned} \tag{37}$$

Daraus sieht man sofort, daß man die Größen $L_{ijk}^{\cdot 11}, \cdots, L_{ij,kl}^{\cdot 2}$ schrittweise nach der Vorschrift des allgemeinen Existenzsatzes bestimmen kann. Die Gleichungen (34) besitzen also Lösungen L_i. Da keine skalaren Gleichungen zwischen den L_i vorhanden sind, kann man die Anfangswerte $L_i(y^0)$ willkürlich vorschreiben. Es gibt also $3r^2$ Funktionen $L_{ki}(y)$, $M_{ki}(y)$, $N_{ki}(y)$ mit den gewünschten Eigenschaften. Damit ist die Umkehrung bewiesen.

Da die d-Operation bei biholomorphen Transformationen invariant ist und die ϑ_{ik} eindeutig bestimmt sind, führt die biholomorphe Transformation (9), die das Formensystem Ω_{1k}, Ω_{2k} in $\overline{\Omega}_{1k}$, $\overline{\Omega}_{2k}$ überführt, gleichzeitig die Formen ϑ_{ik} in die entsprechend gebildeten Formen $\overline{\vartheta}_{ik}$ über. Also sind die zwei Dreigewebe

$$\Omega_{\varrho k} = 0; \qquad \overline{\Omega}_{\varrho k} = 0$$

in den Räumen $(x_1, \cdots, x_{2r})$, $(\overline{x}_1, \cdots, \overline{x}_{2r})$ dann und nur dann topologisch äquivalent, wenn es eine biholomorphe Transformation (9) gibt, die das Formensystem Ω_{1k}, Ω_{2k}, ϑ_{ik} in $\overline{\Omega}_{1k}$, $\overline{\Omega}_{2k}$, $\overline{\vartheta}_{ik}$ überführt.

Aus der allgemeinen Invariantentheorie für n Pfaffsche Formen in n Variablen folgt der Satz:

Das vollständige Invariantensystem eines Dreigewebes besteht aus den Funktionen $c_{ik}^{(m)}$, $f_{ilm}^{(k)}$ *und ihren kovarianten Ableitungen.*

Dabei ist mit dem Ausdruck: vollständiges Invariantensystem gemeint, daß es möglich ist, durch Aufstellung der sämtlichen Invarianten bis zu einer gewissen Ordnung die topologische Äquivalenz oder Nicht-Äquivalenz zweier Gewebe nach einem durchführbaren Verfahren zu entscheiden.

Insbesondere sind alle Gewebe mit

$$c_{ik}^{(m)} = f_{ilm}^{(k)} = 0 \tag{38}$$

topologisch äquivalent. Da das Gewebe aus drei Parallelenbündeln

$$\begin{aligned} x_1 &= \text{konst.}, \cdots\cdots, x_r = \text{konst.},\\ x_{r+1} &= \text{konst.}, \cdots\cdots, x_{2r} = \text{konst.},\\ x_1 + x_{r+1} &= \text{konst.}, \cdots\cdots, x_r + x_{2r} = \text{konst.} \end{aligned}$$

den Bedingungen (38) genügt, ist ein Gewebe, dessen sämtliche Invarianten Null sind, topologisch auf drei Parallelenbündel abbildbar.

§ 2. Die Konfigurationen (S), (P), (D).

In diesem Paragraphen wollen wir als Anwendung der allgemeinen Invariantentheorie die invarianten Formulierungen einiger geometrischen Eigenschaften von Geweben bestimmen. Bei diesen Eigenschaften handelt es sich um das Bestehen gewisser Konfigurationen, die zuerst von THOMSEN vom axiomatischen Standpunkt studiert worden sind. Wenn man in der folgenden Figur die drei Scharen mit *I*, *II*, *III* bezeichnet und die Mannigfaltigkeiten durch Geraden repräsentiert, so sind die betreffenden Konfigurationen die folgenden:

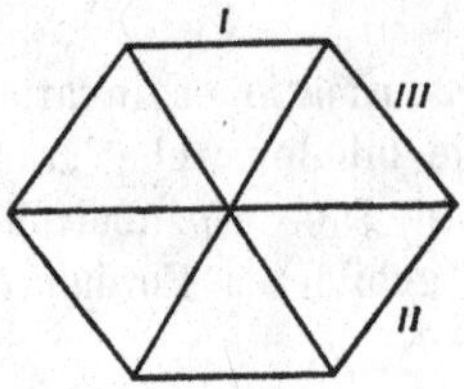

(S)

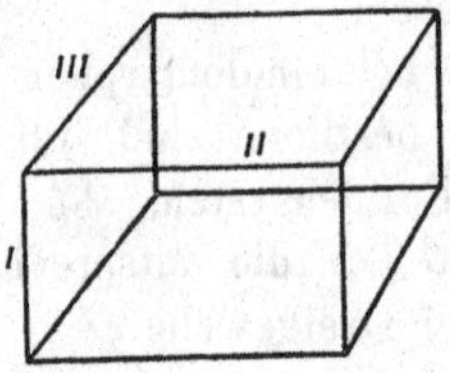

(P)

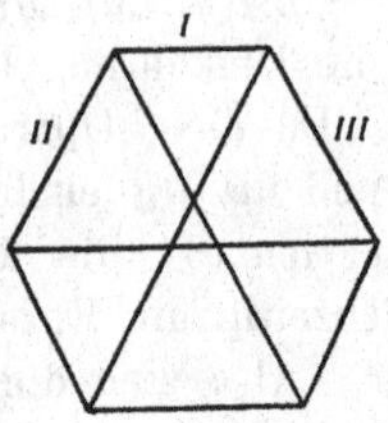

(D)

Diese Konfigurationen haben auch in unserem Falle einen Sinn, weil zwei r-dimensionale Mannigfaltigkeiten im R_{2r} sich in einem Punkte schneiden. Nach THOMSEN bezeichnen wir diese Konfigurationen der Reihe nach mit (S), (P), (D). Genau gesprochen verlangen wir: Wählt man die Punkte, die die Freiheitsgrade $3\,r$ bzw. $5\,r$, $4\,r$ der Konfigurationen bestimmen, hinreichend nahe beieinander, so schließen sich die Figuren in der gezeichneten Weise. Es ist klar, daß das Bestehen einer solchen Konfiguration eine topologische Eigenschaft des Gewebes bedeutet.

THOMSEN hat bewiesen: *Aus* (P) *folgt* (S); *aus* (D) *folgt* (P) *und somit auch* (S).

Stellen wir zunächst die Bedingungen für das Bestehen von (S) auf! Wir denken uns eine topologische Transformation ausgeführt, die die ersten zwei Scharen in die Scharen von Parallelenbündeln

$$(39)\qquad \begin{aligned} x_1 &= \text{konst.}, \cdots\cdots, x_r = \text{konst.}; \\ x_{r+1} &= \text{konst.}, \cdots\cdots, x_{2r} = \text{konst.} \end{aligned}$$

überführt. Die dritte Schar kann man durch die Gleichungen

$$(40)\qquad \varphi_1(x_1, \cdots, x_{2r}) = \text{konst.}, \cdots\cdots, \varphi_r(x_1, \cdots, x_{2r}) = \text{konst.}$$

definieren.

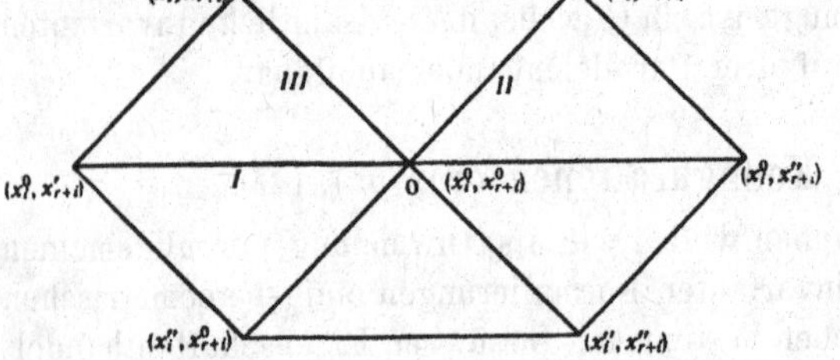

In der nebenstehenden Konfiguration (S) sind die Punkte mit Koordinaten versehen, wobei die Gleichungen (39) für die ersten zwei Scharen berücksichtigt sind.

Außerdem hat man

$$(41)\qquad \begin{aligned} &1.\quad \varphi_k(x_i^0, x_{r+i}^0) = \varphi_k(x_i', x_{r+i}'),\\ &2.\quad \varphi_k(x_i^0, x_{r+i}^0) = \varphi_k(x_i'', x_{r+i}''),\\ &3.\quad \varphi_k(x_i^0, x_{r+i}'') = \varphi_k(x_i', x_{r+i}^0),\\ &4.\quad \varphi_k(x_i^0, x_{r+i}') = \varphi_k(x_i'', x_{r+i}^0). \end{aligned}$$

Das Bestehen von (S) heißt nun: Das Gleichungssystem (41) ist lösbar mit beliebig vorgeschriebenen x_i^0, x_{r+i}^0, x_i'.

Aus (41_3) lassen sich die x_{r+i}'' als Funktionen von x_i^0, x_{r+i}^0, x_i' ausdrücken. Ebenso sind nach (41_4) die x_{r+i}' Funktionen von x_i^0, x_{r+i}^0, x_i'', und zwar sind es analytisch dieselben Funktionen. Trägt man diese Werte in (41_1) und (41_2) ein, so erhält man $2r$ Gleichungen, die mit willkürlichen x_i^0, x_{r+i}^0, x_i' lösbar sein sollen. D. h. aber, daß das Gleichungssystem, das durch Elimination von x_{r+i}' zwischen (41_1), (41_4) entsteht, symmetrisch in x_i', x_i'' sein muß.

Wir setzen

$$(42)\qquad \Delta x_i = x_i' - x_i^0, \qquad \Delta x_{r+i} = x_{r+i}' - x_{r+i}^0, \qquad \Delta y_i = x_i'' - x_i^0,$$

und entwickeln die Funktionen φ_k in der Nähe des Punktes $(x_1^0, \cdots, x_{2r}^0)$. Dann lauten die Gleichungen (41_1), (41_4) etwa so:

$$(43)\qquad \begin{aligned} &\frac{\partial\varphi_k}{\partial x_i}\Delta x_i + \frac{\partial\varphi_k}{\partial x_{r+i}}\Delta x_{r+i} + \tfrac{1}{2}\frac{\partial^2\varphi_k}{\partial x_i\,\partial x_j}\Delta x_i\,\Delta x_j + \frac{\partial^2\varphi_k}{\partial x_i\,\partial x_{r+j}}\Delta x_i\Delta x_{r+j}\\ &\quad + \tfrac{1}{2}\frac{\partial^2\varphi_k}{\partial x_{r+i}\,\partial x_{r+j}}\Delta x_{r+i}\,\Delta x_{r+j} + \tfrac{1}{6}\frac{\partial^3\varphi_k}{\partial x_i\,\partial x_j\,\partial x_l}\Delta x_i\,\Delta x_j\,\Delta x_l\\ &\quad + \tfrac{1}{2}\frac{\partial^3\varphi_k}{\partial x_i\,\partial x_j\,\partial x_{r+l}}\Delta x_i\,\Delta x_j\,\Delta x_{r+l}\\ &\quad + \tfrac{1}{2}\frac{\partial^3\varphi_k}{\partial x_i\,\partial x_{r+j}\,\partial x_{r+l}}\Delta x_i\,\Delta x_{r+j}\,\Delta x_{r+l}\\ &\quad + \tfrac{1}{6}\frac{\partial^3\varphi_k}{\partial x_{r+i}\,\partial x_{r+j}\,\partial x_{r+l}}\Delta x_{r+i}\,\Delta x_{r+j}\,\Delta x_{r+l} = 0,\\ &\frac{\partial\varphi_k}{\partial x_{r+i}}\Delta x_{r+i} + \tfrac{1}{2}\frac{\partial^2\varphi_k}{\partial x_{r+i}\,\partial x_{r+j}}\Delta x_{r+i}\,\Delta x_{r+j}\\ &\quad + \tfrac{1}{6}\frac{\partial^3\varphi_k}{\partial x_{r+i}\,\partial x_{r+j}\,\partial x_{r+l}}\Delta x_{r+i}\,\Delta x_{r+j}\,\Delta x_{r+l} - \frac{\partial\varphi_k}{\partial x_i}\Delta y_i\\ &\quad - \tfrac{1}{2}\frac{\partial^2\varphi_k}{\partial x_i\,\partial x_j}\Delta y_i\,\Delta y_j - \tfrac{1}{6}\frac{\partial^3\varphi_k}{\partial x_i\,\partial x_j\,\partial x_l}\Delta y_i\,\Delta y_j\,\Delta y_l = 0, \end{aligned}$$

wobei die Glieder von vierter oder höherer Ordnung in den Δx_i, Δx_{r+i}, Δy_i vernachlässigt sind und für die partiellen Ableitungen von φ_k ihre Werte im Punkte $(x_1^0, \cdots, x_{2r}^0)$ genommen werden müssen.

Da die Funktionaldeterminanten

$$\frac{\partial(\varphi_1, \cdots, \varphi_r)}{\partial(x_1, \cdots, x_r)} \neq 0, \qquad \frac{\partial(\varphi_1, \cdots, \varphi_r)}{\partial(x_{r+1}, \cdots, x_{2r})} \neq 0,$$

sind, kann man die durch

$$\begin{aligned} \frac{\partial \varphi_i}{\partial x_k} C_{kl} &= \delta_{il}, & C_{ik} \frac{\partial \varphi_k}{\partial x_l} &= \delta_{il}, \\ \frac{\partial \varphi_i}{\partial x_{r+k}} D_{kl} &= \delta_{il}, & D_{ik} \frac{\partial \varphi_k}{\partial x_{r+l}} &= \delta_{il} \end{aligned} \tag{44}$$

definierten Größen C_{ik}, D_{ik} einführen. Subtrahiert man die Gleichungen (43) voneinander, so erhält man

$$\begin{aligned} &\frac{\partial \varphi_k}{\partial x_i}(\Delta x_i + \Delta y_i) + \tfrac{1}{2} \frac{\partial^2 \varphi_k}{\partial x_i \partial x_j}(\Delta x_i \Delta x_j + \Delta y_i \Delta y_j) \\ &+ \tfrac{1}{6} \frac{\partial^3 \varphi_k}{\partial x_i \partial x_j \partial x_l}(\Delta x_i \Delta x_j \Delta x_l + \Delta y_i \Delta y_j \Delta y_l) \\ &+ \frac{\partial^2 \varphi_k}{\partial x_i \partial x_{r+j}} \Delta x_i \Delta x_{r+j} + \tfrac{1}{2} \frac{\partial^3 \varphi_k}{\partial x_i \partial x_j \partial x_{r+l}} \Delta x_i \Delta x_j \Delta x_{r+l} \\ &+ \tfrac{1}{2} \frac{\partial^3 \varphi_k}{\partial x_i \partial x_{r+j} \partial x_{r+l}} \Delta x_i \Delta x_{r+j} \Delta x_{r+l} = 0. \end{aligned} \tag{45}$$

Aus der zweiten Gleichung in (43) bekommt man bis auf Glieder dritter Ordnung

$$\begin{aligned} \Delta x_{r+k} &= \frac{\partial \varphi_m}{\partial x_i} D_{km} \Delta y_i \\ &+ \tfrac{1}{2}\left(\frac{\partial^2 \varphi_k}{\partial x_i \partial x_j} - \frac{\partial^2 \varphi_m}{\partial x_{r+p} \partial x_{r+q}} \frac{\partial \varphi_s}{\partial x_i} \cdot \frac{\partial \varphi_n}{\partial x_j} D_{ps} D_{qn} D_{km}\right) \Delta y_i \Delta y_j. \end{aligned} \tag{46}$$

Setzt man diese Werte von Δx_{r+k} in (45) ein, so kann man die Δy_k durch Δx_k ausdrücken. Das Resultat ist bis auf Glieder dritter Ordnung

$$\Delta y_k = -\Delta x_k + \left(-\frac{\partial^2 \varphi_m}{\partial x_i \partial x_j} C_{km} + \frac{\partial^2 \varphi_m}{\partial x_j \partial x_{r+p}} \frac{\partial \varphi_q}{\partial x_i} D_{pq} C_{km}\right) \Delta x_i \Delta x_j. \tag{47}$$

Diese Werte von Δy_k befriedigen auch die Gleichungen, die man bekommt, indem man die Werte von Δx_{r+k} aus (46) in (45) einsetzt und danach die $\Delta x_k, \Delta y_k$ miteinander vertauscht. Dadurch bekommen wir r Gleichungen zwischen den Δx_k, die identisch befriedigt sein sollen. Aus den Gliedern dritter Ordnung in Δx_k bekommen wir die Bedingungen

$$B_{npqs} + B_{nqsp} + B_{nspq} + B_{npsq} + B_{nqps} + B_{nsqp} = 0, \tag{48}$$

wo

$$
\begin{aligned}
B_{npqs} = & \frac{\partial^3 \varphi_n}{\partial x_i\, \partial x_{r+j}\, \partial x_{r+k}} C_{iq} D_{jp} D_{ks} - \frac{\partial^3 \varphi_n}{\partial x_i\, \partial x_j\, \partial x_{r+k}} C_{iq} C_{jp} D_{ks} \\
& - \frac{\partial^2 \varphi_n}{\partial x_i\, \partial x_{r+j}} \frac{\partial^2 \varphi_m}{\partial x_{r+l}\, \partial x_{r+k}} C_{iq} D_{lp} D_{jm} D_{ks} \\
(49) \qquad & + \frac{\partial^2 \varphi_n}{\partial x_i\, \partial x_{r+j}} \frac{\partial^2 \varphi_m}{\partial x_l\, \partial x_k} C_{im} C_{lp} C_{kq} D_{js} \\
& + \frac{\partial^2 \varphi_n}{\partial x_i\, \partial x_{r+j}} \frac{\partial^2 \varphi_m}{\partial x_l\, \partial x_{r+k}} C_{iq} C_{lp} D_{jm} D_{ks} \\
& - \frac{\partial^2 \varphi_n}{\partial x_i\, \partial x_{r+j}} \frac{\partial^2 \varphi_m}{\partial x_l\, \partial x_{r+k}} C_{im} C_{lq} D_{kp} D_{js}.
\end{aligned}
$$

Da sich die Konfiguration (S) in jedem Punkt $(x_1, \cdots, x_{2r})$ schließt, gelten die Bedingungen (48) für beliebige x_i, x_{r+i}.

Wir wollen nun die Bedingung (48) in invarianter Form ausdrücken. Das Gewebe (39), (40) läßt sich auch durch die Differentialgleichungen

$$\omega_{1k} = 0, \quad \omega_{2k} = 0, \quad \omega_{3k} = 0$$

definieren, wobei

$$
\begin{aligned}
& \omega_{1k} = \frac{\partial \varphi_k}{\partial x_i} d x_i, \\
(50) \qquad & \omega_{2k} = \frac{\partial \varphi_k}{\partial x_{r+i}} d x_{r+i}, \\
& \omega_{3k} = -d\varphi_k = -\frac{\partial \varphi_k}{\partial x_i} d x_i - \frac{\partial \varphi_k}{\partial x_{r+i}} d x_{r+i};
\end{aligned}
$$

folglich bestehen die Relationen

$$\omega_{1k} + \omega_{2k} + \omega_{3k} = 0.$$

In den Gleichungen (11) haben wir wegen

$$d(\omega_{1k} + \omega_{2k}) = d(d\varphi_k) = 0$$

die Beziehungen

$$
(50\text{a}) \qquad
\begin{aligned}
& \alpha_{kij} = \delta_{kij} = 0, \\
& \beta_{kij} = \gamma_{kji}.
\end{aligned}
$$

Die Ableitungsgleichungen lauten also

$$
(51) \qquad
\begin{aligned}
& d\,\omega_{1k} = \beta_{kij}\, \omega_{2j}\, \omega_{1i}, \\
& d\,\omega_{2k} = \beta_{kji}\, \omega_{1j}\, \omega_{2i}.
\end{aligned}
$$

Dabei steht β_{kij} mit φ_k in der folgenden Beziehung

$$(52) \qquad \beta_{kij} = \frac{\partial^2 \varphi_k}{\partial x_l\, \partial x_{r+m}} C_{li} D_{mj}.$$

Leitet man die Gleichungen (51) ab, so erhält man wegen der linearen Unabhängigkeit der ω_{1k}, ω_{2k} die Gleichungen

$$(53)\qquad \begin{aligned} \beta^{(1)}_{kil,m} - \beta^{(1)}_{kml,i} + \beta_{kip}\,\beta_{pml} - \beta_{kmp}\,\beta_{pil} &= 0,\\ \beta^{(2)}_{kli,m} - \beta^{(2)}_{klm,i} + \beta_{kpi}\,\beta_{plm} - \beta_{kpm}\,\beta_{pli} &= 0. \end{aligned}$$

Nach (53) lassen sich die A_{npqs} aus (30) so vereinfachen:

$$(54)\qquad A_{npqs} = \beta^{(2)}_{nqs,p} - \beta^{(1)}_{nqs,p}.$$

Da nach Definition (28)

$$(55)\qquad \begin{aligned} \beta^{(1)}_{nqs,p} &= C_{ip}\frac{\partial \beta_{nqs}}{\partial x_i},\\ \beta^{(2)}_{nqs,p} &= D_{ip}\frac{\partial \beta_{nqs}}{\partial x_{r+i}} \end{aligned}$$

ist, folgt unter Beachtung von (49), (52) und der weiteren Beziehungen

$$(56)\qquad \begin{aligned} \frac{\partial C_{ik}}{\partial x_m} &= -\frac{\partial^2 \varphi_p}{\partial x_q\,\partial x_m} C_{ip}\,C_{qk},\\ \frac{\partial C_{ik}}{\partial x_{r+m}} &= -\frac{\partial^2 \varphi_p}{\partial x_q\,\partial x_{r+m}} C_{ip}\,C_{qk}; \end{aligned}$$

$$(57)\qquad \begin{aligned} \frac{\partial D_{ik}}{\partial x_m} &= -\frac{\partial^2 \varphi_p}{\partial x_{r+q}\,\partial x_m} D_{ip}\,D_{qk},\\ \frac{\partial D_{ik}}{\partial x_{r+m}} &= -\frac{\partial^2 \varphi_p}{\partial x_{r+q}\,\partial x_{r+m}} D_{ip}\,D_{qk}, \end{aligned}$$

daß

$$A_{npqs} = B_{npqs}$$

ist. Die Bedingungen (48) sind also gleichwertig mit den Bedingungen

$$(58)\qquad f^{(k)}_{ilm} + f^{(k)}_{lmi} + f^{(k)}_{mil} + f^{(k)}_{iml} + f^{(k)}_{lim} + f^{(k)}_{mli} = 0.$$

Daher gilt: *Für das Bestehen der Sechseckkonfiguration* (S) *sind die Bedingungen* (58) *notwendig.*

Um zu beweisen, daß die Bedingungen (58) für (S) auch hinreichend sind, beweisen wir den folgenden Satz: *Wenn die Bedingungen* (58) *erfüllt sind, so gibt es zu beliebiger Richtung auf einer beliebigen Mannigfaltigkeit der drei Scharen eine Fläche* (*d. h. zweidimensionale Mannigfaltigkeit*) *mit folgenden Eigenschaften:*

1. *Sie schneidet die Mannigfaltigkeit in einer Kurve längs dieser Richtung.*
2. *Sie schneidet jede Mannigfaltigkeit der drei Scharen in einer Kurve, wenn sie mit ihr einen Punkt gemein hat.*
3. *Auf ihr gibt es somit drei Kurvenscharen, nämlich die Scharen ihrer Schnittkurven mit den drei Gewebescharen. Diese Kurvenscharen bilden ein Kurvensechseckgewebe.*

Nach diesem Satz kann man in der Konstruktion der Konfiguration (S) die durch die Anfangspunkte O und P gehende Fläche nehmen, welche immer existiert, wenn O und P genügend nah sind. Wegen der Eigenschaft 2 schneiden die betreffenden Mannigfaltigkeiten die Fläche in Kurven. Da die Schnittkurvenscharen auf der Fläche ein Kurvensechseckgewebe bilden, gilt die Konfiguration (S) auch für unser Dreigewebe von r-dimensionalen Mannigfaltigkeiten.

Eine Flächenschar des Raumes R_{2r}, die durch jeden Punkt eine und nur eine Fläche schickt, definiert man durch ein vollständig integrierbares Pfaffsches System der Form

$$(59)\qquad \psi_1 = 0, \cdots, \psi_{r-1} = 0, \qquad \psi_{r+1} = 0, \cdots, \psi_{2r-1} = 0,$$

wobei die ψ Pfaffsche Formen in x_k, x_{r+k} bedeuten und

$$(60)\qquad \begin{aligned} d\psi_t &\equiv 0 \pmod{\psi}, \\ d\psi_{r+t} &\equiv 0 \pmod{\psi}, \end{aligned} \qquad t = 1, 2, \cdots, r-1 \text{[6]})$$

gilt. Jedes ψ ist eine Linearkombination von Ω_{1k}, Ω_{2k}. Um die Eigenschaft 2 des Satzes erfüllen zu können, verlangen wir, daß das System (59) mit jedem der drei Systeme

$$\Omega_{1k} = 0; \qquad \Omega_{2k} = 0; \qquad \Omega_{1k} + \Omega_{2k} = 0$$

ein nicht-triviales Lösungssystem besitzt, wenn man diese Systeme als lineare Gleichungen in den Ω auffaßt. Nach Ersetzung der ψ durch geeignete Linearkombinationen kann man erreichen, daß

$$(61)\qquad \begin{aligned} \psi_t &= a_{tk}\,\Omega_{1k}, \\ \psi_{r+t} &= a_{tk}\,\Omega_{2k} \end{aligned}$$

wird, wo a_{tk} Funktionen von x_k, x_{r+k}, u_{kj} sind.

Wir nehmen zwei Hilfsformen hinzu:

$$(62)\qquad \begin{aligned} \psi_r &= b_k\,\Omega_{1k}, \\ \psi_{2r} &= b_k\,\Omega_{2k}, \end{aligned}$$

wo die b_k der Einfachheit halber als Konstanten angenommen werden mit der Zusatzforderung

$$(63)\qquad a = \begin{vmatrix} a_{11} & \cdots & a_{1r} \\ \cdot & \cdot & \cdot \\ a_{r-1,1} & \cdots & a_{r-1,r} \\ b_1 & \cdots & b_r \end{vmatrix} \neq 0.$$

Jetzt denken wir uns die Ω_{1k}, Ω_{2k} nach den ψ_k, ψ_{r+k} aufgelöst, also

$$(64)\qquad \begin{aligned} \Omega_{1k} &= A_{kj}\,\psi_j, \\ \Omega_{2k} &= A_{kj}\,\psi_{r+j}, \end{aligned}$$

[6]) Ausnahmsweise lassen wir vorläufig den Index t von 1 bis $r-1$ laufen.

wobei insbesondere

$$(65)\quad A_k = A_{kr} = \frac{(-1)^{k+r}}{a} \begin{vmatrix} a_{11} & \cdots & a_{1k-1} & a_{1k+1} & \cdots & a_{1r} \\ \cdot & \cdot & \cdot & \cdot & \cdot & \cdot \\ a_{r-1,1} & \cdots & a_{r-1,k-1}, & a_{r-1,k+1} & \cdots & a_{r-1,r} \end{vmatrix}$$

gesetzt sei.

Nach Ableiten von (61) folgt

$$(66)\quad \begin{aligned} d\psi_t &= (d a_{tk} - a_{tj}\vartheta_{kj})\,\Omega_{1k} - a_{tk}\,c_{jk}^{(l)}\,\Omega_{1j}\,\Omega_{1l}, \\ d\psi_{r+t} &= (d a_{tk} - a_{tj}\vartheta_{kj})\,\Omega_{2k} + a_{tk}\,c_{jk}^{(l)}\,\Omega_{2j}\,\Omega_{2l}. \end{aligned}$$

Man kann daher setzen:

$$(67)\quad d a_{tk} = \overset{(1)}{a}_{tk,m}\,\Omega_{1m} + \overset{(2)}{a}_{tk,m}\,\Omega_{2m} + a_{tj}\,\vartheta_{kj},$$

und dann drückt sich die vollständige Integrierbarkeit des Systems (59) so aus:

$$(67\,\mathrm{a})\quad \begin{aligned} \overset{(1)}{a}_{tk,m}\,A_k\,A_m &= 0, \\ \overset{(2)}{a}_{tk,m}\,A_k\,A_m &= 0. \end{aligned}$$

Auf einer Integralfläche des Systems (59) kann man die drei Schnittkurvenscharen als durch die Gleichungen

$$(68)\quad \psi_r = 0, \qquad \psi_{2r} = 0, \qquad \psi_r + \psi_{2r} = 0$$

definiert denken. Aus (62) erhält man, da $\psi_1 = \cdots = \psi_{r-1} = \psi_{r+1} = \cdots = \psi_{2r-1} = 0$ gesetzt werden muß,

$$(69)\quad \begin{aligned} d\psi_r &= b_k\,d\,\Omega_{1k} \equiv b_k\,A_j\,\psi_r\,\vartheta_{jk}, \\ d\psi_{2r} &= b_k\,d\,\Omega_{2k} \equiv b_k\,A_j\,\psi_{2r}\,\vartheta_{jk}, \end{aligned}$$

wobei diese und die folgenden Kongruenzen alle nach dem Modul $(\psi_1, \cdots, \psi_{r-1}, \psi_{r+1}, \cdots, \psi_{2r-1})$ genommen sind. Ich behaupte:

Die notwendige und hinreichende Bedingung, daß das Kurvendreigewebe (68) *auf jeder Integralfläche des Systems* (59) *ein Sechseckgewebe bildet, ist*

$$(70)\quad d(b_k\,A_j\,\vartheta_{jk}) \equiv 0.$$

Zum Beweise, daß diese Bedingung notwendig ist, denken wir uns solche Koordinaten $x_1, \cdots, x_{2r}$ eingeführt, daß

$$(71)\quad x_1 = \text{konst.}, \cdots, x_{r-1} = \text{konst.}, \quad x_{r+1} = \text{konst.}, \cdots, x_{2r-1} = \text{konst.}$$

die Gleichungen jener Flächenschar sind, während

$$x_1 = \text{konst.}, \cdots, x_r = \text{konst.},$$

bzw.

$$x_{r+1} = \text{konst.}, \cdots, x_{2r} = \text{konst.},$$

$$\varphi_1(x_1, \cdots, x_{2r}) = \text{konst.}, \cdots, \varphi_r(x_1, \cdots, x_{2r}) = \text{konst.}$$

die Gleichungen der drei r-dimensionalen Scharen sind. Wenn nun diese Scharen auf allen jenen Flächen Sechseckgewebe ausschneiden, kann man die x noch so gewählt denken, daß die Funktionen φ die Variablen x_r und x_{2r} nur in der Kombination $x_r + x_{2r}$ enthalten. Dann gilt

(72)
$$\psi_r \equiv \varrho_1 \, d x_r, \qquad \psi_{2r} \equiv \varrho_2 \, d x_{2r},$$
$$\psi_r + \psi_{2r} \equiv \varrho_3 (d x_r + d x_{2r}) \qquad (\varrho \text{ skalar}),$$

woraus durch Differentiation wegen der vollständigen Integrabilität von (59) folgt:

(73)
$$d \psi_r \equiv \frac{d \varrho_1}{\varrho_1} \psi_r, \qquad d \psi_{2r} \equiv \frac{d \varrho_2}{\varrho_2} \psi_{2r},$$
$$d \psi_r + d \psi_{2r} \equiv \frac{d \varrho_3}{\varrho_3} (\psi_r + \psi_{2r}).$$

Daraus ergibt sich

$$\frac{d \varrho_1}{\varrho_1} \equiv \frac{d \varrho_2}{\varrho_2} \equiv \frac{d \varrho_3}{\varrho_3}$$

und durch Vergleich mit (69) findet man

(74)
$$b_k \, A_j \, \vartheta_{jk} \equiv - \frac{d \varrho_1}{\varrho_1}$$

was (70) zur Folge hat.

Es sei umgekehrt (70) erfüllt. Die Koordinaten x denken wir wieder so gewählt, daß (71) unsere Flächenschar darstellt. Halten wir $x_1, \cdots, x_{r-1}, x_{r+1}, \cdots, x_{2r-1}$ konstant, d. h. üben wir die Substitution

$$\begin{aligned} x_1 &= a_1 = \text{konst.}, \\ &\cdots\cdots\cdots \\ x_{r-1} &= a_{r-1} = \text{konst.}, \\ x_{r+1} &= a_{r+1} = \text{konst.}, \\ &\cdots\cdots\cdots \\ x_{2r-1} &= a_{2r-1} = \text{konst.} \end{aligned}$$

auf (69), (70) aus, so gehen diese Kongruenzen in gewöhnliche Gleichungen für die substituierten Formen (die wir durch Überstreichen kennzeichnen wollen) über. Man hat also

$$d(\overline{b_k \, A_j \, \vartheta_{jk}}) = 0, \quad \text{d. h.} \quad \overline{b_k \, A_j \, \vartheta_{jk}} = - dv(x_r, x_{2r}, a, u),$$

und daher

(75)
$$d \overline{\psi}_r = dv \, \overline{\psi}_r, \qquad d \overline{\psi}_{2r} = dv \, \overline{\psi}_{2r}.$$

Hiernach ist sowohl $\overline{\psi}_r = 0$ wie $\overline{\psi}_{2r} = 0$ vollständig integrierbar und es gibt somit Funktionen F_1, F_2, y_1, y_2 von x_r, x_{2r}, den a und den u so, daß

(76)
$$\overline{\psi}_r = F_1 \, d y_1, \qquad \overline{\psi}_{2r} = F_2 \, d y_2$$

gilt. Aus (75) und (76) ergibt sich

$$(77)\qquad \begin{aligned} dF_1\, dy_1 &= F_1\, dv\, dy_1, \\ dF_2\, dy_2 &= F_2\, dv\, dy_2. \end{aligned}$$

Es folgt daraus

$$(78)\qquad F_1 = e^v f_1(y_1), \qquad F_2 = e^v f_2(y_2).$$

Also ist das Gewebe topologisch äquivalent mit dem Gewebe

$$dz_1 = 0, \quad dz_2 = 0, \quad dz_1 + dz_2 = 0$$

in der (z_1, z_2)-Ebene, in welches es durch die Transformation

$$z_1 = \int f_1(y_1)\, dy_1, \qquad z_2 = \int f_2(y_2)\, dy_2$$

übergeführt wird.

Nach dem eben Bewiesenen gelten für eine Flächenschar (59), (61) mit den Eigenschaften 2 und 3 die Bedingungen (67), (70), welche dazu auch hinreichend sind. Da die A_k die Richtungen auf den Gewebemannigfaltigkeiten bestimmen, wird die verlangte Eigenschaft 1 erfüllt werden und also der genannte Satz bewiesen sein, wenn man zeigt, daß man mit Benutzung der Bedingungen (58) die a_{tk} als Funktionen der x_k und u_{kj} so bestimmen kann, daß sie den Bedingungen (67), (67a), (70) genügen und die A_k willkürlich vorgegebene Anfangswerte annehmen.

Die linke Seite von (70) ist gleich

$$d(b_k A_j \vartheta_{jk}) \equiv b_k (dA_j + A_l \vartheta_{lj}) \vartheta_{jk} + b_k f^{(k)}_{jlm} A_j A_l A_m \psi_r \psi_{2r}.$$

Nach (58) ist das letzte Glied Null. Mit dem Ansatz

$$(79)\qquad dA_k = A^{\cdot 1}_{kj} \Omega_{1j} + A^{\cdot 2}_{kj} \Omega_{2j} - A_j \vartheta_{jk}$$

und den Zusatzforderungen

$$(80)\qquad \begin{aligned} A^{\cdot 1}_{kj} A_j &= 0, \\ A^{\cdot 2}_{kj} A_j &= 0 \end{aligned}$$

sind die Bedingungen (70) erfüllt. Das ursprüngliche System läßt sich also durch das System ersetzen, das aus den Gleichungen (67), (67a), (79), (80) und

$$(81)\qquad a_{tk} A_k = 0$$

besteht und die Variablen $a_{tk}, A_k, x_k, x_{r+k}, u_{kj}$ enthält.

Ich behaupte: *Wenn das Teilsystem* (79), (80) *in den Variablen* A_k, x_k, x_{r+k}, u_{kj} *ein Lösungssystem* $A_k(x, u)$ *hat, kann man dazu gewisse Funktionen* $a_{tk}(x, u)$ *hinzufügen, so daß ein Lösungssystem* $A_k(x, u)$, $a_{tk}(x, u)$ *des ganzen Systems entsteht.* Übrigens ist dazu nur die Auflösung skalarer Gleichungen nötig.

Zum Beweis brauchen wir den folgenden Hilfssatz:

Es seien H_k r Funktionen in den Variablen x_k, x_{r+k}, u_{kj}. Diese Funktionen genügen einem Differentialgleichungssystem der Gestalt

$$dH_k = H_{kj}^{\cdot 1}\, \Omega_{1j} + H_{kj}^{\cdot 2}\, \Omega_{2j} + H_j\, \vartheta_{kj} \tag{82}$$

dann und nur dann, wenn $H_k = v_{kj} F_j(x_1, \cdots, x_{2r})$ gesetzt werden kann. Sie genügen dem System

$$dH_k = H_{kj}^{\cdot 1}\, \Omega_{1j} + H_{kj}^{\cdot 2}\, \Omega_{2j} - H_j\, \vartheta_{jk} \tag{83}$$

dann und nur dann, wenn $H_k = u_{jk} F_j(x_1, \cdots, x_{2r})$ gesetzt werden kann. In beiden Fällen bedeuten die F_j r Funktionen in den Variablen x_k, x_{r+k} allein.

Der Beweis für den Hilfssatz ist einfach. Man beachtet nämlich die Identitäten:

$$\begin{aligned} d(u_{jk} G_j) &= u_{jk}\, dG_j - G_j\, u_{jl}\, \vartheta_{lk} + \text{Kombination von } \Omega, \\ d(u_{jk} G_k) &= u_{jk}\, dG_k - G_k\, u_{jl}\, \vartheta_{lk} + \text{Kombination von } \Omega, \\ d(v_{kj} G_j) &= v_{kj}\, dG_j + G_j\, v_{lj}\, \vartheta_{kl} + \text{Kombination von } \Omega, \\ d(v_{kj} G_k) &= v_{kj}\, dG_k + G_k\, v_{lj}\, \vartheta_{kl} + \text{Kombination von } \Omega, \end{aligned} \tag{84}$$

welche aus (24) leicht folgen. Wenn H_k z. B. von der Form $H_k = v_{kj} F_j(x_1, \cdots, x_{2r})$ sind, so ist nach der dritten Gleichung von (84) klar, daß die H_k dem System (82) genügen. Umgekehrt, wenn die H_k dem System (82) genügen, dann folgt aus der zweiten Gleichung von (84), daß $d(u_{jk} H_k)$ eine Linearkombination von Ω_{1k}, Ω_{2k} ist, d. h. die $u_{jk} H_k$ sind Funktionen von x_k, x_{r+k} allein und H_k ist von der Form $H_k = v_{kj} F_j(x_1, \cdots, x_{2r})$. Der zweite Teil des Satzes ist mittels der ersten und vierten Gleichung von (84) genau durch dieselbe Überlegung zu beweisen.

Nun seien $A_k(x, u)$ r Funktionen, die dem System (79), (80) genügen. Nach dem Hilfssatz sind die A_k von der Form

$$A_k = u_{jk} F_j(x_1, \cdots, x_{2r}). \tag{85}$$

Man bestimme $r(r-1)$ Funktionen $H_{tj}(x_1, \cdots, x_{2r})$, $t = 1, 2, \cdots, r-1$, so, daß

$$H_{tj} F_j = 0 \tag{86}$$

und die Matrix

$$\begin{pmatrix} H_{11} & \cdots & H_{1r} \\ \cdot & \cdot & \cdot \\ H_{r-1,1} & \cdots & H_{r-1,r} \end{pmatrix}$$

vom Rang $r-1$ sei. Wir nehmen dann die Funktionen A nach (85) und setzen ferner

$$a_{tk} = v_{kj} H_{tj}(x_1, \cdots, x_{2r}). \tag{87}$$

Nach Voraussetzung sind (79), (80) erfüllt. Die Gleichung (81) ist eine Folgerung von (86). (67) ist wegen des Hilfssatzes erfüllt. (67a) folgt durch Ableiten von (81), denn es gelten die Relationen:

$$
(88) \qquad \begin{aligned} a^{(1)}_{tk,m} A_k + a_{tk} A^{\cdot 1}_{km} &= 0, \\ a^{(2)}_{tk,m} A_k + a_{tk} A^{\cdot 2}_{km} &= 0. \end{aligned}
$$

Es bleibt also zu zeigen, daß das System (79), (80) Lösungen A_k hat, deren Anfangswerte sich willkürlich vorschreiben lassen.

Das Differentialideal des Systems (79), (80) besteht aus den Gleichungen (79), (80) und ihren abgeleiteten Gleichungen. Man mache den Ansatz

$$
(89) \qquad \begin{aligned} d A^{\cdot 1}_{kj} &= A^{\cdot 11}_{kjl} \Omega_{1l} + A^{\cdot 12}_{kjl} \Omega_{2l} + A^{\cdot 1}_{kj,lm} \vartheta_{lm}, \\ d A^{\cdot 2}_{kj} &= A^{\cdot 21}_{kjl} \Omega_{1l} + A^{\cdot 22}_{kjl} \Omega_{2l} + A^{\cdot 2}_{kj,lm} \vartheta_{lm}, \end{aligned}
$$

und leite die Gleichungen (79), (80) ab. Dann erhält man zur Bestimmung der Größen $A^{\cdot 11}_{kjl}, \cdots, A^{\cdot 2}_{kj,lm}$ die folgenden Gleichungen:

$$
(90) \qquad \begin{aligned} A^{\cdot 11}_{kjl} - A^{\cdot 11}_{klj} &= 2 A^{\cdot 1}_{km} c^{(j)}_{lm}, \\ A^{\cdot 12}_{kjl} - A^{\cdot 21}_{klj} &= - A_m f^{(k)}_{mjl}, \\ A^{\cdot 22}_{kjl} - A^{\cdot 22}_{klj} &= -2 A^{\cdot 2}_{km} c^{(j)}_{lm}, \\ A^{\cdot 1}_{kj,lm} &= \delta_{jl} A^{\cdot 1}_{km} - \delta_{km} A^{\cdot 1}_{lj}, \\ A^{\cdot 2}_{kj,lm} &= \delta_{jl} A^{\cdot 2}_{km} - \delta_{km} A^{\cdot 2}_{lj}, \\ A_j A^{\cdot 11}_{kjl} + A^{\cdot 1}_{kj} A^{\cdot 1}_{jl} &= 0, \\ A_j A^{\cdot 12}_{kjl} + A^{\cdot 1}_{kj} A^{\cdot 2}_{jl} &= 0, \\ A_j A^{\cdot 21}_{kjl} + A^{\cdot 2}_{kj} A^{\cdot 1}_{jl} &= 0, \\ A_j A^{\cdot 22}_{kjl} + A^{\cdot 2}_{kj} A^{\cdot 2}_{jl} &= 0, \\ A_j A^{\cdot 1}_{kj,lm} - A_l A^{\cdot 1}_{km} &= 0, \\ A_j A^{\cdot 2}_{kj,lm} - A_l A^{\cdot 2}_{km} &= 0. \end{aligned}
$$

Nun nehmen wir die Basisformen in der Reihenfolge $\Omega_{11}, \cdots, \Omega_{1r-1}, \Omega_{21}, \cdots, \Omega_{2r-1}, \Omega_{1r}, \Omega_{2r}, \vartheta_{11}, \cdots, \vartheta_{rr}$. Für das Integral-$E_i$, das durch

$$
\begin{aligned} \Omega_{1i+1} = \cdots = \Omega_{1r-1} = \Omega_{21} = \cdots = \Omega_{2r-1} = \Omega_{1r} = \Omega_{2r} \\ = \vartheta_{11} = \cdots = \vartheta_{rr} = 0 \end{aligned}
$$

bestimmt ist ($i \leqq r-1$), kommen zur Bestimmung der Unbekannten $A^{\cdot 11}_{kji}, A^{\cdot 21}_{kji}$ die folgenden Gleichungen in Betracht:

$$
\begin{aligned} &(90_1) \quad \text{mit } j, l \leqq i, \\ &(90_6) \quad \text{mit } l = i, \\ &(90_8) \quad \text{mit } l = i. \end{aligned}
$$

Die ersten Gleichungen bestimmen $A_{kji}^{\cdot 11}$ für $j < i$, die zweiten $A_{kri}^{\cdot 11}$, und die dritten $A_{kri}^{\cdot 21}$.

Bei dem Integral-E_{r-1+i} $(i \leqq r-1)$:

$$\Omega_{2,i+1} = \cdots = \Omega_{2r-1} = \Omega_{1r} = \Omega_{2r} = \vartheta_{11} = \cdots = \vartheta_{rr} = 0$$

haben wir die Unbekannten $A_{kji}^{\cdot 12}$, $A_{kji}^{\cdot 22}$, und es kommen die Gleichungen

(90_2) mit $j \leqq r-1, \quad l = i$,
(90_3) mit $j, \quad l \leqq i$,
(90_7) mit $l = i$,
(90_9) mit $l = i$.

in Betracht. Die ersten Gleichungen bestimmen $A_{kji}^{\cdot 12}$, $j \leqq r-1$, die zweiten $A_{kji}^{\cdot 22}$, $j < i$, die dritten $A_{kri}^{\cdot 12}$, und die vierten $A_{kri}^{\cdot 22}$.

Bei dem Integral-E_{2r-1}:

$$\Omega_{2r} = \vartheta_{11} = \cdots = \vartheta_{rr} = 0$$

haben wir die Unbekannten $A_{kjr}^{\cdot 11}$, $A_{kjr}^{\cdot 21}$ und die Gleichungen

(90_1) mit $l = r$,
(90_2) mit $l \leqq r-1$, $j = r$,
(90_6) mit $l = r$,
(90_8) mit $l = r$.

Die ersten Gleichungen bestimmen $A_{kjr}^{\cdot 11}$, $j \leqq r-1$, die zweiten $A_{kjr}^{\cdot 21}$, $j \leqq r-1$, die dritten $A_{krr}^{\cdot 11}$, und die vierten $A_{krr}^{\cdot 21}$.

Bei dem Integral-E_{2r}:

$$\vartheta_{11} = \cdots = \vartheta_{rr} = 0$$

haben wir die Unbekannten $A_{kjr}^{\cdot 12}$, $A_{kjr}^{\cdot 22}$ und die Gleichungen

(90_2) mit $l = r$,
(90_3) mit $l = r$,
(90_7) mit $l = r$,
(90_9) mit $l = r$.

Die ersten Gleichungen bestimmen $A_{kjr}^{\cdot 12}$, die zweiten $A_{kjr}^{\cdot 22}$, $j \leqq r-1$, und die vierten $A_{krr}^{\cdot 22}$. Andererseits sind die dritten Gleichungen Folgerungen aus den früheren Gleichungen und den Bedingungen (58). Es folgt nämlich (die Indizes t, t' laufen von 1 bis $r-1$!):

$$\begin{aligned} A_j A_{kjr}^{\cdot 12} + A_{kj}^{\cdot 1} A_{jr}^{\cdot 2} \\ &= A_j A_{krj}^{\cdot 21} - A_j A_m f_{mjr}^{(k)} + A_{kj}^{\cdot 1} A_{jr}^{\cdot 2} \\ &= -\frac{1}{A_r} A_j A_t A_{ktj}^{\cdot 21} - A_j A_m f_{mjr}^{(k)} + A_{kj}^{\cdot 1} A_{jr}^{\cdot 2} \end{aligned}$$

$$
\begin{aligned}
&= -\frac{1}{A_r}\{A_{t'}A_t A_{ktt'}^{\cdot 21} + A_t A_r A_{krt}^{\cdot 12} + A_t A_m A_r f_{mrt}^{\cdot (k)} \\
&\qquad + A_j A_m A_r f_{mjr}^{(k)} - A_r A_{kj}^{\cdot 1} A_{jr}^{\cdot 2}\} \\
&= -\frac{1}{A_r}\{A_{t'}A_t A_{ktt'}^{\cdot 21} - A_{t'} A_t A_{kt't}^{\cdot 12} - A_{kj}^{\cdot 1} A_{jt}^{\cdot 2} A_t + A_t A_m A_r f_{mrt}^{\cdot (k)} \\
&\qquad + A_j A_m A_r f_{mjr}^{(k)} - A_r A_{kj}^{\cdot 1} A_{jr}^{\cdot 2}\} \\
&= -\frac{1}{A_r}\{A_{t'}A_t A_m f_{mt't}^{(k)} + A_t A_m A_r f_{mrt}^{(k)} + A_j A_m A_r f_{mjr}^{(k)}\} \\
&= -\frac{1}{A_r} A_j A_l A_m f_{mjl}^{(k)} \\
&= 0.
\end{aligned}
$$

Bei einem durch

$$\vartheta_{kj} = \cdots = \vartheta_{rr} = 0$$

definierten Integralelement ist klar, daß die Unbekannten sich nach der Vorschrift des Existenzsatzes bestimmen lassen.

Daher hat das System (79), (80) Lösungen A_k, und weil keine skalare Beziehung zwischen den A_k aus den Integrabilitätsbedingungen des Systems (79), (80) entsteht, lassen sich die Anfangswerte $A_k(x_l^0, u_{ij}^0)$ willkürlich vorschreiben, wie zu beweisen war.

Schließlich haben wir also den Satz:

Die notwendige und hinreichende Bedingung für das Bestehen der Sechseckkonfiguration (*S*) *ist das Erfülltsein der Bedingungen* (58).

Betrachten wir nun die Reidemeistersche Konfiguration (*P*). Diese Konfiguration hat eine naheliegende geometrische Bedeutung. Man nimmt zwei Mannigfaltigkeiten *III* und *III'* aus der dritten Schar. Die Mannigfaltigkeit aus der ersten Schar durch einen Punkt *P* von *III* schneidet *III'* in *Q* und die von der zweiten Schar durch *Q* schneidet *III* in einem neuen Punkt P^*. Durch diese Konstruktion ist eine Punkttransformation $P \to P^*$ auf *III* definiert. Wenn *III'* auf der Gesamtheit aller Mannigfaltigkeiten der dritten Schar variiert, bekommen wir auf *III* eine r-parametrige Schar von Transformationen. Es läßt sich leicht zeigen, daß die Konfiguration (*P*) dann und nur dann gilt, wenn die so definierte r-parametrige Schar von Transformationen auf jeder Mannigfaltigkeit der dritten Schar eine Gruppe bildet. Übrigens besteht die Thomsensche Konfiguration (*D*) genau dann, wenn die Gruppe abelsch ist.

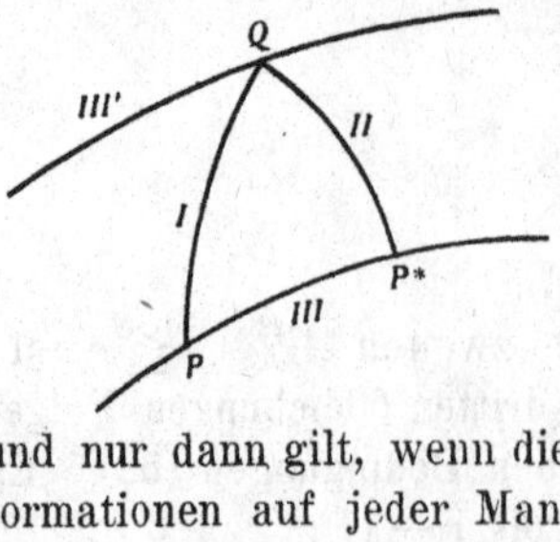

Um die Bedingungen für (*P*) aufzustellen, beachten wir zunächst, daß (*S*) eine Folgerung von (*P*) ist. Es gelten also die Bedingungen (58). Nach dem oben bewiesenen Satz sind die Flächenscharen (59) vorhanden und die Schnittkurvenscharen auf diesen Flächen bilden Sechseckgewebe. Für Kurvengewebe sind aber die Konfigurationen (S) und (P) äquivalent. D. h. die eben definierte Schar von Transformationen ist aus eingliedrigen Gruppen zusammengesetzt. Um die notwendigen und hinreichenden Bedingungen für (P) zu bekommen, braucht man nur weiter zu fordern, daß die Klammerausdrücke der infinitesimalen Transformationen dieser eingliedrigen Gruppen sich als Linearkombinationen der infinitesimalen Transformationen selbst mit konstanten Koeffizienten darstellen lassen.

Wir berechnen also die Symbole der betreffenden infinitesimalen Transformationen. Denken wir uns das Gewebe durch (39), (40) gegeben und betrachten wir die $\Delta\varphi_k$ als Konstante, so sind die Änderungen der Koordinaten Δx_k, Δx_{r+k} bis auf Größen zweiter Ordnung durch die folgenden Gleichungen gegeben:

$$\begin{aligned} \Delta\varphi_i &= \frac{\partial\varphi_i}{\partial x_{r+k}}\Delta x_{r+k}, \\ \frac{\partial\varphi_i}{\partial x_k}\Delta x_k &+ \frac{\partial\varphi_i}{\partial x_{r+k}}\Delta x_{r+k} = 0 \end{aligned} \tag{91}$$

oder, aufgelöst nach Δx_k, Δx_{r+k},

$$\begin{aligned} \Delta x_k &= -C_{ki}\,\Delta\varphi_i, \\ \Delta x_{r+k} &= D_{ki}\,\Delta\varphi_i. \end{aligned} \tag{92}$$

Die infinitesimale Transformation lautet also:

$$\begin{aligned} H(F) &= \Delta x_k\frac{\partial F}{\partial x_k} + \Delta x_{r+k}\frac{\partial F}{\partial x_{r+k}} \\ &= \left(-C_{ki}\frac{\partial F}{\partial x_k} + D_{ki}\frac{\partial F}{\Delta x_{r+k}}\right)\Delta\varphi_i. \end{aligned} \tag{93}$$

Man führt nun $2r$ Operatoren ein durch den Ansatz:

$$dF = X_k(F)\,\omega_{1k} + Y_k(F)\,\omega_{2k} \tag{94}$$

(vgl. (50)), wo F eine beliebige Funktion von x_k, x_{r+k} bedeutet. Dann ergibt sich

$$\begin{aligned} X_k(F) &= C_{ik}\frac{\partial F}{\partial x_i}, \\ Y_k(F) &= D_{ik}\frac{\partial F}{\partial x_{r+i}} \end{aligned} \tag{95}$$

und

$$H(F) = (Y_k(F) - X_k(F))\,\Delta\varphi_k. \tag{96}$$

Unter Benutzung von (51) bekommt man durch Differenzieren von (94)

$$\begin{aligned} (X_i, X_k) &= 0, \qquad (Y_i, Y_k) = 0, \\ (X_i, Y_k) &= \beta_{jik}(X_j - Y_j) \end{aligned} \tag{97}$$

und daraus
$$(Y_i - X_i,\ Y_k - X_k) = (\beta_{jik} - \beta_{jki})(Y_j - X_j).$$

Für das Bestehen der Reidemeisterschen Konfiguration (P) ist es nun notwendig und hinreichend, daß (58) erfüllt ist und ferner, daß $\beta_{jik} - \beta_{jki}$ auf jeder Mannigfaltigkeit der dritten Schar konstant bleibt. Die letzte Bedingung drückt man so aus:
$$\ni d(\beta_{jik} - \beta_{jki}) \equiv 0 \ (\text{mod. } \omega_{1k} + \omega_{2k}).$$
Daraus folgt wegen (54)
$$A_{npqs} = A_{npsq}$$
oder
$$f_{ilm}^{(k)} = f_{iml}^{(k)}. \tag{98}$$
Daher:

Die notwendigen und hinreichenden Bedingungen für das Bestehen der Reidemeisterschen Konfiguration (P) sind
$$\begin{gathered} f_{ilm}^{(k)} = f_{iml}^{(k)}, \\ f_{ilm}^{(k)} + f_{lmi}^{(k)} + f_{mil}^{(k)} = 0. \end{gathered} \tag{99}$$

Soll auch die Thomsensche Konfiguration gelten, so muß die betreffende Gruppe kommutativ sein. Es folgt also
$$\beta_{jik} = \beta_{jki},$$
und somit nach (50a) und (25)
$$c_{ik}^{(m)} = 0.$$
Dann ist nach (32) auch
$$f_{ilm}^{(k)} = f_{lim}^{(k)} = f_{mli}^{(k)}$$
und aus der Sechseckbedingung erhält man
$$f_{ilm}^{(k)} = 0.$$

Die notwendige und hinreichende Bedingung für das Bestehen der Thomsenschen Konfiguration (D) ist daher das Verschwinden aller Invarianten $c_{ik}^{(m)}$, $f_{ilm}^{(k)}$, woraus folgt, *daß in diesem Falle das Gewebe mit drei Parallelenbündeln linearer Räume topologisch äquivalent ist.*

§ 3. Beispiele.

Aus dem Vorherstehenden ist ersichtlich, daß die Bedingungen für (S), (P), (D) wesentlich verschieden lauten. Um ihre Verschiedenheit zu sichern, müßte man z. B. zeigen, daß (99) keine Folgerung von (58) und den Integrierbarkeitsbedingungen (32) ist. Am einfachsten gibt man konkrete Beispiele, wo sich beim Ausrechnen herausstellt, daß (58) erfüllt ist, dagegen (99) nicht.

Bol hat im R_4 das folgende Beispiel gegeben[7]):

$$\begin{array}{lll} \text{I.} & x_1 = \text{konst.}, & x_2 = \text{konst.}, \\ \text{II.} & x_3 = \text{konst.}, & x_4 = \text{konst.}, \\ \text{III.} & \varphi_1 = \dfrac{x_1 + x_3}{x_2 + x_4} = \text{konst.}, & \varphi_2 = \dfrac{x_1 - x_3}{x_2 - x_4} = \text{konst.} \end{array}$$

Die Größen β_{kij} in den Formeln (51) sind in diesem Falle

$$\begin{aligned} \beta_{111} &= \frac{(x_2 - x_4)(x_2 + x_4)}{x_2 x_3 - x_1 x_4}, & \beta_{112} &= \frac{(x_2 - x_4)^2}{2(x_2 x_3 - x_1 x_4)}, \\ \beta_{121} &= -\frac{(x_2 - x_4)^2}{2(x_2 x_3 - x_1 x_4)}, & \beta_{122} &= 0, \\ \beta_{211} &= 0, & \beta_{212} &= \frac{(x_2 + x_4)^2}{2(x_2 x_3 - x_1 x_4)}, \\ \beta_{221} &= -\frac{(x_2 + x_4)^2}{2(x_2 x_3 - x_1 x_4)}, & \beta_{222} &= -\frac{(x_2 - x_4)(x_2 + x_4)}{x_2 x_3 - x_1 x_4}. \end{aligned}$$

Nach (54) bekommen wir

$$\begin{gathered} A_{1111} = A_{1122} = A_{1211} = A_{2111} = 0, \\ A_{1222} = A_{2122} = A_{2211} = A_{2222} = 0, \\ A_{1112} = -A_{1121}, \quad A_{1221} = -A_{1212}, \\ A_{2112} = -A_{2121}, \quad A_{2212} = -A_{2221}. \end{gathered}$$

Da z. B. $A_{1112} \neq 0$, gilt (P) nicht. Die Bedingungen (58) sind aber erfüllt, d. h. die Konfiguration (S) besteht.

Eine leichte Überlegung zeigt, daß für das Gewebe im R_{2r}:

$$\begin{array}{ll} \text{I.} & x_1 = \text{konst.}, \quad x_2 = \text{konst.}, \quad x_3 = \text{konst.}, \cdots, x_r = \text{konst.}, \\ \text{II.} & x_{r+1} = \text{konst.}, \quad x_{r+2} = \text{konst.}, \quad x_{r+3} = \text{konst.}, \cdots, x_{2r} = \text{konst.}, \\ \text{III.} & \begin{array}{l} \varphi_1 = \dfrac{x_1 + x_{r+1}}{x_2 + x_{r+2}} = \text{konst.}, \quad \varphi_2 = \dfrac{x_1 - x_{r+1}}{x_2 - x_{r+2}} = \text{konst.}, \\ \varphi_3 = x_3 + x_{r+3} = \text{konst.}, \cdots, \varphi_r = x_r + x_{2r} = \text{konst.}, \end{array} \end{array}$$

ebenfalls die Sechseckkonfiguration (S), nicht aber (P) gilt.

Als zweites Beispiel nehmen wir im R_{2r} das Gewebe, wo jede der drei Scharen aus den R_r durch einen festen R_{r-1} besteht und diese drei R_{r-1} nicht in einer Hyperebene liegen. Nach einer geeigneten projektiven Transformation kann man die drei Scharen in die folgenden überführen:

$$\begin{array}{ll} \text{I.} & x_1 = \text{konst.}, \quad x_2 = \text{konst.}, \cdots, x_r = \text{konst.}, \\ \text{II.} & x_{r+1} = \text{konst.}, \quad x_{r+2} = \text{konst.}, \cdots, x_{2r} = \text{konst.}, \end{array}$$

---

[7]) Bol, a. a. O., S. 448.

III.
$$\varphi_1 = \frac{x_1 + x_{r+1}}{x_1 + \cdots + x_r} = \text{konst.},$$
$$\varphi_2 = \frac{x_2 + x_{r+2}}{x_1 + \cdots + x_r} = \text{konst.}, \cdots, \varphi_r = \frac{x_r + x_{2r}}{x_1 + \cdots + x_r} = \text{konst.}$$

Setzt man

$$z = x_1 + x_2 + \cdots + x_r,$$
$$C_i = \sum_{l=1}^{r} C_{li},$$

so findet man in diesem Fall

$$\beta_{kij} = -\frac{\delta_{jk}}{z} C_i,$$

also

$$\beta_{kij} = 0, \quad j \neq k,$$
$$\beta_{kik} = -\frac{1}{z} C_i.$$

Da

$$\frac{\partial \varphi_i}{\partial x_k} = \frac{\delta_{ik}}{z} - \frac{\varphi_i}{z},$$

folgt

$$\frac{C_{il}}{z} - \frac{\varphi_i}{z} C_l = \delta_{il}.$$

Summiert man diese Gleichungen von 1 bis r, so erhält man

$$\frac{1}{z}(1 - \varphi_1 - \varphi_2 - \cdots - \varphi_r) C_l = 1,$$

folglich

$$C_l = \frac{z}{1 - \varphi_1 - \varphi_2 - \cdots - \varphi_r}$$

Man findet schließlich

$$\beta_{kik} = \frac{1}{\varphi_1 + \cdots + \varphi_r - 1}, \qquad \beta_{kij} = 0, \quad j \neq k.$$

Da die β_{kik} Funktionen von φ_k allein sind, haben wir

$$A_{npqs} = 0,$$

d. h. es besteht die Reidemeistersche Konfiguration (P). Andererseits gilt (T) nicht, da

$$\beta_{kik} - \beta_{kki} = \beta_{kik} \neq 0, \quad \text{wenn } i \neq k.$$

Hamburg, den 28. August 1935.

SUR LA POSSIBILITÉ DE PLONGER UN ESPACE A CONNEXION PROJECTIVE DONNÉ DANS UN ESPACE PROJECTIF;

PAR M. SHIING-SHEN CHERN (1).

1. Le théorème sur la possibilité de plonger un espace riemannien général à n dimensions dans un espace euclidien à $\frac{n(n+1)}{2}$ dimensions a été énoncé par Schläfli et a été rigoureusement démontré par MM. Janet et Cartan (2). Dans cette Note nous voulons démontrer le théorème analogue dans la géométrie des espaces à connexion projective.

Un espace à connexion projective à n dimensions est défini par une famille de repères $AA_1, \ldots, A_n$ (3), attachés aux différents points A de l'espace, avec une loi de déplacement infinitésimal d'un repère à un repère infiniment voisin. D'autre part, étant donnée une variété V_n à n dimensions dans un espace projectif à N dimensions, nous pouvons attacher à chaque point de la variété, un espace linéaire E_{N-n-1} à $N-n-1$ dimensions sans points communs avec l'espace tangent (T) à V_n en ce point. De E_{N-n-1} nous pouvons projeter l'espace tangent (T) sur un espace tangent (T') infiniment voisin en joignant un point de (T) avec E_{N-n-1} par un espace linéaire à $N-n$ dimensions, dont l'intersection avec (T') donne le point correspondant sur (T'). De cette manière, on a défini une connexion projective sur la variété V_n. Un espace à connexion projective à n dimensions étant donné à l'avance, notre problème est la recherche, dans un espace projectif

(1) Research Fellow of the China Foundation. Je remercie cordialement M. Élie Cartan pour ses précieux conseils et critiques.

(2) M. JANET, *Annales Soc. Pol. Math.*, t. 5, 1926, p. 38-43; E. CARTAN, *même recueil*, t. 6, 1927, p. 1-7.

(3) E. CARTAN, *Leçons sur la théorie des espaces à connexion projective*, IIe Partie (Paris, Gauthier-Villars, 1937).

SHIING-SHEN CHERN

à N dimensions (avec N suffisamment grand), de la possibilité de trouver une variété à n dimensions et de définir sur elle une connexion projective comme ci-dessus, de sorte qu'elle soit isomorphe avec la connexion projective donnée. Nous verrons qu'il faut que l'espace donné satisfasse à des conditions nécessaires et que si ces conditions sont satisfaites, la valeur de N est plus grande que la valeur correspondante dans le cas des espaces riemanniens.

2. Considérons maintenant un espace à connexion projective à n dimensions, rapporté à un système quelconque de repères $AA_1, \ldots, A_n$. Le déplacement infinitésimal du repère est donné par les équations

$$(1)\quad \begin{cases} dA = \varpi^1 A_1 + \ldots + \varpi^n A_n, \\ dA_1 = \varpi_1^0 A + \varpi_1^1 A_1 + \ldots + \varpi_1^n A_n, \\ \ldots\ldots\ldots\ldots\ldots\ldots\ldots\ldots \\ dA_n = \varpi_n^0 A + \varpi_n^1 A_1 + \ldots + \varpi_n^n A_n, \end{cases}$$

où les ϖ^i, ϖ_i^0, ϖ_i^j $(i, j = 1, \ldots, n)$ sont des formes différentielles linéaires des coordonnées $u^1, \ldots, u^n$ de l'espace. Les équations de structure de l'espace sont de la forme

$$(2)\quad \begin{cases} (\varpi^i)' = \sum_{k=1}^{n} [\varpi^k \varpi_k^i] - \frac{1}{2} \sum_{k,h=1}^{n} R_{0kh}^i [\varpi^k \varpi^h], \\ (\varpi_i^j)' = [\varpi_i^0 \varpi^j] + \sum_{k=1}^{n} [\varpi_i^k \varpi_k^j] - \delta_i^j \sum_{k=1}^{n} [\varpi^k \varpi_k^0] - \frac{1}{2} \sum_{k,h=1}^{n} R_{ikh}^j [\varpi^k \varpi^h], \\ (\varpi_i^0)' = \sum_{k=1}^{n} [\varpi_i^k \varpi_k^0] - \frac{1}{2} \sum_{k,h=1}^{n} R_{ikh}^0 [\varpi^k \varpi^h], \end{cases}$$

où les R_{0kh}^i, R_{ikh}^j, R_{ikh}^0 sont les composantes du tenseur de courbure et de torsion et satisfont aux relations

$$(3)\quad \begin{cases} R_{0kh}^i + R_{0hk}^i = 0, \\ R_{ikh}^j + R_{ihk}^j = 0 \qquad (i, j, h, k = 1, \ldots, n), \\ R_{ikh}^0 + R_{ihk}^0 = 0. \end{cases}$$

L'espace est dit sans torsion, si

$$R_{0kh}^i = 0.$$

Dans cé cas, on a aussi les relations

$$R^j_{ikh} + R^j_{khi} + R^j_{hik} = 0. \tag{3'}$$

Prenons dans un espace ordinaire projectif à N dimensions un système de repères $BB_1, \ldots, B_N$ complètement indéterminés, dépendant par suite de $N(N+2)$ paramètres arbitraires. Soient

$$\text{(4)}\quad \begin{cases} dB = \omega^1 B_1 + \ldots + \omega^N B_N, \\ dB_1 = \omega_1^0 B + \omega_1^1 B_1 + \ldots + \omega_1^N B_N, \\ \ldots\ldots\ldots\ldots\ldots\ldots\ldots\ldots \\ dB_N = \omega_N^0 B + \omega_N^1 B_1 + \ldots + \omega_N^N B_N, \end{cases}$$

les équations du déplacement infinitésimal du repère. On a les équations de structure de l'espace projectif

$$\text{(5)}\quad \begin{cases} (\omega^i)' = \sum_{k=1}^{N} [\omega^k \omega_k^i], \\ (\omega_i^j)' = [\omega_i^0 \omega^j] + \sum_{k=1}^{N} [\omega_i^k \omega_k^j] - \delta_i^j \sum_{k=1}^{N} [\omega^k \omega_k^0], \\ (\omega_i^0)' = \sum_{k=1}^{N} [\omega_i^k \omega_k^0] \qquad (i, j = 1, \ldots, N). \end{cases}$$

Soit B le point courant sur la variété à n dimensions qu'on veut chercher et soit $BB_1, \ldots, B_n$ le plan tangent en B. Soit aussi l'espace linéaire $B_{n+1}, \ldots, B_N$ à $N - n - 1$ dimensions, l'espace qu'on attache au point B. Le théorème en question revient à l'étude du système d'équations de Pfaff

$$\text{(I)}\quad \begin{cases} \omega^1 = \varpi^1, \quad \ldots, \quad \omega^n = \varpi^n, \\ \omega^{n+1} = \ldots = \omega^N = 0, \\ \omega_i^0 = \varpi_i^0, \quad \omega_i^j = \varpi_i^j \quad (i, j = 1, \ldots, n), \end{cases}$$

et, en particulier, à la recherche de ses variétés intégrales à n dimensions sur lesquelles les formes $\varpi^1, \ldots, \varpi^n$ soient indépendantes. Le système (I) est un système de $N + n(n+1)$ équations à $N(N+2) + n$ variables.

Dérivons extérieurement les n premières équations de (I) et tenons compte des équations (I) elles-mêmes. Puisque les formes $\varpi^1, \ldots, \varpi^n$ seraient indépendantes sur la variété intégrale à chercher, on obtient

$$R^i_{0kh} = 0.$$

Par conséquent, il faut que l'espace à connexion projective donné soit sans torsion.

Les dérivées extérieures des équations (I) donnent encore

$$\text{(II)}\quad \left\{\begin{aligned} &\sum_{k=1}^{n} [\varpi^k \omega_k^\alpha] = 0 \qquad (\alpha = n+1, \ldots, N), \\ &\sum_{\lambda=n+1}^{N} [\omega_i^\lambda \omega_\lambda^0] = -\frac{1}{2} \sum_{k,h=1}^{n} R^0_{ikh} [\varpi^k \varpi^h] \qquad (i = 1, \ldots, n), \\ &\sum_{\lambda=n+1}^{N} [\omega_i^\lambda \omega_\lambda^j] = -\frac{1}{2} \sum_{k,h=1}^{n} R^j_{ikh} [\varpi^k \varpi^h] \qquad (i, j = 1, \ldots, n). \end{aligned}\right.$$

D'où l'on a

$$\sum_{i=1}^{n} \sum_{\lambda=n+1}^{N} [\varpi^i \omega_i^\lambda \omega_\lambda^0] = -\frac{1}{2} \sum_{i,k,h=1}^{n} R^0_{ikh} [\varpi^i \varpi^k \varpi^h] = 0,$$

et, par suite,

$$(6) \qquad R^0_{ikh} + R^0_{khi} + R^0_{hik} = 0,$$

qui sont encore des conditions nécessaires.

Cela étant, nous allons démontrer que si l'espace à connexion projective donné est sans torsion et que si les conditions (6) sont vérifiées, le système (I) est en involution par rapport à $\varpi^1, \ldots, \varpi$ pour les valeurs suivantes de N :

$$N = \frac{n(n+1)}{2} + \frac{n-1}{2}, \qquad \text{si } n \text{ est impair.}$$

$$N = \frac{n(n+1)}{2} + \frac{n}{2}, \qquad \text{si } n \text{ est pair.}$$

3. C'est au système (I) que nous allons appliquer la théorie des systèmes de Pfaff en involution. Le système (I) contenant $N(N+2)+n$ variables, considérons un point arbitraire dans l'espace de ces $N(N+2)+n$ variables. Un élément linéaire intégral issu de ce point est défini par les valeurs des paramètres directeurs non tous nuls

$$\varpi^i, \quad \omega_\alpha^0, \quad \omega_i^\alpha, \quad \omega_\alpha^i \qquad (i = 1, \ldots, n;\ \alpha = n+1, \ldots, N).$$

Supposons, sans restreindre la généralité, que tout élément intégral à s dimensions, E_s ($s = 1, \ldots, n$), est défini par les équations

$$(7)\qquad \begin{cases} \omega_\alpha^0 = l_{\alpha 1}^0 \varpi^1 + \ldots + l_{\alpha s}^0 \varpi^s, \\ \omega_i^\alpha = l_{i1}^\alpha \varpi^1 + \ldots + l_{is}^\alpha \varpi^s, \\ \omega_\alpha^i = l_{\alpha 1}^i \varpi^1 + \ldots + l_{\alpha s}^i \varpi^s, \\ \varpi^{s+1} = \ldots = \varpi^n = 0. \end{cases}$$

Pour que les équations (7) définissent un élément intégral à s dimensions, les équations (II) doivent être vérifiées; cela donne

$$(8)\qquad \begin{cases} l_{rs}^\alpha - l_{sr}^\alpha = 0, \\ \displaystyle\sum_{\lambda=n+1}^{N} l_{ir}^\lambda l_{\lambda s}^0 - \sum_{\lambda=n+1}^{N} l_{is}^\lambda l_{\lambda r}^0 + R_{irs}^0 = 0, \\ \displaystyle\sum_{\lambda=n+1}^{N} l_{ir}^\lambda l_{\lambda s}^j - \sum_{\lambda=n+1}^{N} l_{is}^\lambda l_{\lambda r}^j + R_{irs}^j = 0 \\ (i, j = 1, \ldots, n;\ r = 1, \ldots, s-1). \end{cases}$$

Pour la démonstration de l'existence, il faut montrer que par tout élément intégral arbitraire E_{s-1}, il passe au moins un élément intégral E_s ($s = 2, \ldots, n$). Autrement dit, il faut montrer que le système d'équations (8) peut être résolu par rapport à $l_{\alpha s}^0$, l_{is}^α, $l_{\alpha s}^i$ ($s = 2, \ldots, n$) sans introduire aucune relation nouvelle entre $l_{\alpha r}^0$, l_{ir}^α, $l_{\alpha r}^i$ ($r = 1, \ldots, s-1$).

Nous regarderons les quantités $l_{\alpha i}^0$, l_{ij}^α, $l_{\alpha i}^j$, pour i, j fixes et pour α variable de $n+1$ à N, comme les projections respectives des vecteurs $\vec{l}_i^{\,0}$, $\vec{l}_{ij}$, $\vec{l}_i^{\,j}$ dans un espace auxiliaire à $N - n$ dimensions. Les sommes $\sum_\lambda l_{ir}^\lambda l_{\lambda s}^0$, $\sum_\lambda l_{ir}^\lambda l_{\lambda s}^j$ peuvent être regardées comme les produits scalaires $(\vec{l}_{ir}\vec{l}_s^{\,0})$, $(\vec{l}_{ir}\vec{l}_s^{\,j})$ des paires de vecteurs $\vec{l}_{ir}$, $\vec{l}_s^{\,0}$ et $\vec{l}_{ir}$, $\vec{l}_s^{\,j}$ respectivement. Cela étant, nous pouvons écrire les équations (8) sous la forme

$$(9)\qquad \begin{cases} \vec{l}_{rs} - \vec{l}_{sr} = 0, \\ (\vec{l}_{ir}\vec{l}_s^{\,0}) - (\vec{l}_{is}\vec{l}_r^{\,0}) + R_{irs}^0 = 0, \\ (\vec{l}_{ir}\vec{l}_s^{\,j}) - (\vec{l}_{is}\vec{l}_r^{\,j}) + R_{irs}^j = 0 \\ (r = 1, \ldots, s-1;\ i, j = 1, \ldots, n). \end{cases}$$

Pour chaque valeur de $s = 2, \ldots, n$ il faut montrer que le système (9) peut être résolu par rapport aux $2n+1$ vecteurs inconnus $\vec{l}_{is}, \vec{l}_s^{\,j}, \vec{l}_s^{\,0}$.

Prenons d'abord $s = 2$. Les équations (9) donnent

$$
(10) \qquad \begin{cases} \vec{l}_{12} = \vec{l}_{21}, \\ (\vec{l}_{i1}\vec{l}_2^{\,0}) - (\vec{l}_{i2}\vec{l}_1^{\,0}) + R^0_{i12} = 0, \\ (\vec{l}_{i1}\vec{l}_2^{\,j}) - (\vec{l}_{i2}\vec{l}_1^{\,j}) + R^j_{i12} = 0. \end{cases}
$$

La première équation détermine le vecteur $\vec{l}_{12}$. Les deux dernières équations forment un système de $n(n+1)$ équations aux $2n$ vecteurs inconnus $\vec{l}_{22}, \ldots, \vec{l}_{n2}, \vec{l}_2^{\,0}, \vec{l}_2^{\,i}$. Ce système est compatible si toutes les équations du système d'équations homogènes correspondantes, à savoir

$$
(\vec{l}_{11}\vec{x}^{\,\alpha}) = 0, \qquad (\vec{l}_{i1}\vec{x}^{\,\alpha}) - (\vec{l}_1^{\,\alpha}\vec{x}_i) = 0
$$
$$
(\alpha = 0, 1, \ldots, n;\ i = 2, \ldots, n),
$$

sont indépendantes. En effet, nous voyons facilement que c'est le cas puisque les coefficients sont les composantes des vecteurs $\vec{l}_{j1}$, $\vec{l}_1^{\,0}, \vec{l}_1^{\,i}$, qui sont arbitraires.

Cela étant, considérons les équations (9) pour une valeur générale de $s = 2, \ldots, n$. Les premières équations déterminent les vecteurs $\vec{l}_{rs}$ pour $r = 1, \ldots, s-1$. Ces vecteurs étant déterminés, nous avons dans les deux dernières équations les $2n - s + 2$ vecteurs inconnus $\vec{l}_s^{\,0}$, $\vec{l}_{is}$ $(i = s, \ldots, n)$, $\vec{l}_s^{\,j}$ $(j = 1, \ldots, n)$, qui donnent $(N-n)(2n-s+2)$ composantes inconnues. Pour calculer le nombre de ces équations indépendantes, remarquons que pour $i = 1, \ldots, s-1$ la deuxième équation de (9) s'écrit

$$
(11) \qquad (\vec{l}_{ir}\vec{l}_s^{\,0}) - (\vec{l}_{si}\vec{l}_r^{\,0}) + R^0_{irs} = 0.
$$

En vertu des relations (6) cette équation est une conséquence des équations

$$
(12) \qquad \begin{cases} (\vec{l}_{sr}\vec{l}_i^{\,0}) - (\vec{l}_{ri}\vec{l}_s^{\,0}) + R^0_{rsi} = 0, \\ (\vec{l}_{si}\vec{l}_r^{\,0}) - (\vec{l}_{sr}\vec{l}_i^{\,0}) + R^0_{sir} = 0, \end{cases}
$$

dont la deuxième équation de (12) est au signe près, l'équation (11), si l'on y permute i et r et la troisième équation est déjà vérifiée par la détermination de l'élément intégral à i ou à r dimensions suivant qu'on a $r < i$ ou $i < r$. La même propriété est vraie pour les dernières équations dans (9) en vertu des relations (3'). Donc on peut supprimer les équations dépendantes dans (9) et écrire les équations qui restent sous la forme

$$(13)\quad \left\{\begin{aligned} &(\vec{l}_{ir}\vec{l}_s^0) - (\vec{l}_{si}\vec{l}_r^0) + R^0_{irs} = 0, \\ &(\vec{l}_{ir}\vec{l}_s^j) - (\vec{l}_{si}\vec{l}_r^j) + R^j_{irs} = 0 \\ &\quad (i \leqq r = 1, \ldots, s-1;\ j = 1, \ldots, n); \\ &(\vec{l}_{ir}\vec{l}_s^0) - (\vec{l}_{is}\vec{l}_r^0) + R^0_{irs} = 0, \\ &(\vec{l}_{ir}\vec{l}_s^j) - (\vec{l}_{is}\vec{l}_r^j) + R^j_{irs} = 0 \\ &\quad (r = 1, \ldots, s-1;\ i = s, \ldots, n;\ j = 1, \ldots, n). \end{aligned}\right.$$

Le nombre de ces équations est égal à

$$p(s) = (n+1)\frac{s(s-1)}{2} + (n+1)(s-1)(n-s+1) = \frac{(n+1)(s-1)(2n-s+2)}{2}.$$

La différence du nombre des inconnus et du nombre des équations est alors

$$\begin{aligned} d(s) &= (N-n)(2n-s+2) - \frac{(n+1)(s-1)(2n-s+2)}{2} \\ &= \frac{(2n-s+2)(2N-ns-n-s+1)}{2}. \end{aligned}$$

Pour $s = n$, on a

$$d(n) = 0 \qquad \text{pour } n \text{ impair,}$$
$$d(n) = \frac{n+2}{2} \qquad \text{pour } n \text{ pair.}$$

Comme $d(s)$ diminue quand s augmente, il est clair que $d(s) > 0$ pour $s = 2, \ldots, n-1$. Dans le système (13) nous avons donc plus d'inconnus que d'équations.

Le système (13) est compatible, si toutes les équations du système d'équations homogènes correspondantes, à savoir

$$(13')\quad \left\{\begin{aligned} &(\vec{l}_{ir}\vec{x}^\alpha) = 0 && (i \leqq r = 1, \ldots, s-1;\ \alpha = 0, 1, \ldots, n), \\ &(\vec{l}_{ir}\vec{x}^\alpha) - (\vec{l}_r^\alpha \vec{x}_i) = 0 && \left(\begin{matrix} i = s, \ldots, n;\ r = 1, \ldots, s-1 \\ \alpha = 0, 1, \ldots, n \end{matrix}\right), \end{aligned}\right.$$

sont indépendantes. Nous le montrons par induction.

Les équations du système (13') étant indépendantes pour $s = 2$, supposons qu'elles sont indépendantes pour s. Les équations correspondantes à (13') pour $s+1$ s'écrivent

$$(14)\quad \begin{cases} (\vec{l}_{ir}\vec{x}^{\alpha}) = 0 & (i \leqq r = 1, \ldots, s-1), \\ (\vec{l}_{ir}\vec{x}^{\alpha}) - (\vec{l}_r^{\alpha}\vec{x}_i) = 0 & (i = s+1, \ldots, n;\ r = 1, \ldots, s-1), \\ (\vec{l}_{is}\vec{x}^{\alpha}) = 0 & (i = 1, \ldots, s), \\ (\vec{l}_{is}\vec{x}^{\alpha}) - (\vec{l}_s^{\alpha}\vec{x}_i) = 0 & (i = s+1, \ldots, n), \end{cases}$$

où $\alpha = 0, 1, \ldots, n$ pour toutes les équations.

Les équations des deux premières lignes du système sont indépendantes d'après l'hypothèse pour s. Comme les coefficients des deux dernières équations de (14) sont les composantes des vecteurs $\vec{l}_{is}(i = 1, \ldots, n)$, $\vec{l}_s^{\alpha}(\alpha = 0, 1, \ldots, n)$, qui sont liés seulement par les relations (13), on voit manifestement qu'en général on peut les choisir de manière que les équations (14_3), (14_4) sont indépendantes de (14_1), (14_2). Les équations du système (14) sont donc toutes indépendantes.

Donc nous avons démontré que le système (13) est compatible pour chaque valeur de s. Le théorème énoncé est alors démontré.

Pour n pair la solution générale dépend de $\frac{n+2}{2}$ fonctions arbitraires de n arguments. Pour n impair la solution générale dépend de $\frac{(n+1)(n+3)}{2}$ fonctions arbitraires de $n-1$ arguments.

4. Il serait intéressant de calculer le degré d'arbitraire (nombre de fonctions arbitraires de n variables) qui entre dans l'espace à connexion projective V_n, construit suivant le procédé considéré dans l'espace projectif à N dimensions. Pour définir V_n il faut donner d'abord la variété et puis à chaque point de la variété un espace linéaire à $N-n-1$ dimensions. Pour cela on a besoin de $(N-n)+(N-n)(n+1) = (N-n)(n+2)$ fonctions de n arguments. Donc les espaces à connexion projective considérés dépendent de $(N-n)(n+2)$ fonctions arbitraires de n arguments pour n impair, et de $(N-n)(n+2) - \frac{n+2}{2}$ pour n pair. Dans tous les cas, ce nombre est égal à

$$\frac{(n-1)(n+1)(n+2)}{2}.$$

Nous pouvons vérifier ce résultat en calculant ce nombre par un autre procédé. Un espace à connexion projective peut être défini par une famille de repères naturels, qui sont complètement déterminés en tous les points. Pour cette famille de repères, on a

$$(15)\qquad \varpi^i = du^i, \qquad \varpi_i^j = \Sigma \Pi_{ik}^j du^k, \qquad \varpi_i^0 = \Sigma \Pi_{ik}^0 du^k, \qquad \varpi_1^1 + \ldots + \varpi_n^n = 0.$$

Les conditions que l'espace soit sans torsion sont :

$$(16)\qquad \Pi_{kh}^i = \Pi_{hk}^i.$$

De plus, on peut vérifier que

$$(17)\qquad R_{ijk}^0 = \frac{\partial \Pi_{ij}^0}{\partial u^k} - \frac{\partial \Pi_{ik}^0}{\partial u^j} + \Sigma \Pi_{ij}^h \Pi_{hk}^0 - \Sigma \Pi_{ik}^h \Pi_{hj}^0.$$

Les conditions (6) sont équivalentes à

$$(18)\qquad \frac{\partial}{\partial u^k}(\Pi_{ij}^0 - \Pi_{ji}^0) + \frac{\partial}{\partial u^i}(\Pi_{jk}^0 - \Pi_{kj}^0) + \frac{\partial}{\partial u^j}(\Pi_{ki}^0 - \Pi_{ik}^0) = 0.$$

Ce système d'équations différentielles partielles est équivalent au système

$$\sum_{i,j,k} \left[\frac{\partial(\Pi_{ij}^0 - \Pi_{ji}^0)}{\partial u^k} du^i \, du^j \, du^k \right] = 0,$$

ou à

$$(19)\qquad \sum_{i,j} [d(\Pi_{ij}^0 - \Pi_{ji}^0) \, du^i \, du^j] = 0,$$

qui est en involution. En regardant les fonctions $\Pi_{ij}^0 - \Pi_{ji}^0$ comme les variables dépendantes, la solution générale du système (18) dépend de $n-1$ fonctions arbitraires de n arguments. Parmi les fonctions Π_{ik}^0, on a alors $n - 1 + \frac{n(n+1)}{2}$ fonctions arbitraires de n arguments. Des fonctions Π_{kh}^i, on a $\frac{n^2(n+1)}{2} - n$, qui sont arbitraires. Puisqu'on peut effectuer sur les u^i un changement de coordonnées, le nombre des fonctions arbitraires de n arguments, dont dépend l'espace à connexion projective considéré, est égal à

$$\left\{ n - 1 + \frac{n(n+1)}{2} \right\} + \left\{ \frac{n^2(n+1)}{2} - n \right\} - n = \frac{(n-1)(n+1)(n+2)}{2},$$

en accord avec le résultat trouvé par l'autre procédé.

(Extrait du *Bulletin des Sciences mathématiques*
2ᵉ série, t. LXI, août 1937.)

THE GEOMETRY OF THE DIFFERENTIAL EQUATION
$y''' = F(x, y, y', y'')$.

By SHIING-SHEN CHERN (陳省身)

Department of Mathematics

(Received June 15, 1939)

ABSTRACT

From the geometric object $y''' = F(x, y, y', y'')$ in the plane we can define either a normal conformal connection with the integral curves of the differential equation as elements or a generalized geometry with a certain five-parameter group as fundamental group. The former case is more important and is characterized by the fact that the condition of contact of neighboring integral curves of the differential equation is given by an equation of Monge of the second order. The solution of the problem depends on its formulation as a problem of equivalence of systems of Pfaffian forms.

Introduction

Cartan has proved[1] that to an ordinary differential equation of the second order

$$(1) \qquad y'' = F(x, y, y')$$

in the plane, where y' and y'' denote the first and second derivatives of y with respect to x, it is possible to define in the plane a projective connection having the integral curves of (1) as its geodesics. Naturally, the definition of this connection is invariant under the group of point transformations in the plane.

The purpose of this paper is to study the geometry in the plane as arised from an ordinary differential equation of the third order

$$(2) \qquad y''' = F(x, y, y', y''),$$

1) E. Cartan, Sur les variétés â connexion projective, Bull. Soc. Math. de France, *52*, 205.

where the invariance under the group of contact transformations is required. One may conject that the fundamental group of the geometry so defined will be the ten parameter group G_{10} of circle transformations of Lie in the plane, as the group which leaves invariant the differential equation $y''' = 0$ is the group of contact transformations leaving unaltered a system of parabolas, group which is similar to G_{10}. But in reality the problem is more complicated. We want to show that this is the case when a certain relative invariant I vanishes. The condition $I = 0$ has a simple geometrical meaning according to a theorem due to Wünschmann. By interpreting the integral curves of (2) as the points in a space of three dimensions, the connection so defined is a normal connection in the sense of Cartan[2]. In the general case $I \neq 0$, it is also possible to define a generalized geometry in the plane with the elements of contact of the second order x, y, y', y'' as the elements of the space and with a certain five-parameter group as its fundamental group. The main results of this paper have been announced in a previous note[3] in the "Comptes Rendus de l'Académie des Sciences à Paris."

The paper is divided into three parts: (1) Formulation of the problem as a problem of equivalence of Pfaffian systems; (2) The case $I = 0$ and its geometrical meaning; (3) The case $I \neq 0$.

The author wishes to express his thanks to Professor Elie Cartan in Paris under whose direction the present study was carried out. He is also indebted to Professor F. Engel in Giessen, Germany for many valuable communications.

§1. Formulation of the Problem

By regarding the y', y'' as independent variables, the equation (2) may be written as a Pfaffian system

$$\left\{\begin{aligned} dy'' - F\,dx &= 0, \\ dy' - y''\,dx &= 0, \\ dy - y'\,dx &= 0. \end{aligned}\right. \tag{3}$$

2) E. Cartan, Les espaces à connexion conforme, Annales Soc. Pol. de Math., 1923, pp. 171-221.

3) Shiing-shen Chern. Sur la géométrie d'une équation différentielle du troisième ordre, Comptes Rendus, 1937, 1227.

We introduce the Pfaffian forms

$$(4)\quad \begin{cases} \omega_1 = \alpha \left\{ dy'' - F\,dx + \beta\,(dy - y'\,dx) + \gamma\,(dy' - y''\,dx) \right\}, \\ \omega_2 = \lambda\,(dy - y'\,dx), \\ \omega_3 = \mu \left\{ dy' - y''\,dx + \nu\,(dy - y'\,dx) \right\}, \\ \omega_4 = u \left\{ dx + v\,(dy - y'\,dx) + w\,(dy' - y''\,dx) \right\}, \end{cases}$$

in which the $\alpha, \beta, \gamma, \lambda, \mu, \nu, u, v, w$ are nine auxiliary variables.

Before defining the generalized geometry in the plane we propose to study of problem of equivalence of the Pfaffian system (3) and a new system

$$(5)\quad \begin{cases} d\bar{y}'' - \bar{F}\,d\bar{x} = 0, \\ d\bar{y}' - \bar{y}''\,d\bar{x} = 0, \\ d\bar{y} - \bar{y}'\,d\bar{x} = 0 \end{cases}$$

under the group of contact transformations. If $\bar{\omega}_1, \bar{\omega}_2, \bar{\omega}_3, \bar{\omega}_4$ are four Pfaffian forms formed from the system (5) with the new auxiliary variables $\bar{\alpha}, \bar{\beta}, \bar{\gamma}, \bar{\lambda}, \bar{\mu}, \bar{\nu}, \bar{u}, \bar{v}, \bar{w}$ in the same way as the ω's are formed from (3), we see that a necessary and sufficient condition that the two systems (3) and (5) be equivalent under the group of contact transformations is that there be a transformation in the variables

$$x, y, y'\, y''; \alpha, \beta, \gamma, \lambda, \mu, \nu, u, v, w$$

leaving the ω's invariant:

$$(6)\qquad \bar{\omega}_1 = \omega_1, \quad \bar{\omega}_2 = \omega_2, \quad \bar{\omega}_3 = \omega_3, \quad \bar{\omega}_4 = \omega_4.$$

The essential feature of the method of equivalence developed by Cartan[4] consists of reducing the number of auxiliary variables by imposing on them relations of an invariant nature (i.e., invariant under the group of contact transformations). This can be done by forming the exterior derivatives of the Pfaffian forms ω_i. As first conditions we suppose

4) E. Cartan, Les sous-groupes des groupes continus, Annales Ecole Norm. Sup. 25 (1908), Chap. I.

$$
(7)\quad \begin{aligned}
\omega_2' &\equiv -[\omega_3\,\omega_4] \quad (\text{mod. } \omega_2),\\
\omega_3' &\equiv -[\omega_1\,\omega_4] \quad (\text{mod. } \omega_2, \omega_3),
\end{aligned}
$$

which are equivalent to the relations

$$\lambda\,[dx\;dy'] = \mu u\,[dx\;dy'],$$

$$\mu\,[dx\;dy''] = \alpha u\,[dx\;dy''],$$

thus giving

$$\lambda = \alpha u^2, \quad \mu = \alpha u.$$

By noticing that the system

$$\omega_1 = 0, \quad \omega_2 = 0, \quad \omega_3 = 0$$

is completely integrable and by taking account of the relation (7) we see that the exterior derivatives of ω_i are necessarily of the form

$$
(8)\quad \begin{cases}
\omega_1' = [w_{11}\,\omega_1] + [w_{12}\,\omega_2] + [w_{13}\,\omega_3],\\
\omega_2' = \qquad\qquad\quad [w_{22}\,\omega_2] - [\omega_3\,\omega_4],\\
\omega_3' = [w_{32}\,\omega_2] + [w_{33}\,\omega_3] - [\omega_1\,\omega_4],\\
\omega_4' = [w_{42}\,\omega_2] + [w_{43}\,\omega_3] + [w_{44}\,\omega_4],
\end{cases}
$$

In these equations the w's are new Pfaffian forms such that

$$
(9)\quad \begin{cases}
w_{11} \equiv \dfrac{d\alpha}{\alpha}, \; w_{12} \equiv \dfrac{1}{u^2}(d\beta - \nu d\gamma), \; w_{13} \equiv \dfrac{1}{u}\,d\gamma,\\
w_{22} \equiv \dfrac{d\alpha}{\alpha} + 2\,\dfrac{du}{u}, \; w_{32} \equiv \dfrac{1}{u}\,d\nu, \; w_{33} \equiv \dfrac{d\alpha}{\alpha} + \dfrac{du}{u}, \; (\text{mod. } \omega_i)\\
w_{42} \equiv \dfrac{1}{\alpha u}(dv - \nu dw), \; w_{43} \equiv \dfrac{1}{\alpha}\,dw, \; w_{44} \equiv \dfrac{du}{u}.
\end{cases}
$$

The $w_{11}, \ldots, w_{44}$ are not completely determined by the equations (8). In fact, we can add to them suitable linear combinations of ω_i such that the form of the equations (8) is still preserved. In particular, the w_{11}, w_{22}, w_{33}, w_{44} can be changed as follows:

$$(10)\qquad \begin{cases} w_{11} = \overline{w}_{11} + a_1\,\omega_1 + a_2\,\omega_2 + a_3\,\omega_3, \\ w_{22} = \overline{w}_{22} + b_2\,\omega_2, \\ w_{33} = \overline{w}_{33} + c_2\,\omega_2 + c_3\,\omega_3, \\ w_{44} = \overline{w}_{44} + e_2\,\omega_2 + e_3\,\omega_3 + e_4\,\omega_4, \end{cases}$$

where the $a_1, a_2, a_3, b_2, c_2, c_3, e_2, e_3, e_4$ are arbitrary functions of the variables.

From (9) we have

$$\begin{aligned} w_{11} + 2w_{44} - w_{22} &\equiv 0, \\ w_{11} + \ \ w_{44} - w_{33} &\equiv 0 \end{aligned} \qquad (\text{mod. } \omega_i).$$

Hence it is possible to have the equations

$$(11)\qquad w_{22} = w_{11} + 2w_{44}, \quad w_{33} = w_{11} + w_{44},$$

if we can choose a_1 and e_4 in the equations (10) such that the coefficients of ω_1 and ω_4 in the two sides of the equations (11) be equal. To see that we can arrive at the equations (11) it is necessary to find the coefficient of $[\omega_1\ \omega_4]$ in ω'_1, those of $[\omega_1\ \omega_2]$, $[\omega_2\ \omega_4]$ in ω'_2, those of $[\omega_1\ \omega_3]$, $[\omega_3\ \omega_4]$ in ω'_3, and that of $[\omega_1\ \omega_4]$ in ω'_4. From the relations

$$\omega_1' \equiv \left[\frac{d\alpha}{\alpha}\ \omega_1\right] + \alpha\left(\frac{\partial F}{\partial y''} + \gamma\right)[dx\ dy''] \ (\text{mod. } \omega_2, \omega_3),$$

$$[\omega_1\,\omega_4] \equiv -\,\alpha u\,[dx\ dy''] \quad (\text{mod. } \omega_2, \omega_3),$$

we see that the coefficient of $[\omega_1\,\omega_4]$ in ω_1' is $-\dfrac{1}{u}\left(\dfrac{\partial F}{\partial y''} + \gamma\right)$. In the same manner we find:

$$\text{Coefficient of } [\omega_1\,\omega_2] \text{ in } \omega_2' = 0,$$

$$\text{,, \quad of } [\omega_2\,\omega_4] \text{ in } \omega_2' = \frac{\nu}{u},$$

$$\text{,, \quad of } [\omega_1\,\omega_3] \text{ in } \omega_3' = \frac{w}{\alpha},$$

$$\text{,, \quad of } [\omega_3\,\omega_4] \text{ in } \omega_3' = \frac{1}{u}(\gamma - \nu),$$

$$\text{,, \quad of } [\omega_1\,\omega_4] \text{ in } \omega_4' = -\frac{w}{\alpha}.$$

The equations (11) then give

$$a_1 = \frac{2w}{\alpha}, \quad e_4 = \frac{1}{u}\left(\nu - 2\gamma - \frac{\partial F}{\partial y''}\right),$$

$$\nu = \gamma + \frac{1}{3}\frac{\partial F}{\partial y''}.$$

This shows that not only a_1 and e_4 but also ν is determined.

The auxiliary variables ν and γ being connected by the above equation, it is again possible to add any linear combination of ω_1. ω_2. ω_3 to $w_{.3}$ without changing the form of the equation (8) and (11). We can thus arrive at the relation

$$w_{13} = w_{32}, \tag{12}$$

if the coefficients of ω_4 in both sides of the equation be equal. In order to have these coefficients it is necessary to calculate the coefficient of $[\omega_3\ \omega_4]$ in ω'_1 and that of $[\omega_2\ \omega_4]$ in ω'_3. As above, we find:

$$\text{Coefficient of } [\omega_3\,\omega_4] \text{ in } \omega_1' = \frac{1}{u^2}\left\{\gamma\left(\frac{\partial F}{\partial y''} + \gamma\right) - \left(\frac{\partial F}{\partial y'} + \beta\right)\right\}.$$

$$\text{,, of } [\omega_2\,\omega_4] \text{ in } \omega_3' = \frac{1}{u^2}(\nu^2 - \gamma\nu + \beta).$$

The equation (12) then gives

$$2\beta = \gamma^2 + \frac{2}{3}\gamma\frac{\partial F}{\partial y''} - \frac{\partial F}{\partial y'} - \frac{1}{9}\left(\frac{\partial F}{\partial y''}\right)^2 + \frac{1}{3}\frac{d}{dx}\frac{\partial F}{\partial y''},$$

where we have introduced the notation

$$\frac{d}{dx}\Phi(x, y, y', y'') = \frac{\partial \Phi}{\partial x} + \frac{\partial \Phi}{\partial y}y' + \frac{\partial \Phi}{\partial y'}y'' + \frac{\partial \Phi}{\partial y''}F, \tag{13}$$

Φ being any function of x, y, y', y''.

By determining β in this way, we easily see that w_{12} is then a linear combination of ω_i and that the coefficient of $[\omega_2\ \omega_4]$ in ω_1' is an invariant. Its expression is of the form I/u^3, where

$$\begin{aligned} I = &-\frac{\partial F}{\partial y} - \frac{1}{3}\frac{\partial F}{\partial y'}\frac{\partial F}{\partial y''} - \frac{2}{27}\left(\frac{\partial F}{\partial y''}\right)^3 + \frac{1}{2}\frac{d}{dx}\frac{\partial F}{\partial y'} \\ &+ \frac{1}{3}\frac{\partial F}{\partial y''}\frac{d}{dx}\frac{\partial F}{\partial y''} - \frac{1}{6}\frac{d^2}{dx^2}\frac{dF}{\partial y''}. \end{aligned} \tag{14}$$

The function I is a relative invariant of the differential equation

(2). We shall divide our discussion into two cases, according as $I=0$ or $I\neq 0$.

§2. The Case $I = 0$

With a little change of notation we write the equations (8) in the form

$$(15)\quad \begin{cases} \omega_1' = [w_1\omega_1] \qquad\qquad + [w_2\omega_3], \\ \omega_2' = \qquad [w_1\omega_2] + 2[w_3\omega_2] - [\omega_3\omega_4], \\ \omega_3' = \qquad [w_2\omega_2] + [w_1\omega_3] + [w_3\omega_3] - [\omega_1\omega_4], \\ \omega_4' = \qquad [w_4\omega_2] + [w_5\omega_3] + [w_3\omega_4], \end{cases}$$

where the w_i are five new Pfaffian forms linearly independent between them and independent of ω_i. The most general transformation on the w_i which preserves the equations (15) is of the form

$$(16)\quad \begin{cases} w_1 = \overline{w}_1, \quad w_2 = \overline{w}_2 + p\omega_3, \quad w_3 = \overline{w}_3 + p\omega_2, \\ w_4 = \overline{w}_4 + q\omega_2 + r\omega_3 + p\omega_4, \quad w_5 = \overline{w}_5 + r\omega_2 + s\omega_3, \end{cases}$$

where the p, q, r, s are arbitrary functions of the variables, which we take as new auxiliary variables.

To find the expressions for the exterior derivatives of w_i we apply the theorem of Poincaré to the equations (15). Putting

$$(17)\quad \begin{cases} \pi_1 = (w_1)' - [w_2\omega_4] - 2[w_5\omega_1] - [w_4\omega_3], \\ \pi_2 = (w_2)' - [w_2 w_3] - [w_4\omega_1], \\ \pi_3 = (w_3)' + [w_2\omega_4] + [w_5\omega_1], \end{cases}$$

we get, from the identities

$$(\omega_1')' = 0, \quad (\omega_2')' = 0, \quad (\omega_3')' = 0,$$

the relations

$$(18)\quad \begin{cases} [\pi_1\omega_1] + [\pi_2\omega_3] = 0, \\ [\pi_1\omega_2] + 2[\pi_3\omega_2] = 0, \\ [\pi_2\omega_2] + [\pi_1\omega_3] + [\pi_3\omega_3] = 0. \end{cases}$$

It follows that

$$[\pi_1 \omega_1 \omega_2] = 0, \quad [\pi_1 \omega_2 \omega_3] = 0, \quad [\pi_1 \omega_1 \omega_3] = 0,$$

so that π_1 is of the form

$$\pi_1 = a\,[\omega_2 \omega_3] + b\,[\omega_3 \omega_1] + c\,[\omega_1 \omega_2].$$

It is possible to modify the Pfaffian forms w_4 and w_5 to have

$$a = b = c = 0.$$

The most general transformation on the w_i which preserves the equations (15) and the equation

$$\omega_1' = [w_2 \omega_4] + 2\,[w_5 \omega_1] + [w_4 \omega_3] \tag{19}$$

is then

$$\begin{cases} w_1 = \overline{w}_1, \quad w_2 = \overline{w}_2 + p\omega_3, \\ w_3 = \overline{w}_3, + p\omega_2, \quad w_4 = \overline{w}_4 + p\omega_4, \quad w_5 = \overline{w}_5. \end{cases} \tag{20}$$

We take p as a new auxiliary variable.

From (18) we have then

$$[\pi_2 \omega_3] = 0,$$

which gives

$$\pi_2 \equiv 0 \quad (\text{mod. } \omega_3).$$

Since w_2' contains the term $[dp\, \omega_3]$, we can choose a Pfaffian form w_6, linearly independent from the ω_i and from $w_1, \ldots, w_5$, such that we have

$$w_2' = [w_2 w_3] + [w_4 \omega_1] + [w_6 \omega_3], \tag{21}$$

where the Pfaffian form w_6 is determined up to an additive term of ω_3. The relations (18) then give

$$[\pi_3 \omega_2] = 0, \quad [w_6 \omega_3 \omega_2] + [\pi_3 \omega_3] = 0,$$

which can be written

$$[(\pi_3 - w_6 \omega_2)\, \omega_2] = 0,$$
$$[(\pi_3 - w_6 \omega_2)\, \omega_3] = 0.$$

This shows that π_3 is of the form

$$\pi_3 = [w_6\,\omega_2] + a\,[\omega_2\,\omega_3].$$

We can choose w_6 to make $a=0$. Then we have

$$w_3' = -[w_2\,\omega_4] - [w_5\,\omega_1] + [w_6\,\omega_2]. \tag{22}$$

The equations (21) and (22) completely determine the Pfaffian form w_6.

It remains to find the expressions for the exterior derivatives of w_4, w_5, w_6. For this purpose we apply the theorem of Poincaré to the last equation of (15) and to the equations (19), (21), (22). Put

$$\begin{cases} \pi_4 = w_4' + [w_1\,w_4] + [w_2\,w_5] + [w_3\,w_4] - [w_6\,\omega_4], \\ \pi_5 = w_5' + [w_1\,w_5] - [w_4\,\omega_4], \\ \pi_6 = w_6' + [w_1\,w_6] + [w_2\,w_4] + 2\,[w_3\,w_6]. \end{cases} \tag{23}$$

We get

$$\begin{cases} 2\,[\pi_5\,\omega_1] + [\pi_4\,\omega_3] = 0, \\ [\pi_4\,\omega_2] + [\pi_5\,\omega_3] = 0, \\ [\pi_4\,\omega_1] + [\pi_6\,\omega_3] = 0, \\ [\pi_5\,\omega_1] - [\pi_6\,\omega_2] = 0. \end{cases} \tag{24}$$

It follows that

$$[\pi_5\,\omega_1\,\omega_2] = [\pi_5\,\omega_2\,\omega_3] = [\pi_5\,\omega_3\,\omega_1] = 0,$$

so that π_5 is of the form

$$\pi_5 = a\,[\omega_2\,\omega_3] + b\,[\omega_3\,\omega_1] + c\,[\omega_1\,\omega_2].$$

The relations (24) then show that π_4 and π_6 are of the form

$$\pi_4 = e\,[\omega_2\,\omega_3] - c\,[\omega_3\,\omega_1] - 2a\,[\omega_1\,\omega_2],$$
$$\pi_6 = f\,[\omega_2\,\omega_3] + a\,[\omega_3\,\omega_1] - e\,[\omega_1\,\omega_2].$$

Summarizing the above results, we get the

Theorem. *If the relative invariant I of the differential equa-*

tion (2) is zero, the problem of equivalence of the differential equation with respect to the group of contact transformations in the plane depends on the fundamental equations

$$
(25)\quad \left\{
\begin{aligned}
\omega_1' &= [w_1\,\omega_1] + [w_2\,\omega_3],\\
\omega' &= [w_1\,\omega_2] + 2\,[w_3\,\omega_2] - [\omega_3\,\omega_4],\\
\omega_3' &= [w_2\,\omega_2] + [w_1\,\omega_3] + [w_3\,\omega_3] - [\omega_1\,\omega_4],\\
\omega_4' &= [w_4\,\omega_2] + [w_5\,\omega_3] + [w_3\,\omega_4],\\
w_1' &= [w_2\,\omega_4] + 2\,[w_5\,\omega_1] + [w_4\,\omega_3],\\
w_2' &= [w_2\,w_3] + [w_4\,\omega_1] + [w_6\,\omega_3],\\
w_3' &= -\,[w_2\,\omega_4] - [w_5\,\omega_1] + [w_6\,\omega_2],\\
w_4' &= -\,[w_1\,w_4] - [w_2\,w_5] - [w_3\,w_4] + [w_6\,\omega_4] + e\,[\omega_2\,\omega_3]\\
&\qquad - c\,[\omega_3\,\omega_1] - 2a\,[\omega_1\,\omega_2],\\
w_5' &= -\,[w_1\,w_5] + [w_4\,\omega_4] + a\,[\omega_2\,\omega_3] + b\,[\omega_3\,\omega_1] + c[\omega_1\,\omega_2],\\
w_6' &= -\,[w_1\,w_6] - [w_2\,w_4] - 2\,[w_3\,w_6] + f\,[\omega_2\,\omega_3]\\
&\qquad + a\,[\omega_3\,\omega_1] - e\,[\omega_1\,\omega_2].
\end{aligned}
\right.
$$

The complete system of invariants consists of the five fundamental invariants a,b,c,e,f and of their covariant derivatives. If all the invariants a,b,c,e,f are zero, the equation is reducible to $y'''=0$ by a contact transformation.

The above formal discussion is capable of geometrical interpretation. By introducing six auxiliary variables, we have attached ten Pfaffian forms $\omega_1, \ldots\ldots \omega_4$, w_1, ..., w_6 to the differential equation (2) in a way invariant under the group of contact transformations in the plane. When all the invariants a,b,c,e,f are zero, as in the case of the differential equation

$$y''' = 0,$$

the equations (25) are the equations of structure of a group in ten parameters. As proved by Lie, this group is similar to the group G_{10} of transformations of contact in the plane which carry circles into circles. Hence in the general case we may interpret the equations (25) as the equations of structure of a generalized space with G_{10} as its fundamental group and with the equations

$$\omega_1 = \omega_2 = \omega_3 = 0$$

as the equations of the elements of the space. The elements are thus the integral curves of the differential equation (2).

By interpreting the integral curves of (2) as the points in a space of three dimensions, the equations (25) with

$$a = b = c = e = f = 0$$

are the equations of structure of the group of conformal transformations in space. The equation of the cone of isotropic directions at a point is

$$\omega_3^2 - 2\,\omega_1\,\omega_2 = 0. \tag{26}$$

In fact, by introducing the operations d and δ with

$$\omega_1(\delta) = \omega_2(\delta) = \omega_3(\delta) = 0,$$

and writing ω_i, u_i, e_4, g_i for $\omega_i\ (d)$, $w_i\ (d)$, $\omega_4(\delta)$, $w_i(\delta)$ respectively, we get, from (25),

$$\begin{cases} \delta\omega_1 = g_1\,\omega_1 + g_2\,\omega_3, \\ \delta\omega_2 = g_1\,\omega_2 + 2g_3\,\omega_2 + g_4\,\omega_3, \\ \delta\omega_3 = g_2\,\omega_2 + g_1\,\omega_3 + g_3\,\omega_3 + g_4\,\omega_1. \end{cases} \tag{27}$$

These equations, when applied to (26), immediately show that the cone (26) remains invariant under a change of reference about the point.

From the above considerations it follows that the equations (25) are the equations of structure of a space with conformal connection. Further we observe that the Pfaffian forms which enter in the expressions for ω_1', ω_2', ω_3' are ω_4, w_1, w_2, w_3 and that their exterior derivatives have the same expressions as in the case of an ordinary conformal space. This shows that our conformal space. This shows that our conformal connection is what Cartan called normal[5]. We have thus the

5) Cartan, Connexion conforme, loc. cit., pp. 195-197.

Theorem. *If the relative invariant I of the given differential equation of the third order is zero, we can define in the space of the integral curves of the differential equation, a normal conformal connection having the equations of structure (25). This definition is invariant under the group of contact transformations in the plane.*

The condition $I=0$ has a simple geometrical meaning, as given by the following theorem due to K. Wünschmann[6] : *The condition $I=0$ is a necessary and sufficient condition that the condition of contact of two neighboring integral curves of the differential equation (2) be given by an equation of Monge of the second order.*

In fact, when $I=0$, the equation (26) is exactly the condition of contact of neighboring integral curves.

§3. The Case $I \neq 0$.

In the case $I \neq 0$ we can choose u as a function of x, y, y', y'' such that the coefficient of $[\omega_2 \omega_4]$ in ω_1' is equal to 1. Then w_{44} is a linear combination of ω_i and the coefficient of $[\omega_1 \omega_4]$ in ω_4' is an invariant. It is easily found that this invariant is

$$\frac{1}{\alpha u}\frac{\partial u}{\partial y''}-\frac{w}{\alpha}.$$

We can reduce it to zero by choosing

$$w=\frac{1}{u}\frac{\partial u}{\partial y''}. \tag{28}$$

In the auxiliary variables there remain only three arbitrary ones, which are α, γ, v. Of the ω's and the w's there remain seven independent ones ω_1, ω_2, ω_3, ω_4, w_{11}, w_{13}, w_{42} while the others are linear combinations of them. On taking account of the relations (11) and (12), we can write the equations (8) in the form

6) K. Wünschmann, Über Beruhrungsbedingungen bei Differentialgleichungen, Dissertation Greifswald 1905, or Enzyklopädie der Math. Wiss., Bd. III, Teil 3D7, pp. 490-492.

$$(29)\begin{cases} \omega_1' = [w_{11}\,\omega_1] + [w_{13}\,\omega_3] + [\omega_2\,\omega_4] + a\,[\omega_1\,\omega_2] + b\,[\omega_2\,\omega_3], \\ \omega_2' = [w_{11}\,\omega_2] - [\omega_3\,\omega_4] - 2a_1\,[\omega_2\,\omega_3] - 2a_2\,[\omega_2\,\omega_4], \\ \omega_3' = [w_{13}\,\omega_2] + [w_{11}\,\omega_3] - [\omega_1\,\omega_4] + a_3\,[\omega_2\,\omega_3] + a_2\,[\omega_3\,\omega_4], \\ \omega_4' = [w_{42}\,\omega_2] + b_1\,[\omega_1\,\omega_3] + b_2\,[\omega_2\,\omega_3] + a_3\,[\omega_2\,\omega_4] \\ \qquad + b_3\,[\omega_3\,\omega_4], \end{cases}$$

where the a's and the b's are functions of x, y, y', y'', α, γ, v.

Let us calculate a_2 and b_3. We find easily

$$a_2 = \frac{1}{u}\left(-\gamma - \frac{2}{3}\frac{\partial F}{\partial y''} + \frac{1}{u}\frac{du}{dx}\right),$$

$$b_3 = \frac{1}{\alpha u}\left(-v + \frac{1}{u}\frac{\partial u}{\partial y'} - \frac{w}{u}\frac{du}{dx} - \frac{dw}{dx}\right).$$

These coefficients can be reduced to zero if we choose

$$(30)\quad\begin{cases} \gamma = -\dfrac{2}{3}\dfrac{\partial F}{\partial y''} + \dfrac{1}{u}\dfrac{du}{dx}, \\ v = \dfrac{1}{u}\dfrac{\partial u}{\partial y'} - \dfrac{1}{u^2}\dfrac{du}{dx} - \dfrac{d}{dx}\left(\dfrac{1}{u}\dfrac{\partial u}{\partial y''}\right) \end{cases}$$

Then both w_{13} and w_{42} become linear combinations of ω_i. On writing w for w_{11} we can write the equations (29) in the form

$$(31)\begin{cases} \omega_1' = [w\,\omega_1] + [\omega_2\,\omega_4] + a\,[\omega_1\,\omega_2] + b\,[\omega_1\,\omega_3] + c\,[\omega_2\,\omega_3] \\ \qquad + e\,[\omega_3\,\omega_4] \\ \omega_2' = [w\,\omega_2] - [\omega_3\,\omega_4] + f\,[\omega_2\,\omega_3], \\ \omega_3' = [w\,\omega_2] - [\omega_1\,\omega_4] + b\,[\omega_1\,\omega_2] + g\,[\omega_2\,\omega_3] + e\,[\omega_2\,\omega_4], \\ \omega_4' = h\,[\omega_1\,\omega_2] + i\,[\omega_1\,\omega_3] + j\,[\omega_2\,\omega_3] + k\,[\omega_2\,\omega_4]. \end{cases}$$

In these equations we regard α as an auxiliary variable and the Pfaffian form w is linearly independent of ω_i. It is possible to choose w to make the coefficients a and b of (31) zero. The resulting equations are then of the form

$$
(32)\quad \begin{cases}
\omega_1' = [w\,\omega_1] + [\omega_2\,\omega_4] + a\,[\omega_2\,\omega_3] + b\,[\omega_3\,\omega_4], \\
\omega_2' = [w\,\omega_2] - [\omega_3\,\omega_4] + c\,[\omega_2\,\omega_3], \\
\omega_3' = [w\,\omega_3] - [\omega_1\,\omega_4] + e\,[\omega_1\,\omega_2] + f\,[\omega_2\,\omega_3] + b\,[\omega_2\,\omega_4], \\
\omega_4' = g\,[\omega_1\,\omega_2] + h\,[\omega_1\,\omega_3] + i\,[\omega_2\,\omega_3] + j\,[\omega_2\,\omega_4],
\end{cases}
$$

where the $a, b, \ldots, j$ do not denote the same functions as those in (31). By the form of the equations (32) the Pfaffian form w is completely determined.

On applying the theorem of Poincaré to the equations (32) we can show that w' is of the form

$$
(33)\quad \begin{aligned} w' = k\,[\omega_1\,\omega_2] + l\,[\omega_1\,\omega_3] + m\,[\omega_1\,\omega_4] + n\,[\omega_2\,\omega_3] \\ + p\,[\omega_2\,\omega_4] + q\,[\omega_3\,\omega_4], \end{aligned}
$$

where we have the relations

$$
(34)\quad m = e = -\tfrac{1}{2}c, \quad p = a + \tfrac{1}{2}bc.
$$

We have thus the

Theorem. *If the relative invariant $I \neq 0$, the complete system of invariants of the differential equation (2) consists of the functions $a, b, \cdots\cdots, q$ in (32), (33) and of their covariant derivatives. When all the invariants $a, b, \cdots\cdots, q$ are zero, the equation is reducible to*

$$
(35)\quad y''' + y = 0
$$

by a contact transformation.

In fact, when all the invariants $a, b, \ldots, q$ are zero, the equations (32), (33) are the equations of structure of a five-parameter group G_5 in the plane. G_5 is evidently the group which leaves invariant the integral curves of (35). Its finite equations are

$$
(36)\quad \begin{cases}
x^* = x + a_1, \\
y^* = a_2\,y + a_3\,e^{-x} + a_4\,e^{\frac{1}{2}x}\cos\left(\frac{\sqrt{3}}{2}x\right) + a_5\,e^{\frac{1}{2}x}\sin\left(\frac{\sqrt{3}}{2}x\right).
\end{cases}
$$

Hence in the general case we have defined in the plane a generlized geometry with the elements of contact of the second order x, y, y', y'' as the elements of the space and with G_5 as the fundamental group.

This result can also be interpreted in the space R_3 whose points are the integral curves of (2). In the case of the differential equation (35) we put

$$(37)\qquad \begin{cases} u_1 = e^{x}(y - y' + y''), \\ u_2 = e^{\varepsilon x}(\varepsilon^2 y - \varepsilon y' + y''), \\ u_3 = e^{\varepsilon^2 x}(\varepsilon y - \varepsilon^2 y' + y''), \end{cases}$$

where ε denotes an imaginary cube root of unity. The condition of contact of neighboring integral curves is given by the equation

$$(37)\qquad du_1^{\varepsilon^2} du_2\, du_3^{\varepsilon} = 1.$$

The variables u_1, u_2, u_3 can be considered as a system of coordinates in R_3 so that at every point of R_3 we have attached a cone (38). The geometry considered is a geometry in R_3 in which there is given a system of cones (38). The finite equations of its fundamental group are

$$(39)\qquad \begin{aligned} x^* &= a_1 x, \\ u_1^* &= a_1 a_2 u_1 + a_3, \\ u_2^* &= a_1^{\varepsilon} a_2 u_2 + a_4, \\ u_3^* &= a_1^{\varepsilon^2} a_2 u_3 + a_5. \end{aligned}$$

In the general case, we have at each point of R_3 a cone whose equation expresses the condition of contact of neighboring integral curves of (2). The geometry is a generalized geometry with the fundamental group (39). The elements of the space are formed by a point and a generator of the cone attached to that point.

ANNALS OF MATHEMATICS
Vol. 43, No. 1, January, 1942

ON INTEGRAL GEOMETRY IN KLEIN SPACES

BY SHIING-SHEN CHERN

(Received September 3, 1940)

The classical results in integral geometry found by Crofton, Poincaré, Cartan, and recently developed by Blaschke[1] and his school, are mostly restricted to Euclidean spaces. It is the object of this paper to give the fundamental concepts of integral geometry in a general space of Klein, by which we mean a number space of n dimensions with a transitive r-parameter group G_r of transformations. The discussion is mainly based on Cartan's theory of Lie's groups.[2]

The paper is divided into three sections. In §1 we give a brief summary of Cartan's theory of Lie's groups with some results concerning the measures of geometrical elements. In §2 we define the incidence of the geometrical elements of different fields, which plays a fundamental rôle in all subsequent discussions. As applications of the general notions we give in the last section two formulas which are respectively generalizations of the well-known formulas of Crofton and Cauchy.

1. Some Fundamental Notions in Klein Spaces

Let G_r be an abstract r-parameter group of Lie[3] with the parameters $a^1, \cdots, a^r$, so that G_r denotes an r-dimensional space and $a^1, \cdots, a^r$ are the coordinates in the space. If S_a denotes a point in G_r with the coordinates $a^1, \cdots, a^r$, there are defined in G_r two simply-transitive groups of transformations

$$(1)\qquad \begin{aligned} S_a &\rightarrow S_c S_a ,\\ S_a &\rightarrow S_a S_c , \end{aligned}$$

called respectively the first and the second groups of parameters. There exists one, and only one, set of r linearly independent Pfaffian forms

$$\omega^1(a, da), \cdots, \omega^r(a, da)$$

invariant under the first group of parameters. They are determined up to a linear transformation with constant coefficients and have the property that any

[1] Cf. W. Blaschke, *Vorlesungen über Integralgeometrie*, Bd. I Leipzig 1936; Bd. II Leipzig 1937. These books will be cited respectively as I. G. I and I. G. II.

[2] Cf. E. Cartan, *La théorie des groupes finis et continus et la géométrie différentielle traitée par la méthode du repère mobile*, Paris 1937. This book will be cited as Cartan, Théorie des groupes.

[3] For the definition of an abstract group of Lie we may follow that given by Cartan in his book "*La théorie des groupes finis et continus et l'analysis situs*," Mémorial des Sciences Mathématiques, Fasc. XLII, Paris 1930.

exterior differential form[4] of degree p invariant under the first group of parameters is of the form

$$\sum_{i_1,\cdots,i_p} A_{i_1\cdots i_p}[\omega^{i_1}\cdots\omega^{i_p}], \tag{2}$$

where $A_{i_1\cdots i_p}$ are constants. According to Cartan, we call $\omega^1, \cdots, \omega^r$ the relative components of G_r. They satisfy the equations of structure of Maurer-Cartan

$$(\omega^i)' = \sum_{j,k=1}^{r} c^i_{jk}[\omega^j\omega^k], \qquad c^i_{jk} + c^i_{kj} = 0, \qquad i = 1, \cdots, r, \tag{3}$$

where c^i_{jk} are the constants of structure of G_r.

Now let g be a subgroup of G_r. This subgroup g and its left-hand cosets Sg fill up the whole space G_r and have the property that no two of them can coincide without being identical. Thus the varieties Sg are the integral varieties of a completely integrable Pfaffian system. Owing to the invariance of the totality of the cosets Sg with respect to the first group of parameters, the left-hand members of the Pfaffian system are linear combinations of $\omega^1, \cdots, \omega^r$ with constant coefficients. We may suppose the system to be

$$\omega^1 = 0, \cdots, \omega^n = 0. \tag{4}$$

The theorem of Frobenius then gives

$$c^i_{jk} = 0, \qquad i = 1, \cdots, n; \qquad j, k = n + 1, \cdots, r, \tag{5}$$

which are the necessary and sufficient conditions for the system (4) to be completely integrable. We say that the subgroup g defines a field of geometrical elements such that each element of the field is represented by a left-hand coset of g or by an integral variety of (4).

We may interpret the geometrical elements defined by (4) as the points of a space E. The first group of parameters becomes then a group of transformations in E, so that, with some necessary assumptions on the analyticity of the equations of the transformations, E is a space in the sense of Klein. It is clear that any space of Klein can be obtained this way. In fact, we need only take G_r to be the group of transformations in the space and g the subgroup of G_r leaving invariant a fixed point.

Let $x^1(a^1, \cdots, a^r), \cdots, x^n(a^1, \cdots, a^r)$ be n independent first integrals of (4), so that $dx^1, \cdots, dx^n$ are n linearly independent combinations of $\omega^1, \cdots, \omega^n$. Then we have

$$[dx^1 \cdots dx^n] = \Delta(a^1, \cdots, a^r)[\omega^1 \cdots \omega^n], \qquad \Delta \neq 0. \tag{6}$$

By the measure of a domain D of elements in E we shall mean an n-tuple integral of the form

$$J = \int_D f(x^1, \cdots, x^n)[dx^1 \cdots dx^n] \tag{7}$$

[4] For the notions of exterior differential forms and exterior derivation cf. E. Cartan, *Leçons sur les invariants intégraux*, Chap. VI, VII, Paris 1922.

extended over D such that its value is invariant under the group of transformations in E, i.e., under the first group of parameters. Since this property holds for all domains D, we see from (6) that it is expressed by

$$f\Delta = \text{constant}.$$

Hence *a necessary and sufficient condition for the elements of E to possess a measure is that the function $\Delta(a)$ in* (6) *be a function of $x^1, \cdots, x^n$ only.* The measure is then defined up to a constant factor and is given by

$$J = \int_D [\omega^1 \cdots \omega^n]. \tag{8}$$

The condition for the existence of a measure can also be expressed in terms of the constants of structure. Let d and δ be two operations such that d denotes a displacement in the set of elements in E and δ a displacement on an integral variety of (4). The condition for the existence of measure can be written as

$$\delta[\omega^1 \cdots \omega^n] = 0.$$

But we have

$$\omega^1(\delta) = \cdots = \omega^n(\delta) = 0.$$

The equations of structure (3) then give

$$\delta\omega^i(d) = 2 \sum_{k=1}^{n} \sum_{j=n+1}^{r} c^i_{jk}\omega^j(\delta)\omega^k(d), \qquad i = 1, \cdots, n.$$

From the last equations we get

$$\delta[\omega^1 \cdots \omega^n] = 2 \sum_{j=n+1}^{r} \sum_{i=1}^{n} c^i_{ji}\omega^j(\delta)[\omega^1 \cdots \omega^n].$$

Therefore *a necessary and sufficient condition for the elements defined by* (4) *to possess a measure is*

$$\sum_{i=1}^{n} c^i_{ji} = 0, \qquad j = n+1, \cdots, r. \tag{9}$$

2. Definition of the Incidence of Elements of Different Fields

Consider two fields of geometrical elements, defined respectively by (4) and

$$w^1 = 0, \cdots, w^m = 0, \tag{10}$$

where w^i $(i = 1, \cdots, m)$ are linear combinations of the relative components with constant coefficients. Each of the systems (4) and (10) being completely integrable, the same must be true of the combined system

$$\omega^1 = 0, \cdots, \omega^n = 0, \qquad w^1 = 0, \cdots, w^m = 0. \tag{11}$$

Of the last system suppose s of them ($s > m$, $s > n$) be linearly independent, so that there exist $m + n - s$ relations of the form

$$\sum_{i=1}^{n} b_{\lambda i}\omega^i + \sum_{j=1}^{m} c_{\lambda j}w^j = 0, \qquad \lambda = 1, \cdots, m + n - s, \tag{12}$$

where $b_{\lambda i}$, $c_{\lambda j}$ are constants. The integral varieties of (11) are of dimension $r - s$. Denoting the integral varieties of (4), (10), (11) respectively by V_{r-n}, V_{r-m}, V_{r-s}, we see that *a V_{r-s} lies completely on a V_{r-n} (or V_{r-m}) if and only if it has a point in common with V_{r-n} (or V_{r-m}). It follows that a V_{r-n} and a V_{r-m} have a V_{r-s} in common if and only if they have a point in common.*

By virtue of the above properties we define *an element N of* (4) *and an element M of* (10) *to be incident when their corresponding integral varieties V_{r-n}, V_{r-m} have a V_{r-s} in common.*

Consider the elements M incident with a given element N. In the group space G_r each of the integral varieties V_{r-m} corresponding to M cuts V_{r-n} in a V_{r-s}. The totality of V_{r-s} on a fixed V_{r-n} depends on $s - n$ parameters. Therefore *the elements of the field* (10) *incident with a given element N of* (4) *depend on $s - n$ parameters.* The same property remains naturally true when the two fields are interchanged.

It is important to notice that this definition of incidence includes the ordinary notions of incidence in Euclidean geometry, affine geometry, projective geometry, etc. as particular cases. Take, for instance, the group of motions in the Euclidean plane

$$\begin{cases} x^* = x \cos c - y \sin c + a, \\ y^* = x \sin c + y \cos c + b, \end{cases} \tag{13}$$

where a, b, c are parameters. The relative components are

$$\begin{cases} \omega^1 = \cos c\,da + \sin c\,db, \\ \omega^2 = -\sin c\,da + \cos c\,db, \\ \omega^3 = dc. \end{cases} \tag{14}$$

As the equations of points and straight lines we may take respectively

$$\omega^1 = 0, \quad \omega^2 = 0 \tag{15}$$

and

$$\omega^1 = 0, \quad \omega^3 = 0. \tag{16}$$

In the three-dimensional group space (a, b, c) the elements of the field (15) are represented by lines parallel to the c-axis and those of (16), by the lines

$$\cos c \cdot a + \sin c \cdot b = \text{const.}, \qquad c = \text{const.}$$

By interpreting a, b as the Cartesian coordinates in the plane and the equation

$$\cos c \cdot a + \sin c \cdot b = \text{const.}$$

as the equation of line in Hesse's normal form, we see immediately that our definition of incidence coincides with the notion of incidence in the ordinary sense.

Before concluding this section, we want to remark that, instead of using the integral varieties in the group space to characterize the elements of a field, we may employ the geometrically more intuitive idea of families of frames (in French "repère").[5] But it is sufficient for our purpose to restrict ourselves to one point of view.

3. The Generalized Crofton's Formula and Cauchy's Formula

Let us first consider the geometry in the Euclidean plane. It is well known that the measures for points and lines exist and that they are given respectively by[6]

$$\int [\omega^1 \omega^2] \quad \text{and} \quad \int [\omega^1 \omega^3].$$

Crofton's formula in its simplest form asserts that the measure of the lines incident with the points of a curve is equal to a constant multiple of the length of the curve, each line being counted as many times as the number of its points of intersection with the curve.

To generalize this formula, we must first have the generalized notion of the length of a curve for a p-dimensional variety of points under an arbitrary group of transformations of Lie. For this purpose, take a p-dimensional variety V_p formed by the elements of the field (4). Let $u^1, \cdots, u^p$ be the parameters on V_p. Then $\omega^1, \cdots, \omega^n$ are linear combinations of $du^1, \cdots, du^p$ on V_p. On eliminating $du^1, \cdots, du^p$, we may express $\omega^{p+1}, \cdots, \omega^n$ as linear combinations of $\omega^1, \cdots, \omega^p$ in the form

$$\omega^\alpha = \sum_{k=1}^{p} \xi_k^\alpha \omega^k, \qquad \alpha = p+1, \cdots, n,$$

where ξ_k^α are functions of the parameters $a^1, \cdots, a^r$ of the group and of $u^1, \cdots, u^p$. By the method of moving frames of Cartan, it is in general possible (by supposing the variety V_p to be sufficiently general) to determine some or all of the ξ_k^α to be constants or functions of the parameters $u^1, \cdots, u^p$ such that the determination is invariant under transformations of the group G_r. If, up to a certain step in the determination of ξ_k^α, the exterior differential forms of degree p

$$[\omega^{i_1} \cdots \omega^{i_p}],$$

where $i_1, \cdots, i_p$ run over all combinations of $1, \cdots, n$, depend on $u^1, \cdots, u^p$ only and differ from each other only by constant factors, we say that the variety V_p possesses a *p-dimensional area*, which is equal to the integral of any one of the

[5] Cf. Cartan, *Théorie des groupes*, Chap. V, or H. Weyl, *The Classical Groups*, Princeton 1939, pp. 16–17.

[6] Blaschke, I. G. I, pp. 5–7.

differential forms and is thus determined up to a constant factor. It is of course possible that the area of V_p does not exist, as in the case that V_p is a quadric in projective space. But in the cases which usually occur to us, the p-dimensional area of V_p under G_r exists.

In the case $p = 1$ the p-dimensional area so defined leads to the Pick's invariant of a curve,[7] which includes the affine arc, projective arc, etc. as particular cases. Pick has proved that *if a group of transformations of order r in the xy-plane transforms transitively the elements of contact of order r − 2*

$$x, y, y' = \frac{dy}{dx}, \cdots, y^{(r-2)} = \frac{d^{r-2}y}{dx^{r-2}},$$

then a plane curve possesses an intrinsic parameter invariant under the group considered. This intrinsic parameter is called the invariant of Pick. Its existence is an easy consequence of our preceding discussions. In fact, it is only necessary to take $x, y, y', \cdots, y^{(r-2)}$ to be parameters of the group and

$$\omega^1 = A^1(x, y, \cdots, y^{(r-2)})\, dx + B^1(x, y, \cdots, y^{(r-2)})\, dy = 0,$$
$$\omega^2 = A^2(x, y, \cdots, y^{(r-2)})\, dx + B^2(x, y, \cdots, y^{(r-2)})\, dy = 0$$

to be the equations of points. Along a curve we have $dy = y'dx$, so that ω^2 is a constant multiple of ω^1, and the integral

$$\int \omega^1$$

gives the intrinsic parameter of the curve.

Another important particular case of the notion of p-dimensional area is the case that $p = s - m$ and that V_p consists of all the elements of the field (4) incident with a fixed element of the field (10). In this case, the relations between $\omega^1, \cdots, \omega^n$ are obtained from (12) by setting $w^1 = \cdots = w^m = 0$ and are therefore

$$\sum_{i=1}^{n} b_{\lambda i}\omega^i = 0, \qquad \lambda = 1, \cdots, m + n - s.$$

Since $b_{\lambda i}$ are constants, we may take as the p-dimensional area of this V_p the integral

$$\int [\omega^1 \cdots \omega^p].$$

If the value of this integral over V_p is finite, we call it *the measure of the elements of* (4) *about a fixed element of* (10).

With these preparations we can give the generalized Crofton's formula in the following theorem:

[7] Cf., for example, G. Kowalewski, *Allgemeine Natürliche Geometrie und Liesche Transformationsgruppen*, Berlin 1931, pp. 106–110.

Let two fields of elements M, N be defined respectively by the equations (10), (4). *Let $p = m + n - s$ and let V_p be a variety of p dimensions formed by the elements of M. If the measure of the elements of N incident with the elements of V_p and the p-dimensional area F of V_p both exist and are finite, then*

$$\int [\omega^1 \cdots \omega^n] = cF, \qquad c = \text{constant}, \tag{17}$$

where the integral is extended over the elements of N incident with the elements of V_p, each element being counted as many times as the number of elements of V_p with which it is incident.

To prove this theorem, consider the $p = m + n - s$ relations (12). Since the relations are independent with respect to $\omega^1, \cdots, \omega^n$, we may solve them in terms of $\omega^1, \cdots, \omega^p$, obtaining

$$\omega^\lambda = \sum_{j=p+1}^{n} e^\lambda_j \omega^j + \sum_{k=1}^{m} f^\lambda_k w^k, \qquad \lambda = 1, \cdots, p,$$

where e^λ_j, f^λ_k are constants. Setting

$$\bar{\omega}^\lambda = \omega^\lambda - \sum_{j=p+1}^{n} e^\lambda_j \omega^j, \qquad \lambda = 1, \cdots, m + n - s,$$

and denoting $\bar{\omega}^\lambda$ again by ω^λ, we have

$$\omega^\lambda = \sum_{k=1}^{m} f^\lambda_k w^k, \qquad \lambda = 1, \cdots, m + n - s. \tag{18}$$

The proof of our theorem then depends on a new form of the expression $[\omega^1 \cdots \omega^n]$. We apply first the method of moving frames of Cartan to V_p. Since by hypothesis the p-dimensional area of V_p exists we can attach to each element of V_p a frame such that within the p-parameter family of frames so attached the exterior differential forms

$$[w^{i_1} \cdots w^{i_p}]$$

where $i_1, \cdots, i_p$ run over all combinations of $1, \cdots, m$, differ from each other only by constant factors. In the group space $G_r V_p$ is represented by a p-parameter family of left-handed cosets with respect to a subgroup h. To the frames attached to the elements of V_p there corresponds in G_r a p-dimensional variety W_p such that on every coset corresponding to an element of V_p there lies one and only one point of W_p. Now every integral variety Sh of (10) is cut by the integral varieties of (4) incident to it in the $(r - s)$-dimensional integral varieties V_{r-s} of (11). The totality of V_{r-s} on Sh depends on $s - m$ parameters. To define a system of coordinates for the V_{r-s} on Sh we may first set up a system of coordinates $\lambda^1, \cdots, \lambda^{s-m}$ for the V_{r-s} on h. By the transformation $S_a \to SS_a$ (S being fixed) the variety h is carried to Sh and we define the coordinates of a V_{r-s} on h to be the coordinates of its image on Sh.

Let $u^1, \cdots, u^p$ be the parameters on V_p. As the coordinates of an element of the field N incident with an element of V_p we may take $u^1, \cdots, u^p, \lambda^1, \cdots, \lambda^{s-m}$. Now the expression $[\omega^1 \cdots \omega^n]$ depends only on the elements of the field N under consideration. This property, called by Blaschke the property of invariance of choice ("Wahlinvarianz"), may be interpreted as follows: Choose on each integral variety of (4) a point and compute the relative components $\omega^1, \cdots, \omega^n$ on the n-dimensional variety so obtained. The resulting expression is independent of the choice of these points.

This being clear, we may chose the points on the integral varieties of (4) to be on the V_{r-s} of the integral varieties corresponding to the elements of V_p. Its totality forms a variety of $p + s - m = n$ dimensions, which we call V_n. To find the relative components on V_n notice that the relative components $\omega^i(a, da)$ $(i = 1, \cdots, r)$ are the parameters of the infinitesimal transformation $S_a^{-1}S_{a+da}$ and that by choosing the infinitesimal transformations X_if $(i = 1, \cdots, r)$ such that $X_{m+1}f, \cdots, X_rf$ generate the subgroup h leaving invariant a fixed element 0 of the field M, the infinitesimal transformation $S_a^{-1}S_{a+da}$ carrying 0 to a neighboring element P is of the form

$$w^1X_1f + \cdots + w^mX_mf + \omega^{m+1}X_{m+1}f + \cdots + \omega^rX_rf,$$

where $w^1, \cdots, w^m$ are exactly the left-hand members of the system (10).

Let $\theta^1, \cdots, \theta^m$ denote the relative components $w^1, \cdots, w^m$ on W_p, $\bar{\omega}^{p+1}, \cdots, \bar{\omega}^n$ the relative components $\omega^{p+1}, \cdots, \omega^n$ on Sh, and let $w^1, \cdots, w^m, \omega^{p+1}, \cdots, \omega^n$ denote these components on V_n. As the independent components on V_n we may then take p independent forms among the $\theta^1, \cdots, \theta^m$, and $\bar{\omega}^{p+1}, \cdots, \bar{\omega}^n$, the latter being Pfaffian forms in $\lambda^1, \cdots, \lambda^{s-m}$ only. Now every point of V_n is of the form

$$S_a = S_bR,$$

where S_b and R belong respectively to W_p and h. By writing

$$S_{a+da} = S_{b+db}R^1, \qquad R^1 \in h,$$

we get

$$S_a^{-1}S_{a+da} = R^{-1}S_b^{-1}S_{b+db}R \cdot R^{-1}R^1$$

Since $R^{-1}R^1$ belongs to h, it produces no effect on 0. The relative components $w^1, \cdots, w^m$ are thus transformed according to the transformations of the adjoint group which correspond to transformations of h. As R transforms between themselves the infinitesimal transformations of h, we have

$$(19) \qquad \begin{cases} w^1 = a_1^1\theta^1 + \cdots + a_m^1\theta^m, \\ \cdots\cdots\cdots\cdots\cdots\cdots\cdots\cdots \\ w^m = a_1^m\theta^1 + \cdots + a_m^m\theta^m, \end{cases}$$

where $a_k^i(i, k = 1, \cdots, m)$ are functions of $\lambda^1, \cdots, \lambda^{s-m}$.

On the other hand, for the components $\omega^{p+1}, \cdots, \omega^n$ on V_n we must have

$$
(20)\quad \begin{cases} \omega^{m+n-s+1} = \bar{\omega}^{m+n-s+1} + b_1^{m+n-s+1}\theta^1 + \cdots + b_m^{m+n-s+1}\theta^m, \\ \cdots\cdots\cdots\cdots\cdots\cdots\cdots\cdots \\ \omega^n = \bar{\omega}^n + b_1^n\theta^1 + \cdots + b_m^n\theta^m, \end{cases}
$$

since $\omega^i (i = p + 1, \cdots, n)$ reduce to $\bar{\omega}^i$ when $u^1, \cdots, u^p$ are constants. Thus we get

$$[\omega^{p+1} \cdots \omega^n] \equiv [\bar{\omega}^{p+1} \cdots \bar{\omega}^n], \qquad (\text{mod } \theta^1, \cdots, \theta^m).$$

Finally, from (18) and (19), we find

$$(21)\qquad [\omega^1 \cdots \omega^n] = f(\lambda^1, \cdots, \lambda^{s-m})[\bar{\omega}^{p+1} \cdots \bar{\omega}^n d\Omega],$$

where $d\Omega$ denotes the element of the p-dimensional area of V_p.

It is easy to finish the proof of our theorem from the last formula. In fact, evaluating the integral

$$\int f(\lambda^1, \cdots, \lambda^{s-m})[\bar{\omega}^{p+1} \cdots \bar{\omega}^n \, d\Omega]$$

by first holding an element of V_p fixed, we get a constant which is the same for all elements of V_p. Integration over V_p then gives the p-dimensional area of V_p.

From this proof we observe that even when the measure of the elements of N incident with the elements of V_p becomes infinite, there still exists a generalized formula of Crofton. In fact, we may make the following restriction: On each variety of $s - m$ dimensions V_{s-m} consisting of the elements of (4) incident with a fixed element of (10) take a fixed domain D such that the integral

$$\int f(\lambda^1, \cdots, \lambda^{s-m})[\bar{\omega}^{p+1} \cdots \bar{\omega}^n]$$

is finite and constant for all elements of (10). Then the measure of the elements of (4) incident with the elements of V_p and belonging to the domain D at each element of incidence is finite and will be given by Crofton's formula (17).

We now come to generalize the formula of Cauchy. Consider a surface S in Euclidean space and the planes E intersecting the surface. If $\dot{E}$ denotes the density of the planes and U_E the perimeter of the curve of intersection of the surface by the plane, then the formula of Cauchy is[8]

$$(22)\qquad \int U_E \dot{E} = \frac{\pi^2}{2} F,$$

where F denotes the area of S and the integral in the left-hand side is extended over all planes intersecting S.

To give a generalization of this formula take a variety V_p of p dimensions formed by the elements of the field M. As in the proof of Crofton's formula (17)

[8] Blaschke, I. G. II, p. 73.

let $u^1, \cdots, u^p$ be the parameters on V_p and let us attach to each element of V_p a frame. Almost all the arguments used above are valid, including the formulas (19), (20). The only difference is that the variety, formally denoted by V_n, is now of $p + s - m$ dimensions. This variety V will be cut by an integral variety V_{r-n} of (4) in a variety $V_{p+s-m-n}$ of $p + s - m - n$ dimensions (supposing $p > m + n - s$). Through each point of $V_{p+s-m-n}$ there passes an integral variety of (10). Its totality consists of all the elements of V_p incident with V_{r-n}. To find the independent components on $V_{p+s-m-n}$ it is only necessary to set $\omega^\lambda = 0$ in (18). The independent Pfaffian forms in the equations so obtained

$$\sum_{k=1}^{m} f_k^\lambda w^k = 0, \qquad \lambda = 1, \cdots, m + n - s$$

then give the independent components on $V_{p+s-m-n}$. The generalized Cauchy's formula states that *the integral over all elements of* (4) *incident with* V_p *of a* $(p + s - m - n)$*-dimensional integral invariant of* $V_{p+s-m-n}$ *of the form*

$$\int \sum C_{i_1 \cdots i_{p+s-m-n}} [w^{i_1} \cdots w^{i_{p+s-m-n}}], \tag{23}$$

where $C_{i_1 \cdots i_{p+s-m-n}}$ *are functions of* $\lambda^1, \cdots, \lambda^{s-m}$ *only, is equal to a constant multiple of the p-dimensional area of* V_p*, provided that both quantities are finite.*

It is only necessary to prove the formula for the case when the sum (23) contains one term. Suppose it be

$$\int C[w^1 \cdots w^{p+s-m-n}].$$

By making use of (18), (20), we get

$$C[w^1 \cdots w^{p+s-m-n} \omega^1 \cdots \omega^n] = F(\lambda^1, \cdots, \lambda^{s-m})[\theta^1 \cdots \theta^p \bar{\omega}^{m+n-s+1} \cdots \bar{\omega}^n],$$

where F is a function of $\lambda^1, \cdots, \lambda^{s-m}$. The generalized Cauchy's formula as stated above is then an easy consequence of the relation obtained.

It may be helpful to give an example of the above general discussions. Consider the three-dimensional Euclidean space with its group of motions. Instead of introducing the group space we may take a fixed right-hand rectangular trihedral T_0 in space and take all such trihedrals T as the elements of the space. Then the motions in space and the trihedrals T are in one-to-one correspondence such that to T corresponds the motion carrying T_0 to T. As the parameters of the group we may take the parameters of T. Let P be the origin of T and $\vec{e}_1, \vec{e}_2, \vec{e}_3$ the three unit vectors along the axes of T. Then the relative components of the group are defined by the equations

$$\begin{cases} dP = \overset{1}{\omega}\vec{e}_1 + \overset{2}{\omega}\vec{e}_2 + \overset{3}{\omega}\vec{e}_3, \\ d\vec{e}_i = \overset{1}{\omega_i}\vec{e}_1 + \overset{2}{\omega_i}\vec{e}_2 + \overset{3}{\omega_i}\vec{e}_3, \end{cases} \qquad \omega_j^i + \omega \ = 0, \qquad (i, j = 1, 2, 3). \tag{24}$$

The equations of structure of the group are

$$(25)\qquad \begin{cases} (\omega^i)' = \sum_{k=1}^{3} [\omega^k \omega_k^i], \\ (\omega_k^i)' = \sum_{j=1}^{3} [\omega_k^j \omega_j^i], \end{cases} \qquad i, k = 1, 2, 3.$$

As the coset in the group space corresponding to a point we take the family of trihedrals having the point as origin. As the coset corresponding to a straight line we take the trihedrals whose origin is on the line and whose vector $\vec{e}_3$ is along the line. Finally, the coset corresponding to a plane is formed by the trihedrals with origin on the plane and with the third unit vector $\vec{e}_3$ perpendicular to the plane. By these definitions, the equations of the points, lines, and planes are respectively

$$(26)\qquad \begin{cases} \omega^1 = 0, & \omega^2 = 0, & \omega^3 = 0; \\ \omega^1 = 0, & \omega^2 = 0, & \omega_3^1 = 0, & \omega_3^2 = 0; \\ \omega^3 = 0, & \omega_3^1 = 0, & \omega_3^2 = 0. \end{cases}$$

For all these three fields of elements the measures exist, as may be verified by using the criterion (9). They are then given by the integrals

$$(27)\qquad \int [\omega^1 \omega^2 \omega^3], \qquad \int [\omega^1 \omega^2 \omega_3^1 \omega_3^2], \qquad \int [\omega^3 \omega_3^1 \omega_3^2].$$

Consider now the fields of points and lines and denote them by M, N respectively. Then

$$m = 3, \qquad n = 4, \qquad s = 5.$$

Put

$$w^1 = \omega^1, \qquad w^2 = \omega^2, \qquad w^3 = \omega^3.$$

The relations (18) become

$$\omega^1 = w^1, \qquad \omega^2 = w^2.$$

To obtain the formula of Crofton in this case take a surface Σ and attach to each point of the surface a trihedral $P\vec{v}_1\vec{v}_2\vec{v}_3$ with origin at this point and with $\vec{v}_3$ normal to the surface. Choose the trihedral $P\vec{e}_1\vec{e}_2\vec{e}_3$ attached to a line through P such that $\vec{e}_1$ is on the plane $P\vec{v}_1\vec{v}_2$. Then we have

$$(28)\qquad \begin{cases} \vec{v}_1 = -\sin\psi\,\vec{e}_1 - \cos\psi\cos\varphi\,\vec{e}_2 + \cos\psi\sin\varphi\,\vec{e}_3, \\ \vec{v}_2 = \cos\psi\,\vec{e}_1 - \sin\psi\cos\varphi\,\vec{e}_2 + \sin\psi\sin\varphi\,\vec{e}_3, \\ \vec{v}_3 = \sin\varphi\,\vec{e}_2 + \cos\varphi\,\vec{e}_3 \end{cases}$$

where φ, ψ are the parameters of the lines through P, φ being the angle between $\vec{e}_3$ and $\vec{v}_3$. From

$$dP = \theta^1 \vec{v}_1 + \theta^2 \vec{v}_2 = w^1 \vec{e}_1 + w^2 \vec{e}_2 + w^3 \vec{e}_3,$$

we get

$$\begin{aligned} \omega^1 &= -\theta^1 \sin \psi + \theta^2 \cos \psi, \\ \omega^2 &= -\theta^1 \cos \psi \cos \varphi - \theta^2 \sin \psi \cos \varphi. \end{aligned} \tag{29}$$

It follows that

$$[\omega^1 \omega^2 \omega_3^1 \omega_3^2] = \cos \varphi [\bar{\omega}_3^1 \bar{\omega}_3^2 \theta^1 \theta^2]. \tag{30}$$

This is a particular case of the formula (21). The classical proof of Crofton's formula is based on this relation.

Next, take the fields of points and planes and denote them by M, N respectively. Then

$$m = 3, \quad n = 3, \quad s = 5.$$

Put

$$w^1 = \omega^1, \quad w^2 = \omega^2, \quad w^3 = \omega^3.$$

The set of relations (18) consists only of one equation

$$\omega^3 = w^3.$$

As by (29), we get

$$w^3 = \theta^1 \cos \psi \sin \varphi + \theta^2 \sin \psi \sin \varphi. \tag{31}$$

Along the curve of intersection of the plane $P\vec{e}_1\vec{e}_2$ with Σ, we have

$$w^3 = 0$$

and

$$dP = w^1 \vec{e}_1$$

so that $\vec{e}_1$ is the tangent and w^1 the element of arc. Then we get

$$[w^1 \omega^3 \omega_3^1 \omega_3^2] = [w^1 w^3 \bar{\omega}_3^1 \bar{\omega}_3^2] = -\sin \varphi [\theta^1 \theta^2 \bar{\omega}_3^1 \bar{\omega}_3^2]. \tag{32}$$

From this relation we derive easily Cauchy's formula.

TSING HUA UNIVERSITY
KUNMING, CHINA

On the Invariants of Contact of Curves in a Projective Space of *N* Dimensions and Their Geometrical Interpretation

By Shiing-shen Chern
Department of Mathematics, National Tsing Hua University
(Received September 15, 1941)
(*from Science Record Vol. 1, Nos. 1–2, pp. 11–15, August 1942*)

The invariants of contact of curves have played an important rôle in the projective theory of curves and surfaces, owing to the fact that so far the simplest geometrical interpretations of various projective differential invariants are given in terms of them. In the case of plane curves the geometrical meanings of the invariants of contact themselves are given by C. Segre and B. Segre in two very interesting theorems.[1] We shall indicate in this note a method which allows us to find such invariants and to derive their geometrical meanings in a natural way. The method is of wide application, but we restrict ourselves, for the sake of simplicity, to the case of a pair of curves in a projective space of n dimensions.

Let C and $\bar{C}$ be two curves having at their common point O the same osculating linear spaces of dimensions $1, 2, \ldots, k$, $1 \leqslant k \leqslant n-1$. We attach to O a projective frame $AA_1 \ldots A_n$ such that A coincides with O, AA_1 with the common tangent at O, and $AA_1 \ldots A_m$ the common osculating m-flat at O, $m = 1, 2, \ldots, k$. Referred to this frame the equations of C and $\bar{C}$ are respectively

$$x_i = b_{i2}x_1^2 + b_{i3}x_1^3 + \cdots \qquad i = 2, \ldots, n.$$

and

$$x_i = c_{i2}x_1^2 + c_{i3}x_1^3 + \cdots \qquad i = 2, \ldots, n.$$

where the coefficients b_{ij} satisfy the conditions

$$\begin{cases} b_{32} = b_{42} = \cdots = b_{n2} = 0, \\ b_{43} = b_{53} = \cdots = b_{n3} = 0, \\ \cdots\cdots\cdots\cdots\cdots\cdots \\ b_{k,k-1} = \cdots = b_{n,k-1} = 0, \\ b_{k+1,k} = \cdots = b_{nk} = 0, \end{cases}$$

and c_{ij} satisfy similar conditions. The transition of the projective frame $AA_1 \ldots A_n$ to a projective frame $\bar{A}\bar{A}_1 \ldots \bar{A}_n$ having the same properties is given by

$$\begin{cases} \bar{A} = A, \\ \bar{A}_1 = a_{10}A + a_{11}A_1, \\ \cdots\cdots\cdots\cdots\cdots\cdots\cdots\cdots \\ \bar{A}_k = a_{k0}A + a_{k1}A_1 + \cdots + a_{kk}A_k, \\ \bar{A}_{k+1} = a_{k+1,0}A + a_{k+1,1}A_1 + \cdots + a_{k+1,n}A_n, \\ \cdots\cdots\cdots\cdots\cdots\cdots\cdots\cdots \\ \bar{A}_n = a_{n0}A + a_{n1}A_1 + \cdots + a_{nn}A_n, \end{cases}$$

where the coefficients a_{ij} are arbitrary. On denoting the coefficients in the equations of C and $\bar{C}$ referred to the new frame $\bar{A}\bar{A} \ldots \bar{A}_n$ by dashes, we find

$$\bar{b}_{kk} = \frac{(a_{11})^k}{a_{kk}} b_{kk},$$

$$\bar{c}_{kk} = \frac{(a_{11})^k}{a_{kk}} c_{kk},$$

It follows that $J_k = c_{kk}/b_{kk}$ is a projective invariant of the two curves. Similarly, we may show that for $k = n-1$, $J_n = c_{nn}/b_{nn}$ is also a projective invariant. For $k = n-1$ these invariants are given by *B.* Segre,[2] while for $k \leqslant n-2$ they are given by B. Su.[3] The author[4] has, on the other hand, arrived at the invariants

$$I_k = \frac{J_k}{J_{k-1}}, \quad I_n = \frac{J_n}{J_{n-1}}$$

by making use of Cartan's method of moving frames.

From the above derivation of the invariants in question it is easy to give their geometrical meanings by generalizing the theorems of *C.* Segre and B. Segre. In fact, the points M of C and the points $\bar{M}$ of $\bar{C}$ may be set in one-to-one correspondence by the condition that the line $M\bar{M}$ meets the $(n-2)$—flat $A_2 \ldots A_n$. Let P_m be the point where $M\bar{M}$ meets $AA_1 \ldots A_{m-1}A_{m+1} \ldots A_n$, $m = 0, 1, \ldots, n$. Then $(M\bar{M}, P_{k-1}P_k)$ approaches I_k as M tends to A. In the case $k = n-1$, the limit of $(M\bar{M}, P_{n-1}P_n)$ is I_n. On remarking that the $(n-2)$—flat $A_2 \ldots A_n$ is arbitrary, we get the following generalization of the theorem of *C.* Segre:

Let C and $\bar{C}$ be two curves having at their common point O the same osculating linear spaces of dimensions $1, 2, \ldots, k$, $k \leqslant n-1$. Take an arbitrary $(n-2)$—flat S_{n-2} not intersecting the common tangent. Let $AA_1, AA_1A_2, \ldots, AA_1 \ldots A_k$ be the common osculating flats of $1, 2, \ldots, k$ dimensions at O such that A coincides with O and that $A_2, \ldots, A_k$ belong to S_{n-2}, and let $A_{k+1}, \ldots, A_n$ be $n-k$ further points in S_{n-2} linearly independent from $A_2, \ldots, A_k$. We set the points M of C and the points $\bar{M}$ of $\bar{C}$ in correspondence by the *condition that* $M\overline{M}$ *meets* S_{n-2}. *If* P_m *is the point of intersection of* $M\overline{M}$ *and* $AA_1 \ldots A_{m-1} A_{m+1} \ldots A_n$, $m = 0, 1, \ldots, n$, *then, as* M *approaches* O,

$$\lim \left(M\overline{M}, P_{k-1}P_k\right) = I_k,$$
$$\lim \left(M\overline{M}, P_{n-1}P_n\right) = I_n, \quad \text{for} \quad k = n-1,$$
$$\lim \left(M\overline{M}, P_0P_k\right) = J_k,$$
$$\lim \left(M\overline{M}, P_0P_n\right) = J_n, \quad \text{for} \quad k = n-1,$$

These limits are independent of the choice of S_{n-2} *and of* $A_2, \ldots, A_n$, *provided that they satisfy the conditions mentioned above.*

A second geometrical interpretation of these invariants is given by the following generalization of the theorem of B. Segre:

Let S_{n-2} *be an arbitrary* $(n-2)$*—flat in space not intersecting the common tangent and let* S_{n-1} *be an arbitrary hyperplane through* S_{n-2} *but not containing* O. *Consider the family of hyperquadrics containing* S_{n-2}. *Let* Σ_2 *and* $\overline{\Sigma}_2$ *be the two hyperquadrics of the family having a contact of the second order at* O *with* C *and* $\overline{C}$ *respectively, with the further property that their tangent hyperplane at any point of* S_{n-2} *is* S_{n-1}. *Each of these hyperquadrics is a quadric hypercone of the* $(n-2)$*-nd kind and each generates a family depending on* $n-2$ *parameters. The hyperquadrics of the two families can be set into one-to-one correspondence by the condition that the intersection of corresponding hyperquadrics consists of two* $(n-2)$*—flats each counted twice. Every pair of corresponding hyperquadrics has then a well-known projective invariant, (namely the cross ratio of the two hyperquadrics and the two hypercones of the* $(n-1)$*st or the nth kind of the pencil determined by them), which is a constant and is equal to* $I_2 = J_2$. *To give a geometrical meaning to* $I_m (m = 3, \ldots, n)$ *denote by* S_{m-3} *the* $(m-3)$*—flat common to* S_{n-2} *and the osculating* $(m-1)$*—flat of the curves at* O. *Suppose* S_{n-2} *be formed by* S_{m-3} *and an arbitrary* $(n-m)$*—flat* S_{n-m} *in* S_{n-2}, *and let* S_{n-1} *be an arbitrary hyperplane through* S_{n-2}. *We consider the family of hyperquadrics having the properties:* 1) *they contain* S_{n-3} *and the* $(m-2)$*—flat joining* O *to* S_{m-3}, 2) *their tangent hyperplane at a point of* S_{n-m} *is* S_{n-1}. *From this family we select two sub-families of hyperquadrics* Σ_m *and* $\overline{\Sigma}_m$ *having a contact of the mth order at* O *with* C *and* $\overline{C}$ *respectively. The hyperquadrics* Σ_m *and* $\overline{\Sigma}_m$ *(hypercones of the* $(n-3)$*rd kind) depend on* $n-m$ *parameters and can be set into one-to-one correspondence by the condition that the intersection of corresponding hyperquadrics consists of four* $(n-2)$*—flats. Then any pair of corresponding hyperquadrics possesses a projective invariant which is equal to* I_m.

The proof of this theorem rests essentially on a particular choice of the frame of reference. Let us take the general case $m \geqslant 3$. When the flat spaces $S_{n-2}, S_{n-1}, S_{n-m}$ in question are given arbitrarily, satisfying the conditions in the theorem, we may evidently choose the frame of reference $AA_1, \ldots, A_n$ such that the following conditions are satisfied:

1) A coincides with O,
2) $AA_1 \ldots A_r$ is the common osculating r-flat of the curves at O, $r = 1, \ldots, m$.
3) $A_2 \ldots A_n$ is S_{n-2},
4) $A_1 \ldots A_n$ is S_{n-1},
5) $A_m \ldots A_n$ is S_{n-m}.

Then the $(m-3)$—flat S_{m-3} in the theorem is $A_2 \ldots A_{m-1}$. By defining the non-homogeneous coordinates $x_1, \ldots, x_n$ of a point M with respect to the frame

$AA_1 \ldots A_n$ by the relation

$$M = A + x_1A_1 + \cdots + x_nA_n,$$

we find the equation of the family of hyperquadrics having the properties 1), 2) of the theorem to be

$$2\alpha x_1 + \beta x_1^2 + 2(\lambda_m x_m + \cdots + \lambda_n x_n) + 2x_1(\mu_2 x_2 + \cdots + \mu_{m-1}x_{m-1}) = 0$$

where $\alpha, \beta, \lambda_m, \ldots, \lambda_n, \mu_2, \ldots, \mu_{m-1}$ are parameters. It follows that the equation of the family of hyperquadrics Σ_m is

$$\lambda_{m+1}x_{m+1} + \cdots + \lambda_n x_n + \left(x_m - \frac{b_{mm}}{b_{m-1,m-1}}x_1x_{m-1}\right) = 0,$$

where $\lambda_{m+1}, \ldots, \lambda_n$ are parameters. Similarly, the equation of $\overline{\Sigma}_m$ is

$$\lambda'_{m+1}x_{m+1} + \cdots + \lambda'_n x_n + \left(x_m - \frac{c_{mm}}{c_{m-1,m-1}}x_1x_{m-1}\right) = 0,$$

$\lambda'_{m+1}, \ldots, \lambda'_n$ being parameters. In order that these two hyperquadrics intersect in four $(n-2)$—flats, it is necessary and sufficient that

$$\lambda'_{m+1} = \lambda_{m+1}, \ldots, \lambda'_n = \lambda_n.$$

When these conditions are satisfied, we may change the frame of reference such that the equations of Σ_m and $\overline{\Sigma}_m$ take the forms

$$x_m - \frac{b_{mm}}{b_{m-1,m-1}}x_1x_{m-1} = 0,$$

$$x_m - \frac{c_{mm}}{c_{m-1,m-1}}x_1x_{m-1} = 0,$$

From these two equations the last statement of the theorem follows.

References

1. Segre, C. *Atti R. Accad. Lincei*, Serie 5, 6_2(1897), 168–175.
   Also Segre, B. *Atti. R. Accad. Lincci*, Serie 6, 9(1929), 975–977.
2. Segre, B. *Atti. R. Accad. Lincei*, Serie 6, 22(1935), 392–399.
3. Su, B. *Science Record*, in press.
4. The relations of the relative components of moving frames to the invariants of contact are given in the author's paper, "Sur les invariants de contact dans la geométrie projective différentielle", which has been presented by Prof. T. Levi-Civita to the "Académie Pontificale des Sciences" in Rome and will be published in Italy.

Reprinted from the Proceedings of the NATIONAL ACADEMY OF SCIENCES,
Vol. 29, No. 1, pp. 38–43. January, 1943

A GENERALIZATION OF THE PROJECTIVE GEOMETRY OF LINEAR SPACES

BY SHIING-SHEN CHERN

TSING HUA UNIVERSITY AND ACADEMIA SINICA

Communicated November 2, 1942

The projective geometry in a space of n dimensions may be briefly described as the geometry of the points and the straight lines in the space. For $n > 2$ the straight lines can be replaced by the linear spaces of a fixed dimension r, $1 \leqq r \leqq n - 1$. We have, in fact, the theorem that a one-to-one point transformation in the space, which carries the linear spaces of dimension r into themselves, is a projective transformation. The number of parameters on which the linear spaces of dimension r depend is equal to $N = (r + 1)(n - r)$. The geometry in a space of n dimensions in which there is given a family of r-dimensional varieties depending on N parameters is therefore in a certain sense a generalization of projective geometry. In this note we shall show that in such a space a projective connection can be defined. The geometry of paths[1] is a particular case of this geometry for $r = 1$, while the case $r = n - 1$ has been studied by M. Hachtroudi[2]

Let $x^1, \ldots, x^n$ be the coördinates in the space and let the family of varieties be defined by a completely integrable Pfaffian system of the form[3]

$$\left.\begin{aligned} dx^i - p^i_\alpha dx^\alpha &= 0, \\ dp^i_\alpha - r^i_{\alpha\beta} dx^\beta &= 0, \end{aligned}\right\} \qquad (1)$$

where $r^i_{\alpha\beta}$ are functions of x^α, x^i, p^i_α. Before defining the projective connection in question, we shall develop the invariant theory of the family of varieties under non-singular point transformations

$$\left.\begin{aligned} x^{1*} &= x^1(x^1, \ldots, x^n) \\ &\cdots\cdots\cdots\cdots\cdots\cdots \\ x^{n*} &= x^n(x^1, \ldots, x^n). \end{aligned}\right\} \qquad (2)$$

The left-hand members of (1) are not invariant Pfaffian forms. To derive invariants or invariant Pfaffian forms of the family of varieties the following device will be repeatedly applied: We adjoin to the variables x^α, x^i, p^i_α the new variables u^i_j, u^α_β, u^α_j, $u^{i\beta}_{\alpha j}$, $u^i_{\alpha j}$. Then the Pfaffian forms

$$\begin{aligned} \omega^\alpha &= u^\alpha_\beta dx^\beta + u^\alpha_j(dx^j - p^j_\beta dx^\beta), \\ \omega^i &= u^i_j(dx^j - p^j_\beta dx^\beta), \\ \omega^i_\alpha &= u^{i\beta}_{\alpha j}(dp^j_\beta - r^j_{\beta\gamma} dx^\gamma) + u^i_{\alpha j}(dx^j - p^j_\beta dx^\beta) \end{aligned} \qquad (3)$$

are invariant (in the space of all the variables). The number of the new variables can sometimes be reduced by invariant conditions. In fact, we shall suppose

$$(\omega^i)' \equiv [\omega^\alpha \omega^i_\alpha], \text{ mod. } \omega^j, \tag{4}$$

which are equivalent to the conditions

$$u^{i\beta}_{\alpha j} = u^i_j v^\beta_\alpha, \tag{5}$$

v^β_α being defined by the relations

$$u^\gamma_\alpha v^\beta_\gamma = v^\gamma_\alpha u^\beta_\gamma = \delta^\beta_\alpha. \tag{6}$$

Suppose the conditions (4) or (5) be satisfied. Then we can write, by introducing the new Pfaffian forms φ^α_β, φ^α_j, φ^i_j, the exterior derivatives of ω^α, ω^i in the forms

$$\left.\begin{aligned}(\omega^\alpha)' &= [\varphi^\alpha_\beta \omega^\beta] + [\varphi^\alpha_j \omega^j],\\ (\omega^i)' &= [\varphi^i_j \omega^j] + [\omega^\alpha \omega^i_\alpha].\end{aligned}\right\} \tag{7}$$

The fact that the exterior derivatives of $(\omega^\alpha)'$, $(\omega^i)'$ are zero gives

$$\left.\begin{aligned}[(\varphi^{i\prime}_j - \varphi^i_k \varphi^k_j - \varphi^\alpha_j \omega^i_\alpha)\omega^j] - [(\omega^{i\prime}_\alpha + \varphi^\beta_\alpha \omega^i_\beta - \varphi^i_j \omega^j_\alpha)\omega^\alpha] &= 0,\\ [(\varphi^{\alpha\prime}_j - \varphi^\alpha_\beta \varphi^\beta_j - \varphi^\alpha_k \varphi^k_j)\omega^j] + [(\varphi^{\alpha\prime}_\beta - \varphi^\alpha_\gamma \varphi^\gamma_\beta + \varphi^\alpha_j \omega^j_\beta)\omega^\beta] &= 0.\end{aligned}\right\} \tag{8}$$

The first equation shows that the expression $(\omega^i_\alpha)' + [\varphi^\beta_\alpha \omega^i_\beta] - [\varphi^i_j \omega^j_\alpha]$ contains ω^j or ω^β in each of its terms. Since the system

$$\omega^i = 0, \quad \omega^i_\alpha = 0 \tag{9}$$

is equivalent to (1) and is hence completely integrable, we see that we can put

$$(\omega^i_\alpha)' = -[\varphi^\beta_\alpha \omega^i_\beta] + [\varphi^i_j \omega^j_\alpha] + [\varphi^i_{\alpha j} \omega^j] + Q^{i\beta}_{\alpha\gamma j}[\omega^j_\beta \omega^\gamma], \tag{10}$$

where $\varphi^i_{\alpha j}$ are new Pfaffian forms and where

$$Q^{i\beta}_{\alpha\gamma j} = Q^{i\beta}_{\gamma\alpha j}. \tag{11}$$

Let us see whether $Q^{i\beta}_{\alpha\gamma j}$ are invariants. For this purpose notice that the Pfaffian forms φ^α_β, φ^i_j in (7) are determined up to the transformation

$$\left.\begin{aligned}\varphi^{\alpha *}_\beta &= \varphi^\alpha_\beta + a^\alpha_{\beta\gamma}\omega^\gamma + a^\alpha_{\beta j}\omega^j,\ a^\alpha_{\beta\gamma} = a^\alpha_{\gamma\beta},\\ \varphi^{i *}_j &= \varphi^i_j + a^i_{jk}\omega^k,\ a^i_{jk} = a^i_{kj},\end{aligned}\right\} \tag{12}$$

where the a's are arbitrary. When φ^α_β, φ^i_j are replaced, respectively, by $\varphi^{\alpha *}_\beta$, $\varphi^{i *}_j$, the equation (10) retains its form, while $Q^{i\beta}_{\alpha\gamma j}$ are replaced by $Q^{i\beta *}_{\alpha\gamma j}$, related to $Q^{i\beta}_{\alpha\gamma j}$ as follows:

$$Q^{i\beta *}_{\alpha\gamma j} = Q^{i\beta}_{\alpha\gamma j} - \delta^i_j a^\beta_{\alpha\gamma}.$$

This shows that $Q^{i\beta}_{\alpha\gamma j}$ are not invariants. We can, however, modify them such that the conditions

$$Q^{j\beta}_{\alpha\gamma j} = 0 \tag{13}$$

are satisfied. Under these conditions $Q^{i\beta}_{\alpha\gamma j}$ are invariants.

If we carry out the calculation of the expression for $Q^{i\beta}_{\alpha\beta j}$, we see that, by a proper choice of $u^i_{\alpha j}$, the conditions

$$Q^{i\beta}_{\alpha\beta j} = 0 \tag{14}$$

can be fulfilled. This determines $u^i_{\alpha j}$ in terms of x^α, x^i, p^i_α, u^i_j, u^α_β, and parameters or new variables which we denote by v_α. When $r = 1$ or $r = n - 1$, the conditions (13) and (14) signify that all $Q^{i\beta}_{\alpha\gamma j}$ vanish.

The conditions (13) and (14) have an effect on the expressions for $(\varphi^\alpha_\beta)'$, $(\omega^i_\alpha)'$. To find these expressions we apply exterior differentiation to (10), which gives

$$[\{\delta^i_j(-\varphi^{\beta\prime}_\alpha - \varphi^\gamma_\alpha\varphi^\beta_\gamma) + \delta^\beta_\alpha(\varphi^{i\prime}_j - \varphi^i_k\varphi^k_j) - \pi^{i\beta}_{\alpha\gamma j}\omega^\gamma - \varphi^i_{\alpha j}\omega^\beta + Q^{i\rho}_{\alpha\gamma k}Q^{k\beta}_{\rho\sigma j}\omega^\sigma\omega^\gamma\}\omega^j_\beta] \equiv 0, \text{ mod. } \omega^k,$$

where

$$\pi^{i\beta}_{\alpha\gamma j} = dQ^{i\beta}_{\alpha\gamma j} - Q^{k\beta}_{\alpha\gamma j}\varphi^i_k + Q^{i\beta}_{\alpha\gamma k}\varphi^k_j - Q^{i\rho}_{\alpha\gamma j}\varphi^\beta_\rho + Q^{i\beta}_{\rho\gamma j}\varphi^\rho_\alpha + Q^{i\beta}_{\alpha\rho j}\varphi^\rho_\gamma.$$

From (8) we get respectively

$$(\varphi^i_j)' - [\varphi^i_k\varphi^k_j] \equiv [\varphi^\alpha_j\omega^i_\alpha] - [\varphi^i_{\alpha j}\omega^\alpha], \text{ mod. } \omega^k,$$
$$(\varphi^\alpha_\beta)' - [\varphi^\alpha_\gamma\varphi^\gamma_\beta] + [\varphi^\alpha_j\omega^j_\beta] \equiv [\theta^\alpha_{\beta\gamma}\omega^\gamma], \text{ mod. } \omega^k,$$

where $\theta^\alpha_{\beta\gamma}$ are newly introduced Pfaffian forms. Substituting these expressions into the last equation, we shall get

$$\delta^i_j\theta^\beta_{\alpha\gamma} + \delta^\beta_\alpha\varphi^i_{\gamma j} + \delta^\beta_\gamma\varphi^i_{\alpha j} + \pi^{i\beta}_{\alpha\gamma j} \equiv 0, \quad \text{mod. } \omega^\alpha, \omega^i, \omega^i_\alpha.$$

The conditions (13) and (14) have as consequences

$$\pi^{j\beta}_{\alpha\gamma j} = 0, \quad \pi^{i\beta}_{\alpha\beta j} = 0,$$

so that the above equation gives, respectively,

$$(n - r)\theta^\beta_{\alpha\gamma} + \delta^\beta_\alpha\varphi_\gamma + \delta^\beta_\gamma\varphi_\alpha \equiv 0, \quad \text{mod. } \omega^\alpha, \omega^i, \omega^i_\alpha,$$

$$\varphi^i_{\alpha j} \equiv \frac{1}{n - r}\,\delta^i_j\varphi_\alpha, \quad \text{mod. } \omega^\alpha, \omega^i, \omega^i_\alpha,$$

where φ_α is an abbreviation of $\varphi^j_{\alpha j}$. It follows that the exterior derivatives of ω^i_α, φ^α_β are of the forms

$$(\omega^i_\alpha)' = -[\varphi^\beta_\alpha\omega^i_\beta] + [\varphi^i_j\omega^j_\alpha] + \frac{1}{n - r}[\varphi_\alpha\omega^i] + \Omega^i_\alpha, \tag{15}$$

$$(\varphi_\beta^\alpha)' \equiv [\varphi_\gamma^\alpha \varphi_\beta^\gamma] - [\varphi_j^\alpha \omega_\beta^j] - \frac{1}{n-r}\delta_\beta^\alpha[\varphi_\gamma\omega^\gamma] - \frac{1}{n-r}[\varphi_\beta\omega^\alpha] + R_{\beta\gamma\rho}^\alpha[\omega^\rho\omega^\gamma] + R_{\beta\gamma j}^{\alpha\rho}[\omega_\rho^j\omega^\gamma], \text{ mod. } \omega^i, \tag{16}$$

where

$$\Omega_\alpha^i = Q_{\alpha j\beta}^i[\omega^\beta\omega^j] + Q_{\alpha jk}^i[\omega^k\omega^j] + Q_{\alpha jk}^{i\beta}[\omega_\beta^k\omega^j] + Q_{\alpha\gamma j}^{i\beta}[\omega_\beta^j\omega^\gamma]. \tag{17}$$

It is to be noticed that the Pfaffian forms φ_β^α, φ_j^i, φ_j^α, φ_α introduced above are not invariant, because they are not completely determined by the equations (7), (15). These Pfaffian forms are determined up to a transformation which can easily be written down. By investigating the effect of this transformation on the coefficients in (15), (16), it is not difficult to show that we can choose the φ's such that we have

$$Q_{\alpha j\beta}^j = 0, \quad Q_{\alpha jk}^{j\beta} = 0, \quad R_{\beta\gamma j}^{\alpha\gamma} = 0, \quad Q_{\beta jk}^{i\beta} = 0. \tag{18}$$

Under these conditions the Pfaffian forms in question are determined up to the transformation

$$\left.\begin{aligned}
\varphi_\beta^{\alpha*} &= \varphi_\beta^\alpha + \frac{1}{n-r+1}\delta_\beta^\alpha a_j\omega^j,\\
\varphi_j^{\alpha*} &= \varphi_j^\alpha + \frac{1}{n-r+1}a_j\omega^\alpha + a_{jk}^\alpha\omega^k, \ a_{jk}^\alpha = a_{kj}^\alpha,\\
\varphi_j^{i*} &= \varphi_j^i + \frac{1}{n-r+1}(\delta_j^i a_k\omega^k + a_j\omega^i),\\
\overset{*}{\varphi}_\alpha &= \varphi_\alpha + a_{\alpha j}\omega^j + \frac{n-r}{n-r+1}a_j\omega_\alpha^j,
\end{aligned}\right\} \tag{19}$$

where the a's are arbitrary.

We now take v_α, a_j to be new variables and adjoin them to the set of variables x^α, x^i, p_α^i, u_j^i, u_β^α, u_j^α. All these $n(n+2)$ variables then form the set which we shall deal with and by invariance we shall mean the invariance under a general transformation in these variables. It is in this sense that the Pfaffian forms φ_β^α, φ_j^i are invariant. In order to find a set of linearly independent invariant Pfaffian forms, whose number is equal to the number of variables, we form the exterior derivatives of φ_β^α, φ_j^i. From (8), (10) we see that we can set

$$\left.\begin{aligned}
&(\varphi_j^i)' - [\varphi_k^i\varphi_j^k] - [\varphi_j^\alpha\omega_\alpha^i] + \frac{1}{n-r}\delta_j^i[\varphi_\alpha\omega^\alpha] - Q_{\alpha j\beta}^i[\omega^\alpha\omega^\beta] - \\
&\qquad Q_{\alpha jk}^i[\omega^\alpha\omega^k] - Q_{\alpha jk}^{i\beta}[\omega^\alpha\omega_\beta^k] = [\theta_{jk}^i\omega^k],\\
&(\varphi_\beta^\alpha)' - [\varphi_\gamma^\alpha\varphi_\beta^\gamma] + [\varphi_j^\alpha\omega_\beta^j] + \frac{1}{n-r}\delta_\beta^\alpha[\varphi_\gamma\omega^\gamma] + \frac{1}{n-r}[\varphi_\beta\omega^\alpha] - \\
&\qquad R_{\beta\gamma\rho}^\alpha[\omega^\rho\omega^\gamma] - R_{\beta\gamma j}^{\alpha\rho}[\omega_\rho^j\omega^\gamma] = [\theta_{\beta j}^\alpha\omega^j],
\end{aligned}\right\} \tag{20}$$

where θ^i_{jk}, $\theta^\alpha_{\beta j}$ are newly introduced Pfaffian forms, of which the former ones are subjected to the conditions

$$[\theta^i_{jk}\omega^j\omega^k] = 0. \tag{21}$$

We form the equation obtained by exterior differentiation of (15). A complete analysis of that equation will give the following two consequences:

$$\delta^i_k\theta^\beta_{\beta j} - r\theta^i_{kj} - \frac{1}{n-r}\,\delta^i_j(\theta^\beta_{\beta k} - r\theta_k) \equiv 0, \quad \text{mod. } \omega^\alpha, \omega^i, \omega^i_\alpha; \tag{22}$$

$$\varphi_\alpha{}' \equiv -[\varphi^\beta_\alpha\varphi_\beta] - [\theta^\beta_{\alpha k}\omega^k_\beta] + [\theta_k\omega^k_\alpha] - Q^{j\beta}_{\alpha\gamma k}\,[\varphi^\gamma_j\omega^k_\beta] + U^{\ \beta}_{\alpha\gamma k}[\omega^k_\beta\omega^\gamma], \text{mod. } \omega^i + U_{\alpha\beta\gamma}[\omega^\beta\omega^\gamma] \tag{23}$$

where $\theta_k = \theta^j_{kj}$. On the other hand, by exterior differentiation of the second equation of (20), we can get

$$Q^{k\alpha}_{\beta\sigma j}\varphi^\sigma_k + \delta^\alpha_\beta\theta^\gamma_{\gamma j} - (r+1)\delta^\alpha_\beta\theta_j + (n+1)\theta^\alpha_{\beta j} \equiv 0, \text{ mod. } \omega^\beta, \omega^i, \omega^i_\alpha. \tag{24}$$

From (22) and (24) we then find

$$\theta^\alpha_{\beta j} \equiv \frac{1}{n-r+1}\,\delta^\alpha_\beta\theta_j - \frac{1}{n+1}\,Q^{k\alpha}_{\beta\sigma j}\varphi^\sigma_k, \text{ mod. } \omega^\alpha, \omega^i, \omega^i_\alpha, \tag{25}$$

$$\theta^i_{kj} \equiv \frac{1}{n-r+1}\,(\delta^i_j\theta_k + \delta^i_k\theta_j), \text{ mod. } \omega^\alpha, \omega^i, \omega^i_\alpha. \tag{26}$$

In this way we have introduced the Pfaffian forms θ_k, which involve da_k and are linearly independent from ω^α, ω^i, ω^i_α, φ^α_β, φ^i_j, φ^α_j, φ_α. The total number of linearly independent Pfaffian forms is now equal to the number of variables. But of these forms φ^α_j, φ_α, θ_k are not yet invariant. To derive invariant Pfaffian forms from them, further conditions are necessary. We put, according to (25), (26),

$$\left.\begin{aligned}
[\theta^i_{jk}\omega^k] &= \frac{1}{n-r+1}\,(\delta^i_j[\theta_k\omega^k] + [\theta_j\omega^i]) + S^i_{jkl}[\omega^l\omega^k] + \\
&\qquad S^i_{jk\gamma}[\omega^\gamma\omega^k] + S^{i\gamma}_{jkl}[\omega^l_\gamma\omega^k], \\
[\theta^\alpha_{\beta j}\omega^j] &= -\frac{1}{n+1}\,Q^{k\alpha}_{\beta\sigma j}[\varphi^\sigma_k\omega^j] + \frac{1}{n-r+1}\,\delta^\alpha_\beta[\theta_j\omega^j] + \\
&\qquad R^\alpha_{\beta j\gamma}[\omega^\gamma\omega^j] + R^\alpha_{\beta jk}[\omega^k\omega^j] + R^{\alpha\gamma}_{\beta jk}[\omega^k_\gamma\omega^j].
\end{aligned}\right\} \tag{27}$$

It is then easy to show that under the conditions

$$S^j_{kj\gamma} = 0,\ S^{j\gamma}_{kjl} = 0,\ R^\beta_{\beta j\gamma} = 0,\ R^{\beta\alpha}_{\beta jk} + R^{\beta\alpha}_{\beta kj} = 0 \tag{28}$$

the Pfaffian forms φ^α_j, φ_α are completely determined. The uniqueness of θ_k depends on the expression for $(\varphi^\alpha_j)'$. We find

$$(\varphi_j^\alpha)' = [\varphi_k^\alpha \varphi_j^k] + [\varphi_\beta^\alpha \varphi_j^\beta] + \frac{1}{n-r+1} [\theta_j \omega^\alpha] - \frac{1}{n+1} Q_{\beta\sigma j}^{k\alpha} [\varphi_k^\sigma \omega^\beta] -$$
$$\frac{1}{2} (Q_{\beta jk}^{l\alpha} + Q_{\beta kj}^{l\alpha}) [\varphi_l^\beta \omega^k] - \frac{n-r+1}{2(n+1)} (R_{\beta\rho k}^{\beta\alpha} [\varphi_j^\rho \omega^k] + R_{\beta\rho j}^{\beta\alpha} [\varphi_k^\rho \omega^k]) +$$
$$R_{\beta j\gamma}^{\alpha} [\omega^\gamma \omega^\beta] + R_{\beta jk}^{\alpha\gamma} [\omega_\gamma^k \omega^\beta] + T_{jk\beta}^{\alpha} [\omega^\beta \omega^k] + T_{jkl}^{\alpha} [\omega^l \omega^k] + T_{jki}^{\alpha\beta} [\omega_\beta^i \omega^k]. \quad (29)$$

The conditions

$$T_{jk\beta}^{\beta} = 0 \quad (30)$$

determine θ_k completely.

From the complete determination of φ_j^α, φ_α, θ_k follows their invariance. We have therefore a set of linearly independent invariant Pfaffian forms whose number is equal to the number of variables. A necessary and sufficient condition for two families of varieties to be equivalent is that a transformation in all our variables exists, which carries one set of Pfaffian forms to the other. Hence the problem of equivalence is solved.

If the given family of varieties is the family of linear spaces, our Pfaffian forms are those which define the infinitesimal transformation between two neighboring projective frames. In the general case they define a projective connection. The actual calculation of the components of the projective connection in terms of the coördinates of the space offers no essential difficulty.

[1] Eisenhart, L. P., *Non-Riemannian Geometry*, New York, 1927.

[2] Hachtroudi, M., *Les espaces d'éléments à connexion projective normale*, Paris, 1937.

[3] We agree that Greek indices run from 1 to r and that Latin indices run from $r + 1$ to n.

Annals of Mathematics
Vol. 45, No. 4, October, 1944

A SIMPLE INTRINSIC PROOF OF THE GAUSS-BONNET FORMULA FOR CLOSED RIEMANNIAN MANIFOLDS

By Shiing-shen Chern

(Received November 26, 1943)

Introduction

C. B. Allendoerfer[1] and W. Fenchel[2] have independently given a generalization of the classical formula of Gauss-Bonnet to a closed orientable Riemannian manifold which can be imbedded in a euclidean space. Recently, Allendoerfer and André Weil[3] extended the formula to a closed Riemannian polyhedron and proved in particular its validity in the case of a general closed Riemannian manifold. In their proof use is still made of the imbedding of a Riemannian cell in a euclidean space. The object of this paper is to offer a direct intrinsic proof of the formula by making use of the theory of vector fields in differentiable manifolds.

The underlying idea of the present proof is very simple, so that a brief summary might be helpful. Let R^n be a closed orientable Riemannian manifold of an even dimension n. According to details to be given below, we define in R^n an intrinsic exterior differential form Ω of degree n, which is of course equal to a scalar invariant of R^n multiplied by the volume element. The formula of Gauss-Bonnet in question asserts that the integral of this differential form over R^n is equal to the Euler-Poincaré characteristic χ of R^n. To prove this we pass from the manifold R^n to the manifold M^{2n-1} of $2n - 1$ dimensions formed by the unit vectors of R^n.[4] In M^{2n-1} we show that Ω is equal to the exterior derivative of a differential form Π of degree $n - 1$. By defining a continuous field of unit vectors over R^n with isolated singular points, we get, as its image in M^{2n-1}, a submanifold V^n of dimension n, and the integral of Ω over R^n is equal to the same integral over V^n. The application of the theorem of Stokes shows that the latter is equal to the integral of Π over the boundary of V^n. Now, the boundary of V^n corresponds exactly to the singular points of the vector field defined in R^n, the sum of whose indices is, by a well-known theorem, equal to χ. With such an interpretation the integral of Π over the boundary of V^n can be evaluated and is easily proved to be equal to χ.

The method can of course be applied to derive other formulas of the same type and, with suitable modifications, to deduce the Gauss-Bonnet formula for a Riemannian polyhedron. We publish this proof, because it is in the present case that the main ideas of our method are most clear. Further results will be given in a forthcoming paper.

§1. Résumé of some fundamental formulas in Riemannian Geometry

Let R^n be a closed orientable differentiable manifold[5] of an even dimension $n = 2p$ and class $r \geqq 4$. In R^n suppose a Riemannian metric be defined, with the fundamental tensor g_{ij}, whose components we suppose to be of class 3. Since we are to deal with multiple integrals, it seems convenient to follow Cartan's treatment of Riemannian Geometry,[6] with the theory of exterior differential forms, instead of the ordinary tensor analysis, playing the dominant rôle. The differential forms which occur below are exterior differential forms.

According to Cartan we attach to each point P of R^n a set of n mutually perpendicular unit vectors $\mathfrak{e}_1, \cdots, \mathfrak{e}_n$, with a certain orientation. Such a figure $P\mathfrak{e}_1 \cdots \mathfrak{e}_n$ is called a frame. A vector $\mathfrak{v}$ of the tangent space of R^n at P can be referred to the frame at P, thus

$$\mathfrak{v} = u_i \mathfrak{e}_i, \tag{1}$$

where the index i runs from 1 to n and repeated indices imply summation. The law of infinitesimal displacement of tangent spaces, as defined by the parallelism of Levi-Civita, is given by equations of the form

$$\begin{cases} dP = \omega_i \mathfrak{e}_i, \\ d\mathfrak{e}_i = \omega_{ij} \mathfrak{e}_j, \qquad \omega_{ij} + \omega_{ji} = 0 \end{cases} \tag{2}$$

where ω_i, ω_{ij} are Pfaffian forms. These Pfaffian forms satisfy the following "equations of structure":

$$\begin{cases} d\omega_i = \omega_j \omega_{ji}, \\ d\omega_{ij} = -\omega_{ik}\omega_{jk} + \Omega_{ij}, \qquad \Omega_{ij} + \Omega_{ji} = 0. \end{cases} \tag{3}$$

In (3) Ω_{ij} are exterior quadratic differential forms and give the curvature properties of the space.

The forms Ω_{ij} satisfy a system of equations obtained by applying to (3) the theorem that the exterior derivatives of the left-hand members are zero. The equations are

$$\begin{cases} \omega_j \Omega_{ji} = 0, \\ d\Omega_{ij} - \omega_{jk}\Omega_{ik} + \omega_{ik}\Omega_{jk} = 0, \end{cases} \tag{4}$$

and are called the Bianchi identities.

For the following it is useful to know how the Ω_{ij} behave when the frame $\mathfrak{e}_1 \cdots \mathfrak{e}_n$ undergoes a proper orthogonal transformation. In a neighborhood of P in which the same system of coordinates is valid let $\mathfrak{e}_1 \cdots \mathfrak{e}_n$ be changed to $\mathfrak{e}_1^* \cdots \mathfrak{e}_n^*$ according to the proper orthogonal transformation:

$$\mathfrak{e}_i^* = a_{ij}\mathfrak{e}_j \tag{5}$$

or

$$\mathfrak{e}_i = a_{ji}\mathfrak{e}_j^*, \tag{5'}$$

where (a_{ij}) is a proper orthogonal matrix, whose elements a_{ij} are functions of the coordinates. Suppose Ω_{ij}^* be formed from the frames $P\mathfrak{e}_1^* \cdots \mathfrak{e}_n^*$ in the same way as Ω_{ij} are formed from $P\mathfrak{e}_1 \cdots \mathfrak{e}_n$. Then we easily find

$$\Omega_{ij}^* = a_{ik}a_{jl}\Omega_{kl}. \tag{6}$$

From (6) we deduce an immediate consequence. Let $\epsilon_{i_1 \cdots i_n}$ be a symbol which is equal to $+1$ or -1 according as $i_1, \cdots, i_n$ form an even or odd permutation of $1, \cdots, n$, and is otherwise zero. Since our space R^n is of even dimension $n = 2p$, we can construct the sum

$$\Omega = (-1)^{p-1} \frac{1}{2^{2p} \pi^p p!} \epsilon_{i_1 \cdots i_{2p}} \Omega_{i_1 i_2} \Omega_{i_3 i_4} \cdots \Omega_{i_{2p-1} i_{2p}}, \tag{7}$$

where each index runs from 1 to n. Using (6), we see that Ω remains invariant under a change of frame (5) and is therefore intrinsic. This intrinsic differential form Ω is of degree n and is thus a multiple of $\omega_1 \cdots \omega_n$. As the latter product (being the volume element of the space) is also intrinsic, we can write

$$\Omega = I\omega_1 \cdots \omega_n, \tag{8}$$

where the coefficient I is a scalar invariant of the Riemannian manifold.

With all these preparations we shall write the formula of Gauss-Bonnet in the following form

$$\int_{R^n} \Omega = \chi, \tag{9}$$

χ being the Euler-Poincaré characteristic of R^n.

§2. The space of unit vectors and a formula for Ω

From the Riemannian manifold R^n we pass now to the manifold M^{2n-1} of dimension $2n - 1$ formed by its unit vectors. M^{2n-1} is a closed differentiable manifold of class $r - 1$. As its local coordinates we may of course take the local coordinates of R^n and the components u_i of the vector $\mathfrak{v}$ in (1), subjected to the condition

$$u_i u_i = 1. \tag{1'}$$

If θ_i are the components of $d\mathfrak{v}$ with respect to the frame $\mathfrak{e}_1 \cdots \mathfrak{e}_n$, we have

$$d\mathfrak{v} = \theta_i \mathfrak{e}_i, \tag{10}$$

where

$$\theta_i = du_i + u_j \omega_{ji} \tag{11}$$

and

$$u_i \theta_i = 0. \tag{12}$$

From (11) we get, by differentiation,

$$d\theta_i = \theta_j \omega_{ji} + u_j \Omega_{ji}. \tag{13}$$

As to the effect of a change of frame (5) on the components u_i, θ_i, it is evidently given by the equations

$$u_i^* = a_{ij} u_j, \qquad \theta_i^* = a_{ij} \theta_j. \tag{14}$$

We now construct the following two sets of differential forms:

$$\Phi_k = \epsilon_{i_1 \cdots i_{2p}} u_{i_1} \theta_{i_2} \cdots \theta_{i_{2p-2k}} \Omega_{i_{2p-2k+1} i_{2p-2k+2}} \cdots \Omega_{i_{2p-1} i_{2p}}, \quad k = 0, 1, \cdots, p-1, \tag{15}$$

$$\Psi_k = \epsilon_{i_1 \cdots i_{2p}} \Omega_{i_1 i_2} \theta_{i_3} \cdots \theta_{i_{2p-2k}} \Omega_{i_{2p-2k+1} i_{2p-2k+2}} \cdots \Omega_{i_{2p-1} i_{2p}}, \quad k = 0, 1, \cdots, p-1. \tag{16}$$

The forms Φ_k are of degree $2p - 1$ and Ψ_k of degree $2p$, and we remark that Ψ_{p-1} differs from Ω only by a numerical factor. Using (6) and (14), we see that Φ_k and Ψ_k are intrinsic and are therefore defined over the entire Riemannian manifold R^n.

We shall prove the following recurrent relation:

$$d\Phi_k = \Psi_{k-1} + \frac{2p - 2k - 1}{2(k+1)} \Psi_k, \qquad k = 0, 1, \cdots, p-1, \tag{17}$$

where we define $\Psi_{-1} = 0$. Using the property of skew-symmetry of the symbol $\epsilon_{i_1 \cdots i_{2p}}$ in its indices, we can write

$$\begin{aligned} d\Phi_k = {} & \epsilon_{(i)} du_{i_1} \theta_{i_2} \cdots \theta_{i_{2p-2k}} \Omega_{i_{2p-2k+1} i_{2p-2k+2}} \cdots \Omega_{i_{2p-1} i_{2p}} \\ & + (2p - 2k - 1) \epsilon_{(i)} u_{i_1} d\theta_{i_2} \theta_{i_3} \cdots \theta_{i_{2p-2k}} \Omega_{i_{2p-2k+1} i_{2p-2k+2}} \cdots \Omega_{i_{2p-1} i_{2p}} \\ & - k \epsilon_{(i)} u_{i_1} \theta_{i_2} \cdots \theta_{i_{2p-2k}} d\Omega_{i_{2p-2k+1} i_{2p-2k+2}} \Omega_{i_{2p-2k+3} i_{2p-2k+4}} \cdots \Omega_{i_{2p-1} i_{2p}}, \end{aligned}$$

where $\epsilon_{(i)}$ is an abbreviation of $\epsilon_{i_1 \cdots i_{2p}}$. For the derivatives du_i, $d\theta_i$, $d\Omega_{ij}$ we can substitute their expressions from (11), (13), and (4). The resulting expression for $d\Phi_k$ will then consist of terms of two kinds, those involving ω_{ij} and those not. We collect the terms not involving ω_{ij}, which are

$$\Psi_{k-1} + (2p - 2k - 1) \epsilon_{(i)} u_{i_1} u_j \Omega_{j i_2} \theta_{i_3} \cdots \theta_{i_{2p-2k}} \Omega_{i_{2p-2k+1} i_{2p-2k+2}} \cdots \Omega_{i_{2p-1} i_{2p}}. \tag{18}$$

This expression is obviously intrinsic. Its difference with $d\Phi_k$ is an expression which contains a factor ω_{ij} in each of its terms.

We shall show that this difference is zero. In fact, let P be an arbitrary but fixed point of R^n. In a neighborhood of P we can choose a family of frames $e_1 \cdots e_n$ such that at P,

$$\omega_{ij} = 0.$$

(This process is "equivalent" to the use of geodesic coordinates in tensor notation.) Hence, for this particular family of frames, the expressions (18) and $d\Phi_k$ are equal at P. It follows that they are identical, since both expressions are intrinsic and the point P is arbitrary.

To transform the expression (18) we shall introduce the abbreviations

$$\begin{cases} P_k = \epsilon_{(i)} u_{i_1}^2 \Omega_{i_1 i_2} \theta_{i_3} \cdots \theta_{i_{2p-2k}} \Omega_{i_{2p-2k+1} i_{2p-2k+2}} \cdots \Omega_{i_{2p-1} i_{2p}}, \\ \Sigma_k = \epsilon_{(i)} u_{i_1} u_{i_3} \Omega_{i_3 i_2} \theta_{i_3} \cdots \theta_{i_{2p-2k}} \Omega_{i_{2p-2k+1} i_{2p-2k+2}} \cdots \Omega_{i_{2p-1} i_{2p}}, \\ T_k = \epsilon_{(i)} u_{i_3}^2 \Omega_{i_1 i_2} \theta_{i_3} \cdots \theta_{i_{2p-2k}} \Omega_{i_{2p-2k+1} i_{2p-2k+2}} \cdots \Omega_{i_{2p-1} i_{2p}}, \end{cases} \tag{19}$$

which are forms of degree $2p$. Owing to the relations (1′) and (12) there are some simple relations between these forms and Ψ_k. In fact, we can write

$$P_k = \epsilon_{(i)}(1 - u_{i_2}^2 - u_{i_3}^2 - \cdots - u_{i_{2p}}^2)\Omega_{i_1i_2}\theta_{i_3} \cdots \theta_{i_{2p-2k}}\Omega_{i_{2p-2k+1}i_{2p-2k+2}} \cdots \Omega_{i_{2p-1}i_{2p}}$$
$$= \Psi_k - P_k - 2(p - k - 1)T_k - 2kP_k,$$

which gives

$$\Psi_k = 2(k + 1)P_k + 2(p - k - 1)T_k. \tag{20}$$

Again, we have

$$\Sigma_k = \epsilon_{(i)}\, u_{i_1}\Omega_{i_3i_2}(-u_{i_1}\theta_{i_1} - u_{i_2}\theta_{i_2} - u_{i_4}\theta_{i_4} - \cdots - u_{i_{2p}}\theta_{i_{2p}})\theta_{i_4} \cdots$$
$$\theta_{i_{2p-2k}}\Omega_{i_{2p-2k+1}i_{2p-2k+2}} \cdots \Omega_{i_{2p-1}i_{2p}}$$
$$= T_k - (2k + 1)\Sigma_k,$$

and hence

$$T_k = 2(k + 1)\Sigma_k. \tag{21}$$

The expression (18) for $d\Phi_k$ therefore becomes

$$d\Phi_k = \Psi_{k-1} + (2p - 2k - 1)\{P_k + 2(p - k - 1)\Sigma_k\}, \qquad k = 0, 1, \cdots, p - 1.$$

Using (20) and (21), we get the desired formula (17).

From (17) we can solve Ψ_k in terms of $d\Phi_0, d\Phi_1, \cdots, d\Phi_k$. The result is easily found to be

$$\psi_k = \sum_{m=0}^{k} (-1)^m \frac{2^{m+1}(k+1)k \cdots (k-m+1)}{(2p-2k-1)(2p-2k+1)\cdots(2p-2k+2m-1)}\, d\Phi_{k-m}, \tag{22}$$
$$k = 0, 1, \cdots, p - 1.$$

In particular, it follows that Ω is the exterior derivative of a form Π:

$$\Omega = (-1)^{p-1}\frac{1}{2^{2p}\pi^p p!}\Psi_{p-1} = d\Pi, \tag{23}$$

where

$$\Pi = \frac{1}{\pi^p}\sum_{m=0}^{p-1}(-1)^m \frac{1}{1\cdot 3 \cdots (2p - 2m - 1)m!\,2^{p+m}}\Phi_m. \tag{24}$$

§3. Proof of the Gauss-Bonnet formula

Basing on the formula (24) we shall give a proof of the formula (9), under the assumption that R^n is a closed orientable Riemannian manifold.

We define in R^n a continuous field of unit vectors with a point 0 of R^n as the only singular point.[7] By a well-known theorem the index of the field at 0 is equal to χ, the Euler-Poincaré characteristic of R^n. This vector field defines in M^{2n-1} a submanifold V^n, which has as boundary χZ, where Z is the $(n - 1)$-

dimensional cycle formed by all the unit vectors through 0. The integral of Ω over R^n is evidently equal to the same over V^n. Applying Stokes's theorem, we get therefore

$$\int_{R^n} \Omega = \int_{V^n} \Omega = \chi \int_Z \Pi = \chi \frac{1}{1\cdot 3 \cdots (2p-1)2^p \pi^p} \int_Z \Phi_0 . \tag{25}$$

From the definition of Φ_0 we have

$$\Phi_0 = (2p-1)! \sum_{i=1}^{n} (-1)^i \theta_1 \cdots \theta_{i-1} u_i \theta_{i+1} \cdots \theta_{2p} . \tag{26}$$

The last sum is evidently the volume element of the $(2p-1)$-dimensional unit sphere. Therefore

$$\int_Z \Phi_0 = (2p-1)! \frac{2\pi^p}{(p-1)!} .$$

Substituting this into (25), we get the formula (9).

INSTITUTE FOR ADVANCED STUDY, PRINCETON, N. J. AND
TSING HUA UNIVERSITY, KUNMING, CHINA.

REFERENCES

1. ALLENDOERFER, C. B., *The Euler number of a Riemann manifold*, Amer. J. Math., 62 (1940), 243–248.
2. FENCHEL, W., *On total curvatures of Riemannian manifolds I*, Jour. London Math. Soc., 15 (1940), 15–22.
3. ALLENDOERFER, C. B., AND ANDRÉ WEIL, *The Gauss-Bonnet theorem for Riemannian polyhedra*, Trans. Amer. Math. Soc., 53 (1943), 101–129.
4. For its definition and topology see, for instance, E. STIEFEL, *Richtungsfelder und Fernparallelismus in n-dimensionalen Mannigfaltigkeiten*, Comm. Math. Helv., 8 (1936), 3–51.
5. WHITNEY, H., *Differentiable manifolds*, Annals of Math., 37 (1936), 645–680.
6. CARTAN, E., *Leçons sur la géométrie des espaces de Riemann*, Paris 1928.
7. ALEXANDROFF-HOPF, Topologie I, 550.

ANNALS OF MATHEMATICS
Vol. 46, No. 4, October, 1945

ON THE CURVATURA INTEGRA IN A RIEMANNIAN MANIFOLD

BY SHIING-SHEN CHERN

(Received May 23, 1945)

Introduction

In a previous paper [1] we have given an intrinsic proof of the formula of Allendoerfer-Weil which generalizes to Riemannian manifolds of n dimensions the classical formula of Gauss-Bonnet for $n = 2$. The main idea of the proof is to draw into consideration the manifold of unit tangent vectors which is intrinsically associated to the Riemannian manifold. Denoting by R^n the Riemannian manifold of dimension n and by M^{2n-1} the manifold of dimension $2n - 1$ of its unit tangent vectors, our proof has led, in the case that n is even, to an intrinsic differential form of degree $n - 1$ (which we denoted by Π) in M^{2n-1}. We shall introduce in this paper a differential form of the same nature for both even and odd dimensional Riemannian manifolds. We find that this differential form bears a close relation to the "Curvatura Integra" of a submanifold in a Riemannian manifold, because it will be proved that its integral over a closed submanifold of R^n is equal to the Euler-Poincaré characteristic of the submanifold. The method can be carried over to deduce relations between relative topological invariants of a submanifold of the manifold and differential invariants derived from the imbedding, and some remarks are to be added to this effect.

§1. Definition of the Intrinsic Differential Form in M^{2n-1}

Let R^n be an orientable Riemannian manifold of dimension n and class $\geqq 3$. For a résumé of the fundamental formulas in Riemannian Geometry we refer to §1 of the paper quoted above.

Let M^{2n-1} be the manifold of the unit tangent vectors of R^n. To a unit tangent vector we attach a frame $Pe_1 \cdots e_n$ such that it is the vector e_n through P The frame $Pe_1 \cdots e_n$ is determined up to the transformation

$$e_\alpha^* = \sum_\beta a_{\alpha\beta} e_\beta \tag{1}$$

where $(a_{\alpha\beta})$ is a proper orthogonal matrix of order $n - 1$ and where, as well as throughout the whole section, we shall fix the ranges of the indices α, β to be from 1 to $n - 1$. Since the manifold of frames over R^n is locally a topological product, we can, to a region in M^{2n-1} the points of which have their local coordinates expressed as differentiable functions of certain parameters, attach the frames $Pe_1 \cdots e_n$ which depend differentiably (with the same class) on the same parameters. From the family of frames we construct the forms ω_i, $\omega_{ij} = -\omega_{ji}$, Ω_{ij} according to the equations

$$\begin{aligned} dP &= \sum_i \omega_i e_i , \\ de_i &= \sum_j \omega_{ij} e_j , \\ \Omega_{ij} &= d\omega_{ij} - \sum_k \omega_{ik}\omega_{kj} , \end{aligned} \tag{2}$$

it being agreed that the indices i, j, k range from 1 to n. From the forms ω_i, ω_{ij}, Ω_{ij} we construct by exterior multiplication differential forms of higher degree. If we change the frames $Pe_1 \cdots e_n$ into the frames $Pe_1^* \cdots e_n^*(e_n^* = e_n)$ according to the equations (1), where $a_{\alpha\beta}$ are differentiable functions of the local parameters, and denote by $\overset{*}{\omega}_i$, $\overset{*}{\omega}_{ij}$, $\overset{*}{\Omega}_{ij}$ the forms constructed from $Pe_1^* \cdots e_n^*$ as the same forms without asterisks are constructed from the frames $Pe_1 \cdots e_n$, we shall have

$$
\begin{aligned}
\overset{*}{\omega}_{\alpha n} &= \sum_{\beta} a_{\alpha\beta}\,\omega_{\beta n}\,, \\
\overset{*}{\Omega}_{\alpha\beta} &= \sum_{\rho,\sigma=1}^{n-1} a_{\alpha\rho}\,a_{\beta\sigma}\,\Omega_{\rho\sigma}\,, \\
\overset{*}{\Omega}_{\alpha n} &= \sum_{\beta} a_{\alpha\beta}\,\Omega_{\beta n}\,, \\
\overset{*}{\Omega}_{nn} &= \Omega_{nn}\,.
\end{aligned}
\tag{3}
$$

A differential form constructed from the frames $Pe_1 \cdots e_n$ will be a differential form in M^{2n-1}, if it remains invariant under the transformations (1), (3).

To apply this remark, let us put

$$
\begin{aligned}
\Phi_k &= \sum \epsilon_{\alpha_1\cdots\alpha_{n-1}}\,\Omega_{\alpha_1\alpha_2} \cdots \Omega_{\alpha_{2k-1}\alpha_{2k}}\,\omega_{\alpha_{2k+1}n} \cdots \omega_{\alpha_{n-1}n}\,, \\
\Psi_k &= 2(k+1) \sum \epsilon_{\alpha_1\cdots\alpha_{n-1}}\,\Omega_{\alpha_1\alpha_2} \cdots \Omega_{\alpha_{2k-1}\alpha_{2k}}\,\Omega_{\alpha_{2k+1}n}\,\omega_{\alpha_{2k+2}n} \cdots \omega_{\alpha_{n-1}n}\,,
\end{aligned}
\tag{4}
$$

where $\epsilon_{\alpha_1 \cdots \alpha_{n-1}}$ is the Kronecker index which is equal to $+1$ or -1 according as $\alpha_1, \cdots, \alpha_{n-1}$ constitute an even or odd permutation of $1, \cdots, n-1$, and is otherwise zero, and where the summation is extended over all the indices α_1, $\cdots, \alpha_{n-1}$. These forms are defined for $k = 0, 1, \cdots, \left[\frac{n}{2}\right] - 1$, where $\left[\frac{n}{2}\right]$ denotes the largest integer $\leqq \frac{n}{2}$. Furthermore, when n is odd, $\Phi_{[\frac{1}{2}n]}$ is also defined. It will be convenient to define by convention

$$
\Psi_{-1} = \Psi_{[\frac{1}{2}n]} = 0. \tag{5}
$$

Under the transformations (1), (3) each of the forms in (4) is multiplied by the value of the determinant $| a_{\alpha\beta} |$, which is $+1$. Hence they are differential forms in M^{2n-1}. We remark that Φ_k is of degree $n-1$ and Ψ_k is of degree n. When n is even, they reduce to the forms of the same notation introduced in our previous paper.

The exterior derivative $d\Phi_k$ is a differential form in M^{2n-1}, and is equal to

$$
\begin{aligned}
d\Phi_k = k \sum & \epsilon_{\alpha_1\cdots\alpha_{n-1}}\,d\Omega_{\alpha_1\alpha_2}\,\Omega_{\alpha_3\alpha_4} \cdots \Omega_{\alpha_{2k-1}\alpha_{2k}}\,\omega_{\alpha_{2k+1}n} \cdots \omega_{\alpha_{n-1}n} \\
&+ (n-2k-1) \sum \epsilon_{\alpha_1\cdots\alpha_{n-1}}\,\Omega_{\alpha_1\alpha_2} \cdots \Omega_{\alpha_{2k-1}\alpha_{2k}}\,d\omega_{\alpha_{2k+1}n}\,\omega_{\alpha_{2k+2}n} \cdots \omega_{\alpha_{n-1}n}\,.
\end{aligned}
$$

In substituting the expressions for $d\Omega_{\alpha_1\alpha_2}$, $d\omega_{\alpha_{2k+1}n}$ into this equation, the terms involving $\omega_{\alpha\beta}$ will cancel each other, because $d\Phi_k$ is a differential form in M^{2n-1}.

Hence we immediately get

$$d\Phi_k = \Psi_{k-1} + \frac{n - 2k - 1}{2(k + 1)} \Psi_k. \tag{6}$$

Solving for Ψ_k, we get

$$\Psi_k = d\Theta_k, \qquad k = 0, 1, \cdots, \left[\frac{n}{2}\right] - 1, \tag{7}$$

where

$$\Theta_k = \sum_{\lambda=0}^{k} (-1)^{k-\lambda} \frac{(2k + 2) \cdots (2\lambda + 2)}{(n - 2\lambda - 1) \cdots (n - 2k - 1)} \Phi\lambda, \qquad k = 0, 1, \cdots, \left[\frac{n}{2}\right] - 1. \tag{8}$$

If n is even, say $= 2p$, then we have

$$d\Theta_{p-1} = \Psi_{p-1}$$

where

$$\Psi_{p-1} = n \sum \epsilon_{\alpha_1 \cdots \alpha_{n-1}} \Omega_{\alpha_1\alpha_2} \cdots \Omega_{\alpha_{n-1}n} = \sum \epsilon_{i_1 \cdots i_n} \Omega_{i_1 i_2} \cdots \Omega_{i_{n-1}i_n}.$$

If n is odd, say $= 2q + 1$, then

$$d\Theta_{q-1} = \Psi_{q-1}.$$

But in this case we have also

$$d\Phi_q = \Psi_{q-1},$$

so that

$$d(\Theta_{q-1} - \Phi_q) = 0.$$

We define*

$$\Pi = \begin{cases} \dfrac{1}{\pi^p} \displaystyle\sum_{\lambda=0}^{p-1} (-1)^\lambda \frac{1}{1 \cdot 3 \cdots (2p - 2\lambda - 1) \cdot 2^{p+\lambda} \cdot \lambda!} \Phi_\lambda, & \text{if } n = 2p \text{ is even}, \\ \dfrac{1}{2^{2q+1} \pi^q q!} \displaystyle\sum_{\lambda=0}^{q} (-1)^{\lambda+1} \binom{q}{\lambda} \Phi_\lambda, & \text{if } n = 2q + 1 \text{ is odd}, \end{cases} \tag{9}$$

or, for a formula covering both cases,

$$\Pi = \frac{(-1)^n}{2^n \pi^{\frac{1}{2}(n-1)}} \sum_{\lambda=0}^{[\frac{1}{2}(n-1)]} (-1)^\lambda \frac{1}{\lambda! \Gamma(\frac{1}{2}(n - 2\lambda + 1))} \Phi_\lambda, \tag{9a}$$

* Our present form Ω differs, in the case of even n, from the corresponding one in our previous paper by a sign. There are several reasons which indicate that the present choice is the appropriate one.

and

$$(10)\quad \Omega = \begin{cases} (-1)^p \dfrac{1}{2^{2p}\pi^p p!} \sum \epsilon_{i_1 \cdots i_n} \Omega_{i_1 i_2} \cdots \Omega_{i_{n-1} i_n}, & \text{if } n = 2p \text{ is even} \\ 0 & , \text{if } n \text{ is odd.} \end{cases}$$

Our foregoing relations can then be summarized in the formula

$$(11)\qquad -d\Pi = \Omega.$$

We remark that Π is a differential form of degree $n - 1$ in M^{2n-1}.

Over a simplicial chain of dimension $n - 1$ in M^{2n-1} whose simplexes are covered by coordinate neighborhoods of M^{2n-1} the integral of Π is defined.

§2. Remarks on the Formula of Allendoerfer-Weil

As we have shown before, the formula (11) leads immediately to a proof of the formula of Allendoerfer-Weil. We shall, however, add here a few remarks.

Let O be a point of R^n, and let $O\mathfrak{e}_1^0 \cdots \mathfrak{e}_n^0$ be a frame with origin at O. A point P of R^n sufficiently near to O is determined by the direction cosines λ^i (referred to $O\mathfrak{e}_1^0 \cdots \mathfrak{e}_n^0$) of the tangent of the geodesic joining O to P and the geodesic distance $s = OP$. The coordinates x^i of P defined by

$$(12)\qquad x^i = s\lambda^i$$

are called the normal coordinates. In a neighborhood of O defined by $s \leqq R$ we shall employ s, λ^i to be the local coordinates, where

$$(13)\qquad \sum_i (\lambda^i)^2 = 1.$$

As the components of a vector $\mathfrak{v}$ through P we shall take the components referred to $O\mathfrak{e}_1^0 \cdots \mathfrak{e}_n^0$ of the vector at O obtained by transporting $\mathfrak{v}$ parallelly along the geodesic OP.

In the neighborhood $s \leqq R$ of O let a field of unit vectors $\mathfrak{v}$ be given, whose components are differentiable functions of the normal coordinates x^i, except possibly at O. The forms Φ_k, $k \geqq 1$, being at least of degree two in dx^i, there exists a constant M such that

$$\left| \int_S \Phi_k \right| < Ms, \qquad k \geqq 1,$$

where S is the geodesic hypersphere of radius s about O. Let I be the index of the vector field at O, which is possibly a singular point. By Kronecker's formula we have

$$(14)\qquad I = \frac{1}{O_{n-1}} \int_S \omega_{1n} \cdots \omega_{n-1,n},$$

where O_{n-1} denotes the area of the unit hypersphere of dimension $n-1$ and is given by

$$O_{n-1} = \frac{2\pi^{\frac{1}{2}n}}{\Gamma(\frac{1}{2}n)} . \tag{15}$$

It follows that there exists a constant M_1 such that

$$\left| I - (-1)^n \int_S \Pi \right| < M_1 s$$

or that

$$I = (-1)^n \lim_{s \to 0} \int_S \Pi . \tag{16}$$

Let the Riemannian manifold R^n be closed. It is well-known and is also easy to prove directly that it is possible to define in R^n a continuous vector field with a finite number of singular points. Draw about each singular point a small geodesic hypersphere. The vector field at points not belonging to the interior of these geodesic hyperspheres defines a chain in M^{2n-1} over which Ω can be integrated. From (11) and (16) we get, by applying the formula of Stokes,

$$\int_{R^n} \Omega = (-1)^n I , \tag{17}$$

where I is the sum of indices of the vector field. Hence the sum of indices of the singular points of a vector field is independent of the choice of the field, provided that their number is finite. By the construction of a particular vector field, as was done by Stiefel and Whitney [2], we get the formula

$$\int_{R^n} \Omega = (-1)^n I = (-1)^n \chi(R^n), \tag{18}$$

where $\chi(R^n)$ is the Euler-Poincaré characteristic of R^n. In particular, it follows that $\chi(R^n) = 0$ if n is odd.

The same idea can be applied to derive the formula of Allendoerfer-Weil for differentiable polyhedra. Let P^n be a differentiable polyhedron whose boundary ∂P^n is a differentiable submanifold imbedded in R^n. Let ∂P^n be orientable and therefore two-sided. To each point of ∂P^n we attach the inner unit normal vector to ∂P^n, the totality of which defines a submanifold of dimension $n-1$ in M^{2n-1}. The integral of Π over this submanifold we shall denote simply by $\int_{\partial P^n} \Pi$. Then the formula of Allendoerfer-Weil for a differentiable polyhedron P^n is

$$\int_{P^n} \Omega = -\int_{\partial P^n} \Pi + \chi'(P^n), \tag{19}$$

where $\chi'(P^n)$ is the inner Euler-Poincaré characteristic of P^n

To prove the formula (19), we notice that the field of unit normal vectors on ∂P^n can be extended continuously into the whole polyhedron P^n, with the possible exception of a finite number of singular points. Application of the formula of Stokes gives then

$$\int_{P^n} \Omega = -\int_{\partial P^n} \Pi + (-1)^n J,$$

where J is the sum of indices at these singular points. That $J = (-1)^n \chi'(P^n)$ follows from a well-known theorem in topology [3]. It would also be possible to deduce this theorem if we carry out the construction of Stiefel-Whitney for polyhedra and verify in an elementary way that $J = (-1)^n\chi'(P^n)$ for a particular vector field.

§3. A New Integral Formula

Let R^m be a closed orientable differentiable (of class $\geqq 3$) submanifold of dimension $m \leqq n - 2$ imbedded in R^n. The unit normal vectors to R^m at a point of R^m depend on $n - m - 1$ parameters and their totality defines a submanifold of dimension $n - 1$ in M^{2n-1}. Denote by $\int_{R^m} \Pi$ the integral of Π over this submanifold. Our formula to be proved is then

$$-\int_{R^m} \Pi = \chi(R^m), \tag{20}$$

where the right-hand member stands for the Euler-Poincaré characteristic of R^m, which is zero if m is odd.

As a preparation to the proof we need the formulas for the differential geometry of R^m imbedded in R^n. At a point P of R^m we choose the frames $Pe_1 \cdots e_n$ such that $e_1, \cdots, e_m$ are the tangent vectors to R^m. We now restrict ourselves on the submanifold R^m and agree on the following ranges of indices

$$1 \leqq \alpha, \beta \leqq m, \qquad m + 1 \leqq r, s \leqq n, \qquad 1 \leqq A, B \leqq n - 1.$$

By our choice of the frames we have

$$\omega_r = 0,$$

and hence, by exterior differentiation,

$$\sum_\alpha \omega_{r\alpha}\, \omega_\alpha = 0$$

which allows us to put

$$\omega_{r\alpha} = \sum_\beta A_{r\alpha\beta}\, \omega_\beta \tag{21}$$

with

$$A_{r\alpha\beta} = A_{r\beta\alpha}. \tag{22}$$

Consequently, the fundamental formulas for the Riemannian Geometry on R^m, as induced by the Riemannian metric of R^n, are

$$(23)\qquad \begin{aligned} d\omega_\alpha &= \sum_\beta \omega_\beta \omega_{\beta\alpha}, \\ d\omega_{\alpha\beta} &= \sum_{\gamma=1}^{m} \omega_{\alpha\gamma}\omega_{\gamma\beta} + \tilde{\Omega}_{\alpha\beta}, \end{aligned}$$

where

$$(24)\qquad \tilde{\Omega}_{\alpha\beta} = \Omega_{\alpha\beta} + \sum_r \omega_{\alpha r}\omega_{r\beta}$$

To evaluate the integral on the left-hand side of (20) we introduce a differentiable family of frames $P\mathfrak{a}_1 \cdots \mathfrak{a}_n$ in a neighborhood of R^m, satisfying the condition that $\mathfrak{a}_\alpha = \mathfrak{e}_\alpha$ and that exactly one of the frames has the origin P. The relation between the vectors $\mathfrak{a}_{m+1}, \cdots, \mathfrak{a}_n$ and $\mathfrak{e}_{m+1}, \cdots, \mathfrak{e}_n$ is then given by the equations

$$(25)\qquad \mathfrak{e}_r = \sum_s u_{rs}\mathfrak{a}_s,$$

where u_{rs} are the elements of a proper orthogonal matrix. In particular, the quantities $u_{nr} = u_r$ may be regarded as local coordinates of the vector $\mathfrak{e}_n$ with respect to this family of frames. We now get all the normal vectors to R^m at P by letting u_r vary over all values such that $\sum_r (u_r)^2 = 1$. The forms $\omega_{n\alpha}$, ω_{nr} which occur in Π can be calculated according to the formulas

$$(26)\qquad \begin{aligned} \omega_{n\alpha} &= d\mathfrak{e}_n \cdot \mathfrak{e}_\alpha = \sum u_r \theta_{r\alpha}, \\ \omega_{nr} &= d\mathfrak{e}_n \cdot \mathfrak{e}_r = \sum_s du_s \cdot u_{rs} + \sum_{s,t=m+1}^{n} u_s u_{rt} \theta_{st}, \end{aligned}$$

where the product of vectors is the scalar product and where we define

$$(27)\qquad \theta_{ij} = d\mathfrak{a}_i \cdot \mathfrak{a}_j.$$

It is evident that

$$\Phi_k = 0, \qquad 2k > m.$$

For $k \leqq m/2$ we have by definition

$$\Phi_k = \sum \epsilon_{A_1 \cdots A_{n-1}} \Omega_{A_1 A_2} \cdots \Omega_{A_{2k-1}A_{2k}} \omega_{A_{2k+1}n} \cdots \omega_{A_{n-1}n}.$$

Each term of this sum is of degree m in the differentials of the local coordinates on R^m and of degree $n - m - 1$ in the differentials du_r. It follows that the non-vanishing terms are the terms where the indices $m + 1, \cdots, n - 1$ occur among $A_{2k+1}, \cdots, A_{n-1}$. We can therefore write

$$\begin{aligned} \Phi_k &= (-1)^{n-1} \frac{(n-2k-1)!}{(m-2k)!} \sum \epsilon_{\alpha_1 \cdots \alpha_m} \Omega_{\alpha_1\alpha_2} \cdots \Omega_{\alpha_{2k-1}\alpha_{2k}} \omega_{n\alpha_{2k+1}} \cdots \\ &\qquad\qquad \omega_{n\alpha_m} \omega_{n,m+1} \cdots \omega_{n,n-1} \\ &= (-1)^{n-m-1} \frac{(n-2k-1)!}{(m-2k)!} \sum \epsilon_{\alpha_1 \cdots \alpha_m} \Omega_{\alpha_1\alpha_2} \cdots \Omega_{\alpha_{2k-1}\alpha_{2k}} \Big(\sum_r u_r \theta_{\alpha_{2k+1} r}\Big) \cdots \\ &\qquad\qquad \Big(\sum_r u_r \theta_{\alpha_m r}\Big) \Lambda_{n-m-1}, \end{aligned}$$

where Λ_{n-m-1} is the surface element of a unit hypersphere of dimension $n - m - 1$.

The integration of Φ_k over R^m is then carried out by iteration. In fact, we shall keep a point of R^m fixed and integrate over all the unit normal vectors through that point. This leads us to the consideration of integrals of the form

$$\int u_{m+1}^{\lambda_{m+1}} \cdots u_n^{\lambda_n} \Lambda_{n-m-1}$$

over the unit hypersphere of dimension $n - m - 1$. It is clear that the integral is not zero, only when all the exponents $\lambda_{m+1}, \cdots, \lambda_n$ are even. But for the integrals obtained from Φ_k we have $\sum \lambda_r = m - 2k$. It follows that, if m is odd, we shall have

$$\int_{R^m} \Phi_k = 0, \qquad 0 \leqq k \leqq \frac{m}{2},$$

and hence

$$\int_{R^m} \Pi = 0.$$

This proves the formula (20) for the case that m is odd.

More interesting is naturally the case that m is even, which we are going to suppose from now on. It was proved that [4]

$$\int u_{m+1}^{2\lambda_{m+1}} \cdots u_n^{2\lambda_n} \Lambda_{n-m-1} = \frac{2\lambda_{m+1}) \cdots 2\lambda_n) O_{n-m-1}}{(n-m)(n-m+2) \cdots (n-m+2\lambda_{m+1}+\cdots+2\lambda_n-2)}, \tag{28}$$

where the symbol in the numerator is defined by

$$0) = 1, \quad 2\lambda) = 1.3 \cdots (2\lambda - 1). \tag{29}$$

To evaluate the integral of Φ_k over R^m we have to expand the product

$$\left(\sum_r u_r \theta_{\alpha_{2k+1} r}\right) \cdots \left(\sum_r u_r \theta_{\alpha_m r}\right).$$

We introduce the notation

$$\Delta(k; \lambda_{m+1}, \cdots, \lambda_n) = \sum \epsilon_{\alpha_1 \cdots \alpha_m} \Omega_{\alpha_1 \alpha_2} \cdots \Omega_{\alpha_{2k-1} \alpha_{2k}} \bigstar \tag{30}$$

where the last symbol stands for a product of θ's, whose first indices are $\alpha_{2k+1}, \cdots, \alpha_m$ respectively and whose second indices are respectively $2\lambda_{m+1}(m+1)$'s, $2\lambda_{m+2}(m+2)$'s, and finally $2\lambda_n$ n's. Let it be remembered that $\Delta(k; \lambda_{m+1}, \cdots, \lambda_n)$ is a differential form of degree m in R^m. Expanding Φ_k and using (28), we shall get

$$(31)\quad \int_{R^m} \Phi_k = (-1)^{n-1} \frac{(n-2k-1)!\,O_{n-m-1}}{2^{\frac{1}{2}m-k}(n-m)(n-m+2)\cdots(n-2k-2)} \cdot \sum_{\lambda_{m+1}+\cdots+\lambda_n=\frac{1}{2}m-k} \frac{1}{\lambda_{m+1}!\cdots\lambda_n!} \int_{R^m} \Delta(k;\lambda_{m+1},\cdots,\lambda_n),$$

where the summation is extended over all $\lambda_r \geqq 0$, whose sum is $\frac{m}{2} - k$.

It is now to be remarked that for the curvature forms $\tilde{\Omega}_{\alpha\beta}$ of the Riemannian metric on R^m we have to substitute $\theta_{\alpha r}$ for $\omega_{\alpha r}$ in the expressions (24). $\tilde{\Omega}$ being the form on R^m whose integral over R^m is equal to the Euler-Poincaré characteristic $\chi(R^m)$ by the Allendoerfer-Weil formula, we have

$$\tilde{\Omega} = (-1)^{\frac{1}{2}m} \frac{1}{2^m \pi^{\frac{1}{2}m}(\frac{1}{2}m)!} \sum \epsilon_{\alpha_1\cdots\alpha_m}(\Omega_{\alpha_1\alpha_2} - \sum \theta_{\alpha_1 r}\theta_{\alpha_2 r}) \cdots (\Omega_{\alpha_{m-1}\alpha_m} - \sum \theta_{\alpha_{m-1} r}\theta_{\alpha_m r})$$

or, by expansion,

$$(32)\quad \tilde{\Omega} = \frac{1}{2^m \pi^{\frac{1}{2}m}} \sum_{k=0}^{\frac{1}{2}m} (-1)^k \frac{1}{k!} \sum_{\lambda_{m+1}+\cdots+\lambda_n=\frac{1}{2}m-k} \frac{1}{\lambda_{m+1}!\cdots\lambda_n!} \Delta(k;\lambda_{m+1},\cdots,\lambda_n).$$

By a straightforward calculation which we shall omit here, we get from (9), (31), (32), and (18) the desired formula (20).

So far we have assumed that $m \leqq n-2$, that is, that R^m is not a hypersurface of R^n. In case $m = n-1$ the unit normal vectors of $R^{n-1} = R^m$ in R^n are, under our present assumptions concerning orientability, divided into two disjoint families. It is possible to maintain the formula (20) by making suitable conventions. In fact, we suppose that the integrals $\int_{R^{n-1}} \Pi$ over the families of inward and outward unit normal vectors are taken over the oppositely oriented manifold R^{n-1}. Then we have

$$\int_{(R^{n-1})^-} \Pi = (-1)^{n-1} \int_{(R^{n-1})^+} \Pi,$$

where the integrals at the left and right hand sides are over the families of inward and outward normals respectively. If n is even, we have

$$\int_{R^{n-1}} \Pi = \int_{(R^{n-1})^-} \Pi + \int_{(R^{n-1})^+} \Pi = 0.$$

If n is odd, we have

$$\int_{R^{n-1}} \Pi = 2\int_{(R^{n-1})^-} \Pi = -\chi(R^{n-1}).$$

Both cases can be considered as included in the formula (20). In particular, if n is odd and if R^{n-1} is the boundary ∂P^n of a polyhedron P^n, we have also, by (19),

$$\int_{(R^{n-1})^-} \Pi = \chi'(P^n).$$

Comparing the two equations, we get

$$-\chi(\partial P^n) = 2\chi'(P^n),$$

which asserts that the inner Euler-Poincaré characteristic of a polyhedron in an odd-dimensional manifold is $-\frac{1}{2}$ times the Euler-Poincaré characteristic of its boundary, a well-known result in the topology of odd-dimensional manifolds.

It is interesting to remark in passing that, so far as the writer is aware, the formula (20) seems not known even for the Euclidean space.

§4. Fields of Normal Vectors

We consider the case that R^{2n} is an even-dimensional orientable Riemannian manifold of class $\geqq 3$ and R^n a closed orientable submanifold of the same class imbedded in R^{2n}. By considering normal vector fields over R^n, Whitney [5] has defined a topological invariant of R^n in R^{2n}, which is the sum of indices at the singular points of a normal vector field (with a finite number of singular points) over R^n. Let us denote by ψ this invariant of Whitney.

To prepare for the study of this invariant we make use of the discussions at the beginning of §3. To each point P of R^n we attach the frames $P\mathfrak{e}_1 \cdots \mathfrak{e}_{2n}$ such that $\mathfrak{e}_1, \cdots, \mathfrak{e}_n$ are tangent vectors to R^n at P. Then we have, in particular,

$$d\omega_{ij} = \sum \omega_{ik}\omega_{kj} + \Theta_{ij}, \tag{33}$$

where

$$\Theta_{ij} = \Omega_{ij} - \sum_{\alpha=1}^{n} \omega_{i\alpha}\,\omega_{j\alpha} \tag{34}$$

the indices i, j running from $n + 1$ to $2n$. The differential forms Θ_{ij} are exterior quadratic differential forms depending on the imbedding of R^n in R^{2n}. They give what is essentially known as the Gaussian torsion of R^n in R^{2n}. We put, similar to (10),

$$\Theta = \begin{cases} (-1)^p \dfrac{1}{2^{2p}\,\pi^p\,p!} \sum \epsilon_{i_1 \cdots i_n} \Theta_{i_1 i_2} \cdots \Theta_{i_{n-1} i_n}, & \text{if } n = 2p \text{ is even}, \\ 0, & \text{if } n \text{ is odd}. \end{cases} \tag{35}$$

With these preparations we are able to state the following theorems:

1. *If R^n is a closed orientable submanifold imbedded in an orientable Riemannian manifold R^{2n}, the Whitney invariant ψ is given by*

$$\psi = \int_{R^n} \Theta. \tag{36}$$

2. *It is always possible to define a continuous normal vector field over a closed orientable odd-dimensional differentiable submanifold (of class* $\geqq 3$*) imbedded in an orientable differentiable manifold of twice its dimension.*

The first theorem can be proved in the same way as the formula of Allendoerfer-Weil. We shall give a proof of the second theorem.

For this purpose we take a simplicial decomposition of our submanifold R^n and denote its simplexes by σ_i^n, $i = 1, \cdots, m$. We assume the decomposition to be so fine that each σ_i^n lies in a coordinate neighborhood of R^n. According to a known property on the decomposition of a pseudo-manifold [6], the simplexes σ_i^n can be arranged in an order, say $\sigma_1^n, \cdots, \sigma_m^n$, such that σ_k^n, $k < m$, contains at least an $(n - 1)$-dimensional side which is not incident to $\sigma_1^n, \cdots, \sigma_{k-1}^n$. We then define a continuous normal vector field by induction on k. It is obviously possible to define a continuous normal vector field over σ_1^n. Suppose that such a field is defined over $\sigma_1^n + \cdots + \sigma_{k-1}^n$. The simplex σ_k^n has in common with $\sigma_1^n + \cdots + \sigma_{k-1}^n$ at most simplexes of dimension $n - 1$ and there exists, when $k < m$, at least one boundary simplex of dimension $n - 1$ of σ_k^n which does not belong to $\sigma_1^n + \cdots + \sigma_{k-1}^n$. It follows that the subset of σ_k^n at which the vector field is defined is contractible to a point in σ_k^n. By a well-known exten sion theorem [7], the vector field can be extended throughout σ_k^n, $k < m$. In the final step $k = m$ the extension of the vector field throughout σ_m^n will lead possibly to a singular point in σ_m^n. Hence it is possible to define a continuous normal vector field over R^n with exactly one singular point, the index at which is equal to the Whitney invariant ψ. If n is odd, we have, by (36), $\psi = 0$, and the singular point can be removed. This proves our theorem.

INSTITUTE FOR ADVANCED STUDY, PRINCETON, NEW JERSEY, AND
TSING HUA UNIVERSITY, KUNMING, CHINA.

REFERENCES

[1] S. CHERN, A simple intrinsic proof of the Gauss-Bonnet formula for closed Riemannian manifolds, these Annals, Vol. 45, pp. 747–752 (1944).
Also C. B. ALLENDOERFER AND A. WEIL, The Gauss-Bonnet theorem for Riemannian polyhedra, Trans. Amer. Math. Soc. vol. 53, pp. 101–129 (1943).
[2] E. STIEFEL, Richtungsfelder und Fernparallelismus in n-dimensionalen Mannigfaltigkeiten, Comm. Math. Helv. Vol. 8, pp. 339–340 (1936). H. Whitney, On the topology of differentiable manifolds, University of Michigan Conference 1941, pp. 127–129.
[3] S. LEFSCHETZ, Topology 1930, p. 272.
[4] H. WEYL, On the volume of tubes, Amer. J. Math., Vol. 61, p. 465 (1939).
[5] H. WHITNEY, loc. cit., pp. 126–141.
[6] P. ALEXANDROFF UND H. HOPF, Topologie I, p. 550.
[7] P. ALEXANDROFF UND H. HOPF, loc. cit., p. 501.

ANNALS OF MATHEMATICS
Vol. 47, No. 1, January, 1946

CHARACTERISTIC CLASSES OF HERMITIAN MANIFOLDS

BY SHIING-SHEN CHERN

(Received July 10, 1945)

INTRODUCTION

In recent years the works of Stiefel,[1] Whitney,[2] Pontrjagin,[3] Steenrod,[4] Feldbau,[5] Ehresmann,[6] etc. have added considerably to our knowledge of the topology of manifolds with a differentiable structure, by introducing the notion of so-called fibre bundles. The topological invariants thus introduced on a manifold, called the characteristic cohomology classes, are to a certain extent susceptible of characterization, at least in the case of Riemannian manifolds,[7] by means of the local geometry. Of these characterizations the generalized Gauss-Bonnet formula of Allendoerfer-Weil[8] is probably the most notable example.

In the works quoted above, special emphasis has been laid on the sphere bundles, because they are the fibre bundles which arise naturally from manifolds with a differentiable structure. Of equal importance are the manifolds with a complex analytic structure which play an important rôle in the theory of analytic functions of several complex variables and in algebraic geometry. The present paper will be devoted to a study of the fibre bundles of the complex tangent vectors of complex manifolds and their characteristic classes in the sense of Pontrjagin. It will be shown that there are certain basic classes from which all the other characteristic classes can be obtained by operations of the cohomology ring. These basic classes are then identified with the classes obtained by generalizing Stiefel-Whitney's classes to complex vectors. In the sense of de Rham the cohomology classes can be expressed by exact exterior differential forms which are everywhere regular on the (real) manifold. It is then shown that, in case the manifold carries an Hermitian metric, these differential forms can be constructed from the metric in a simple way. This means that the characteristic classes are completely determined by the local structure of the Hermitian metric. This result also includes the formula of Allendoerfer-Weil and can be regarded as a generalization of that formula.

Concerning the relations between the characteristic classes of a complex manifold and an Hermitian metric defined on it, the problem is completely solved by the above results. It is to be remarked that corresponding questions for Rie-

[1] STIEFEL, [24]. The number in the bracket refers to the bibliography at the end of the paper.
[2] WHITNEY, [29], [30].
[3] PONTRJAGIN, [18], [19].
[4] STEENROD, [21], [22].
[5] FELDBAU, [12].
[6] EHRESMANN, [10].
[7] CHERN, [5], [6], [7].
[8] ALLENDOERFER-WEIL, [2].

mannian manifolds remain open. Roughly speaking, the difficulty in the real case lies in the existence of finite homotopy groups of certain real manifolds, namely the manifolds formed by the ordered sets of linearly independent vectors of a finite-dimensional vector space.

The paper is divided into five chapters. In Chapter I we consider the fibre bundles which include the bundles of tangent complex vectors of a complex manifold and which are called complex sphere bundles. To a given base space a complex sphere bundle can be defined by a continuous mapping of the base space into a complex Grassmann manifold and it is shown that this is the most general way of generating a complex sphere bundle. We take the Grassmann manifold to be that in a complex vector space of sufficiently high dimension and define a characteristic cohomology class in the base space to be the inverse image under this mapping of a cohomology class of the Grassmann manifold. We are therefore led to the study of the cocycles or cycles on a complex Grassmann manifold, a problem treated exhaustively by Ehresmann.[9] A close examination of Ehresmann's results is therefore made in Chapter II, in the light of the problems which concern us here. In fact, we are only interested in the cocycles of the Grassmann manifold which are of dimension not greater than the dimension of the base space. If the Grassmann manifold is that of the linear spaces of n (complex) dimensions in a linear vector space of $n + N$ dimensions, there are on it n basic cocycles such that all other cocycles of dimension $\leqq 2n$ can be obtained from them by operations of the cohomology ring. The cycles corresponding to these cocycles are determined and geometrically interpreted. In Chapter III we identify the images of these cocycles in the base space with the cocycles obtained by generalizing the Stiefel-Whitney invariants to complex vectors. A new definition of these cocycles is given, which is important for applications to differential geometry in the large. Chapter IV is devoted to the study of a complex manifold with an Hermitian metric. It is proved that the n basic cocycles in question can be characterized in a simple way in terms of differential forms constructed from the Hermitian metric. These results are then applied in Chapter V to the complex projective space with the elliptic Hermitian metric. Classical formulas of Cartan[10] and Wirtinger[11] are derived from our formulas as particular cases.

Chapter I

Complex Sphere Bundles and Their Imbedding

1. The complex sphere

Various definitions have been given of a fibre bundle. For definiteness we shall adopt the one of Steenrod[12] and follow his terminology. We are, however, going to restrict the kind of fibre bundles under consideration.

[9] Ehresmann, [8].
[10] Cartan, [3].
[11] Wirtinger, [31].
[12] Steenrod, [22].

Let $E(n; C)$ be a complex vector space of n dimensions,[13] whose vectors will be denoted by small German letters. In $E(n; C)$ suppose a positive definite Hermitian form be given, which, in terms of a suitable base, has the expression

$$\mathfrak{z}\bar{\mathfrak{z}} = \sum_{i=1}^{n} z^i \bar{z}^i, \tag{1}$$

where z^i are the components of the vector $\mathfrak{z}$ in terms of the base and the bar denotes the operation of taking the complex conjugate. A vector $\mathfrak{z}$ such that $\mathfrak{z}\bar{\mathfrak{z}} = 1$ is called a unit vector. The group of linear transformations

$$\mathfrak{z}^{*i} = \sum_{j=1}^{n} a_j^i \mathfrak{z}^j, \qquad i = 1, \cdots, n, \tag{2}$$

which leaves the form (1) unaltered is the unitary group and will be denoted by $U(n; C)$. We shall call the complex sphere $S(n; C)$ the manifold of all the unit vectors of $E(n; C)$. It is homeomorphic to the real sphere of topological dimension $2n - 1$. The letter C in these notations will be dropped, when there is no danger of confusion.

In this paper we shall be concerned with fibre bundles such that the fibres are homeomorphic to the complex sphere $S(n)$ and that the group in each fibre is the unitary group $U(n)$. Such a fibre bundle is called a *complex sphere bundle.*

The most important complex sphere bundle is obtained from the consideration of the complex tangent vectors of a complex manifold $M(n)$ of complex dimension n and topological dimension $2n$. By a complex manifold $M(n)$ we shall mean a connected Hausdorff space which satisfies the following conditions:

1) It is covered by a finite or denumerable set of neighborhoods each of which is homeomorphic to the interior of the polycylinder

$$|z^i| < 1, \qquad i = 1, \cdots, n,$$

in the space of n complex variables, so that z^i can be taken as local coordinates of $M(n)$.

2) In a region in which two local coordinate systems z^i and z^{*i} overlap the coordinates of the same point are connected by the relations

$$z^{*i} = f^i(z^1, \cdots, z^n), \tag{3}$$

where f^i are analytic functions.

It follows from this definition that the notions of $M(n)$ which are expressed in terms of local coordinates but which remain invariant under the transformations (3) have an intrinsic meaning in $M(n)$. This is in particular true of a *tangent vector* at a point, which we define in the usual way as an object which has n com-

[13] Throughout this paper we shall mean by dimension the complex dimension. The dimension of a manifold in the sense of topology will be called the topological dimension, which is twice the complex dimension. The dimensions of simplexes, chains, cycles, homology groups, etc., are understood in the sense of topology, so long as there is no danger of confusion.

ponents Z^i in each local coordinate system and whose components Z^i, Z^{*i} in two local coordinate systems z^i, z^{*i} are transformed according to the equations

$$Z^{*i} = \sum_{k=1}^{n} \frac{\partial z^{*i}}{\partial z^k} Z^k. \tag{4}$$

It is clear that the space of the tangent vectors at a point is homeomorphic to $E(n)$. We consider the non-zero tangent vectors and call two such vectors equivalent if their components Z^i, W^i with respect to the same local coordinate system satisfy the conditions

$$W^i = \rho Z^i,$$

where ρ is a positive real quantity. This relation remains unchanged under transformation of local coordinates. Also it is an equivalence relation in the sense of algebra, being reflexive, symmetric, and transitive. Hence the non-zero tangent vectors can be divided by means of this equivalence relation into mutually disjoint classes. We call such a class of non-zero tangent vectors a direction. With a natural topology the space of directions at a point is homeomorphic to the complex sphere $S(n)$. Furthermore, by using the so-called unitarian trick in group theory, it is easy to verify that the manifold of all directions at the points of $M(n)$ is a complex sphere bundle with $M(n)$ as the base space It will be called the tangent bundle of $M(n)$.

Although all the results in this chapter will be formulated for general complex sphere bundles, it is the particular case of the tangent bundle of a complex manifold that justifies the study of complex sphere bundles.

2. The Grassmann manifold and the imbedding theorems

Consider the space $E(n + N;C)$ and the linear subspaces of $E(n + N;C)$ of dimension n. The manifold of all such linear subspaces is called a Grassmann manifold and will be denoted by $H(n, N;C)$ or simply $H(n, N)$. It is of dimension nN. The unit vectors of the complex sphere $S(n + N)$ in $E(n + N)$, which belong to a linear subspace $E(n)$ of dimension n, constitute a complex sphere $S(n)$, to be denoted by $S(n + N) \cap E(n)$.

Now let B be a finite polyhedron in the sense of combinatorial topology and let f be a continuous mapping of B into $H(n, N)$. From the mapping f we can define a complex sphere bundle $\mathfrak{F}$ with B as base space as follows: $\mathfrak{F}$ consists of the points $(b, \mathfrak{v})$ of the topological product $B \times S(n + N)$ such that $\mathfrak{v} \in f(b) \cap S(n + N)$, and the projection π of $\mathfrak{F}$ onto B is defined by $\pi(b, \mathfrak{v}) = b$. It is easy to verify that $\mathfrak{F}$ is a complex sphere bundle over B, which we shall call the *induced bundle* over B.

The importance of the notion of complex sphere bundles induced by the mapping of the base space into a Grassmann manifold is justified by the following theorems:

THEOREM 1. *To every bundle $\mathfrak{F}$ of complex spheres $S(n)$ over a finite polyhedron B of topological dimension d there exists a continuous mapping f of B into $H(n, N)$ with $N \geq d/2$, such that $\mathfrak{F}$ is equivalent to the bundle induced by f.*

THEOREM 2. *Let $\mathfrak{F}_1$ and $\mathfrak{F}_2$ be two bundles of complex spheres $S(n)$ over a finite polyhedron B of topological dimension d induced by the mappings f_1, f_2 respectively of B into $H(n, N)$, $N \geqq d/2$. The bundles $\mathfrak{F}_1$ and $\mathfrak{F}_2$ are equivalent when and only when the mappings f_1 and f_2 are homotopic.*

Similar theorems for real sphere bundles are known.[14] It follows from these theorems that to a class of equivalent bundles of complex spheres $S(n)$ over a finite polyhedron B of topological dimension d corresponds a class of homotopic mappings of B into the Grassmann manifold $H(n, N)$, where N is an integer satisfying $2N \geqq d$. This class of mappings induces a homomorphism h of the cohomology groups of dimension $\leqq d$ of $H(n, N)$ into the cohomology groups of the same dimension of B. A cohomology class of B which is the image under h of a cohomology class of $H(n, N)$ is called a *characteristic cohomology class* or simply a *characteristic class* and each of its cocycles is called a *characteristic cocycle.*

3. Proofs of Theorems 1 and 2

The proofs of the Theorems 1 and 2 do not differ essentially from the real case. We shall therefore restrict ourselves to a brief description of the general procedure. We need the following two lemmas:

LEMMA 1. (Covering homotopy theorem)[15] *Let $\mathfrak{F}$ be a fibre space over a base space B, which is a compact metric space. Let S be a compact topological space and let I be the unit interval. Suppose a mapping $h(S \times I) \subset \mathfrak{F}$ be given having the property: There exists a mapping $H(S \times 0) \subset \mathfrak{F}$ such that*

$$h(p \times 0) = \pi H(p \times 0), \qquad p \in S,$$

where π is the projection of $\mathfrak{F}$ into B. Then there exists a mapping $H(S \times I) \subset \mathfrak{F}$ such that

$$h(p \times t) = \pi H(p \times t), \qquad p \in S, \qquad 0 \leqq t \leqq 1.$$

From Lemma 1 follows the lemma:

LEMMA 2 (Feldbau).[16] *Let $\mathfrak{F}$ be a fibre bundle over a compact metric base space B. If B is contractible to a point, then $\mathfrak{F}$ is equivalent to the topological product of B and one of its fibres F.*

To prove Theorem 1 we are going to define the mapping f whose existence was asserted by the theorem. We take a simplicial decomposition of B which is so fine that each simplex lies in a neighborhood, and denote by σ_i^k, $i = 1, \cdots, \alpha_k$, $k = 0, 1, \cdots, d$, its simplexes. We denote as usual by π the projection of $\mathfrak{F}$ onto B. Our purpose is to define a mapping $f(B) \subset H(n, H)$ and a mapping $f^*(\mathfrak{F}) \subset B \times S(n + N)$ such that

[14] WHITNEY, [29]; STEENROD, [22].

[15] HUREWICZ-STEENROD, [15]. The theorem is given in various papers, in slightly different versions.

[16] FELDBAU, [12].

(5) $$f^*(p) \in \pi(p) \times \{f(\pi(p)) \cap S(n+N)\}$$

and that for a fixed $\pi(p)$ the mapping $f^*(p)$ is a homeomorphism preserving the scalar product. The definition of these mappings is given by induction on the dimension of the simplexes of B. The images $f(\sigma_i^0) \in H(n, H)$, $i = 1, \cdots, \sigma_0$, are defined in an arbitrary way and it is clear how $f^*(p)$ can be defined for all $p \in \mathfrak{F}$ such that $\pi(p) = \sigma_i^0$, $i = 1, \cdots, \alpha_0$. We suppose the mappings be defined over the $(k-1)$-dimensional skeleton of B and consider any simplex σ^k of dimension k. We take a neighborhood U which contains σ^k and decompose the set $\pi^{-1}(U)$ into a topological product of U and a complex sphere $S_0(n)$. Then we can define n mappings $\varphi_i(\sigma^k) \subset \mathfrak{F}$, $i = 1, \cdots, n$, such that: 1) $\pi\varphi_i(p) = p$, $p \in \sigma^k$; 2) $\varphi_i(p)$, $\varphi_j(p)$, $i \neq j$, are orthogonal vectors on the complex sphere $\pi^{-1}(p)$. We proceed to define by induction $f^*(\varphi_i(p)) = p \times q_i$, which will satisfy the condition that q_i, q_j for $i \neq j$ are orthogonal vectors of $S(n+N)$. By hypothesis, $f^*(\varphi_1(p))$ is defined for all $p \in \partial\sigma^k$.[17] Since $\partial\sigma^k$ is topologically a sphere of topological dimension $k - 1 \leqq d - 1 \leqq 2N - 1 < 2(n+N) - 1$, which is the topological dimension of the complex sphere $S(n+N)$, and since $\pi(p)$, $p \in \varphi_1(\sigma^k)$, is the cell σ^k, it follows that $f^*(\varphi_1(p))$, $p \in \partial\sigma^k$, is contractible in $B \times S(n+N)$. This means that there is a continuous mapping $g(\partial\sigma^k \times t) \subset B \times S(n+N)$, $0 \leqq t \leqq 1$, such that the following conditions are satisfied: 1) $g(\partial\sigma^k \times 0)$ is a point; 2) $g(\partial\sigma^k \times 1)$ is identical with $f^*(\varphi_1(p))$. On the other hand, we can introduce in σ^k the "polar coordinates" ρ, p, where $0 \leqq \rho \leqq 1$ and $p \in \partial\sigma^k$. For a point of σ^k having the coordinates ρ, p we define

$$f^*(\varphi_1(\rho, p)) = g(p \times \rho).$$

Suppose now that

$$f^*(\varphi_1(p)) = p \times q_1, \cdots, f^*(\varphi_{i-1}(p)) = p \times q_{i-1}, \qquad p \in \sigma^k,$$

are defined, such that q_k, q_j, $k, j = 1, \cdots, i-1$, $k \neq j$, are orthogonal vectors of $S(n+N)$. To define $f^*(\varphi_i(p))$ we consider on $S(n+N)$ the complex spheres $S(n+N-i+1)$ whose vectors are orthogonal to $q_1, \cdots, q_{i-1}$. These complex spheres $S(n+N-i+1)$, depending on p, constitute a complex sphere bundle over the simplex σ^k. By Lemma 2, it is a topological product of σ^k and a complex sphere $S_0(n+N-i+1)$. By induction hypothesis, the boundary $\partial\sigma^k$ is mapped into $S_0(n+N-i+1)$, by means of the vectors $q_i \in S(n+N-i+1)(p)$. Since the topological dimension $k-1$ of $\partial\sigma^k$ is smaller than the topological dimension $2(n+N-i)+1$ of $S_0(n+N-i+1)$, the map is contractible and the mapping of $\partial\sigma^k$ can be extended continuously throughout σ^k. It follows that a mapping $h(\sigma^k) \subset S_0(n+N-i+1)$ and hence a mapping $h_1(\sigma^k) \subset S(n+N)$ can be defined such that

$$h_1(p) \in S(n+N-i+1)(p), \qquad p \in \sigma^k.$$

We then define $f^*(\varphi_i(p)) = p \times h_1(p) = p \times q_i$, $p \in \sigma^k$. Clearly the vector q_i is orthogonal to $q_1, \cdots, q_{i-1}$.

[17] We shall make use of the notation $\partial\sigma^k$ to denote both the combinatorial and the set-theoretical boundary of the simplex σ^k, as the meaning will be clear by context.

To complete the induction on the dimension k let $p^* \in \mathfrak{F}$ such that $\pi(p^*) = p \in \sigma^k$. Then p^* has n components $u_1, \cdots, u_n$ with respect to $\varphi_1(p), \cdots, \varphi_n(p)$. We define $f(p)$ to be the linear space of n dimensions of $E(n + N)$ which contains the complex sphere determined by $q_1, \cdots, q_n$ and

$$f^*(p^*) = p \times q,$$

where q belongs to $f(p) \cap S(n + N)$ and has the components $u_1, \cdots, u_n$ with respect to $q_1, \cdots, q_n$.

Thus our induction is complete and it is easily seen that the mappings f and f^* fulfill our desired conditions. It is also clear that the complex sphere bundle induced by the mapping $f(B) \subset H(n, N)$ is equivalent to $\mathfrak{F}$. This proves our Theorem 1.

Concerning Theorem 2 it is not difficult to prove that $\mathfrak{F}_1$ and $\mathfrak{F}_2$ are equivalent if f_1 and f_2 are homotopic. The converse is proved by defining a mapping $f(B \times I) \subset H(n, H)$, with $f(B \times 0)$ and $f(B \times 1)$ coinciding with the given mappings f_1 and f_2 respectively. Because of the equivalence of $\mathfrak{F}_1$ and $\mathfrak{F}_2$ a complex sphere bundle can be defined over $B \times I$ in an obvious way. The rest of the argument consists of defining the mapping $f(B \times I)$ by an extension process analogous to the proof of Theorem 1. We shall omit the details here.

CHAPTER II

STUDY OF THE COCYCLES ON A COMPLEX GRASSMANN MANIFOLD

1. Summary of some known results

Let $H(n, N)$ be the Grassmann manifold of n-dimensional linear subspaces in $E(n, N)$. Our main purpose in this chapter is to give a homology base for the cocycles of dimension $\leqq 2n$ of $H(n, N)$. It is to be remarked that, $H(n, N)$ being a manifold of topological dimension $2nN$, there corresponds to each cycle of dimension s a cocycle of dimension $2nN - s$, and vice versa.

There are two different ways to describe the cocycles of $H(n, N)$, which are both useful to our purpose.

To explain the first method let $0 \leqq \varphi(i) \leqq N$, $1 \leqq i \leqq n$, be a non-decreasing integral-valued function. Let L_i, $1 \leqq i \leqq n$, be a linear vector space of dimension $i + \varphi(i)$ in $E(n + N)$, such that

$$L_1 \subset L_2 \subset \cdots \subset L_n\,.$$

Let $Z(\varphi(i))$ be the set of all n-dimensional linear spaces $X(n)$ such that

$$\dim\,(X(n) \cap L_i) \geqq i, \qquad i = 1, \cdots, n,$$

where the notation in the parenthesis denotes the linear space common to $X(n)$ and L_i. $Z(\varphi(i))$ is called a Schubert variety in algebraic geometry. It is a pseudo-manifold of dimension $s = \sum_{i=1}^n \varphi(i)$ and carries an integral cycle of dimension $2s$ of $H(n, N)$. Concerning the significance of the Schubert varieties for the topology of Grassmann manifolds the following theorem was proved by Ehresmann:[18]

[18] EHRESMANN, [8], p. 418.

THEOREM 3. *The Grassmann manifold $H(n, N)$ has no torsion coefficients and has all its Betti numbers of odd dimension equal to zero. Its Betti number of dimension $2s$ is equal to the number of distinct non-decreasing integral-valued functions $\varphi(i)$, $1 \leqq i \leqq n$, such that $\sum_{i=1}^{n} \varphi(i) = s$. The integral cycles carried by the corresponding Schubert varieties $Z(\varphi(I))$ constitute a homology base for the Betti group of dimension $2s$.*

The second method to describe the cocycles of $H(n, N)$ is by means of differential forms. Let $X(n) \in H(n, H)$ and let $\mathfrak{e}_1, \cdots, \mathfrak{e}_n$ be n vectors in $X(n)$ such that

$$\mathfrak{e}_i \cdot \bar{\mathfrak{e}}_j = \delta_{ij}, \qquad 1 \leqq i, j \leqq n.$$

To these n vectors we add N further vectors $\mathfrak{e}_{n+1}, \cdots, \mathfrak{e}_{n+N}$ satisfying the conditions

$$\mathfrak{e}_A \cdot \bar{\mathfrak{e}}_B = \delta_{AB}, \qquad 1 \leqq A, B \leqq n + N. \tag{6}$$

When there is a differentiable family of the vectors $\mathfrak{e}_1, \cdots, \mathfrak{e}_{n+N}$, we put

$$\theta_{AB} = d\mathfrak{e}_A \cdot \bar{\mathfrak{e}}_B, \tag{7}$$

which are linear differential forms satisfying the conditions

$$\theta_{AB} + \bar{\theta}_{BA} = 0. \tag{8}$$

Among θ_{AB} the forms

$$\theta_{ir}, \qquad \bar{\theta}_{ir}, \qquad 1 \leqq i \leqq n, \qquad n + 1 \leqq r \leqq n + N,$$

constitute a set of $2nN$ linearly independent forms at each point of $H(n, N)$. Let Θ be a form in θ_{ir}, $\bar{\theta}_{ir}$ with constant coefficients. Θ is called an invariant form if it remains unchanged under the groups of transformations

$$\theta_{ir}^* = \sum_{j=1}^{n} a_{ij} \theta_{jr}, \tag{9}$$

$$\theta_{ir}^* = \sum_{s=n+1}^{n+N} b_{rs} \theta_{is}, \tag{10}$$

where (a_{ij}), (b_{rs}) are arbitrary unitary matrices. It is called exact, if $d\Theta = 0$. It is well-known that on a differentiable manifold of class two a cocycle with rational coefficients can be expressed by an exact differential form, and conversely. For convenience we shall therefore call an exact differential form a cocycle. Then we have the following theorem of E. Cartan:[19]

THEOREM 4. *Every invariant form of $H(n, N)$ is exact. The Betti number of dimension $2s$ of $H(n, N)$ is equal to the number of linearly independent (with constant coefficients) invariant differential forms of degree $2s$. The set of these forms constitutes a cohomology base of dimension $2s$.*

[19] CARTAN, [3]; EHRESMANN, [8], p. 409.

2. The basic forms

Let r be an integer between 1 and n. For reasons which will be clear later we shall be particularly interested in the n cycles Z_r, $r = 1, \cdots, n$, carried by the Schubert varieties defined by the functions

$$\begin{aligned} \varphi_r(i) &= N - 1, & i &= 1, \cdots, n - r + 1, \\ \varphi_r(i) &= N, & i &= n - r + 2, \cdots, n. \end{aligned} \tag{11}$$

The cycle Z_r is of dimension $2(Nn - n + r - 1)$. We shall find the invariant differential form which gives the cocycle of dimension $2(n - r + 1)$ corresponding to Z_r.

For this purpose we put

$$\Theta_{ij} = \sum_{s=n+1}^{n+N} \theta_{is}\,\theta_{sj}, \qquad 1 \leqq i, j \leqq n, \tag{12}$$

and

$$\Phi_r = \frac{1}{(2\pi\sqrt{-1})^{n-r+1}(n - r + 1)!} \sum \delta(i_1 \cdots i_{n-r+1}; j_1 \cdots j_{n-r+1}) \cdot \Theta_{i_1 j_1} \cdots \Theta_{i_{n-r+1} j_{n-r+1}}, \tag{13}$$

where $\delta(i_1 \cdots i_{n-r+1}; j_1 \cdots j_{n-r+1})$ is zero except when $j_1, \cdots, j_{n-r+1}$ form a permutation of $i_1, \cdots, i_{n-r+1}$, in which case it is $+1$ or -1 according as the permutation is even or odd, and where the summation is extended over all indices $i_1, \cdots, i_{n-r+1}$ from 1 to n. It is easy to verify that Φ_r is an invariant form on $H(n, N)$. Our problem is then solved by the following theorem:

THEOREM 5. *The invariant differential form Φ_r defines a cocycle of dimension $2(n - r + 1)$ on $H(n, N)$, which corresponds to the cycle Z_r in the sense that, for any cycle ζ of dimension $2(n - r + 1)$, the relation*

$$KI(\zeta_1, Z_r) = \int_\zeta \Phi_r \tag{14}$$

holds,[20] *whenever both sides are defined.*

To prove this theorem we notice that both sides of the equation (14) are linear in ζ, so that it is sufficient to prove (14) for the cycles of a homology base of dimension $2(n - r + 1)$. By Theorem 3, these are the cycles carried by the Schubert varieties $Z(\varphi(i))$ such that

$$\sum_{i=1}^{n} \varphi(i) = n - r + 1.$$

Since the function $\varphi(i)$ is non-decreasing, we must have

$$\varphi(1) = \cdots = \varphi(r - 1) = 0. \tag{15}$$

[20] The notation KI, due to Lefschetz, means the Kronecker index (or the intersection number).

Let ζ_k, $k = 1, \cdots, m$, be the cycles of a homology base of dimension $2(n - r + 1)$ defined in this way, and let ζ_1 be the cycle defined by

(16) $$\varphi(1) = \cdots = \varphi(r-1) = 0, \qquad \varphi(r) = \cdots = \varphi(n) = 1.$$

Then for the cycles ζ_k, $k \neq 1$, we must also have

$$\varphi(r) = 0.$$

It is therefore sufficient to prove that

(17) $$KI(\zeta_k, Z_r) = \int_{\zeta_k} \Phi_r, \qquad k = 1, \cdots, m,$$

for the cycles ζ_k which are well chosen so that both sides of (17) are defined.

Let us first assume that $k \neq 1$. By definition, there exists a fixed linear vector space $L(r)$ of dimension r such that any $X(n)$ of the Schubert variety carrying ζ_k satisfies the condition

$$\dim\,(X(n) \cup L(r)) \geqq r,$$

which means that

$$X(n) \supset L(r).$$

On the other hand, any $X(n)$ of the Schubert variety carrying Z_r has the property that it intersects a fixed linear space $L(N + n - r)$ of dimension $N + n - r$ in a linear space of dimension $\geqq n - r + 1$. We take in $E(n + N)$ a frame $\mathfrak{e}_1, \cdots, \mathfrak{e}_{n+N}$ and let $L(r)$ and $L(N + n - r)$ be spanned by the vectors $\mathfrak{e}_1, \cdots, \mathfrak{e}_r$ and $\mathfrak{e}_{r+1}, \cdots, \mathfrak{e}_{n+N}$ respectively. There is clearly no $X(n)$ which contains $\mathfrak{e}_1, \cdots, \mathfrak{e}_r$ and has in common with $L(N + n - r)$ a linear space of dimension $n - r + 1$, which means that

$$KI(Z_r, \zeta_k) = 0, \qquad k \neq 1.$$

To show that the integral on the right-hand side of (17) is also zero, we choose the frame $\mathfrak{e}_1, \cdots, \mathfrak{e}_{n+N}$ such that $\mathfrak{e}_1, \cdots, \mathfrak{e}_r$ belong to $L(r)$ and $\mathfrak{e}_1, \cdots, \mathfrak{e}_n$ belong to $X(n)$. It is obviously possible to choose $\mathfrak{e}_1, \cdots, \mathfrak{e}_r$ to be fixed. Under this choice we have

$$\Theta_{ij} = 0, \qquad 1 \leqq i, j \leqq r,$$

and hence

$$\Phi_r = 0.$$

Thus the relations (17) are proved for $k \neq 1$.

To define ζ_1 we take two linear vector spaces $L(r-1)$, $L(n+1)$ of dimensions $r - 1$, $n + 1$ respectively, such that $L(r-1) \subset L(n+1)$. An element $X(n)$ of ζ_1 is then defined by the conditions

$$L(r-1) \subset X(n) \subset L(n+1).$$

Let $\mathfrak{e}_1, \cdots, \mathfrak{e}_{n+N}$ be a frame in $E(n+N)$. We suppose the linear vector spaces in question to be so chosen that

$$\mathfrak{e}_1, \cdots, \mathfrak{e}_{r-1} \subset L(r-1),$$
$$\mathfrak{e}_1, \cdots, \mathfrak{e}_{n+1} \subset L(n+1),$$
$$\mathfrak{e}_r, \cdots, \mathfrak{e}_{n-1}, \mathfrak{e}_{n-1}, \cdots, \mathfrak{e}_{n+N} \subset L(n+N-r).$$

By this choice $L(n+1)$ and $L(n+N-r)$ have in common a linear vector space of dimension $n-r+1$, namely the one spanned by $\mathfrak{e}_r, \cdots, \mathfrak{e}_{n-1}, \mathfrak{e}_{n+1}$. It follows that Z_r and ζ_1 have in common only one $X(n)$, which is spanned by $\mathfrak{e}_1, \cdots, \mathfrak{e}_{n-1}, \mathfrak{e}_{n+1}$. This intersection is to be counted simply, and we have[21]

$$KI(Z_r, \zeta_1) = 1.$$

It now only remains to evaluate the integral in the right-hand side of (17) for $k = 1$. The linear vector spaces $L(r-1)$ and $L(n+1)$ being fixed, we choose a fixed frame $\mathfrak{a}_1, \cdots, \mathfrak{a}_{n+N}$ in $E(n+N)$ such that $\mathfrak{a}_1, \cdots, \mathfrak{a}_{r-1}$ belong to $L(r-1)$ and $\mathfrak{a}_1, \cdots, \mathfrak{a}_{n+1}$ to $L(n+1)$. If $X(n) \in \zeta_1$, we choose the frame $\mathfrak{e}_1, \cdots, \mathfrak{e}_{n+N}$, whose first n vectors belong to $X(n)$ such that

$$\text{(18)} \quad \begin{aligned} \mathfrak{e}_A &= \mathfrak{a}_A, & A &= 1, \cdots, r-1, n+2, \cdots, n+N, \\ \mathfrak{e}_B &= \sum_C u_{BC}\mathfrak{a}_C, & B, C &= r, \cdots, n+1, \end{aligned}$$

where u_{BC} are the elements of a unitary matrix. It follows that

$$\text{(19)} \quad \begin{aligned} \theta_{ki} &= d\mathfrak{e}_k \cdot \bar{\mathfrak{e}}_i = 0, \\ \theta_{ik} &= -\bar{\theta}_{ki} = 0, \end{aligned} \qquad k = n+2, \cdots, n+N, \quad i = 1, \cdots, n+N,$$

and hence that

$$\Theta_{ij} = \theta_{in+1}\theta_{n+1j}, \qquad 1 \leqq i, j \leqq n.$$

For simplicity we shall write $\mathfrak{e}$ for $\mathfrak{e}_{n+1}$, u_B for $u_{n+1,B}$, and θ_i for $\theta_{n+1,i}$. We remark that $X(n)$ is completely determined by the vector $\mathfrak{e}$, whose components with respect to $\mathfrak{a}_r, \cdots, \mathfrak{a}_{n+1}$ are $u_r, \cdots, u_{n+1}$. Our purpose therefore is to transform the form Φ_r into an exterior differential form in $u_r, \cdots, u_{n+1}$, from which the integration can be carried out. With this purpose in mind, we have

$$\text{(20)} \quad \begin{aligned} \theta_i &= d\mathfrak{e} \cdot \bar{\mathfrak{e}}_i = \left(\sum_{B=r}^{n+1} du_B \mathfrak{a}_B\right)\left(\sum_{C=r}^{n+1} \bar{u}_{iC}\bar{\mathfrak{a}}_C\right) \\ &= \sum_{B=r}^{n+1} du_B \bar{u}_{iB}, \qquad i = r, \cdots, n, \\ \theta_i &= 0, \qquad i = 1, \cdots, r-1, \end{aligned}$$

and

$$\Theta_{ij} = -\bar{\theta}_i\theta_j, \qquad i, j = r, \cdots, n,$$

all other Θ_{ij} being zero. It follows that Φ_r is equal to

[21] Cf. Ehresmann, [8], p. 421.

$$(21)\quad \frac{(2\pi\sqrt{-1})^{r-n-1}}{(n-r+1)!}\cdot\sum\delta(i_1\cdots i_{n-r+1};j_1\cdots j_{n-r+1})\theta_{i_1}\bar{\theta}_{j_1}\cdots\theta_{i_{n-r+1}}\bar{\theta}_{j_{n-r+1}}$$

$$= (-1)^{\frac{1}{2}(n-r)(n-r+1)}\frac{(n-r+1)}{(2\pi\sqrt{-1})^{n-r+1}}\theta_r\cdots\theta_n\bar{\theta}_r\cdots\bar{\theta}_n .$$

From now on we shall agree on the following ranges of indices:

$$\gamma \leqq \alpha, \beta \leqq n, \qquad \gamma \leqq i, j \leqq n+1.$$

From (20) and

$$\theta_{n+1} = \sum_j du_j \bar{u}_j$$

we get, by solving for du_i,

$$(22)\qquad du_i = \sum_\alpha u_{\alpha i}\theta_\alpha + u_i\theta_{n+1} .$$

We notice that

$$(23)\qquad \theta_{n+1} + \bar{\theta}_{n+1} = 0.$$

Consider now the form

$$(24)\quad \Psi_r = \sum_{k=0}^{n-r+1} du_r\cdots du_{r+k-1}\,du_{r+k-1}\cdots du_{n+1}\,d\bar{u}_r\cdots d\bar{u}_{r+k-1}\,d\bar{u}_{r+k+1}\cdots d\bar{u}_{n+1},$$

which we shall prove to differ from Φ_r by a numerical factor. In fact, it is easy to verify, by means of the fact that the matrix (u_{ij}) is unitary, that

$$\Psi_r = \theta_r\cdots\theta_n\bar{\theta}_r\cdots\bar{\theta}_n ,$$

and hence that

$$(25)\qquad \Phi_r = (-1)^{\frac{1}{2}(n-r)(n-r+1)}\frac{(n-r+1)!}{(2\pi\sqrt{-1})^{n-r+1}}\Psi_r .$$

To integrate Φ_r or Ψ_r over ζ_1 let us notice that $X(n)$ will remain unchanged if $\mathfrak{e}$ is replaced by the vector $e^{\sqrt{-1}\rho}\mathfrak{e}$, ρ being real. We can therefore normalize the coordinates $u_r, \cdots, u_{r+1}$ of $X(n)$ by assuming that u_{n+1} is real and positive. Then we have

$$\Psi_r = du_r\cdots du_n\,d\bar{u}_r\cdots d\bar{u}_n ,$$

which is to be integrated over the domain

$$u_{n+1}^2 + u_r\bar{u}_r + \cdots + u_n\bar{u}_n = 1, \qquad u_{n+1} > 0.$$

This integration is then easily achieved. In fact, we put

$$u_r = v_r + \sqrt{-1}\,w_r ,$$

$$\bar{u}_r = v_r - \sqrt{-1}\,w_r .$$

Then

$$\Psi_r = (-1)^{\frac{1}{2}(n-r)(n-r+1)}(-2\sqrt{-1})^{n-r+1}dv_r\,dw_r\cdots dv_n\,dw_n ,$$

and the integral is over the domain D:

$$u_{n+1}^2 + v_r^2 + w_r^2 + \cdots + v_n^2 + w_n^2 = 1, \quad u_{n+1} > 0.$$

But the integral

$$\int_D dw_r\, dv_r \cdots dw_n\, dv_n$$

is the volume of the domain bounded by the unit hypersphere of dimension $2n - 2r + 1$, which is equal to $\pi^{n-r+1}/(n - r + 1)!$. Hence we have

$$\int_{\zeta_1} \Psi_r = (-1)^{\frac{1}{2}(n-r)(n-r+1)} \frac{(2\pi\sqrt{-1})^{n-r+1}}{(n - r + 1)!},$$

and finally,

$$\int_{\zeta_1} \Phi_r = 1.$$

Our Theorem 5 is therefore completely proved.

3. The basis theorem

The importance of the invariant differential forms Φ_r, $1 \leqq r \leqq n$, lies in a theorem we proceed to prove, which asserts that every invariant differential form of degree $\leqq 2n$ in $H(n, N)$ is a polynomial in Φ_r with constant coefficients. The exact statement of our theorem is as follows:

THEOREM 6. *Every invariant differential form of degree $\leqq 2n$ in $H(n, N)$ is a polynomial in Φ_r, $1 \leqq r \leqq n$, with constant coefficients. If the form defines an integral cocycle on $H(n, N)$, the coefficients are integers.*

The theorem follows easily from the so-called first main theorem on vector invariants for the unitary group, which we state as follows:

LEMMA 3. *Let $\mathfrak{v}_1, \cdots, \mathfrak{v}_m$ be a set of vectors in $E(n)$ under transformations of the unitary group $U(n)$. Every integral rational invariant in the components of $\mathfrak{v}_k$, $1 \leqq k \leqq m$, is an integral rational function of the scalar products $\mathfrak{v}_i\bar{\mathfrak{v}}_k$, $1 \leqq i, k \leqq m$.*

It is known that[22] under the unimodular unitary group such an invariant is an integral rational function of the scalar products and of determinants of the form $[\mathfrak{v}_1 \cdots \mathfrak{v}_n]$ or $[\bar{\mathfrak{v}}_1 \cdots \bar{\mathfrak{v}}_n]$. But under a general unitary transformation of determinant $e^{\sqrt{-1}\alpha}$ the determinants $[\mathfrak{v}_1 \cdots \mathfrak{v}_n]$ and $[\bar{\mathfrak{v}}_1 \cdots \bar{\mathfrak{v}}_n]$ will be multiplied by $e^{\sqrt{-1}\alpha}$ and $e^{-\sqrt{-1}\alpha}$ respectively. It follows that an invariant will involve the determinants only in products of the form $[\mathfrak{v}_1 \cdots \mathfrak{v}_n]\cdot[\bar{\mathfrak{v}}_1' \cdots \bar{\mathfrak{v}}_n']$, which can however be expressed as a determinant of scalar products:

$$[\mathfrak{v}_1 \cdots \mathfrak{v}_n][\bar{\mathfrak{v}}_1' \cdots \bar{\mathfrak{v}}_n'] = \begin{vmatrix} \mathfrak{v}_1 \bar{\mathfrak{v}}_1' & \cdots & \mathfrak{v}_1 \bar{\mathfrak{v}}_n' \\ \cdots & \cdots & \cdots \\ \mathfrak{v}_n \bar{\mathfrak{v}}_1' & \cdots & \mathfrak{v}_n \bar{\mathfrak{v}}_n' \end{vmatrix}.$$

Thus the lemma is proved.

[22] WEYL, [28], p. 45.

To prove Theorem 6 let Ψ be an invariant differential form of degree $2s \leqq 2n$ in $H(n, N)$, which is therefore an exterior form in θ_{ir}, $\bar{\theta}_{ir}$, $1 \leqq i \leqq n$, $n+1 \leqq r \leqq n+N$, with constant coefficients. The form Ψ being in particular invariant under the transformation $\theta^*_{ir} = e^{\sqrt{-1}\alpha}\theta_{ir}$, it follows that Ψ, when reduced to its lowest terms, will contain in each term exactly s factors each of θ_{ir} and $\bar{\theta}_{ir}$. Let us fix our attention for the moment to the group (10). We take from Ψ all the terms of the form

$$\text{const. } \theta_{i_1 r_1} \cdots \theta_{i_s r_s} \bar{\theta}_{j_1 t_1} \cdots \bar{\theta}_{j_s t_s},$$

with a fixed set of the indices $i_1, \cdots, i_s, j_1 . \cdots, j_s$, and call their sum Ψ_1. Since the indices $i_1, \cdots, i_s, j_1, \cdots, j_s$ are now fixed, we shall drop them for simplicity.

Now it is well-known that there is an isomorphism between the ring of exterior forms and the ring of multilinear forms with alternating coefficients. To Ψ_1 corresponds, in the complex vector space of N dimensions, an alternating multilinear form of degree $2s$. Since Ψ_1 is invariant under the unitary group (10), the same is true of its corresponding alternating multilinear form. By our Lemma the latter is an integral rational function of the scalar products. It follows by the isomorphism that Ψ_1 can be expressed as a polynomial in sums of the form $\sum_r \theta_{ir}\bar{\theta}_{jr} = -\Theta_{ij}$. Consequently, Ψ is a polynomial in Θ_{ij}, $1 \leqq i, j \leqq n$, with constant coefficients.

Let us now put

$$P_r = \sum \Theta_{i_1 i_2}\Theta_{i_2 i_3} \cdots \Theta_{i_{n-r+1} i_1}, \qquad r = 1, \cdots, n. \tag{26}$$

By the same argument as above, we can prove that Ψ, being also invariant under the group (9), is a polynomial in P_r, $1 \leqq r \leqq n$, with constant coefficients. On the other hand, it is easy to show, by induction on r, that P_r is a polynomial in $\Phi_1, \cdots, \Phi_r$, with constant coefficients. Hence the first part of our theorem is proved.

To prove the second part of the theorem consider the products of the form

$$\Phi_n^{\lambda_1} \cdots \Phi_1^{\lambda_n}, \tag{27}$$

such that

$$\lambda_1 + 2\lambda_2 + \cdots + n\lambda_n = s. \tag{27a}$$

These forms constitute a basis for all invariant differential forms of degree $2s \leqq 2n$ on $H(n, N)$. Since $s \leqq n$, their number is equal to the number of partitions of s as a sum of integral summands. By Theorem 3 this is equal to the Betti number of dimension $2s$ of $H(n, N)$. It follows that the products in (27) are linearly independent, and that every invariant differential form of degree $2s$ of $H(n, N)$ representing an integral cocycle is equal to a linear combination of the products (27) with integral coefficients.

Chapter III

The Basic Characteristic Classes on a Complex Manifold

1. A second definition of the basic characteristic classes

Let M be a complex manifold of dimension n. We consider the complex sphere bundle defined from the tangent vectors of M and imbed it, according to Theorem 1, in a Grassmann manifold $H(n, N)$, $N \geqq n$, by means of a mapping of M into $H(n, N)$. It follows from Theorem 2 that the inverse image of a cohomology class of dimension $\leqq 2n$ of $H(n, N)$ induced by this mapping is an invariant of M (or rather of the analytic structure of M), which we have called a characteristic cohomology class of M. From Theorem 6 we see that of all the characteristic cohomology classes of M those which are inverse images of the cohomology classes of $H(n, N)$ containing the cocycles Φ_r, $1 \leqq r \leqq n$, play a particularly important rôle, because all the others can be obtained from them by operations of the cohomology ring. We therefore call these n classes the basic characteristic classes, the inverse image of the class containing Φ_r being the rth basic class.

Our first aim is to identify these basic classes with the classes obtained by generalizing to complex manifolds the well-known procedure of Stiefel-Whitney.[23]

In order to understand the situation we recall briefly the results of Stiefel-Whitney for real sphere bundles, emphasizing the differences between the real and complex cases. From a bundle of real spheres of dimension $n - 1$ over a polyhedron as base space Stiefel and Whitney considered the fibre bundle over the same base space whose fibre at each point is the manifold $V(n, r)$ of $r(1 \leqq r \leqq n)$ linearly independent points of the real sphere at this point. It was proved that all homology groups of dimension $<n - r$ of $V(n, r)$ vanish and that the homology group $H^{n-r}(V(n, r))$ of dimension $n - r$ of $V(n, r)$ is the free cyclic group or the cyclic group of order 2 according as the following condition is satisfied or not: $n - r$ is even or $r = 1$. To define a generator of $H^{n-r}(V(n, r))$ we take an ordered set of $r - 1$ mutually perpendicular unit vectors $\mathfrak{e}_1, \cdots, \mathfrak{e}_{r-1}$ in the Euclidean space E^n of dimension n which contains the sphere of dimension $n - 1$. The unit vector $\mathfrak{e}_r$ of E^n perpendicular to $\mathfrak{e}_1, \cdots, \mathfrak{e}_{r-1}$ describes a sphere of dimension $n - r$, which, when oriented, defines a cycle ζ_0^{n-r} belonging to one of the generating homology classes of $H^{n-r}(V(n, r))$. We then take a simplicial decomposition K of the base space such that each simplex of K belongs to a neighborhood. Let the simplexes of K be oriented. It is possible to define a continuous mapping of the $(n - r)$-dimensional skeleton of K into the fibre bundle such that each point is mapped into a point belonging to the fibre over it. Let σ^{n-r+1} be a simplex of dimension $n - r + 1$ of K. In a neighborhood containing σ^{n-r+1} the fibre bundle can be resolved into a topological product and can therefore be mapped into one fibre. Since the mapping is defined over the boundary $\partial\sigma^{n-r+1}$, we get a mapping of

[23] Stiefel, [24]; Whitney, [29].

a sphere of dimension $n - r$ into a fibre and hence an element h of $H^{n-r}(V(n, r))$. The delicate point is to get from this element h an integer or a residue class mod. 2. This is possible if a generating element of $H^{n-r}(V(n, r))$ is defined. When $H^{n-r}(V(n, r))$ is cyclic of order two, it has only one generating element, so that no further assumption is necessary. When $H^{n-r}(V(n, r))$ is a free cyclic group, we assume that a continuous field of generating elements of $H^{n-r}(V(n, r))$ can be defined over the whole M, which is possible if M is orientable. The element h is then equal to the generator of $H^{n-r}(V(n, r))$ so defined, multiplied by an integer or a residue class mod. 2. Taking this integer or the residue class mod. 2 as the value of a cochain for the simplex σ^{n-r+1}, we get an integral cochain or a cochain mod. 2. It was proved that the cochain is a cocycle and that its cohomology class is independent of the choice of the mapping from which it is defined. This cohomology class is the class of Stiefel-Whitney. It is to be remarked that the definition can be given under more general conditions, but we shall be satisfied with the above résumé.

The situation is simpler in the case of complex sphere bundles. From a bundle of complex spheres $S(n)$ we consider the fibre bundle $\mathfrak{F}^{(r)}$ over the same base space whose fibre at each point is the manifold $U(n, r)$ of $r(1 \leqq r \leqq n)$ linearly independent vectors in $E(n)$. It can be proved that[24] all homology groups of dimension $<2n - 2r + 1$ of $U(n, r)$ vanish and that the homology group $H^{2n-2r+1}(U(n, r))$ of dimension $2n - 2r + 1$ is a free cyclic group. To define a generator of $H^{2n-2r+1}(U(n, r))$ we take in $E(n)$ an ordered set of $r - 1$ mutually perpendicular unit vectors $\mathfrak{e}_1, \cdots, \mathfrak{e}_{r-1}$. The unit vector $\mathfrak{e}_r$ in $E(n)$ perpendicular to $e_1, \cdots, e_{r-1}$ describes a complex sphere in the $E(n - r + 1)$ perpendicular to $\mathfrak{e}_1, \cdots, \mathfrak{e}_{r-1}$. The complex sphere $S(n - r + 1)$ in $E(n - r + 1)$ is topologically a real sphere of topological dimension $2n - 2r + 1$. Its two orientations define two cycles belonging respectively to the two generating classes of $H^{2n-2r+1}(U(n, r))$. The cycle carried by the oriented real sphere $S(n - r + 1)$ is completely determined by the orientation of $E(n - r + 1)$ considered as a real Euclidean space of topological dimension $2(n - r + 1)$. This orientation is independent of the order of the vectors $\mathfrak{e}_r, \cdots, \mathfrak{e}_n$ in $E(n - r + 1)$. It follows that the fibre bundle $\mathfrak{F}^{(r)}$ is orientable in the sense of Steenrod, which means that there is an isomorphism in the large between the $(2n - 2r + 1)$-dimensional homology groups of the fibres of $\mathfrak{F}^{(r)}$, or that a continuous field of generating elements of $H^{2n-2r+1}(U(n, r))$ can be defined over the whole manifold. The fibre bundle $\mathfrak{F}^{(r)}$ has two opposite orientations and we shall from now on make a definite choice of one of them. Using this "oriented" fibre bundle, we shall be allowed to replace an element of $H^{2n-2r+1}(U(n, r))$ at a point by the integer which, when multiplied by the generating element at this point, is equal to the element in question.

With these explanations understood, we have:

---

[24] EHRESMANN, [9]. This fact is easily proved by making use of the covering homotopy theorem, the complex case being even simpler than the real case.

THEOREM 7. *The* r^{th} *basic characteristic class of a complex manifold* M *of dimension* n *can be defined as follows: Take a simplicial decomposition of* M *each of whose simplexes belongs to a neighborhood, and define over its skeleton of dimension* $2n - 2r + 1$ *a continuous field of ordered sets of* r *linearly independent complex tangent vectors. To each simplex of dimension* $2n - 2r + 2$ *take a point in its interior and consider the manifold of the ordered sets of* r *linearly independent complex tangent vectors at that point. The field on the boundary of the simplex defines a mapping of the boundary into this manifold and hence an element of its* $(2n - 2r + 1)$*-dimensional homotopy or homology group, which is free cyclic. Attach the corresponding integer to the simplex. The cochain so defined is a cocycle and belongs to the* r^{th} *basic class.*

To prove this theorem we take on $H(n, N)$ a definite Schubert variety V defined by the function in (11), which carries a cycle Z, dual to the cocycle Φ_r. The Schubert variety being an algebraic variety on the algebraic variety $H(n, N)$, it follows from the triangulation theorem of algebraic varieties[25] that $H(n, N)$ can be covered by a complex L such that V is a subcomplex. Let L^* be the dual cellular subdivision of L and let f be a mapping of M into L^*, which, according to Theorem 1, induces over M a complex sphere bundle equivalent to the tangent bundle of M. By the theorem on the simplicial approximation of mappings there exists a subdivision of M and a simplicial mapping f_1 of the subdivision into L^* such that f and f_1 are homotopic. By Theorem 2 the complex sphere bundle over M induced by f_1 is equivalent to the tangent bundle of M. For simplicity of notation we can therefore assume f to be a simplicial mapping of M into L^*.

Let K be the skeleton of M of dimension $2(n - r + 1)$. $f(K)$ is a subcomplex of L^* of dimension $\leqq 2(n - r + 1)$. The rth basic cocycle of M is by definition a linear function of the integral chains of dimension $2(n - r + 1)$ of K such that its value at a simplex σ of K is the intersection number of $f(\sigma)$ and V.

To give a description of the Schubert variety V we take in $E(n + N)$ a linear subspace $L(n + N - r)$ of dimension $n + N - r$, and let $L(r)$ be the linear subspace of dimension r which is totally perpendicular to $L(n + N - r)$. Then V consists of all $X(n)$ of $H(n, N)$ satisfying the condition

$$\dim(X(n) \cap L(n + N - r)) \geqq n - r + 1.$$

We take in $L(r)$ r mutually perpendicular unit vectors $\mathfrak{a}_1, \cdots, \mathfrak{a}_r$. To each $X(n)$ not belonging to V the projection of $\mathfrak{a}_1, \cdots, \mathfrak{a}_r$ on $X(n)$ will give r vectors which are linearly independent, and the construction fails exactly for the $X(n)$ belonging to V.

With these preparations consider a simplex σ of K. If the intersection number $KI(f(\sigma), V) = 0$, the above construction will give a continuous field of r linearly independent complex vectors at each point of $f(\sigma)$ and hence also at each point of σ, which shows that the integer attached to σ according to the statement of the theorem is also zero. It remains therefore to consider the case

[25] VAN DER WAERDEN, [26].

that $KI(f(\sigma), V) = \epsilon \neq 0$. In this case $f(\sigma)$ is of dimension $2(n - r + 1)$. Since $f(\sigma)$ and V belong to dual subdivisions, we have $\epsilon = +1$. Let $X_0(n)$ be the linear space of $H(n, N)$ common to V and $f(\sigma)$. The orthogonal projection of $\mathfrak{a}_1, \cdots, \mathfrak{a}_r$ defines on each $X(n) \neq X_0(n)$ of $f(\sigma)$ r linearly independent complex vectors, which, by the resolution of all the $X(n)$ belonging to $f(\sigma)$ into a topological product, are mapped into the manifold μ of all the ordered sets of r linearly independent vectors in $X_0(n)$. Our purpose is to prove that by means of the sets of vectors on the boundary $\delta f(\sigma)$, $\delta f(\sigma)$ is mapped into a cycle belonging to the generator of the $(2n - 2r + 1)st$ homology group of μ. For simplicity of language let us call index the integer m obtained by mapping the field on the boundary $\delta f(\sigma)$ into a fibre $U(n, r)$, the image cycle being in the homology class equal to m times the generator of $H^{2n-2r+1}(U(n, r))$.

Suppose first that a continuous field of ordered sets of n linearly independent vectors $\mathfrak{e}_1, \cdots, \mathfrak{e}_n$ is defined throughout $f(\sigma)$ and let r linearly independent vectors $\mathfrak{f}_1, \cdots, \mathfrak{f}_r$ be defined over $\delta f(\sigma)$ such that

$$\mathfrak{f}_i = \sum_{k=1}^{n} \mathfrak{f}_{ik} \mathfrak{e}_k, \qquad 1 \leqq i \leqq r.$$

Regarding $\mathfrak{e}_1, \cdots, \mathfrak{e}_n$ as fixed, these equations also define a mapping of $\partial f(\sigma)$ into μ. We assert that their indices are equal. In fact, the existence of the field $\mathfrak{e}_1, \cdots, \mathfrak{e}_n$ throughout $f(\sigma)$ provides exactly a deformation of the field $\mathfrak{e}_1, \cdots, \mathfrak{e}_n$ over $\partial f(\sigma)$ into vectors $\mathfrak{e}_1', \cdots, \mathfrak{e}_n'$ which are constant.

We assume that $X_0(n)$ has the property that the orthogonal projections of $a_1, \cdots, a_{r-1}$ onto it are linearly independent, which is possible, after applying a small deformation if necessary. Since $f(\sigma)$ is a simplex, we can define over $f(\sigma)$ a continuous field of n linearly independent vectors, such that in every $X(n)$ the first $r - 1$ of these vectors are the orthogonal projections in $X(n)$ of $\mathfrak{a}_1, \cdots, \mathfrak{a}_{r-1}$ respectively. This continuous field is then deformed into a continuous field $\mathfrak{e}_1, \cdots, \mathfrak{e}_n$ over $f(\sigma)$ such that in each $X(n)$ the vectors $\mathfrak{e}_1, \cdots, \mathfrak{e}_n$ constitute a frame (that is, are mutually perpendicular unit vectors). It is well-known that the deformation can be so chosen that during the deformation the vector subspace determined by the first s vectors $(1 \leqq s \leqq n)$ remains fixed. With these deformations performed, we proceed to study the orthogonal projection $\mathfrak{a}_r^*$ of $\mathfrak{a}_r$ in $X(n)$.

The index of the field of orthogonal projections of $\mathfrak{a}_1, \cdots, \mathfrak{a}_r$ on $\partial f(\sigma)$ is equal to the index of the field $\mathfrak{e}_1, \cdots, \mathfrak{e}_{r-1}, \mathfrak{a}_r^*$ on $\partial f(\sigma)$ and is also equal to the index of the same field, when $\mathfrak{e}_1, \cdots, \mathfrak{e}_{r-1}$ are considered as constant vectors. We also remark that the vector $\mathfrak{a}_r^*$ is linearly independent of $\mathfrak{e}_1, \cdots, \mathfrak{e}_{r-1}$ at every point $\neq X_0(n)$ of $f(\sigma)$.

To show that the index in question is 1 we choose the continuous field of vectors $\mathfrak{e}_{n+1}, \cdots, \mathfrak{e}_{n+N}$ over $f(\sigma)$ such that $\mathfrak{e}_A$, $1 \leqq A \leqq n + N$, is a frame in $E(n + N)$. Then we have

$$\mathfrak{a}_r = \sum_{A=1}^{n+N} u_A \mathfrak{e}_A, \qquad \sum_{A=1}^{n+N} u_A \bar{u}_A = 1,$$

and

$$\mathfrak{a}_r^* = \sum_{i=1}^{n} u_{ri} \mathfrak{e}_i .$$

According to our previous remark we can regard the vectors $\mathfrak{e}_A$, $1 \leqq A \leqq n + N$, as fixed and consider the mapping of $f(\sigma) - X_0(n)$ into $S(n - r)$ defined by the vector whose components with respect to a fixed frame are $u_{r,r+1}, \cdots, u_{rn}$. Thus we see that the index is 1, and Theorem 7 is proved.

It is also possible to introduce from a bundle of complex spheres $S(n)$ the fibre bundles $\mathfrak{F}^{(r)*}$ over the same base space whose fibre at each point is the manifold $U^*(n, r)$ of all ordered sets of $r(1 \leqq r \leqq n)$ mutually perpendicular vectors of $S(n)$. The manifold $U^*(n, r)$ is an absolute retract of $U(n, r)$. Theorem 7 still holds, if we replace everywhere the phrase "ordered sets of r linearly independent complex vectors" by "ordered sets of r mutually perpendicular vectors of $S(n)$", and, naturally also the manifold $U(n, r)$ by $U^*(n, r)$.

2. A third definition of the basic characteristic classes

We suppose in this section that the base space M, which is a complex manifold, is compact. From the tangent bundle $\mathfrak{F}$ over M we construct the fibre bundle $\mathfrak{F}^{(r)*}$ $(1 \leqq r \leqq n)$ as explained at the end of the last section. Then the following theorem gives a third definition of the rth basic characteristic class:

THEOREM 8. *The rth basic characteristic class of M is the cohomology class of M, each of whose cocycles γ has the following property: Under the projection of $\mathfrak{F}^{(r)*}$ into M, γ is mapped into a cocycle γ^*. There exists in $\mathfrak{F}^{(r)*}$ a cochain β^*, such that $\delta\beta^* = \gamma^*$ and that β^* reduces to a fundamental cocycle on each fibre of $\mathfrak{F}^{(r)*}$.*

The last statement in the theorem needs some explanation. Let P be a polyhedron and $Q \subset P$ a closed subpolyhedron of P. If γ is a cochain in P, its *reduced cochain* on Q is the cochain γ' such that $\gamma' \cdot \sigma = \epsilon \gamma \cdot \sigma$, where $\epsilon = 1$ or 0 according as the simplex σ belongs to Q or not. Moreover, the integral cohomology group of dimension $2n - 2r + 1$ of a fibre being free cyclic, a *fundamental cocycle* on the fibre is a cocycle of dimension $2n - 2r + 1$ which belongs to a generator of the cohomology group. It is also understood that the cycles and cocycles are defined in terms of simplicial decompositions of M and $\mathfrak{F}^{(r)*}$. To define the inverse mapping of the cocycles of M into the cocycles of $\mathfrak{F}^{(r)*}$ induced by the projection π of $\mathfrak{F}^{(r)*}$ into M, we therefore take a simplicial approximation π' of π. Let σ^* be a simplex of dimension $2n - 2r + 2$ of $\mathfrak{F}^{(r)*}$. Then we define

$$\gamma^* \cdot \sigma^* = \gamma \cdot \pi'(\sigma^*),$$

if $\pi'(\sigma^*)$ is of dimension $2n - 2r + 2$ and $\gamma^* \cdot \sigma^* = 0$ if $\pi'(\sigma^*)$ is of lower dimension. The cocycle γ^* depends on π', but its cohomology class is independent of it. The theorem asserts that any such cocycle γ^* reduces to a fundamental cocycle on a fibre.

In order to prove Theorem 8 we need the following lemma:

LEMMA 4. *With the notations of Theorem 8 let U be a neighborhood of M and let $\pi^{-1}(U)$ be its complete inverse image in $\mathfrak{F}^{(r)*}$. Let K be a finite complex and L its skeleton of dimension $\leqq 2n - 2r$. If f and g are two continuous mappings into $\pi^{-1}(U)$ of K and L respectively, there is a continuous mapping f^* of K into $\pi^{-1}(U)$ which is homotopic to f and coincides with g on L.*

We denote by I the unit segment $0 \leqq t \leqq 1$ and consider the topological product $K \times I$. To prove Lemma 4 is to define a continuous mapping $f(K \times I) \subset \pi^{-1}(U)$, with $f(K \times 0)$ and $f(L \times 1)$ given. For this purpose we decompose K simplicially and arrange the simplexes of the decomposition in a sequence that every simplex is preceded by its faces. The mapping is then defined by successive extensions over the prisms constructed on the simplexes of the sequence. We resolve $\pi^{-1}(U)$ into the topological product of U and a fixed fibre F_0 and denote by λ the projection of $\pi^{-1}(U)$ onto F_0. Let σ^0 be a vertex of K. Then $f(\sigma^0 \times 0)$ and $f(\sigma^0 \times 1)$ are both defined in $\mathfrak{F}^{(r)*}$, and can be joined by a segment, on which the prism on σ^0 is mapped. Using mathematical induction we suppose f be defined over all prisms on simplexes preceding σ^m of the sequence, and consider σ^m, $m \leqq 2n - 2r$. By hypothesis, the mapping is defined over $\partial(\sigma^m \times I)$, which is topologically a sphere of topological dimension m. The mapping can be extended over $\sigma^m \times I$, if and only if $f(\partial(\sigma^m \times I))$ is homotopic to zero in $\mathfrak{F}^{(r)*}$, that is, by the covering homotopy theorem, if and only if $\lambda f(\partial(\sigma^m \times I))$ is homotopic to zero in F_0. The latter is the case, because the mth homotopy group of F_0 is zero. It follows that f is defined for the subcomplex $L \times I + K \times 0$ of the prism $K \times I$. By a well-known elementary geometric construction[26] f is then extended over $K \times I$. Thus the lemma is proved.

We proceed to prove Theorem 8. Let γ be a cocycle of M belonging to the rth basic class defined by the construction of Theorem 7, with the bundle $\mathfrak{F}^{(r)}$ replaced by $\mathfrak{F}^{(r)*}$. To explain Theorem 7 for this case, we take a simplicial decomposition of M which is so fine that each simplex belongs to a neighborhood of M. Let $K^{2n-2r+1} = K$ be the $(2n - 2r + 1)$-dimensional skeleton of the simplicial decomposition. There exists a continuous mapping Ψ of K into $\mathfrak{F}^{(r)*}$ such that $\pi\Psi(p) = p$ for every $p \in K$. Let σ be a simplex of dimension $2n - 2r + 2$. The mapping Ψ defines a mapping of the boundary $\partial\sigma$ of σ into $\mathfrak{F}^{(r)*}$ and the mapping $\lambda\Psi$ defines a mapping of $\partial\sigma$ into F_0, and hence a cycle of dimension $2n - 2r + 1$ of F_0. Let this cycle be homologous to a multiple $\gamma(\sigma)$ of the generating cycle of dimension $2n - 2r + 1$ of F_0. According to Theorem 7 the cocycle $\gamma(\psi)$ defined by assigning the integer $\gamma(\sigma)$ to σ belongs to the rth basic class. Moreover, it was also proved[27] that to a given cocycle γ of the rth basic class there exists a mapping ψ such that $\gamma(\psi) = \gamma$. We suppose ψ to be so chosen.

Let π' be a simplicial approximation of π and let γ^* be the inverse image of γ under π' as defined above. We shall show that γ^* is the coboundary of an

[26] ALEXANDROFF-HOPF, [1], p. 501.
[27] STEENROD, [21], p. 124.

integral cochain β^* of dimension $2n - 2r + 1$ of $\mathfrak{F}^{(r)*}$. To define β^* let τ^* be a simplex of dimension $2n - 2r + 1$ of $\mathfrak{F}^{(r)*}$, and let $\tau = \pi'(\tau^*)$. The discussion will be divided into two cases, according as τ is of dimension equal to or less than $2n - 2r + 1$.

Suppose τ be of dimension $2n - 2r + 1$. In this case π' establishes a simplicial and therefore topological mapping between τ and τ^*. By Lemma 4 there exists a mapping $\pi''(\tau) \subset \mathfrak{F}^{(r)*}$, which is homotopic to $\pi'^{-1}(\tau)$ and coincides with Ψ on the boundary $\partial\tau$. We then take an oriented sphere of topological dimension $2n - 2r + 1$ and denote by H_1, H_2 its two hemispheres. We map H_1 and H_2 into τ by the mappings h_1 and h_2 of the degrees -1 and $+1$ respectively, such that h_1 and h_2 are identical on the "equator" $H_1 \cap H_2$. A mapping f of the sphere $H_1 + H_2$ into $\mathfrak{F}^{(r)*}$ is then defined by the conditions

$$f(p) = \pi'' h_1(p), \qquad p \in H_1,$$
$$f(p) = \psi h_2(p), \qquad p \in H_2.$$

This mapping f is by construction continuous. Taking its projection λf on F_0, we get an element of the $(2n - 2r + 1)$-dimensional homotopy group of the fibre and hence an integer, on account of the orientability of the bundle $\mathfrak{F}^{(r)*}$. This integer we define to be $\beta^* \cdot \tau^*$. It is to be remarked that $\beta^* \cdot \tau^*$ in general depends on the deformation which carries π'^{-1} to π'', but only one such deformation will be utilized, and $\beta^* \cdot \tau^*$ is thus well defined.

Next let τ be of dimension $< 2n - 2r + 1$. We suppose without loss of generality that $p_0 = \pi(F_0) \in \tau$. Let any simplex τ' of dimension $2n - 2r + 1$ be mapped into τ^* by a non-degenerate orientation-preserving simplicial mapping. By Lemma 4 this mapping is homotopic to a mapping $\pi'''(\tau') \subset \mathfrak{F}^{(r)*}$ such that $\pi'''(\partial\tau') = q_0$, where q_0 is a point of F_0. We identify all the points on $\partial\tau'$, thus getting an oriented sphere of topological dimension $2n - 2r + 1$, which is mapped into F_0 by the mapping $\lambda\pi'''(\tau') \subset F_0$. This mapping defines an element of the $(2n - 2r + 1)$-dimensional homotopy group and hence an integer, which is defined to be $\beta^* \cdot \tau^*$.

It remains to show that the coboundary of the cochain β^* so defined is equal to γ^*. For this purpose let σ^* be a simplex of $\mathfrak{F}^{(r)*}$ of dimension $2n - 2r + 2$. It is sufficient to verify that $\gamma^* \cdot \sigma^* = \delta\beta^* \cdot \sigma^* = \beta^* \cdot \partial\sigma^*$.

Suppose first that $\sigma = \pi'(\sigma^*)$ is of dimension $2n - 2r + 2$. The mapping π'' is defined for each simplex of $\partial\sigma$ and hence for $\partial\sigma$ itself, because it coincides with ψ on the $(2n - 2r)$-dimensional skeleton K^{2n-2r}. We therefore have two mappings, $\lambda\psi$ and $\lambda\pi''$ respectively, of $\partial\sigma$ into F_0 such that they are identical on K^{2n-2r}. Our fibre F_0 being $(2n - 2r)$-simple, this is a situation discussed by Eilenberg,[28] who introduced several cochains, denoted in his notation by $c(\lambda\psi)$, $c(\lambda\pi'')$, $d(\lambda\psi, \lambda\pi'')$ respectively. In our notation they are given by

$$c(\lambda\psi) \cdot \sigma = \gamma^* \cdot \sigma^*,$$
$$c(\lambda\pi'') = 0,$$
$$d(\lambda\psi, \lambda\pi'') \cdot \partial\sigma = \beta^* \cdot \partial\sigma^*,$$

[28] Eilenberg, [11], pp. 235–237.

where the second relation follows from the fact that $\lambda\pi''$ is defined for the simplex σ bounded by $\partial\sigma$. From a theorem of Eilenberg we have

$$\delta d(\lambda\psi, \lambda\pi'') = c(\lambda\psi) - c(\lambda\pi''),$$

or

$$\gamma^* \cdot \sigma^* = \beta^* \cdot \partial\sigma^*,$$

which is to be proved.

Next suppose $\sigma = \pi'(\sigma^*)$ be of dimension $< 2n - 2r + 2$. If each simplex of $\partial\sigma^*$ is mapped by π' into a simplex of dimension $< 2n - 2r + 1$, $\beta^* \cdot \partial\sigma^*$ is clearly zero. The other possibility is that σ is of dimension $2n - 2r + 1$ and that exactly two of the simplexes of $\partial\sigma^*$, say τ_1^* and τ_2^*, are mapped by π' into σ. If τ_1^* and τ_2^* are coherently oriented with the boundary $\partial\sigma^*$, it follows by definition that $\beta^* \cdot (\tau_1^* + \tau_2^*) = 0$. On the other hand, it is not difficult to see that $\beta^* \cdot (\partial\sigma^* - \tau_1^* - \tau_2^*) = 0$. Hence we have $\beta^* \cdot \partial\sigma^* = 0$. We have thus proved that γ^* is the coboundary of a cochain β^*.

To see that β^* reduces on a fibre F_0, we suppose the simplicial decomposition of $\mathfrak{F}^{(r)*}$ so made that F_0 is a subcomplex. Every simplex of F_0 is mapped by π' into a point, so that we have $\gamma^* \cdot \sigma^* = 0$ for every simplex σ^* of dimension $2n - 2r + 2$ of F_0, which shows that β^* reduces to a cocycle on F_0. To show that β^* is the fundamental cocycle on F_0, it is sufficient to show that one is the value of its product with a cycle belonging to the generating homology class of dimension $2n - 2r + 1$ of F_0. For this purpose we take an oriented sphere of topological dimension $2n - 2r + 1$ and map it simplicially into F_0 such that the map belongs to the generator of the $(2n - 2r + 1)$-dimensional homotopy group of F_0. The image of this map defines a cycle of F_0 belonging to the generating homology class of the $(2n - 2r + 1)$-dimensional homology group, and its product with β^* is 1. It follows that β^* reduces to a fundamental cocycle on F_0.

Our Theorem 8 is now completely proved.

3. In terms of differential forms

Consider again the Grassmann manifold $H(n, N)$. We take as point of a new space a linear space $E(n)$ and r vectors $e_{n-r+1}, \cdots, e_n$ belonging to $E(n)$ such that $e_i\bar{e}_j = \delta_{ij}, n - r + 1 \leq i, j \leq n$. This space, to be denoted by $R(r, n, N)$, is clearly a fibre bundle over $H(n, N)$, the projection of a point of $R(r, n, N)$ being the corresponding $E(n)$ and each fibre being homeomorphic to $U^*(n, r)$. This fibre bundle $R(r, n, N)$ is transformed transitively by the unitary group in the space $E(n + N)$. Let W_r be the rth basic class of $H(n, N)$ and $\gamma \in W_r$ be one of its cocycles. γ is mapped by the inverse mapping induced by any simplicial approximation of π into a cocycle having the properties asserted by Theorem 8. In particular, we can take for γ the cocycle defined by the differential form Φ_r. The inverse image of Φ_r in $R(r, n, N)$ under π is a differential form which for convenience we denote by Φ_r. The form Φ_r then defines a cocycle γ^* in $R(r, n, N)$.

From Theorem 8 it follows that[29] there exists a cochain β^* in $R(r, n, N)$ such that $\delta\beta^* = \gamma^*$ and such that β^* reduces to a fundamental cocycle on a fibre.

In order to define β^* in $R(r, n, N)$ by means of a suitably chosen differential form, we shall make use of the following lemma proved by de Rham:[30]

LEMMA 5. *Let M be a compact differentiable manifold of class $\geqq 2$ and let K be a simplicial decomposition of M, whose simplexes are σ_i^p, $i = 1, \cdots, \alpha_p$, $p = 0, 1, \cdots, n$, and whose incidence relations are*

$$\partial \sigma_i^p = \sum_{j=1}^{\alpha_{p-1}} \eta_{ij}^{(p)} \sigma_j^{p-1}$$

Then there exists a set of differential forms

$$\varphi_i^{(p)}, \qquad i = 1, \cdots, \alpha_p, \qquad p = 0, 1, \cdots, n,$$

such that the following conditions are satisfied:

$$1) \qquad \int_{\sigma_j(p)} \varphi_i^{(p)} = \delta_{ij},$$

$$2) \qquad d\varphi_i^{(p)} = \sum \eta_{ji}^{(p+1)} \varphi_j^{(p+1)}.$$

We apply this lemma to the manifold $R(r, n, N)$, and write for simplicity $p = 2n - 2r + 1$. The cochain β^* is defined by definition by a system of equations

$$\beta^* \cdot \sigma_i^p = \lambda_i, \qquad i = 1, \cdots, \alpha_p.$$

Let

$$\omega = \sum_{i=1}^{\alpha_p} \lambda_i \varphi_i^{(p)}.$$

Then we have

$$\int_{\sigma_i^p} \omega = \lambda_i$$

which shows that the differential form ω defines the cochain β^*. By construction we have

$$d\omega = \Phi_r.$$

Now the unitary group $U(n + N)$ in $E(n + N)$ transforms transitively the manifold $R(r, n, N)$. Let s be a transformation of $U(n + N)$. If θ is a differen-

[29] We have tacitly assumed at this point of our discussion that the Theorem 8, proved by a combinatorial construction for the simplicial approximations of the projection π, holds for π itself, when the cocycles are expressed by means of differential forms. It is, however, possible to avoid this assumption by observing that the cochain β^* exists in $R(r, n, N)$ such that $\delta\beta^* = \gamma^*$. That β^* reduces to a fundamental cocycle on a fibre then follows from the very definition of the characteristic cocycle on $H(n, N)$.

[30] DE RAHM, [20], p. 178.

tial form in $R(r, n, N)$, we shall denote by $s\theta$ its transform by the transformation s. We also use the notation $\theta \sim 0$ to denote that θ is derived.

LEMMA 6. *Let β^* be a cochain of dimension $2n - 2r + 1$, whose coboundary is γ^*. Let ω be a differential form which defines β^*. Then $s\omega - \omega \sim 0$.*

First of all, the differential form $s\omega - \omega$ is exact, since we have

$$d(s\omega - \omega) = s\Phi_r - \Phi_r = 0.$$

Let ζ be a cycle of dimension $2n - 2r + 1$ of $R(r, n, N)$. It is sufficient to prove that

$$\int_\zeta s\omega = \int_\zeta \omega.$$

The cycle $s\omega - \omega$ being homologous to zero, let Z^* be the chain it bounds. Then we have

$$\int_\zeta s\omega - \omega = \int_{s\zeta - \zeta} \omega = \int_{\partial Z^*} \omega = \int_{Z^*} \Phi_r.$$

Let Z be the projection in the base space $H(n, N)$ of the chain Z^* in $R(r, n, N)$. The boundary of Z is the projection of $s\zeta - \zeta$. Since the Betti group of dimension $2n - 2r + 1$ of $H(n, N)$ is zero, the projection of ζ bounds in $H(n, N)$ a chain which we shall call Z_1. Then the projection of $s\zeta$ bounds the chain sZ_1, and we have

$$Z \sim sZ_1 - Z_1.$$

It follows that

$$\int_{Z^*} \Phi_r = \int_Z \Phi_r = \int_{Z_1} s\Phi_r - \int_{Z_1} \Phi_r = 0.$$

Thus Lemma 6 is proved.

THEOREM 9. *Under the projection of $R(r, n, N)$ into $H(n, N)$ the differential form Φ_r is mapped by the inverse mapping into $R(r, n, N)$. There exists a differential form π which is invariant under transformations of the unitary group $U(n + N)$ operating in $R(r, n, N)$ and whose exterior derivative $d\pi$ is equal to Φ_r.*

By Lemma 5 we can construct the differential form ω in $R(r, n, N)$ such that

$$d\omega = \Phi_r.$$

For such a differential form ω it follows from Lemma 6 that

$$\delta\omega - \omega \sim 0.$$

Let dv be the invariant volume element of $U(n + N)$ such that the integral of dv over $U(n + N)$ is equal to 1. We put

$$\Pi = \int_{U(n+N)} s\omega \, dv.$$

Then Π is invariant under $U(n + N)$ and we have

$$\lambda\Pi = \int_{U(n+N)} s \cdot d\omega \cdot dv = \Phi_r \int_{U(n+N)} dv = \Phi_r,$$

which proves our theorem.

To a point of $R(r, n, N)$ we now attach the frames $e_1, \cdots, e_{n+N}$ in $E(n + N)$ such that $e_1, \cdots, e_n$ determine the $E(n)$ and that $e_{n-r+1}, \cdots, e_n$ are the vectors in question. For clearness let us agree in the remainder of this section on the following ranges of indices:

$$1 \leqq \alpha, \beta, \gamma \leqq n - r, \qquad n - r + 1 \leqq A, B, C \leqq n,$$
$$n + 1 \leqq i, j, k \leqq n + N.$$

In a neighborhood of $R(r, n, N)$ we can choose a differentiable family of such frames, one attached at each point of the neighborhood. By means of the family of frames the forms $\theta_{A\alpha}, \bar{\theta}_{A\alpha}, \theta_{Ai}, \bar{\theta}_{Ai}, \theta_{AB}$ can be constructed according to the equations (7). They constitute a set of linearly independent linear differential forms at each point of $R(r, n, N)$. Our form π, whose existence was asserted by Theorem 9 and which is invariant under $U(n + N)$, is necessarily a polynomial (in the sense of Grassmann algebra) in the forms of this set, with constant coefficients. On the other hand, the form Π, being itself in $R(r, n, N)$, must be invariant under the transformation

$$\theta^*_{Ai} = \sum_j a_{ij}\theta_{Aj}$$

where a_{ij} are the elements of a unitary matrix. We put

$$\Theta_{AB} = \sum_i \theta_{Ai}\theta_{iB}$$

It follows from the first main theorem on vector invariants of the unitary group that Π is a polynomial in $\theta_{A\alpha}, \bar{\theta}_{A\alpha}, \theta_{AB}, \Theta_{AB}$, with constant coefficients. Moreover, on a fibre, that is, omitting all terms in Θ_{AB}, Π becomes a fundamental cocycle.

All these results can be summarized in the following theorem:

THEOREM 10. *There exists in $R(r, n, N)$ a polynomial Π in $\theta_{A\alpha}, \bar{\theta}_{A\alpha}, \theta_{AB}, \Theta_{AB}$, with constant coefficients, such that $d\Pi = \Phi_r$. When all terms involving Θ_{AB} in Π are omitted, the form defines a fundamental cocycle on a fibre.*

CHAPTER IV

HERMITIAN MANIFOLDS

1. Fundamental formulas of Hermitian Geometry

Let M be a compact complex manifold. M is called an Hermitian manifold, if an intrinsic Hermitian differential form is given throughout the manifold. In each local coordinate system z^i the Hermitian differential form is defined by

(28) $$ds^2 = \sum_{i,j=1}^{n} g_{ij}(z, \bar{z})(dz^i\, d\bar{z}^j), \qquad \bar{g}_{ij} = g_{ji},$$

where, as well as in later formulas, we insert a parenthesis to designate that the multiplication of the differential forms in question is ordinary multiplication. We shall agree, unless otherwise stated, that the indices i, j, k take the values 1 to n.

Our main result in this chapter is to establish that the n basic classes which arise from the analytic structure of a complex manifold, are completely determined by the Hermitian metric, if the manifold in question is an Hermitian manifold. In particular, as we shall see later, the theorem for the class W_1 reduces to the formula of Allendoerfer-Weil, if we interpret the Hermitian metric as a Riemannian metric for the real manifold of $2n$ topological dimensions.

We begin by establishing the fundamental formulas for local Hermitian Geometry.

For this purpose we determine in a neighborhood of M n linear differential forms φ_i in the local coordinates z^i such that

(29) $$ds^2 = \sum_{i=1}^{n} (\varphi_i \bar{\varphi}_i).$$

The forms φ_i are determined up to a unitary transformation:

(30) $$\omega_i = \sum_{i=1}^{n} u_{ij}\varphi_j, \qquad i = 1 \cdots, n,$$

where u_{ij} are the elements of a unitary matrix

$$U = (n_{ij}),$$

and which we take to be independent variables. Let $\mathfrak{e}_i$ be the dual base corresponding to ω_k, so that

$$\omega_i(\mathfrak{e}_k) = \delta_{ik}.$$

From the Hermitian differential form a scalar product of the contravariant vectors can be defined, and we have

$$\mathfrak{e}_i \cdot \bar{\mathfrak{e}}_k = \delta_{ik}.$$

We shall call a frame the figure formed by a point P and n such vectors $\mathfrak{e}_i$. With a natural topology the set of frames constitutes a fibre bundle over M.

The forms ω_i are intrinsically defined in the fibre bundle. By actual calculation in terms of a local coordinate system, we find that their exterior derivatives are of the form

$$d\omega_i = \sum_j \omega_j \omega_{ji} + \sum_{j,k} a_{ijk}\omega_j\omega_k + \sum_{j,k} b_{ijk}\omega_j\bar{\omega}_k, \qquad a_{ijk} + a_{ijk} = 0,$$

where

$$\omega_{ij} = \sum_k \bar{u}_{ik}\, d\bar{u}_{jk},$$

so that

$$\omega_{ij} + \omega_{ji} = 0.$$

But the forms ω_{ij} are defined up to the transformation

$$\omega_{ij} \rightarrow \omega_{ij} + \sum \lambda_{ijk}\omega_k - \sum \bar{\lambda}_{jik}\bar{\omega}_k,$$

where the quantities λ_{ijk} are arbitrary. It is easy to show that there exists one, and only one, set of forms ω_{ij}, such that the following equations are satisfied:

$$(31) \quad d\omega_i = \sum_j \omega_j\omega_{ji} + \sum_{jik} A_{ijk}\omega_j\omega_k, \qquad A_{ijk} + A_{ikj} = 0, \qquad \omega_{ij} + \bar{\omega}_{ji} = 0,$$

From the uniqueness of this set of forms follows the fact that they are intrinsically defined in the fibre bundle. The forms ω_i, ω_{ij} constitute therefore a set of linearly independent linear differential forms in the fibre bundle.

From equations (31) it is possible to draw all the consequences of local Hermitian Geometry. In fact, we put

$$(32) \qquad \Omega_i = \sum_{j,k} A_{ijk}\,\omega_j\omega_k.$$

Exterior differentiation of the first set of equations in (31) will give

$$d\Omega_i - \sum_k \omega_k\Omega_{ki} + \sum_k \Omega_k\omega_{ki} = 0,$$

where we have put

$$(33) \qquad \Omega_{ij} = d\omega_{ij} - \sum_k \omega_{ik}\,\omega_{kj}.$$

We remark that $d\Omega_i + \sum_k \Omega_k\omega_{ki}$ is of the form $\sum_{j,k}\psi_{ijk}\omega_j\omega_k$. It follows that

$$\omega_1 \cdots \omega_n\Omega_{ij} = 0$$

and that we can put

$$\Omega_{ik} = \sum \chi_{ikj}\omega_j.$$

On the other hand, we have

$$\Omega_{ik} + \bar{\Omega}_{ki} = 0$$

or

$$\sum_j \chi_{ikj}\,\omega_j + \sum_j \bar{\chi}_{kij}\,\bar{\omega}_j = 0,$$

which shows that χ_{ikj} is a linear combination of ω_k, $\bar{\omega}_k$:

$$\chi_{ikj} = \sum_{l=1}^{n} a_{ikjl}\,\omega_l + \sum_{l=1}^{n} b_{ijkl}\,\bar{\omega}_l.$$

Substituting this expression of χ_{ikj} into the last equation, we see immediately that a_{ikjl} must be symmetric in the last two indices j, l. It follows that Ω_{ik} is of the form

$$\Omega_{ik} = \sum b_{ikjl}\bar{\omega}_l\omega_j .$$

The equations for the exterior derivatives $d\omega_i$, $d\omega_{sj}$ and the equations obtained therefrom by exterior differentiation we shall call the fundamental equations of local Hermitian Geometry. These equations will now be summarized as follows:

$$\begin{aligned} d\omega_i &= \sum_j \omega_j\omega_{ji} + \Omega_i , \\ d\omega_{ij} &= \sum_k \omega_{ik}\omega_{kj} + \Omega_{ij} , \\ d\Omega_i &+ \sum_j \Omega_j\omega_{ji} - \sum_j \omega_j\Omega_{ji} = 0, \\ d\Omega_{ij} &+ \sum_k \Omega_{ik}\omega_{kj} - \sum_k \omega_{ik}\Omega_{kj} = 0, \\ \Omega_i &= \sum A_{ijk}\omega_j\omega_k , \qquad A_{ijk} + A_{ikj} = 0, \\ \Omega_{ij} &= \sum_{l,m=1}^{n} R_{ij,lm}\bar{\omega}_l\omega_m , \qquad R_{ij,lm} = \bar{R}_{ji,ml} , \\ \omega_{ij} &+ \bar{\omega}_{ji} = 0, \qquad \Omega_{ij} + \bar{\Omega}_{ji} = 0. \end{aligned} \tag{34}$$

In a well-known way the forms ω_i, ω_{ij} can be interpreted as defining an infinitesimal displacement, by means of the equations

$$\begin{aligned} dp &= \sum_i \omega_i \mathfrak{e}_i , \\ d\mathfrak{e}_i &= \sum_j \omega_{ij}\mathfrak{e}_j . \end{aligned} \tag{35}$$

Of importance are the Hermitian metrics satisfying the condition

$$\Omega_i = 0, \tag{36}$$

which will be called Hermitian metrics without torsion. An Hermitian metric without torsion can be characterized by the condition

$$d(\sum_i \omega_i\bar{\omega}_i) = 0, \tag{37}$$

and was studied by E. Kähler.[31] Kähler proved that in this case there exists locally a function $F(z^i, \bar{z}^i)$ such that the metric can be written in the form

$$ds^2 = \sum_{i,k} \frac{\partial^2 F}{\partial z^i\, \partial \bar{z}^k} (dz^i\, d\bar{z}^k). \tag{38}$$

Hermitian metrics without torsion play an important rôle in the theory of automorphic functions of several complex variables.

[31] KÄHLER, [16].

2. Formulas for the basic characteristic classes

As defined in Chapter III there are on M n basic characteristic classes, the rth one ($1 \leqq r \leqq n$) being of dimension $2(n - r + 1)$. We shall show that, if M is an Hermitian manifold, these classes are defined by the local properties of the Hermitian metric. For this purpose we put

$$\Psi_r = \frac{1}{(2\pi\sqrt{-1})^{n-r+1}(n-r+1)!} \cdot \sum \delta(i_1 \cdots i_{n-r+1}\,; j_1 \cdots j_{n-r+1})\Omega_{i_1 j_1} \cdots \Omega_{i_{n-r+1} j_{n-r+1}}, \tag{39}$$

where Ω_{ij} are the forms defined in (34) and where the meaning of the summation has been explained before. Then we have the following theorem, which is the main result of this paper:

THEOREM 11. *The form Ψ_r defined by (39) is the form corresponding to the rth basic characteristic class W_r in the sense that the product of any homology class ζ of dimension $2(n - r + 1)$ with W_r is equal to the integral of Ψ_r over ζ:*

$$\zeta \cdot W_r = \int_\zeta \Psi_r. \tag{40}$$

We first establish the following lemma:

LEMMA 7. *Let Δ be the differential form Π in Theorem 10, with every θ and Θ replaced by the corresponding ω and Ω with the same indices. Then $d\Delta = \Psi_r$.*

We observe that the equations for $d\omega_{ij}$, $d\Omega_{ij}$ are exactly of the same form as the equations for $d\theta_{ij}$, $d\Theta_{ij}$, the only difference being that Θ_{ij} is given by the equation (12). It follows that

$$d\Delta - \Psi_r \equiv 0 \text{ mod. } \Omega_{ij} - \sum_{B=n+1}^{n+N} \theta_{iB}\,\theta_{Bj}.$$

By mathematical induction on the degree of $d\Delta - \Psi_r$ it is easy to show that then

$$d\Delta - \Psi_r = 0.$$

To prove Theorem 11 we make use of the definition of W_r given in Theorem 7, with $\mathfrak{F}^{(r)}$ replaced by $\mathfrak{F}^{(r)*}$. A sufficiently fine simplicial decomposition K of M is taken and a continuous mapping ψ of its $(2n - 2r + 1)$-dimensional skeleton into $\mathfrak{F}^{(r)*}$ is defined, such that the image of every point belongs to the fibre over it. Let $\sigma_A^{2n-2r+2}$ or σ_A, $1 \leqq A \leqq \alpha_{2n-2r+2}$, be the simplexes of dimension $2n - 2r + 2$ of K. By the construction of Theorem 7, an integer $\gamma(\sigma_A)$ is defined for every σ_A, and the corresponding cocycle γ belongs to W_r. It is sufficient to prove that

$$\zeta \cdot \gamma = \int_\zeta \Psi_r. \tag{40a}$$

Both sides of the equation (40a) being linear in ζ, equation (40a) will follow from the relation

$$\gamma(\sigma_A) = \int_{\sigma_A} \Psi_r.$$

But, by Lemma 7,

$$\int_{\sigma_A} \Psi_r = \int_{\psi(\sigma_A)} \Psi_r = \int_{\psi(\sigma_A)} d\Delta = \int_{\partial\psi(\sigma)_A} \Delta.$$

The image $\psi(\sigma_A)$ has a singular point and we see that the last integral is precisely the definition of $\gamma(\sigma_A)$ given in integral form. Hence Theorem 11 is proved.

3. The case $r = 1$ and the formula of Allendoerfer-Weil

In the case $r = 1$ we can take for the cycle ζ in (40) one of the fundamental cycles of the manifold M. Then we have $\zeta \cdot W_1 = \chi$, the Euler-Poincaré characteristic of M. On the other hand, the manifold M can be considered as a real differentiable manifold and the Hermitian metric can be used to define a Riemannian metric in the real manifold. It is to be expected that the formula (40) will then reduce to the formula of Allendoerfer-Weil. We shall show that this is actually the case, if the Hermitian metric is without torsion.

To study the Hermitian metric as a Riemannian metric, we decompose each of the forms ω_i, ω_{ij}, Ω_{ij} into its real and imaginary parts, writing

$$\begin{aligned} \omega_i &= \theta_i + \sqrt{-1}\psi_i, \\ \omega_{ij} &= \theta_{ij} + \sqrt{-1}\psi_{ij}, \\ \Omega_{ij} &= \Theta_{ij} + \sqrt{-1}\Psi_{ij}. \end{aligned} \tag{41}$$

From the last two equations of (34) we have

$$\begin{aligned} \theta_{ij} + \theta_{ji} &= 0, \qquad \psi_{ij} - \psi_{ji} = 0, \\ \Theta_{ij} + \Theta_{ji} &= 0, \qquad \Psi_{ij} - \Psi_{ji} = 0. \end{aligned} \tag{41a}$$

The Hermitian metric can then be written as a Riemannian metric in the form:

$$ds^2 = \sum_i \{(\theta_i)^2 + (\psi_i)^2\}.$$

We get, moreover, by separating the real and imaginary parts of the equations for $d\omega_i$, $d\omega_{ij}$, the following equations:

$$\begin{aligned} d\theta_i &= \sum_j \theta_j\theta_{ji} - \sum_j \psi_j\psi_{ji}, \\ d\psi_i &= \sum_j \theta_j\psi_{ji} + \sum_j \psi_j\theta_{ji}, \end{aligned} \tag{42}$$

and

$$\begin{aligned} d\theta_{ij} &= \sum_k \theta_{ik}\theta_{kj} - \sum_k \psi_{ik}\psi_{kj} + \Theta_{ij}, \\ d\psi_{ij} &= \sum_k \theta_{ik}\psi_{kj} + \sum_k \psi_{ik}\theta_{kj} + \Psi_{ij}. \end{aligned} \tag{43}$$

It follows that, for the Riemannian Geometry of $2n$ dimensions thus obtained, the curvature forms can be conveniently described by the matrix

(44)
$$\begin{pmatrix} \Theta_{ij} & -\Psi_{ij} \\ \Psi_{ij} & \Theta_{ij} \end{pmatrix}$$

or simply by

$$\begin{pmatrix} \Theta & -\Psi \\ \Psi & \Theta \end{pmatrix}.$$

It only remains to compare the integrand of the Allendoerfer-Weil formula calculated from this matrix of curvature forms with the expression Ψ_1 defined in (39).

Let Ω denote the integrand of the Allendoerfer-Weil formula. We observe that $(-2\pi)^n\Omega$ obeys the same expansion rule as a so-called Pfaffian function of the $2n^{\text{th}}$ order,[32] which is an integral rational function in a number of independent variables, whose square is equal to the value of a skew-symmetric determinant of order $2n$. It follows that

$$(2\pi)^{2n}\Omega^2 = \begin{vmatrix} \Theta & -\Psi \\ \Psi & \Theta \end{vmatrix}.$$

On the other hand, we have

$$\Psi_1 = \frac{1}{(2\pi\sqrt{-1})^n}\,|\Omega_{ij}| = \frac{1}{(2\pi\sqrt{-1})^n}\,|\Theta + \sqrt{-1}\Psi|,$$

from which we get, after some reduction,

$$(-1)^n(2\pi)^{2n}\Psi_1^2 = \begin{vmatrix} \Theta + \sqrt{-1}\Psi & 0 \\ 0 & \Theta + \sqrt{-1}\Psi \end{vmatrix}$$
$$= \begin{vmatrix} \Theta + \sqrt{-1}\Psi & 0 \\ \Theta + \sqrt{-1}\Psi & -\Theta + \sqrt{-1}\Psi \end{vmatrix} = (-1)^n \begin{vmatrix} \Theta & -\Psi \\ \Psi & \Theta \end{vmatrix}.$$

Hence we have

$$\Omega^2 = \Psi_1^2,$$

and finally

$$\Omega = \Psi_1,$$

by comparing the coefficients of one of the terms in both sides.

Chapter V

Applications to Elliptic Hermitian Geometry

1. Preliminaries

We are going to make use of the above results to derive some consequences for elliptic Hermitian Geometry.

---

[32] Pascal, [17], pp. 60–64.

The local Hermitian Geometry whose fundamental equations are given by (34) is called elliptic Hermitian, if we have

$$(45)\qquad \begin{aligned} \Omega_{ii} &= \omega_i\bar{\omega}_i + \sum_k \omega_k\bar{\omega}_k, \\ \Omega_{ij} &= \omega_j\bar{\omega}_i, \qquad i \neq j. \end{aligned}$$

This definition owes its origin to the type of geometry studied by G. Fubini,[33] E. Study,[34] and E. Cartan,[35] which we shall prove to satisfy the conditions (45).

The elliptic Hermitian Geometry of Fubini-Study is defined as follows: We consider the complex projective space of n dimensions P_n, with the homogeneous coordinates z^α, where, as well as throughout this section, we shall use the following ranges of our indices:

$$0 \leqq \alpha, \beta, \gamma \leqq n, \qquad 1 \leqq i, j, k \leqq n.$$

In P_n let a positive definite Hermitian form be given:

$$(46)\qquad (z\bar{z}) = \sum_\alpha z^\alpha \bar{z}^{-\alpha},$$

which will serve to define the scalar product of two vectors in the affine space A_{n+1} of $n + 1$ dimensions, with the (non-homogeneous) coordinates z^α. We normalize the coordinates z^α in P_n, such that

$$(47)\qquad (z\bar{z}) = 1.$$

An Hermitian metric in P_n is then defined by the Hermitian form

$$(48)\qquad ds^2 = (dz\, d\bar{z}) - (z\, d\bar{z})\cdot(\bar{z}\, dz).$$

The group of linear transformations in A_{n+1} which leaves the form (46) invariant is the unitary group $U(n + 1)$. We take in A_{n+1} $n + 1$ vectors A_k such that

$$(49)\qquad A_\alpha \bar{A}_\beta = \delta_{\alpha\beta}.$$

For a differentiable family of such sets of vectors we have

$$(50)\qquad dA_\alpha = \sum_\beta \theta_{\alpha\beta} A_\beta,$$

where

$$(51)\qquad \theta_{\alpha\beta} + \bar{\theta}_{\beta\alpha} = 0,$$

and where the forms $\theta_{\alpha\beta}$ satisfy the equations of structure:

$$(52)\qquad d\theta_{\alpha\beta} = \sum_\gamma \theta_{\alpha\gamma}\theta_{\gamma\beta}.$$

---

[33] FUBINI, [13].

[34] STUDY, [25].

[35] CARTAN, [4].

Let B_α be fixed vectors in A_{n+1}, satisfying the equations

$$B_\alpha \bar{B}_\beta = \delta_{\alpha\beta} .$$

The coordinates of a vector A_0 with respect to B_α are defined by the equation

$$A_0 = \sum_\alpha z^\alpha B_\alpha ,$$

from which we find

$$(dA_0\, d\bar{A}_0) - (A_0\, d\bar{A}_0)(\bar{A}_0\, dA_0) = (dz\, d\bar{z}) - (z\, d\bar{z})(\bar{z}\, dz).$$

It follows that, if we regard A_0 as defining the points in P_n, the Hermitian form in (48) can be written as

$$ds^2 = (dA_0\, d\bar{A}_0) - (A_0\, d\bar{A}_0)(\bar{A}_0\, dA_0),$$

and, by (50), as

$$ds^2 = \sum_i (\theta_{0i} \bar{\theta}_{0i}). \tag{53}$$

This proves in particular that the Hermitian form in (48) is positive definite.

To calculate the curvature of the metric (53) we shall make use of the equations (52). Using the notations of local Hermitian Geometry in the last Chapter, we put

$$\omega_i = \theta_{0i} .$$

Then, by (52), we get

$$d\omega_i = d\theta_{0i} = \sum_{j \neq i} \omega_j \theta_{ji} + \omega_i(\theta_{ii} - \theta_{00}).$$

From the uniqueness of the set of forms ω_{ij} satisfying the first and the ninth equations of (34) it follows that

$$\omega_{ij} = \theta_{ij} , \qquad i \neq j,$$

$$\omega_{ii} = \theta_{ii} - \theta_{00} .$$

We find then

$$d\omega_{ij} = \sum_k \omega_{ik} \omega_{kj} + \theta_{i0} \theta_{0j} , \qquad i \neq j,$$

$$d\omega_{ii} = \sum_k \omega_{ik} \omega_{ki} + \theta_{i0} \theta_{0i} + \sum_k \theta_{k0} \theta_{0k} ,$$

and therefore

$$\Omega_{ij} = -\bar{\omega}_i \omega_j , \qquad i \neq j,$$

$$\Omega_{ii} = -\bar{\omega}_i \omega_i - \sum_k \bar{\omega}_k \omega_k ,$$

which are exactly the equations (45). Thus we have proved that it is possible to define in the complex projective space an elliptic Hermitian Geometry.

2. Formulas of Cartan and Wirtinger

When the Hermitian manifold is locally elliptic, it is possible to calculate the forms Ψ_r in (39) more explicitly. In fact, we are going to prove the following theorem:

THEOREM 12. *In a locally elliptic Hermitian manifold let* Λ *be the exterior differential form corresponding to the Hermitian differential form, that is,* $\Lambda = \sum_i \omega_i \bar{\omega}_i$ *in case the given Hermitian form is* $ds^2 = \sum_i (\omega_i \bar{\omega}_i$ *Then we have*

$$\Psi_r = \frac{1}{(2\pi\sqrt{-1})^{n-r+1}} \binom{n+1}{r} \Lambda^{n-r+1}, \qquad 1 \leqq r \leqq n. \tag{54}$$

It is clear that the construction of Λ from ds^2 is independent of the choice of the base linear differential forms ω_i in terms of which ds^2 is expressed.

The theorem is proved by induction on $n - r$. If $n - r = 0$, that is, $r = n$, we have

$$\Psi_n = \frac{1}{2\pi\sqrt{-1}} \sum_i \Lambda_{ii} = \frac{n+1}{2\pi\sqrt{-1}} \Lambda.$$

Suppose the formula (54) be true for $r + 1, \cdots, n$. We have

$$\Psi_r = \frac{1}{(2\pi\sqrt{-1})^{n-r+1}(n-r+1)!} \{\sum \delta(i_1 \cdots i_{n-r}; j_1 \cdots j_{n-r})$$

$$\cdot \Omega_{i_1 j_1} \cdots \Omega_{i_{n-r} j_{n-r}} \sum_k \Omega_{kk} - (n-r) \sum \delta(i_1 \cdots i_{n-r}; j_1 \cdots j_{n-r})$$

$$\cdot \Omega_{i_1 j_1} \cdots \Omega_{i_{n-r-1} j_{n-r-1}} \cdot \sum_k \Omega_{i_{n-r} k} \Omega_{k j_{n-r}} \}.$$

Consider the second sum inside the braces. If $i_{n-r} \neq j_{n-r}$, we can replace the sum $\sum_k \Omega_{i_{n-r} k} \Omega_{k j_{n-r}}$ by $\omega_{j_{n-r}} \bar{\omega}_{i_{n-r}} \Lambda = \Omega_{i_{n-r} j_{n-r}} \Lambda$. If $i_{n-r} = j_{n-r}$, the sum can be replaced by $(\omega_{i_{n-r}} \bar{\omega}_{i_{n-r}} + \Lambda)\Lambda = \Omega_{i_{n-r} i_{n-r}} \Lambda$. By our induction hypothesis we get easily the desired formula (54).

As an application let us determine the n basic classes of the complex projective space of n dimensions. The result is given by the following theorem:

THEOREM 13. *The* r^{th} *basic characteristic class* $(1 \leqq r \leqq n)$ *of a complex projective space of* n *dimensions is dual to the homology class containing the cycle carried by a linear subspace of dimension* $r - 1$ *multiplied by* $\binom{n+1}{r}$.

To prove this theorem we consider the affine space A_{n+1} of dimension $n + 1$ with the coordinates z^α such that the projective space P_n under consideration is the hyperplane at infinity with which A_{n+1} is made into a projective space of $n + 1$ dimensions. As before, z^α are homogeneous coordinates in P_n. Suppose that we have defined at each point $\mathfrak{z}$ of A_{n+1} different from the origin r vectors $\mathfrak{v}_i$, $1 \leqq i \leqq r$, whose components $v_i^0, \cdots, v_i^n$ are linear forms in the coordinates

z^α of $\mathfrak{z}$. Then all the points on the line joining the origin 0 to $\mathfrak{z}$ are projected from 0 into the same point p of P_n and the vectors $\mathfrak{v}_i$, $1 \leqq i \leqq r$, at the points of $0\mathfrak{z}$ are projected from 0 into the same vector of P_n, which we attach to p. It is easy to verify that the r vectors thus defined at each point of P_n are linearly independent when and only when the $r + 1$ vectors $\mathfrak{z}, \mathfrak{v}_1, \cdots, \mathfrak{v}_r$ in A_{n+1} are linearly independent.

We take

$$v_i^\alpha = a_i^\alpha z^\alpha, \qquad 1 \leqq i \leqq r,$$

where a_i^α are constants. The a_i^α can be so chosen that none of the determinants of order $r + 1$ of the matrix

$$\begin{pmatrix} 1 & \cdots & 1 \\ a_1^0 & \cdots & a_1^n \\ \cdots & \cdots & \cdots \\ a_r^0 & \cdots & a_r^n \end{pmatrix}$$

will vanish. It then follows that the vectors $\mathfrak{z}, \mathfrak{v}_1, \cdots, \mathfrak{v}_r$ will be linearly dependent when and only when all products

$$z^{\alpha_0} \cdots z^{\alpha_r} = 0, \qquad 0 \leqq \alpha_0, \cdots, \alpha_r \leqq n,$$

the indices $\alpha_0, \cdots, \alpha_r$ being distinct from each other. This is possible when and only when $n + 1 - r$ of the coordinates z^α will vanish. In other words, for this particular field of r vectors the points of P_n at which the r vectors are linearly dependent are the linear spaces of dimension $r - 1$ defined by setting $n + 1 - r$ of the homogeneous coordinates of P_n to zero. The number of such linear spaces is $\binom{n+1}{r}$ and each of them is to be counted simply. Hence our theorem is proved.

THEOREM 14. (E. Cartan).[36] *Let M be a closed submanifold of topological dimension $2n - 2r + 2$ of the complex projective space of n dimensions P_n in which an elliptic Hermitian Geometry is defined. Let m be the number of points of intersection of M with a generic linear subspace of dimension $r - 1$ of P_n. Then*

$$m = \frac{1}{(2\pi\Gamma - 1^{n-r+1})} \int_m \Lambda^{n-r+1} \tag{55}$$

In particular, if M is an algebraic variety of dimension $n - r + 1$ in P_n, m is its order.

This theorem is an immediate consequence of the Theorems 11, 12, and 13.

Related to these discussions is also a formula due to W. Wirtinger. In Hermitian Geometry, as in Riemannian Geometry, it is common to define an element of volume of topological dimension $2p$ by means of the equation

$$\Delta_{2p} = \pm \sum_{(i_1 \cdots i_p)} \omega_{i_1} \cdots \omega_{i_p} \bar{\omega}_{i_1} \cdots \bar{\omega}_{i_p}, \tag{56}$$

---

[36] CARTAN, [3], p. 206.

the summation being over all the combinations of $i_1, \cdots, i_p$ from 1 to n. Up to a sign Λ_{2p} is equal to $\frac{1}{p!}\Lambda^p$. We define

$$\Delta_{2p} = \frac{1}{p!}\Lambda^p. \tag{57}$$

From (55) follows the theorem:

THEOREM 15. (W. Wirtinger).[37] *In the complex projective space of n dimensions with the elliptic Hermitian metric let V_{2p} be a p-dimensional algebraic variety of order m and volume V. Then*

$$V = \frac{(2\pi)^p}{p!}m. \tag{58}$$

INSTITUTE FOR ADVANCED STUDY, PRINCETON, AND
TSING HUA UNIVERSITY, CHINA.

BIBLIOGRAPHY

1. ALEXANDROFF, P., UND HOPF, H., *Topologie* I, Berlin 1935.
2. ALLENDOERFER, C. B., AND WEIL, A., *The Gauss-Bonnet theorem for Riemannian polyhedra*, Trans. Amer. Math. Soc., Vol. 53 (1943), pp. 101–129.
3. CARTAN, E., *Sur les invariants intégraux de certains espaces homogènes clos et les propriétés topologiques de ces espaces*, Annales Soc. pol. Math., Tome 8 (1929), pp. 181–225.
4. CARTAN, E., *Leçons sur la géométrie projective complexe*, Paris 1931.
5. CHERN, S., *A simple intrinsic proof of the Gauss-Bonnet formula for closed Riemannian manifolds*, Annals of Math., Vol. 45 (1944), pp. 747–752.
6. CHERN, S., *Integral formulas for the characteristic classes of sphere bundles*, Proc. Nat. Acad. Sci., Vol. 30 (1944), pp. 269–273.
7. CHERN, S., *Some new viewpoints in differential geometry in the large*, to appear in Bull. Amer. Math. Soc.
8. EHRESMANN, C., *Sur la topologie de certains espaces homogènes*, Annals of Math., Vol. 35 (1934), pp. 396–443.
9. EHRESMANN, C., *Sur la topologie des groupes simples clos*, C. R. Acad. Sci. Paris, Vol. 208 (1939), pp. 1263–1265.
10. EHRESMANN, C., Various notes on fibre spaces in C. R. Acad. Sci. Paris, Vol. 213 (1941), pp. 762–764; Vol. 214 (1942), pp. 144–147; Vol. 216 (1943), pp. 628–630.
11. EILENBERG, S., *Cohomology and continuous mappings*, Annals of Math., Vol. 41 (1940), pp. 231–251.
12. FELDBAU, J., *Sur la classification des espaces fibrés*, C.R. Acad. Sci. Paris, Vol. 208 (1939), pp. 1621–1623.
13. FUBINI, G., *Sulle metriche definite da una forma Hermitiana*, Instituto Veneto, Vol. 63, 2 (1904), pp. 502–513.
14. HOPF, H., see ALEXANDROFF, P.
15. HUREWICZ, W., AND STEENROD, N., *Homotopy relations in fibre spaces*, Proc. Nat. Acad. Sci., Vol. 27 (1941), pp. 60–64.
16. KÄHLER, E., *Über eine bemerkenstwerte Hermitische Metrik*, Abh. Math. Sem. Hamburg, Vol. 9 (1933), pp. 173–186.
17. PASCAL, E., *Die Determinanten*, Leipzig 1900.
18. PONTRJAGIN, L., *Characteristic cycles on manifolds*, C. R. (Doklady) Acad. Sci. URSS (N. S.), Vol. 35 (1942), pp. 34–37.

---

[37] WIRTINGER, [31].

19. PONTRJAGIN, L., *On some topologic invariants of Riemannian manifolds*, C. R. (Doklady), Acad. Sci. URSS (N. S.), Vol. 43 (1944), pp. 91–94.
20. DE RHAM, G., *Sur l'analysis situs des variétés à n dimensions*, J. Math. pures et appl., Tome 10 (1931), pp. 115–200.
21. STEENROD, N., *Topological methods for the construction of tensor functions*, Annals of Math., Vol. 43 (1942), pp. 116–131.
22. STEENROD, N., *The classification of sphere bundles*, Annals of Math., Vol. 45 (1944), pp. 294–311.
23. STEENROD, N., see HUREWICZ, W.
24. STIEFEL, E., *Richtungsfelder und Fernparallelismus in n-dimensionalen Mannigfultigkeiten*, Comm. Math. Helv., Vol. 8 (1936), pp. 305–343.
25. STUDY, E., *Kürzeste Wege im komplexen Gebiet*, Math. Annalen, Vol. 60 (1905), pp. 321–377.
26. VAN DER WAERDEN, B. L., *Topologische Begründung des Kalküls der abzählenden Geometrie*, Math. Annalen, Vol. 102 (1930), pp. 337–362.
27. WEIL, A., see ALLENDOERFER, C. B.
28. WEYL, H., *The Classical Groups*, Princeton 1939.
29. WHITNEY, H., *Topological properties of differentiable manifolds*, Bull. Amer. Math. Soc., Vol. 43 (1937), pp. 785–805.
30. WHITNEY, H., *On the topology of differentiable manifolds*, Lectures in Topology, pp. 101–141, Michigan 1941.
31. WIRTINGER, W., *Eine Determinantenidentität und ihre Anwendung auf analytische Gebilde in Euklidischer und Hermitischer Massbestimmung*, Monatshefte für Math. u. Physik, Vol. 44 (1936), pp. 343–365.

SUR UNE CLASSE REMARQUABLE DE VARIÉTÉS DANS L'ESPACE PROJECTIF À N DIMENSIONS

Par M. Shiing-shen Chern (陳省身)

Department of Mathematics

(*Received September 6, 1940*)

RÉSUME

On donne dans cette Note une propriété caractéristique d'une classe de variétés dans l'espace projectif à n dimensions, étudiées auparavant par Cartan. Ces variétés fournissent une généralisation des surfaces ayant un réseau. Il existe une transformation entre ces variétés généralisant la transformation bien connue de Laplace.

INTRODUCTION

La géométrie différentielle projective des réseaux conjugués[1] a été l'objet de recherches de bien des géométres. Analytiquement cette théorie est étroitement liée à la théorie de l'équation aux dérivées partielles de Laplace:

$$\frac{\partial^2 x}{\partial u \partial v} = a\frac{\partial x}{\partial u} + b\frac{\partial x}{\partial v} + cx, \tag{1}$$

où a, b, c sont des fonctions de u, v. C'est le problème de l'intégration de l'équation (1) qui a conduit Laplace, Moutard, Goursat de fonder sur cette équation une théorie analytique profonde. Les interprétations géométriques de ces résultats sont dues à Darboux, Koenigs, Guichard, Tzitzéica, Bompiani, etc., en ne citant que les noms les plus importants. Je vais signaler dans cette Note une classe de variétés à p dimensions dans l'espace projectif a n dimensions. qui jouent de propriétés généralisant les propriétés des surfaces sur lesquelles il existe un réseau conjugué. M. Cartan a rencontré ces variétés dans un problème de nature differente.[2] Je me propose donc d'appeler ces variétés les variétés de Cartan.

§1. NOTIONS PRÉLIMINAIRES. EQUATIONS FONDAMENTALES.

Nous définissons d'après Cartan comme repère projectif[3] de l'espace projectif à n dimensions l'ensemble de $n+1$ points analytiques (chacun à $n+1$ composantes)

$A, A_1, \cdots, A_n$, définis à un facteur commun près. Un tel repère dépend de $n(n+2)$ paramtères arbitraires. Les équations du déplacement infinitésimal sont de la forme

$$\begin{aligned} dA &= \omega_{00} A + \omega_1 A_1 + \cdots + \omega_n A_n, \\ dA_1 &= \omega_{10} A + \omega_{11} A_1 + \cdots + \omega_{1n} A_n, \\ &\cdots\cdots\cdots \\ dA_n &= \omega_{n0} A + \omega_{n1} A_1 + \cdots + \omega_{nn} A_n, \end{aligned} \tag{2}$$

où les ω sont des formes de Pfaff à ces paramètres. En introduisant les notions "produit extérieur" et "dérivée extérieure" comme d'habitude, on sait que les ω satisfont aux équations

$$\begin{aligned} \omega'_{00} &= [\omega_1 \omega_{10}] + \cdots + [\omega_n \omega_{n0}], \\ \omega'_i &= [\omega_{00} \omega_i] + [\omega_1 \omega_{1i}] + \cdots + [\omega_n \omega_{ni}], \\ \omega'_{ij} &= [\omega_{i0} \omega_j] + [\omega_{i1} \omega_{1j}] + \cdots + [\omega_{in} \omega_{nj}], \\ \omega'_{i0} &= [\omega_{i0} \omega_{00}] + [\omega_{i1} \omega_{10}] + \cdots + [\omega_{in} \omega_{n0}], \end{aligned} \qquad i, j = 1, \cdots, n, \tag{3}$$

qui s'appellent les équations de structure de l'espace.

Cela étant, considérons dans l'espace une variété V_p à p dimensions[4)] et faisons correspondre à chaque point A de cette variété un repère dont les $n+1$ sommets seront le point A, puis p points $A_1, A_2, \cdots, A_p$ situés dans l'hyperplan tangent, enfin $n-p$ autres points $A_{p+1}, \cdots, A_n$. Pour tout déplacement infinitésimal dans cette famille de repères on a

$$\begin{aligned} &\omega_{p+1} = \cdots = \omega_n = 0, \\ &\omega_{i\alpha} = \sum_{j=1}^{p} g_{ij\alpha} \omega_j, \quad i = 1, \cdots, p;\ \alpha = p+1, \cdots, n, \end{aligned} \tag{4}$$

où les $g_{ij\alpha}$ sont des fonctions des paramètres sur la variété et sont symétriques par rapport aux indices , i, j. Soit q le nombre des formes quadratiques

$$\Phi_\alpha = \omega_1 \omega_{1\alpha} + \cdots + \omega_p \omega_{p\alpha}, \quad \alpha = p+1, \cdots, n, \tag{5}$$

qui sont linéairement indépendantes. On sait qu'on peut choisir le repère attaché au point A de manière que les formes $\Phi_{p+1}, \cdots, \Phi_{p+q}$ soient linéairement indépendantes, les formes $\Phi_{p+q+1}, \cdots, \Phi_n$ étant identiquement nulles. Le réseau linéaire de cônes du second ordre de sommet A défini par l'équation

$$\lambda_{p+1} \Phi_{p+1} + \cdots + \lambda_{p+q} \Phi_{p+q} = 0 \tag{6}$$

est appelé le réseau asymptotique relatif au point A. Il jouit de la propriété que tout cône du réseau est engendré par les tangentes aux courbes de la variété dont le plan osculateur est situé dans une variété plane fixe à $p+q-1$ dimensions qui contient l'hyperplan tangent et est contenue dans l'hyperplan osculateur de V_p.

Cela posé, M. Cartan s'est proposé de déterminer toutes les variétés à p dimensions dont le réseau asymptotique est réductible à la forme[5])

$$\lambda_1 \omega_1^2 + \cdots + \lambda_p \omega_p^2 = 0. \tag{7}$$

Pour une telle variété on pourra évidemment supposer choisi le repère attaché au point A de manière à avoir

$$\Phi_{p+i} = \omega_i^2, \quad i = 1, \cdots, p. \tag{8}$$

Les variétés cherchées peuvent être donc regardées comme les solutions du système de Pfaff:

$$\begin{aligned} &\omega_\alpha = 0, && \alpha = p+1, \cdots, n, \\ &\omega_{i,\, p+i} = \omega_i, && i = 1, \cdots, p, \\ &\omega_{j,\, p+i} = 0, && i \neq j;\ i, j = 1, \cdots, p, \\ &\omega_{i,\, 2p+\lambda} = 0, && i = 1, \cdots, p;\ \lambda = 1, \cdots, n-2p. \end{aligned} \tag{9}$$

La dérivation extérieure de ces équations nous donne alors

$$\begin{aligned} &[\omega_i (\omega_{p+i,\, p+i} - 2\omega_{ii} + \omega_{00})] - \sum_{k \neq i}^{1-p} [\omega_k \omega_{ki}] = 0, \quad i = 1, \cdots, p, \\ &-[\omega_i\, \omega_{ji}] + [\omega_j\, \omega_{p+j,\, p+i}] = 0, \quad i \neq j;\ i, j = 1, \cdots, p, \\ &[\omega_i\, \omega_{p+i,\, 2p+\lambda}] = 0, \quad i = 1, \cdots, p;\ \lambda = 1, \cdots, n-2p. \end{aligned} \tag{10}$$

En appliquant la théorie des systèmes de Pfaff en involution, M. Cartan a démontré que *les variétés cherchées dépendent de $p(p-1)$ fonctions arbitraires de deux arguments*. Nous appellerons une telle variété une variété de Cartan.

Supposons que les indices i, j, k prennent les valeurs $1, \cdots, p$ et qu'ils sont tous distincts. Supposons encore que λ prend les valeurs $1, \cdots, n-2p$. Des equations (10) on déduit

$$\begin{aligned}
\omega_{ji} &= a_{ji}\,\omega_i + b_{ji}\,\omega_j,\\
\omega_{p+j,\,p+i} &= -\,b_{ji}\,\omega_i + c_{ji}\,\omega_j,\\
\omega_{p+i,\,2p+\lambda} &= e_{i\lambda}\,\omega_i,\\
\omega_{p+i,\,p+i} - 2\omega_{ii} + \omega_{00} &= -\sum_j a_{ji}\,\omega_j + c_i\,\omega_i,
\end{aligned} \tag{11}$$

où les coefficients des ω sont des fonctions des paramètres auxquels dépend le repère. En utilisant les équations de structure et tenant compte des équations (11), on a

$$\begin{aligned}
\omega_i' &= [\,\omega_i\,(\,\omega_{ii} - \omega_{00} - \sum_j a_{ji}\,\omega_j\,)\,],\\
\omega_i' &= [\,\omega_i\,(\,a_{ji}\,\omega_{ii} - a_{ji}\,\omega_{jj} - \omega_{j0} - \sum_k a_{jk}\,a_{kj}\,\omega_k\,)\,]\\
&\quad + [\,\omega_j\,(\,b_{ji}\,\omega_{ii} - b_{ji}\,\omega_{jj} + \omega_{p+j,\,i} + \sum_k b_{jk}\,b_{ki}\,\omega_k\,)\,]\\
&\quad - \sum_k b_{jk}\,a_{ki}\,[\,\omega_i\ \omega_j\,].
\end{aligned} \tag{12}$$

On trouve donc, en dérivant la première équation de (11),

$$\begin{aligned}
da_{ji} &= a_{ji}(\,\omega_{jj} - \omega_{00}\,) + \omega_{j0} + \sum_k a_{jk}\,a_{ki}\,\omega_k - a_{ji}\sum_k a_{ki}\,\omega_k\\
&\qquad + f_{ji}\,\omega_i + g_{ji}\,\omega_j,\\
db_{ji} &= b_{ji}(\,-\,\omega_{ii} + 2\,\omega_{jj} - \omega_{00}\,) - \omega_{p+j,\,i} + h_{ji}\,\omega_j\\
&\quad - \sum_{k\neq i,\,j}(\,b_{jk}\,b_{ki} + b_{ji}\,a_{kj}\,)\,\omega_k + (\,g_{ji} - b_{ji}\,a_{ij} - \sum_{k\neq i,\,j} b_{jk}\,a_{ki}\,)\,\omega_i.
\end{aligned} \tag{13}$$

On tire de la première équation de (12) une conséquence importante. C'est que *chacune des équations*

$$\omega_i = 0, \quad i = 1, \cdots, p \tag{14}$$

est complètement intégrable. Géométriquement cela signifie qu'il existe sur la variété p familles de variétés à $p-1$ dimensions tangentes en chacun de leurs points à l'une des variétés planes $AA_1 \cdots A_{i-1}\,A_{i+1} \cdots A_p$, $i = 1, \cdots, p$. Il est facile de voir qu'une telle variété à $p-1$ dimensions est aussi une variété de Cartan.

§2. UNE PROPRIÉTÉ CARACTÉRISTIQUE DES VARIÉTÉS DE CARTAN

Nous dirons *qu'une variété à p dimensions possède un réseau généralisé* ou simplement un réseau s'il existe sur la variété p familles de courbes telles que par tout point de la variété il passe une et une seule courbe de chaque famille et telles que quand l'hyperplan tangent est déplacé le long d'une courbe d'une de ces familles deux hyperplans tangents infiniment voisins se rencontrent à une variété plane à $p-1$ dimensions tangente à toutes les courbes des $p-1$ autres familles passant par le point considéré. Il est clair que cette propriété généralise celle d'un réseau conjugué sur une surface (variété à deux dimensions).

Attachons à chaque point A de la variété un repère $AA_1\cdots A_n$ tel que la variété plane à p dimensions $[AA_1\cdots A_p]$ soit l'hyperplan tangent au point A. Nous avons donc les équations (4). Supposons que les p familles de courbes sont données par les équations

$$\omega_1=0,\ldots, \omega_{i-1}=0, \omega_{i+1}=0,\ldots, \omega_p=0, \; i=1,\ldots, p. \tag{15}$$

Nous allons montrer que *les conditions nécessaires et suffisantes pour qu'une variété à p dimensions ait un réseau généralisé donné par les équations (15) sont*

$$g_{ij\alpha}=0 \quad , \; i\neq j, \; i,j=1,\ldots, p, \; \alpha=p+1,\ldots, n, \tag{16}$$

avec la condition supplémentaire que les fonctions $g_{iip+1},\ldots, g_{iin}$, pour chaque i fixe, ne sont pas toutes nulles.

La condition est manifestement nécessaire. En effect, quand l'hyperplan tangent est déplacé le long de la courbe (15), la condition que le point

$$xA+x_1A_1+\cdots+x_pA_p$$

appartienne à l'hyperplan tangent infiniment voisin est que le point

$$xdA+x_1dA_1+\cdots+x_pdA_p$$

soit une combinaison linéaire de $A, A_1,\ldots, A_p$. Cette condition s'exprime par les équations

$$x_1\omega_{1\alpha}+\cdots+x_p\omega_{p\alpha}=0, \quad \alpha=p+1,\ldots, n,$$

ou

$$\sum_{jk=1}^{p} g_{jk\alpha}x_j\,\omega_k=0, \quad \alpha=p+1,\ldots, n.$$

Pour que ces équations soient identiques à l'équation $x_i = 0$ il faut que les conditions dans notre théorème soient remplies.

Nous supprimerons la démonstration de la suffisance de ces conditions dont les raisonnements sont analogues.

On peut remarquer qu'il suffit de supposer que la condition concernant l'intersection des hyperplans tangents infiniment voisins en les déplaçant le long d'une courbe soit vérifiée pour $p-1$ des familles de courbes sur la variété. La même propriété pour les courbes de la p-ième famille en résulte comme leur conséquence.

Pour une variété ayant un réseau on a donc

$$\Phi_\alpha = \sum_{i=1}^{p} g_{ii\alpha}\omega_i^2 \quad , \quad \alpha = p+1, \ldots, n. \tag{17}$$

Soit q le rang de la matrice

$$\begin{pmatrix} g_{11p+1} & g_{22p+1} & \cdots & g_{ppp+1} \\ \cdots\cdots & & & \\ g_{11n} & g_{22n} & \cdots & g_{ppn} \end{pmatrix} \tag{18}$$

de sorte que q des formes Φ_α soient linéairement indépendantes. Il en résulte que les hyperplans osculateurs de la variété sont à $p+q$ dimensions et que

$$1 \leqq q \leqq p \ , \ q \leqq n-p.$$

Après ces discussions préliminaires on peut énoncer dans le théorème suivant la propriété caractéristique pour les variété de Cartan définies au numéro précédent:

La condition nécessaire et suffisante pour qu'une variété soit une variété de Cartan est qu'il existe sur elle un réseau généralisé et que ses hyperplans osculateurs soient à 2p dimensions.

Il est évident que la condition est nécessaire. Pour voir qu'elle est aussi suffisante remarquons qu'on a

$$dA_i = \omega_{io} A + \omega_{i1} A_1 + \cdots + \omega_{ip} A_p + \omega_i (g_{iip+1} A_{p+1} + \cdots + g_{iin} A_n),$$

$$i = 1, \cdots, p.$$

Dans ces équations les points

$$g_{iip+1} A_{p+1} + \cdots + g_{iin} A_n, \quad i=1,\cdots, p$$

sont linéairement indépendants. On peut modifier le repère attaché au point A en les choisissant comme les points $A_{p+1}, \cdots, A_{2p}$ du repère nouveau. Avec cette famille de repères nouveaux on voit facilement que la variété considérée est une variété de Cartan.

§3. LA TRANSFORMATION GÉNÉRALISÉE DE LAPLACE

Dans ce numéro nous allons étudier une transformation des variétés de Cartan, qui jouit de propriétés analogues à la transformation bien connue de Laplace des réseaux conjugués.

Commençons à démonter le théorème suivant:

Sur chaque tangente AA_i il existe $p-1$ points

$$A_{ij} = A_i - a_{ij} A, \quad i \neq j, \; i, j=1, \cdots, p \tag{19}$$

ayant la propriété que quand la variété à v dimensions $[AA_{i_1} \cdots A_{i_v}]$ est déplacée le long de AA_j, $j \neq i_1, \cdots, i_v$, deux variétés planes à v dimensions infiniment voisines ont en commun la variété plane à $v-1$ dimensions $[A_{i_1 j} \cdots A_{i_v j}]$.

Pour démontrer ce théoréme supposons que l'indice t prend les valeurs $i_1, \cdots, i_v$. Un point de la variété plane $[AA_{i_1} \cdots A_{i_v}]$ est de la forme

$$x A + \sum_t x_t A_t$$

La condition pour qu'il appartienne aussi à la variété plane infiniment voisine de $[AA_{i_1} \cdots A_{i_v}]$ est que le point

$$x\, dA + \sum_t x_t\, dA_t$$

soit une combinaison linéaire de $A, A_{i_1}, \cdots, A_{i_v}$. Lorsque $[AA_{i_1} \cdots A_{i_v}]$ se déplace le long de AA_j, cette condition s'exprime par l'équation

$$x + \sum_t x_t \; a_{tj} = 0.$$

Cela montre que l'intersection de $[AA_{i_1} \cdots A_{i_\nu}]$ et de la variété plane infiniment voisine se compose des points

$$-\sum_t x_t\, a_{ti}\, A + \sum_t x_t\, A_t = \sum_t x_t\, A_{tj},$$

où $x_{i_1}, \cdots, x_{i_\nu}$ sont arbitraires. Donc le théorème est démontré.

Lorsque le point A décrit la variété V_p, chacun des points A_{ji} $(i, j = 1, \cdots, p;\ i \neq j)$ décrit une variété dont les points sont en correspondance biunivoque avec ceux de V_p. Pour ces variété on a la propriété fondamentals suivante: ***La variété décrite par A_{ji} est en général une variété de Cartan.***

Pour démontrer ce théorème on utilise les équations du déplacement infinitésimal (2) et les équations (13). On trouve alors

$$dA_{ji} = (\omega_{jj} - a_{ji}\,\omega_j)\, A_{ji} - f_{ji}\,\omega_i\, A + \sum_{k \neq i,j} (a_{jk} - a_{ji})\,\omega_k\, A_{ki} + \omega_j\, B_{ji}, \quad (20)$$

où l'on a posé

$$B_{ji} = A_{p+j} + \sum_{k \neq j} b_{jk}\, A_k - g_{ji}\, A. \quad (21)$$

En supposant que

$$f_{ji} \neq 0, \quad a_{jk} - a_{ji} \neq 0, \quad k \neq i,$$

on voit que la variété décrite par A_{ji} est à p dimensions, dont l'hyperplan tangent au point A_{ji} est la variété plane $[AA_{1i} \cdots A_{i-1i}\, A_{i+1i} \cdots A_{pi}\, B_{ji}]$. Pour voir si elle est une variété de Cartan calculons dA, dA_{ki} $(k \neq i, j)$, dB_{ji} mod. A, A_{ki} $(k \neq i)$, B_{ji}, c'est-à-dire mod. A, A_k $(k \neq i)$, $A_{p+i} + b_{ji}\, A_i$. Un calcul facile nous donne

$$dA \equiv \omega_i\, A_i,$$

$$dA_{ki} \equiv \omega_k\, B_{ki}, \quad k \neq i, j,$$

$$dB_{ji} \equiv \omega_j \Big(\sum_{k \neq j} c_{jk}\, A_{p+k} + \sum_{\lambda} e_{j\lambda}\, A_{2p+\lambda} - c_j\, b_{ji}\, A_i + h_{ji}\, A_i\Big).$$

Cela montre que la varieété décrite par A_{ji} est bien une variété de Cartan, si les points

$$A_i, B_{ki}, \quad k \neq i, j,$$

$$\sum_{k \neq j} c_{jk} A_{p+k} + \sum_{\lambda} l_{j\lambda} A_{2p+\lambda}$$

sont linéairement indépendants.

En généralisant la notion classique de la transformation de Laplace on peut dire que les variétés décrites par A_{ji}, $i \neq j$; $i=1,\cdots,p$ sont les variétés transformées de Laplace de la variété donnée. Une variété de Cartan possède donc $p(p\text{-}1)$ variété transformées de Laplace qui sont en général aussi les variétés de Cartan. Dans le cas spécial $p=2$ on obtient les deux transformes de Laplace d'un réseau conjugué.

On peut chercher à trouver les transformes de Laplace de la variété décrite par A_{ji}. En ce qui concerne ce problème signalons les résultats suivants dont la démonstration est d'ailleurs facile:

Les tangentes aux courbes du réseau généralisé A_{ji} étant les droites A_{ji} A, A_{ji} A_{ki} $(k \neq i, j)$, A_{ji} B_{ji}, les transformes de Laplace sur A_{ji} A sont A_{jl} $(l \neq i, j)$, A, ceux sur A_{ji} A_{ki} $(k \neq i, j)$ sont

$$f_{ki} A_{ji} - f_{ji} A_{ki}, \quad A_{ki},$$

$$(a_{ki} - a_{kl}) A_{ji} + (a_{jl} - a_{ji}) A_{ki}, \quad l \neq i, j, k,$$

tandis que ceux sur A_{ji} B_{ji} sont donnés par des expressions plus compliquées.

BIBLIOGRAPHIE

1. Les travaux sur la théorie des réseaux sont nombreux. Citons seulement le livre: G. Tzitzéica, *Géométrie différentielle projective des réseaux*, Paris 1924.

2. E. Cartan, *Sur les variétés de courbure constante d'un espace euclidien ou non-euclidien*, Bull. Soc. math. de France, **47**, 125-160 (1919), **48**, 132-208 (1920). Cette Note sera citée comme: Cartan, *variétés*.

3. Cartan, *variétés*, nos. 1, 2.

4. Cartan, *variétés*, nos. 33-38.

5. Cartan, *variétés*, nos. 43-45.

Annals of Mathematics
Vol. 49, No. 2, April, 1948

ON THE MULTIPLICATION IN THE CHARACTERISTIC RING OF A SPHERE BUNDLE

Shiing-shen Chern

(Received June 9, 1947)

Introduction

Perhaps the most important unifying idea in the theory of sphere bundles is the imbedding theorem.[1] It asserts essentially that every sphere bundle whose spheres are of dimension $n - 1$ is equivalent to the bundle induced by mapping the base space M into the manifold $H(n, N)$, the Grassmann manifold of all the n-dimensional linear spaces through a point in an Euclidean space E^{n+N} of dimension $n + N$ ($N \geqq \dim M + 1$). Steenrod[2] proved that two such induced sphere bundles are equivalent if and only if the mappings of M into $H(n, N)$ are homotopic. We take a coefficient ring R and denote by $\mathcal{H}(H(n, N))$ the cohomology ring of $H(n, N)$ relative to R. The class of homotopic mappings of M into $H(n, N)$ induces a definite ring homomorphism of the cohomology ring $\mathcal{H}(H(n, N))$ into the cohomology ring of M. Its image, to be denoted by $C(M, R)$, will be called the *characteristic ring* of the sphere bundle in M. A cohomology class of $C(M, R)$ is called a *characteristic cohomology class.*

The purpose of this paper is to study the multiplication in the characteristic ring, in the sense of Whitney's cup product. Following some indications in Schubert's enumerative geometry (Abzählende Geometrie), we shall define symbols which describe the characteristic cohomology classes. Formulas are then established which enable us to express the product of two characteristic cohomology classes as a sum of these classes. For simplicity we shall restrict ourselves in this paper largely to the case that the coefficient ring is the ring of residue classes of integers modulo 2. The characteristic cohomology classes in our generalized sense of course include as particular cases the Stiefel-Whitney classes. It follows, however, from our product formulas that every characteristic class is a polynomial of the Stiefel-Whitney classes, so long as the coefficient ring mod. 2 is concerned. As application we derive also from our product formulas some necessary conditions for the possibility of imbedding a manifold in an Euclidean space of a given dimension. We prove, for instance, that a real projective space of even dimension n cannot be imbedded in an Euclidean space of dimension $n + 2$, if n is not of the form $n = 2(2^k - 1)$, $k \geqq 1$.

§1 is devoted to the preliminaries. The main part of the paper is §2, in which the product formulas are proved and some consequences drawn. Some applications are given in §3.

---

[1] Whitney, H., *Topological properties of differentiable manifolds*, Bull. Amer. Math. Soc., 43 (1937), 785–805.

[2] Steenrod, N., *The classification of sphere bundles*, these Annals, 45 (1944), 294–311.

§1. Preliminaries

1. Let F be a sphere bundle, with the base space M and with fibres which are spheres of dimension $n - 1$. M is supposed to be a polyhedron. According to the imbedding theorem due to Whitney and Steenrod, F is equivalent to the bundle induced by a (continuous) mapping

$$f: M \to H(n, N), \tag{1}$$

where $H(n, N)$ is the Grassmann manifold of all n-dimensional linear spaces through a fixed point 0 in an Euclidean space E^{n+N} of dimension $n + N$. This mapping f is determined up to a homotopy.

We take a coefficient ring R and denote by $\mathcal{H}(H(n, N), R)$ the cohomology ring of $H(n, N)$, relative to the coefficient ring R. The image $C(M, R)$ in M of $\mathcal{H}(H(n, N), R)$ under the ring homomorphism f^{*-1} induced by f is called the *characteristic ring* of M, relative to the coefficient ring R. A cohomology class of M belonging to $C(M, R)$ will be called a *characteristic cohomology class.*

2. To describe a characteristic cohomology class it is sufficient to give its original in $\mathcal{H}(H(n, N), R)$. For this purpose we shall introduce the Schubert symbols. Let $a_1, \cdots, a_n$ be a sequence of integers such that

$$0 \leqq a_1 \leqq a_2 \leqq \cdots \leqq a_n \leqq N. \tag{2}$$

We take through 0 a sequence of linear spaces

$$L_1 \subset L_2 \subset \cdots \subset L_n, \tag{3}$$

whose dimensions are

$$\dim L_i = a_i + i. \tag{4}$$

In $H(n, N)$ we consider the linear spaces X which satisfy the conditions

$$\dim (X \cap L_i) \geqq i, \quad i = 1, \cdots, n. \tag{5}$$

It was proved by Ehresmann[3] that these linear spaces, minus the ones on the boundary, form an open cell of dimension $\sum_{i=1}^{n} a_i$ in $H(n, N)$ and that every chain in $H(n, N)$ is homologous to a linear combination of these cells, when considered as chains. Such a chain will be denoted by the Schubert symbol $[a_1 \cdots a_n]$.[4]

We shall denote by $\{a_1 \cdots a_n\}$ the cochain whose value is one for $[a_1 \cdots a_n]$ and zero for other chains of that dimension. For convenience of terminology we shall call such chains (and cochains) Schubert chains (and Schubert cochains).

---

[3] Ehresmann, C., *Sur la topologie des certaines variétés algébriques réelles*, Journal de Math. Pures, 104 (1939), 69–100.

[4] This is in fact a modified form of the Schubert symbol.

By orienting the Schubert chains $[a_1 \cdots a_n]$ according to Ehresmann, the boundary relations can be written in the form

$$(6)\qquad \begin{aligned} \partial[a_1 \cdots a_n] &= \sum_{i=1}^{n} \eta_i [a_1 \cdots a_i - 1 \cdots a_n], \\ \delta\{a_1 \cdots a_n\} &= \sum_{i=1}^{n} \zeta_i \{a_1 \cdots a_i + 1 \cdots a_n\}, \end{aligned}$$

where

$$(7)\qquad \begin{aligned} \eta_i &= 0, \quad \text{if} \quad a_i + n + i \quad \text{is odd}, \\ \eta_i &= 2(-1)^{a_1 + \cdots + a_i}, \quad \text{if} \quad a_i + n + i \quad \text{is even}, \\ \zeta_i &= 0, \quad \text{if} \quad a_i + n + i \quad \text{is even}, \\ \zeta_i &= 2(-1)^{a_1 + \cdots + a_i + 1}, \quad \text{if} \quad a_i + n + i \quad \text{is odd}. \end{aligned}$$

Of these relations the first set is essentially the boundary relations given by Ehresmann with a slight change of notation, while the second set follows from the first set by making use of the definition of the cochains $\{a_1 \cdots a_n\}$. It will be understood here that all Schubert symbols which have no sense, that is, for which the inequalities (2) are not satisfied, will be replaced by zero.

It follows from (6) that

$$(8)\qquad \delta\{a_1 \cdots a_n\} = 2B\{a_1 \cdots a_n\},$$

where $B\{a_1 \cdots a_n\}$ is a cocycle of dimension $\sum_{i=1}^{n} a_i + 1$ and order two.

From the above discussion we see that a characteristic class in M is defined by an expression of the form

$$(9)\qquad \gamma = \sum \lambda_i \{a_1 \cdots a_n\}, \qquad \lambda_i \in R,$$

being the image class under f^{*-1} of the class to which γ belongs.

3. The boundary relations (6) lead to the following classification of the cocycles on $H(n, N)$: 1) Cocycles of the first kind, which are cocycles of the form $\{a_1 \cdots a_n\}$ and are not of order two; 2) Cocycles of the second kind, which are cocycles of the form $\{a_1 \cdots a_n\}$ and are of order two; 3) Cocycles of the third kind, which are cocycles of the form $B\{a_1 \cdots a_n\}$, when it contains more than one term. Every cocycle of the third kind is of order two.

We shall prove the following theorem:

THEOREM 1. *Every cocycle on $H(n, N)$ is cohomologous to a linear combination of the cocycles of the three kinds. The number of cocycles of the first kind of dimension r is equal to the r-dimensional Betti number of $H(n, N)$.*

It follows from Ehresmann's work that every cocycle γ on $H(n, N)$ is cohomologous to a linear combination of the Schubert cochains. We shall say that the cochain $\{a_1 \cdots a_n\}$ is of *smaller rank* than $\{b_1 \cdots b_n\}$ (and that $\{b_1 \cdots b_n\}$ is of larger rank than $\{a_1 \cdots a_n\}$), if $a_1 = b_1, \cdots, a_{i-1} = b_{i-1}, a_i < b_i$. It is sufficient for the proof of the first part of this theorem to show that by subtracting from γ a linear combination of the cocycles of the three kinds all the Schubert cochains in the resulting cocycle γ^* are of larger rank than the Schubert cochain of smallest rank in γ. In other words, the operation leads to an increase of the minimum rank and successive applications of it will come to an end.

We first remove from γ all the Schubert cochains which are cocycles. In the cocycle which remains let $\{a_1 \cdots a_n\}$ be the Schubert cochain of minimum rank, and let $\{a_1 \cdots a_i + 1 \cdots a_n\}$ be that of minimum rank in $\delta\{a_1 \cdots a_n\}$. In order that $\delta\gamma = 0$, γ must contain a Schubert cochain of the form $\{a_1 \cdots a_i + 1 \cdots a_h - 1 \cdots a_n\}$, where $i < h$ and where $n + a_h + h$ is even. The symbol $\{a_1 \cdots a_i \cdots a_h - 1 \cdots a_n\}$ has then a sense, and we have

$$\begin{aligned}\delta\{a_1 \cdots a_i \cdots a_h - 1 \cdots a_n\} &= 2B\{a_1 \cdots a_i \cdots a_h - 1 \cdots a_n\}\\ &= \sum_{\alpha=0}^{i-1} \zeta_\alpha\{a_1 \cdots a_\alpha + 1 \cdots a_i \cdots a_h - 1 \cdots a_n\}\\ &\quad \pm 2\{a_1 \cdots a_i + 1 \cdots a_h - 1 \cdots a_n\}\\ &\quad \pm 2\{a_1 \cdots a_i \cdots a_h \cdots a_n\}.\end{aligned}$$

In $B\{a_1 \cdots a_i \cdots a_h - 1 \cdots a_n\}$ all the Schubert cochains, with the exception of $\{a_1 \cdots a_n\}$, are of larger rank than that of $\{a_1 \cdots a_n\}$. If, therefore, a suitable multiple of $B\{a_1 \cdots a_i \cdots a_h - 1 \cdots a_n\}$ is subtracted from γ, all the Schubert cochains in the difference will be of rank larger than that of $\{a_1 \cdots a_n\}$. Thus the minimum rank is increased. This proves the first part of the theorem.

To prove the second part of the theorem we notice that if $B\{a_1 \cdots a_n\}$ contains more than one term, it is of the form

$$\begin{aligned}B\{a_1 \cdots a_n\} &= \pm\{a_1 \cdots a_i + 1 \cdots a_h \cdots a_n\}\\ &\quad \pm \{a_1 \cdots a_i \cdots a_h + 1 \cdots a_n\} + \cdots,\end{aligned}$$

where $a_{i+1} - a_i \geqq 1$, $a_{h+1} - a_h \geqq 1$, and where $n + a_i + i$ and $n + a_h + h$ are both odd. These conditions imply, for instance, that

$$\{a_1 \cdots a_i + 1 \cdots a_h \cdots a_n\}$$

is not a cocycle. In other words, a Schubert cocycle never appears as a term of a cocycle of the third kind. It follows in particular that the cocycles of the first kind are homologically independent with respect to weak homology, and that the number of r-dimensional Schubert cocycles of the first kind is equal to the r-dimensional Betti number of $H(n, N)$.

4. From now on we shall restrict our discussion to the case that the coefficient ring is the ring of residue classes of integers mod. 2. Then every Schubert chain is a cycle and every Schubert cochain a cocycle. We have, moreover, the following intersection formulas:

$$\begin{aligned}&KI([a_1 \cdots a_n], [N - a_n \cdots N - a_1]) = 1,\\ (10)\qquad &KI([a_1 \cdots a_n], [N - b_n \cdots N - b_1]) = 0,\end{aligned}$$

where, in the last formula, the numbers $b_1, \cdots, b_n$ form a sequence distinct from $a_1, \cdots, a_n$, but with $\sum_{i=1}^n a_i = \sum_{i=1}^n b_i$.

It follows that by means of the Kronecker index the cycle $[N - a_n \cdots N - a_1]$ defines a cocycle, which is $\{a_1 \cdots a_n\}$.

§2. The Multiplication Formulas

Our main aim in this section is to establish two formulas which will enable us to express the cup product $\{a_1 \cdots a_n\} \smile \{b_1 \cdots b_n\}$ as a linear combination of Schubert cocycles of dimension $\sum_{i=1}^n (a_i + b_i)$.

1. In order to give the multiplication formulas in question let us write for simplicity $\{h\} = \{o \cdots oh\}$, and in general $\{a_1 \cdots a_k\} = \{o \cdots o\, a_1 \cdots a_k\}$. Then our first formula is

$$\{a_1 \cdots a_n\} \smile \{h\} \backsim \sum \{b_1 \cdots b_n\}, \tag{11}$$

where the summation is extended over all combinations $b_1, \cdots, b_n$, such that

$$\begin{gathered} o \leqq b_1 \leqq b_2 \leqq \cdots \leqq b_n \leqq N, \\ a_i \leqq b_i \leqq a_{i+1}\ (a_{n+1} = N), \qquad i = 1, \cdots n, \\ \sum_{i=1}^n a_i + h = \sum_{i=1}^n b_i. \end{gathered} \tag{12}$$

The second multiplication formula is

$$\{a_1 \cdots a_n\} \backsim \begin{vmatrix} \{a_1\} & \{a_1 - 1\} & \cdots & \{a_1 - \overline{n-1}\} \\ \{a_2 + 1\} & \{a_2\} & \cdots & \{a_2 - \overline{n-2}\} \\ & \cdots & & \\ \{a_n + \overline{n-1}\} & \{a_n + \overline{n-2}\} & \cdots & \{a_n\} \end{vmatrix}, \tag{13}$$

where the determinant expansion is by means of the cup product and where we define by convention: $\{o\} = 1$ and $\{c\} = 0$, if $c < 0$.

It is clear that by means of (11) and (13) we can express the cup product of any two Schubert cocycles as a linear combination of Schubert cocycles.

Formula (13) is an easy consequence of (11). For, by mathematical induction on n, on supposing the formula to be true for $n - 1$, we can expand the determinant in the right-hand side of (13) by Laplace's development according to the first column. This gives, when Δ denotes the determinant,

$$\Delta \backsim \sum_{i=1}^n \{a_i + i - 1\} \smile \{a_1 - 1 \cdots a_{i-1} - 1, a_{i+1}, \cdots a_n\}.$$

By (11) the ith summand, $i = 1, \cdots, n$, is equal to

$$\sum \{b_1 \cdots b_n\}.$$

where

$$o \leqq b_1 \leqq a_1 - 1 \leqq \cdots \leqq b_{i-1} \leqq a_{i-1} - 1 \leqq b_i \leqq a_{i+1} \leqq \cdots \leqq b_{n-1} \leqq a_n \leqq b_n \leqq N.$$

It is not hard to see that these terms cancel with each other, except the term $\{a_1 \cdots a_n\}$. This proves (13), under the assumption that (11) is true.

2. We proceed to give a proof of formula (11). By the remark in no. 4, §1, this formula is equivalent to the intersection formula

$$\begin{aligned}(14)\qquad &[N - a_n \cdots N - a_1] \cdot [N - h\, N \cdots N] \\ &= \sum [N - b_n \cdots N - b_1],\end{aligned}$$

where the left-hand side denotes the intersection of the two Schubert cycles. To prove (14) it is sufficient to prove that

$$(15)\qquad [b_1 \cdots b_n]\cdot[N - a_n \cdots N - a_1]\cdot[N - h\, N \cdots N] = 1,$$

if the b's satisfy the inequalities (12) and that the intersection number is otherwise zero.

The intersection cycle in the left-hand side is zero-dimensional. To determine the Schubert cycles $[b_1 \cdots b_n]$, $[N - a_n \cdots N - a_1]$ in the intersection we take respectively the sequences of linear spaces

$$(16)\qquad \begin{aligned} K_1 \subset K_2 \subset \cdots \subset K_n, \\ L_1 \subset L_2 \subset \cdots \subset L_n, \end{aligned}$$

which are supposed to be in general position. Their dimensions are given by

$$\begin{aligned} \dim K_i &= b_i + i, \\ \dim L_{n-i+1} &= N - a_i + n - i + 1, \qquad i = 1, \cdots, n. \end{aligned}$$

In order that there is through O an n-dimensional linear space

$$X \in [b_1 \cdots b_n]\cdot[N - a_n \cdots N - a_1],$$

we must have

$$\begin{aligned} \dim (X \cap K_i) &\geqq i, \\ \dim (X \cap L_{n-i+1}) &\geqq n - i + 1. \end{aligned}$$

Since these two intersections both belong to X, they intersect in a linear space of dimension $\geqq 1$. It follows that the same is true of K_i and L_{n-i+1}. The K's and L's being in general position, we must have

$$(b_i + i) + (N - a_i + n - i + 1) \geqq N + n + 1$$

or

$$b_i \geqq a_i.$$

These conditions are therefore necessary in order that the intersection cycle in the left-hand side of (15) is not zero.

For further discussion we shall introduce in E^{n+N} the rectangular coordinates $x_1, \cdots, x_n, y_1, \cdots, y_N$. As the equations of K_i and L_i we choose respectively

$$(17)\qquad x_1 = \cdots = x_{n-i} = y_{b_i+1} = \cdots = y_N = 0,$$

$$x_{i+1} = \cdots = x_n = y_1 = \cdots = y_{a_{n-i+1}} = 0. \tag{18}$$

We shall put

$$M_i = K_i \cap L_{n-i+1}, \qquad i = 1, \cdots, n. \tag{19}$$

A linear space X of dimension n through O can be defined by a set of n linearly independent vectors on it, that is, by a matrix of rank n:

$$\begin{pmatrix} x_1^{(1)} & \cdots & x_n^{(1)} & y_1^{(1)} & \cdots & y_N^{(1)} \\ & & \cdots & & & \\ x_1^{(n)} & \cdots & x_n^{(n)} & y_1^{(n)} & \cdots & y_N^{(n)} \end{pmatrix} \tag{20}$$

We introduce an n-dimensional auxiliary space A^n and regard the $n + N$ columns of this matrix as the components of $n + N$ *column vectors* $\mathfrak{x}_1, \cdots, \mathfrak{x}_n; \mathfrak{y}_1, \cdots, \mathfrak{y}_N$ in A^n. An X is thus determined by $n + N$ column vectors which span the whole space A^n and conversely.

We shall prove the following lemma:

LEMMA. *If* $X \epsilon [b_1 \cdots b_n]\cdot[N - a_n \cdots N - a_1]$ *and if* $b_i < a_{i+1}$, *then*

$$\mathfrak{y}_{b_i+1} = \cdots = \mathfrak{y}_{a_i+1} = 0, \qquad i = 1, \cdots, n.$$

In fact, by our choice of the sequences (16) the condition

$$\dim (X \cap K_i) \geqq i$$

is equivalent to the condition that the column vectors

$$\mathfrak{x}_1, \cdots, \mathfrak{x}_{n-i}, \qquad \mathfrak{y}_{b_i+1}, \cdots, \mathfrak{y}_N$$

belong to a linear space of dimension $n - i$, and the condition

$$\dim (X \cap L_{n-i}) \geqq n - i$$

to the condition that the column vectors

$$\mathfrak{x}_{n-i+1}, \cdots, \mathfrak{x}_n, \mathfrak{y}_1, \cdots, \mathfrak{y}_{a_i+1}$$

belong to a linear space of dimension i. If these two sets of vectors have a non-zero vector in common, they together will span a linear space of dimension $\leqq n - 1$, which contradicts our assumption. Thus the lemma is proved.

We shall denote by M the join of $M_1, \cdots, M_n$. Since the non-vanishing coordinates in M_i are $x_{n-i+1}, y_{a_i+1}, \cdots, y_{b_i}$, it follows from our lemma that $X \epsilon [b_1 \cdots b_n]\cdot[N - a_n \cdots N - a_1]$ only when $X \epsilon M$.

Suppose that, for a certain i, $b_i > a_{i+1}$. Then M is of dimension $\leqq h + n - 1$. We can choose a linear space H_0^{N-h+1} of dimension $N - h + 1$ which intersects M in the point O only. Then from $X \epsilon M$ we have $X \cap H_0^{N-h+1} = 0$. This means that

$$[b_1 \cdots b_n] \cdot [N - a_n \cdots N - a_1] \cdot [N - hN \cdots N] = 0.$$

We suppose next that

$$a_i \leqq b_i \leqq a_{i+1}, \qquad i = 1, \cdots, n.$$

Then M is exactly of dimension $h + n$, and M_i, M_j $(i \neq j)$ have only the point O in common. To determine $[N - h\ N \cdots N]$ we choose an H^{N-h+1} defined by the equations

$$y_{a_i+1} = \cdots = y_{b_i} = 0, \qquad i = 1, \cdots, n,$$

$$\frac{x_1}{\lambda_1} = \cdots = \frac{x_n}{\lambda_n}, \qquad \lambda_i \neq 0,$$

so that an $X \in [N - h\ N \cdots N]$ is characterized by the condition that it intersects H^{N-h+1} in a linear space of dimension $\geqq 1$. This H^{N-h+1} intersects M in the linear space of one dimension:

$$\frac{x_1}{\lambda_1} = \cdots = \frac{x_n}{\lambda_n}, \qquad y_\alpha = 0, \qquad \alpha = 1, \cdots, N.$$

From the definition of M_i in (19) it is seen that an

$$X \in [b_1 \cdots b_n] \cdot [N - a_n \cdots N - a_1]$$

intersects M_i in a linear space of dimension $\geqq 1$. Under the assumptions $a_i \leqq b_i \leqq a_{i+1}$, $i = 1, \cdots, n$, such an X can be determined by n non-zero vectors, one each in $M_1, \cdots, M_n$. The corresponding non-zero column vectors to determine X can therefore be supposed to be of the form

$$\mathfrak{x}_i = (0, \cdots, 0, x_i^{(i)}, 0, \cdots, 0), \qquad i = 1, \cdots, n,$$

$$\mathfrak{y}_{\alpha_i} = (0, \cdots, 0, y_{\alpha_i}^{(i)}, 0, \cdots, 0), \qquad \alpha_i = a_i + 1, \cdots, b_i.$$

Since $X \in [b_1 \cdots b_n] \cdot [N - a_n \cdots N - a_1] \cdot [N - h\ N \cdots N]$ contains the vector $(\lambda_1, \cdots, \lambda_n, 0, \cdots, 0)$, there is a linear combination of the n row vectors which is equal to this vector. Let $\mu_1, \cdots, \mu_n$ be the coefficients of the linear combination. Then we have

$$\mu_i x_i^{(i)} = \lambda_i,$$

$$\mu_i y_k^{(i)} = 0, \qquad a_i + 1 \leqq k \leqq b_i, \qquad i = 1, \cdots, n.$$

Since $\lambda_i \neq 0$, the first equation implies $\mu_i \neq 0$. But then we have

$$y_k^{(i)} = 0, \qquad a_i + 1 \leqq k \leqq b_i.$$

It follows that the only linear space belonging to $[b_1 \cdots b_n] \cdot [N - a_n \cdots N - a_1] \cdot [N - h\ N \cdots N]$ is the linear space spanned by the first n coordinate vectors.

It is easy to see that this intersection is to be counted simply, which completes the proof of (15). Combining all these results, we have proved formula (11).

3. For the treatment of the characteristic ring (mod 2) a particularly important rôle will be played by the Schubert cocycles $\{k\}$ and $\{0 \cdots 0\ 1 \cdots 1\}$, where the

last symbol consists of $n - k$ zeros followed by k one's. The image in $C(M)$ of their cohomology classes under the ring homomorphism f^{*-1} will be denoted respectively by W^k and $\overline{W}^k$, $k = 1, \cdots, n$. We remark here that the characteristic classes W^k are exactly the classes of Stiefel-Whitney,[5] introduced by them in their study of continuous $(n - k + 1)$-fields of vectors over M. This can be proved by essentially the same argument which the author used to establish an analogous theorem for complex analytic manifolds.[6] It may be noticed that the superscripts of the characteristic classes W^k and $\overline{W}^k$ denote their dimensions.

We shall denote by $Y(h, k)$ the characteristic class which is the image under f^{*-1} of the class to which $\{1 \cdots 1k\}$ (h one's) belongs. It then follows from (11) that

$$W^i \smile \overline{W}^{r-i} = Y(i-1, r-i+1) + Y(i, r-i), \qquad i = 0, 1, \cdots, r, \tag{21}$$

where we define

$$Y(-1, r+1) = Y(r, 0) = 0, W^0 = \overline{W}^0 = 1. \tag{22}$$

Summing these relations, we get

$$\sum_{i=0}^{r} W^i \smile \overline{W}^{r-i} = 0, \qquad r = 1, \cdots, n. \tag{23}$$

These relations show that $\overline{W}^i$ is a polynomial in W^0 $(=1)$, $W^1, \cdots, W^i$ and that W^i is a polynomial in $\overline{W}^0$ $(=1)$, $\overline{W}^1, \cdots, \overline{W}^i$.

From this remark and formula (13) we get the theorem:

THEOREM 2. *Every characteristic cohomology class* mod 2 *is a polynomial in the Stiefel-Whitney classes* $W^1, \cdots, W^n$ *and is also a polynomial in the classes* $\overline{W}^1, \cdots, \overline{W}^n$.

It is remarked that our notation for W^k and $\overline{W}^k$ agrees with that of Whitney.[7] Whitney introduced an independent variable t and wrote

$$\begin{aligned} W &= \sum_{k=0}^{n} W^k t^k, \\ \overline{W} &= \sum_{k=0}^{N} \overline{W}^k t^k. \end{aligned} \tag{24}$$

With this notation the formulas (23) can be condensed into one formula

$$W\overline{W} = 1. \tag{25}$$

It may be convenient to call W the *Whitney polynomial* of F.

---

[5] STIEFEL, E., *Richtungsfelder und Fernparallelismus in n-dimensionalen Mannigfaltigkeiten*, Comm. Math. Helv., 8 (1936), 305–343; Whitney, loc. cit.

[6] CHERN, S. S., *Characteristic classes of Hermitian manifolds*, these Annals, 47 (1946), 85–121, in particular, p. 101, Theorem 7.

[7] WHITNEY, H., *On the topology of differentiable manifolds*, Lectures in Topology, 101–141, Michigan 1941, in particular, 132–133.

§3. Some Applications

In this section we shall derive some simple consequences of the multiplication formulas. We shall restrict ourselves for this purpose to the particular case of the tangent bundle of a differentiable manifold M of dimension n, about which, however, no assumption on finiteness or orientability is made.

1. From the restriction $k \leqq N$ imposed on the characteristic class $\overline{W}^k$ we have immediately the theorem:

THEOREM 3. *A necessary condition that a differentiable manifold of dimension n can be imbedded in an Euclidean space of dimension $n + r$ is*

$$\overline{W}^{r+1} = \cdots = \overline{W}^n = 0. \tag{26}$$

By means of (23) these conditions can be expressed in terms of W^k. In particular, we get, as a set of necessary conditions for a differentiable manifold of dimension n to be imbeddable in an Euclidean space of dimension $n + 1$,

$$W^k = (W^1)^k, \qquad k = 1, \cdots, n, \tag{27}$$

where the multiplication in the right-hand side is by means of the cup product.

Similarly, a set of necessary conditions for a differentiable manifold of dimension n to be imbeddable in an Euclidean space of dimension $n + 2$ is

$$W^k + W^{k-1}W^1 + W^{k-2}(W^2 + (W^1)^2) = 0, \qquad 3 \leqq k \leqq n, \tag{28}$$

multiplication being again the cup product.

2. We shall apply these conditions to the real projective space P^n of dimension n. In P^n we denote by Q^k ($k = 0, 1, \cdots, n$) the cohomology class, whose dual homology class contains the projective space P^{n-k}. Then we have

$$Q^k \smile Q^h = Q^{k+h}.$$

The Stiefel-Whitney classes of P^n were determined by Stiefel.[8] The results are

$$W^k = \begin{cases} Q^k, & \text{if } \binom{n+1}{k} \text{ is odd} \\ 0, & \text{if } \binom{n+1}{k} \text{ is even.} \end{cases} \tag{29}$$

In our terminology this may be described by saying that the Whitney polynomial of P^n is

$$W = (1 + Q^1t)^{n+1}. \tag{30}$$

We shall study the conditions (on n) under which the equations (28) will be satisfied by P^n. These conditions are: either 1) $W^k \neq 0$, $k = 1, \cdots, n$; or 2) $W^k = 0$, $k = 1, \cdots, n$; or 3) $W^k = 0$, k odd, $W^k \neq 0$, k even.

[8] STIEFEL, E., *Über Richtungsfelder in den projektiven Räumen und einen Satz aus der reellen Algebra*, Comm. Math. Helv., 13, 201–218.

To prove this, suppose $W^1 \neq 0$, so that $W^n \neq 0$. If $W^2 = W^3 = 0$, it would follow from (28) that all $W^k = 0$, $2 \leqq k \leqq n$, which is impossible. Consider then the case $W^2 = 0$, $W^3 \neq 0$. This means that $\binom{n+1}{2}$ is even and $\binom{n+1}{3} = \binom{n+1}{2}\frac{n-1}{3}$ is odd, which is also not possible. It follows that $W^2 \neq 0$. But then $W^2 + (W^1)^2 = 0$, and (28) can be written

$$W^k + W^{k-1} W^1 = 0, \qquad 3 \leqq k \leqq n.$$

From this equation we see that $W^k \neq 0$, $k \geqq 3$.

Suppose next that $W^1 = 0$. Then

$$W^k + W^{k-2} W^2 = 0, \qquad 3 \leqq k \leqq n.$$

If $W^2 = 0$, then all $W^k = 0$, $k = 1, \cdots, n$. If $W^2 \neq 0$, then

$$W^1 = W^3 = \cdots = W^n = 0, \qquad W^2 \neq 0, W^4 \neq 0, \cdots, W^{n-1} \neq 0.$$

This completes the proof of the above necessary conditions.

It follows that if P^n can be imbedded in E^{n+2}, n must be a positive integer satisfying one of the following conditions:

1) $(1 + t)^{n+1} \equiv 1 + t + t^2 + \cdots + t^{n+1}$, mod. 2;

2) $(1 + t)^{n+1} \equiv 1 + t^{n+1}$, mod. 2;

3) $(1 + t)^{n+1} \equiv 1 + t^2 + t^4 + \cdots + t^{n+1}$, mod. 2.

An elementary argument shows that these conditions are respectively equivalent to the conditions that n be one of the forms: 1) $n = 2^k - 2$, $k \geqq 2$; 2) $n = 2^k - 1$, $k \geqq 2$; 3) $n = 2^k - 3$, $k \geqq 3$.

We get therefore the following theorem:

THEOREM 4. *A real projective space P^n of dimension n cannot be imbedded in E^{n+2}, if n is not of one of the forms: 1)$n = 2^k - 2$, $k \geqq 2$; 2) $n = 2^k - 1$, $k \geqq 2$; 3) $n = 2^k - 3$, $k \geqq 3$.*

INSTITUTE OF MATHEMATICS, ACADEMIA SINICA
SHANGHAI, CHINA

Apart from vol. 25, fasc. 3 (1951)
COMMENTARII MATHEMATICI HELVETICI
Printed in Switzerland by Art. Institut Orell Füssli AG. Zürich

A Theorem on Orientable Surfaces in Four-Dimensional Space

By SHIING-SHEN CHERN and E. SPANIER, Chicago

1. *Introduction.* Let M be a closed oriented surface differentiably imbedded in a Euclidean space E of four dimensions. Let G denote the Grassmann manifold of oriented planes through a fixed point O of E. It is well known that G is homeomorphic to the topological product $S_1 \times S_2$ of two 2-spheres. By mapping each point P of M into the oriented plane through O parallel to the oriented tangent plane to M at P, we define a mapping $t: M \to G$. If M, S_1, S_2 denote also the fundamental cycles of the respective manifolds and t_* denotes the homomorphism induced by t, we have

$$t_*(M) \sim u_1 S_1 + u_2 S_2 \ .$$

In a recent paper[1]) Blaschke studied the situation described above by methods of differential geometry and proved that the sum $u_1 + u_2$ equals the Euler characteristic of M. He also asserted that $u_1 = u_2$. The object of this note is to give a proof of this assertion, as well as a new proof of the theorem on $u_1 + u_2$.

2. *Review of some known results on sphere bundles.* Let B be an oriented sphere bundle of d-spheres over a base space X with projection f. The relation between the homology properties of B and X are summarized in the following exact sequence[2]);

$$\cdots \to H^p(X) \overset{f^*}{\to} H^p(B) \overset{\psi}{\to} H^{p-d}(X) \overset{\cup\Omega}{\to} H^{p+1}(X) \to \cdots$$

where each H denotes a cohomology group relative to a coefficient group which is the same for all the terms of the sequence. The homomorphisms that occur in the sequence can be described briefly as follows:

[1]) *Blaschke, W.*, Ann. Mat. Pura Appl. (4) 28, 205—209 (1949).

[2]) *Gysin, W.*, Comm. Math. Helv. 14, 61—122 (1942). — *Chern, S. S. and Spanier, E. H.*, Proc. Nat. Acad. Sci., U. S. A. 36, 248—255 (1950).

f^* is the dual homomorphism induced by the projection f; ψ is a mapping which amounts to „integrating over the fiber"; the third homomorphism is the cup product with the characteristic class Ω (with integer coefficients) of the bundle. From this sequence we see that if, for every coefficient system, the fiber $S^d \not\sim 0$ in B then the unit element 1 of the integral cohomology ring of X is in the image of ψ and $\Omega = 0$.

Let E be oriented. Over the oriented surface $M \subset E$ there are two vector bundles, the *tangent bundle* of tangent vectors and the *normal bundle* of normal vectors. By taking unit vectors we get two bundles of circles over M. According to a theorem of Seifert and Whitney [3]) the characteristic class of the normal bundle is zero. Since this theorem holds in a more general situation and can be proved in a simple way, we state and prove the theorem for the general case [4]).

Theorem. Let M be an orientable manifold imbedded in a Riemann manifold M'. If $M \sim O$ in M', then the characteristic class of the normal bundle of M in M' is zero.

Proof. Let B be a small tube around M. B is then the normal bundle of M. We will show that no fiber S of B bounds in B. Assume that $S = \partial C$ in B mod p for some p. Let D be the set of normal vectors of length $\leqq \epsilon$ having S as boundary. Then $C - D$ is a cycle mod p in M' intersecting M in exactly one point. This is impossible because $M \sim O$ in M'.

The above theorem also follows easily from results of Thom [5]).

3. *Plücker coordinates in G.* Let e_1, e_2, e_3, e_4 be an orthonormal base for E such that $e_1 \wedge e_2 \wedge e_3 \wedge e_4$ [6]) is the orientation of E. If R is any oriented plane of E, let f_1, f_2 be an orthonormal base in R such that $f_1 \wedge f_2$ is the orientation of R. Then

$$f_1 \wedge f_2 = a_{12}\, e_1 \wedge e_2 + a_{23}\, e_2 \wedge e_3 + a_{31}\, e_3 \wedge e_1 + a_{34}\, e_3 \wedge e_4 + a_{14}\, e_1 \wedge e_4 + a_{24}\, e_2 \wedge e_4\,.$$

These "Plücker coordinates" a_{ij} of R are independent of the choice of f_1, f_2 and satisfy the two relations

---

[3]) *Seifert, H.*, Math. Zeitschr. 41 (1936) 1—17. — *Whitney, H.*, Lectures in Topology, Univ. of Mich. Press (1941) 101—141.

[4]) We owe this simple description of the proof to Professor H. Hopf, who also called our attention to the problem settled in this paper.

[5]) *Thom, R.*, C. R. Paris 230, 507—508 (1950).

[6]) The wedge denotes Grassmann multiplication as in Bourbaki, N., Algèbre Multilinéare, Hermann, Paris (1948).

$$a_{12}\,a_{34} + a_{23}\,a_{14} + a_{31}\,a_{24} = 0 \tag{1}$$

$$\Sigma\, a_{ij}^2 = 1\ . \tag{2}$$

Conversely, any set of six real numbers satisfying (1) and (2) are the Plücker coordinates of some oriented plane in E; hence, G is homeomorphic to the subset of six space consisting of a_{ij} such that (1) and (2) hold. We introduce a linear change of coordinates by

$$x_1 = a_{12} + a_{34} \qquad x_2 = a_{23} + a_{14} \qquad x_3 = a_{31} + a_{24}$$

$$y_1 = a_{12} - a_{34} \qquad y_2 = a_{23} - a_{14} \qquad y_3 = a_{31} - a_{24}\ .$$

Then G is homeomorphic to the subset of six space consisting of (x_i, y_j) such that $\Sigma\, x_i^2 = \Sigma\, y_j^2 = 1$.

Let S_1, S_2 be the unit spheres in the x-space and y-space respectively. We orient S_1 and S_2 by the orientations (x_1, x_2, x_3) and (y_1, y_2, y_3) of the x-space and y-space. Let $h\colon\ G \to S_1 \times S_2$ be the homeomorphism defined above using the Plücker coordinates.

Let α; $G \to G$ map each oriented plane R into its normal plane R', oriented so that R, R' determine the given orientation of E. We want to determine the mapping $h\,\alpha\,h^{-1}\colon\ S_1 \times S_2 \to S_1 \times S_2$. If R has Plücker coordinates a_{ij} and R' has Plücker coordinates b_{ij}, it is easy to see that the following equations are satisfied

$$\sum_k a_{ik}\, b_{jk} = 0 \qquad (i \neq j)$$

$$\Sigma\, a_{ij}\, b_{kl} = 1\ ,$$

the last summation being taken over all even permutations of $1, 2, 3, 4$.

It follows from these that $b_{ij} = a_{kl}$, where i, j, k, l is an even permutation of 1, 2, 3, 4. Therefore, we see that

$$h\,\alpha\,h^{-1}(x, y) = (x, -y)$$

where $-y$ denotes the antipodal point to y.

4. *The Theorem.* Let M be a closed oriented surface in E. Let $t\colon\ M \to G$ and $n\colon\ M \to G$ be the maps defined by taking tangent planes and normal planes respectively. It is clear that $t = \alpha\, n$ and $n = \alpha\, t$.

Over G there is a bundle of circles obtained by considering as the fiber over an oriented plane through O the unit circle in that plane. Let Ω denote the characteristic class of this bundle and let Ω_t, Ω_n denote the characteristic classes of the tangent and normal bundles of M. Then

$$t^*\,\Omega = \Omega_t\ , \qquad n^*\,\Omega = \Omega_n\ .$$

3

The bundle of circles over G defined above is the Stiefel manifold V of ordered pairs of orthogonal unit vectors through O in E and is easily seen to be homeomorphic to $S^2 \times S^3$. The following section of Gysin's sequence

$$H^1(V) \xrightarrow{\psi} H^0(G) \xrightarrow{\cup\Omega} H^2(G) \xrightarrow{f^*} H^2(V) \xrightarrow{\psi} H^1(G)$$

shows that Ω is a generator of the kernel of f^* in $H^2(V)$, since $H^1(V)$ and $H^1(G)$ are trivial. To find the kernel of f^* we determine the homomorphism

$$f_* : H_2(V) \to H_2(G)$$

of the second homology groups.

A generating 2-cycle in V is $S^2 \times e_4$. The points z of S^2 can be represented as vectors of the form $z_1 e_1 + z_2 e_2 + z_3 e_3$. Then

$$f\left(\sum_{i=1}^{3} z_i e_i, e_4\right) = \Sigma z_i (e_i \wedge e_4)$$

and so

$$h f(\Sigma z_i e_i, e_4) = (z, -z) .$$

Therefore, we see that $f_*(S^2 \times e_4) = S_1 - S_2$. If S_1^*, S_2^* denote cohomology classes dual to the homology classes S_1, S_2, then the kernel of f^* consists of all elements of the form $u(S_1^* + S_2^*)$ where u is an integer. Orient S_1 and S_2 so that $\Omega = S_1^* + S_2^*$. Orient M so that $\Omega_t \cdot M = \chi_M$ = Euler characteristic of M. Then

$$\Omega_t = t^*(S_1^* + S_2^*) = t^* S_1^* + t^* S_2^*$$

and

$$\Omega_n = n^*(S_1^* + S_2^*) = t^* \alpha^* (S_1^* + S_2^*) = t^* (S_1^* - S_2^*) = t^* S_1^* - t^* S_2^* .$$

Since $\Omega_n = 0$, we see that

$$(t^* S_1^*) \cdot M = (t^* S_2^*) \cdot M = (\tfrac{1}{2}) \chi_M .$$

We summarize the above results in the theorem:

Let M be a closed orientable surface in four space E. Let G be the Grassmann manifold of oriented planes through O in E and let $t: M \to G$ be the map into oriented planes through O parallel to the tangent planes of M. Since G is homeomorphic to $S_1 \times S_2$, we have $t_*(M) = u_1 S_1 + u_2 S_2$. Then S_1, S_2 and M can be oriented so that $u_1 = u_2 = (\tfrac{1}{2}) \chi_M$ where χ_M *is the Euler characteristic of M.*

5. *Remarks.* The above theorem expresses relations between differential topological invariants of surfaces imbedded in Euclidean space

4

and suggests a more general problem. To describe the general situation let $M^k \subset E^{k+l}$ be a manifold of dimension k differentiably imbedded in a Euclidean space of $k+l$ dimensions. Let $G(k, l)$ be the Grassmann manifold of k-dimensional linear spaces through a point O and $G(l, k)$ that of l-dimensional linear spaces through O. There is a natural homeomorphism

$$\alpha: \; G(k, l) \to G(l, k) \; .$$

Using tangent planes and normal planes to M we define mappings

$$t: \; M \to G(k, l) \; , \qquad n: \; M \to G(l, k)$$

such that

$$t = \alpha^{-1} n \; , \qquad n = \alpha t \; .$$

The general problem is to study the relation between the homomorphisms

$$\left.\begin{array}{l} t^* ; \; H^p(G(k, l)) \to H^p(M) \\ n^* ; \; H^p(G(l, k)) \to H^p(M) \end{array}\right\} p = 0, 1, \ldots$$

We hope to study this question on a later occasion.

(Received 31th Mars 1951.)

5

ON THE CHARACTERISTIC CLASSES OF COMPLEX SPHERE BUNDLES AND ALGEBRAIC VARIETIES.* †

By SHIING-SHEN CHERN.

Introduction. In a recent paper [1] Hodge studied the question of identifying, for non-singular algebraic varieties over the complex field, the characteristic classes of complex manifolds [2] with the canonical systems introduced by M. Eger and J. A. Todd.[3] He proved that they are identical up to a sign, when the algebraic variety is the complete intersection of non-singular hypersurfaces in a projective space. His method does not seem to extend to a general algebraic variety. One of the main difficulties lies in the fact that the theory of canonical systems of algebraic varieties has so far been developed only in broad outlines, with the result that very few of their properties are available.

We shall give in this paper a more direct treatment of the problem, by proving that there is an equivalent definition of the characteristic classes, which is valid for algebraic varieties. In order to make the paper as self-contained as possible, let us begin by recalling the original definition of the characteristic classes. We consider a compact complex manifold M_n [4] of complex dimension n, and over M_n consider the bundle $B_{nr}{}^*$ of ordered sets $(e_1, \cdots, e_r)$ of r linearly independent complex vectors with the same origin.[5] The fiber of this bundle is the complex Stiefel manifold V_{nr} of all the ordered sets of r linearly independent complex vectors in a complex vector space of dimension n. It is well-known that V_{nr} is connected and that its first non-vanishing homotopy group is $\pi_{2n-2r+1}(V_{nr})$, the latter being free cyclic. To describe a generator of $\pi_{2n-2r+1}(V_{nr})$ we fix $e_1, \cdots, e_{r-1}$ and let W_{n-r+1}

* Received September 29, 1952.

† This work is done under partial support of the Office of Naval Research.

[1] Hodge [9]. The number refers to the bibliography at the end of the paper.

[2] Chern [1].

[3] Eger [2] and Todd [13].

[4] In general, we shall use a subscript to denote complex dimension and a superscript to denote topological dimension. When the meaning is clear, it will be dropped to simplify notation.

[5] While we shall explain, in so far as possible, the notions which will be utilized, we shall refer to Steenrod [12] as our standard reference on fiber bundles.

be a vector space of dimension $n-r+1$ which contains no non-trivial linear combination of $e_1, \cdots, e_{r-1}$. Let $W_{n-r+1}{}^*$ be the space obtained from W_{n-r+1} by deleting its origin. Then $\pi_{2n-2r+1}(V_{nr})$ and $\pi_{2n-2r+1}(W_{n-r+1}{}^*)$ are naturally isomorphic (under the homomorphism induced by the inclusion mapping $W_{n-r+1}{}^* \subset V_{nr}$). Since W_{n-r+1}, as a complex vector space, is oriented, its orientation determines uniquely a generator of $\pi_{2n-2r+1}(W_{n-r+1}{}^*)$, and hence of $\pi_{2n-2r+1}(V_{nr})$. In other words, the bundle $B_{nr}{}^*$ is orientable. It follows from the theory of obstructions that the primary obstruction of this bundle is a cohomology class C_{n-r+1} of (topological) dimension $2(n-r+1)$ with integer coefficients. We call C_r, $r=1, \cdots, n$, the *r-th characteristic class* of M. If M also denotes the fundamental homology class of the oriented manifold M, the homology class γ_r defined by the cap product $\gamma_r = C_{n-r} \cap M$ is called the *r-th characteristic homology class* of M.

All these considerations apply to the case in which M is a non-singular algebraic variety over the complex field. However, since the obstructions are defined in terms of continuous cross sections over the skeletons of a triangulation of M, it does not follow that the characteristic homology class γ_r contains as representative an algebraic cycle, that is, a cycle in the form of a finite sum $\Sigma\lambda_i V_r{}^i - \Sigma\mu_k V_r{}^k$, where $V_r{}^i$, $V_r{}^k$ are algebraic sub-varieties of dimension r in M and $\lambda_i \geqq 0$, $\mu_k \geqq 0$. A main purpose of this paper is to prove that this is the case.

We proceed to enumerate our results, postponing their proofs for later sections. In the course of our discussion several theorems on the homology theory of fiber bundles and on complex sphere bundles will be used. While they are to some extent known, they are either not easily accessible or not given in a form needed for our purpose. For the sake of completeness such results will be included here.

We consider a fiber bundle $p\colon B \to X$, about which we make once for all the following assumptions: 1) the base space X is a finite polyhedron; 2) the fiber F is a connected finite polyhedron; 3) the fundamental group of X acts trivially on the homology groups of the fibers under consideration. The last assumption makes it possible to use these groups as coefficient groups in the homology of X. Let $r>1$ be such that $\pi_r(F) \neq 0$, $\pi_s(F) = 0$ for all $s<r$. Then it is well-known that $\pi_r(F)$ is isomorphic to the homology group $H_r(F)$ with integer coefficients. The primary obstruction of the bundle is an element of $H^{r+1}(X, H_r(F))$. Its vanishing has an implication described by the following theorem:

THEOREM 1. *Let* $\pi_r(F) \cong H_r(F)$, $r>1$, *be the first non-vanishing* *homotopy group of F. If the primary obstruction vanishes, the injection mapping $l: F \to B$ induces a homomorphism $l^*: H^r(B, H_r(F)) \to H^r(F, H_r(F))$, which is onto.*

The conclusion of this theorem gives information on the "homology position" of a fiber in the bundle. Relative to a coefficient group G the simplest situation is when F is totally non-homologous to zero, i. e., when the homomorphism $l^*: H^r(B, G) \to H^r(F, G)$ is onto for all r. When G is a field, such bundles were studied by Leray and Hirsch, and the cohomology ring of the bundle is found to be isomorphic, in its additive structure, to that of the Cartesian product of the fiber and the base space. Since we are mainly interested in integer coefficients, we need the following strengthened form, due to E. H. Spanier, of the Leray-Hirsch theorem:[6]

THEOREM 2. *Let $l: F \to B$ be the injection of the fiber into the bundle. Relative to a simple coefficient system G suppose there is a homomorphism $\mu: H^r(F; G) \to H^r(B; G)$ such that $l^*\mu$ is the identity automorphism of $H^r(F; G)$ for all $r \geqq 0$. Then $H^r(B; G)$ is isomorphic with $H^r(X \times F; G)$.*

We now describe an important operation in the homology theory of fiber bundles, known as "integration over the fiber." Let $H^r(F)$, $r > 0$, be the last non-vanishing cohomology group of the fiber, so that $H^s(F) = 0$ for all $s > r$. If G is a simple system of coefficient groups (that is, a system of local groups in B on which $\pi_1(B)$ acts trivially), integration over the fiber is a homomorphism

$$\Psi: H^m(B; G) \to H^{m-r}(X; H^r(F; G)). \tag{1}$$

To define Ψ let X be triangulated and let X^k be its k-dimensional skeleton. Put $B_k = p^{-1}(X^k)$. Then it is easy to see that

$$H^m(B_s; G) = 0, \qquad 0 \leqq s \leqq m - r - 1. \tag{2}$$

From the exact sequence of the pair (B_{m-r}, B_{m-r-1}):

$$\cdots \to H^m(B_{m-r}, B_{m-r-1}; G) \xrightarrow{j^*} H^m(B_{m-r}; G) \xrightarrow{i^*} H^m(B_{m-r-1}; G) \xrightarrow{\delta^*} \cdots, \tag{3}$$

it follows that j^* is onto. To an element $u \in H^m(B; G)$ let

$$u' = i^*u \in H^m(B_{m-r}; G)$$

be the image of u under the dual homomorphism of the homomorphism induced by $i: B_{m-r} \to B$. Since j^* is onto, there exists $v \in H^m(B_{m-r}, B_{m-r-1}; G)$

[6] Hirsch [7], Leray [10], pp. 183-184, and Spanier [11].

such that $j^*v = u'$. By an isomorphism which will later be described in more detail, we see that v can be identified with an $(m-r)$-dimensional cochain of X, with coefficients in $H^r(F;G)$. It can be proved that it is a cocycle and that its cohomology class depends only on u. This class is defined to be Ψu.

This operation Ψ has a simple geometrical interpretation, when X, B, F are oriented manifolds. In this case $H^r(F;G)$ is naturally isomorphic to G, so that the coefficient group on the right-hand side of (1) can be replaced by G. Denote also by X, B, F the fundamental homology classes of these manifolds. For $u \varepsilon H^m(B;G)$, $v \varepsilon H^s(X;G)$ we define, by means of cap products, the operations

$$\begin{aligned} \mathcal{D}_B u &= u \cap B, \\ \mathcal{D}_X v &= v \cap X. \end{aligned} \tag{4}$$

Then we have the theorem:

THEOREM 3. *If the spaces X, B, F of a fiber bundle are oriented manifolds and $u \varepsilon H^m(B;G)$, we have*

$$p_* \mathcal{D}_B u = \mathcal{D}_X \Psi u. \tag{5}$$

In other words, integration over the fiber is in this case dual to the homomorphism of homology classes induced by the projection p.

The results we need next center around the theory of complex sphere or vector bundles. As the structural group of the bundle of tangent vectors of M is the general linear group $G(n)$ in n complex variables, there is an associated bundle corresponding to every subgroup of $G(n)$. We realize $G(n)$ as the group of all $n \times n$ non-singular matrices with complex elements. Let $H(n,r)$ be the subgroup of $G(n)$, consisting of all matrices (a_{ik}) for which

$$a_{ik} = 0, \quad 1 \leqq k < i \leqq r; \quad r+1 \leqq i \leqq n, \quad 1 \leqq k \leqq r, \tag{6}$$

that is, of all matrices of the form

$$\begin{pmatrix} a_{11} & \cdots & a_{1r} & a_{1r+1} & \cdots & a_{1n} \\ & \ddots & & & & \\ 0 & & a_{rr} & a_{r,r+1} & \cdots & a_{rn} \\ & & & a_{r+1,r+1} & \cdots & a_{r+1,n} \\ & 0 & & & \cdots & \\ & & & a_{n,r+1} & \cdots & a_{nn} \end{pmatrix}$$

Let $K(n,r)$ be the subgroup of $H(n,r)$ whose matrices satisfy the further conditions $a_{11}=\cdots=a_{rr}=1$. Finally, let $L(n,r)$ be the subgroup of $K(n,r)$ whose matrices satisfy the additional conditions $a_{ik}=0$, $1\leqq i<k\leqq r$. We denote by B_{nr}, $\tilde{B}_{nr}$, and $B_{nr}{}^*$ the associated bundles corresponding to the subgroups $H(n,r)$, $K(n,r)$, and $L(n,r)$ respectively. The bundle $B_{nr}{}^*$ is the one introduced above, whose points are ordered sets of r linearly independent vectors $e_1(x),\cdots,e_r(x)$ with the same origin $x\,\varepsilon\,M$. Similarly, a point of $\tilde{B}_{nr}$ can be identified with a sequence of simple multivectors of the form $e_1(x), e_1(x)\wedge e_2(x),\cdots,e_1(x)\wedge\cdots\wedge e_r(x)\neq 0$, with the same origin $x\,\varepsilon\,M$. To describe the geometrical meaning of the bundle B_{nr} we need the notion of a tangent direction, which is the class of non-zero tangent vectors differing from each other by a non-zero complex factor. All the tangent directions at a point form a complex projective space of dimension $n-1$. A point of B_{nr} can be regarded as a sequence of linear spaces of directions $L_0(x)\subset L_1(x)\subset\cdots\subset L_{r-1}(x)$ in the space of tangent directions at $x\,\varepsilon\,M$, with the subscripts indicating the dimensions of these linear spaces. We notice that these spaces are related by natural projections as follows:

$$B_{nr}{}^* \xrightarrow{m_{nr}} \tilde{B}_{nr} \xrightarrow{p_{nr}} B_{nr} \xrightarrow{q_{nr}} M. \tag{7}$$

Moreover, under these projections every space is a bundle over the spaces which follow it.

It is the bundle

$$q_{nr}\colon B_{nr}\to M, \tag{8}$$

which we are most interested. Since $H(n,r)$ is defined for $1\leqq r\leqq n-1$, we shall assume r to be restricted by these inequalities, thus excluding $r=n$. The fiber F_{nr} of B_{nr} is the space of all sequences of linear subspaces $L_0\subset L_1\subset\cdots\subset L_{r-1}$ in a complex projective space of dimension $n-1$. Its cohomology ring with integer coefficients is generated by r two-dimensional cohomology classes and its first non-vanishing homology group of dimension >0 is $H_2(F_{nr})$, which is free abelian with r generators. Let M' be the $(2n-2r+1)$-dimensional skeleton of M. The bundle $B_{nr}{}^*$ has a cross-section $f\colon M'\to B_{nr}{}^*$ over M'. Then $p_{nr}\circ m_{nr}\circ f$ defines a cross-section of the bundle B_{nr} over M'. Since $2n-2r+1\geqq 3$, the primary obstruction of the bundle B_{nr} is zero. It follows from Theorem 1 that the homomorphism

$$l_{nr}{}^*\colon H^2(B_{nr};H_2(F_{nr}))\to H^2(F_{nr};H_2(F_{nr}))$$

induced by the inclusion mapping $l_{nr}\colon F_{nr}\to B_{nr}$ is onto. Since $H_2(F_{nr})$ is a free abelian group with r generators, the cohomology groups $H^2(F_{nr})$, $H^2(B_{nr})$

11

with integer coefficients can be imbedded isomorphically into $H^2(F_{nr}; H_2(F_{nr}))$, $H^2(B_{nr}; H_2(F_{nr}))$ respectively. It follows that the induced homomorphism $l_{nr}^*: H^2(B_{nr}) \to H^2(F_{nr})$ is also onto.

This induced homomorphism has another geometrical interpretation. In fact, the fiber of the bundle $p_{nr}: \tilde{B}_{nr} \to B_{nr}$ is a Cartesian product of r complex lines, each with the origin deleted. This bundle gives rise in the base space B_{nr} to r 2-dimensional characteristic classes with integer coefficients. It can be seen that their images under l_{nr}^* generate the cohomology ring of F_{nr}. Since l_{nr}^* is multiplicative, we see that the conditions of Theorem 2 are satisfied. This leads to the conclusion that the cohomology groups of B_{nr} are isomorphic to those of $M \times F_{nr}$. In other words, the space B_{nr} has rather simple additive homology properties.

In the study of the bundles in (7) an important tool is the so-called duality theorem. To describe the situation in geometrical terms, we take, over the same base space X, two bundles B_1, B_2 of complex vector spaces of dimensions ν_1, ν_2 respectively and construct a bundle of complex vector spaces of dimension $\nu_1 + \nu_2$ over X by taking as the fiber at a point $x \varepsilon X$ the space spanned by the fibers at x of the given bundles. This bundle will be called the product of B_1 and B_2 and will be denoted by $B_1 \boxtimes B_2$. The question naturally arises as to express the characteristic classes of $B_1 \boxtimes B_2$ in terms of those of B_1 and B_2. To express this relationship we introduce, for a bundle of complex vector spaces of dimension ν, the *characteristic polynomial*

$$(9) \qquad C(t) = \sum_{i=0}^{\nu} C_i t^i, \qquad C_0 = 1.$$

This is a polynomial in an auxiliary variable t, whose coefficients are the characteristic cohomology classes C_i, with the convention that the classes of dimension greater than the topological dimension of X are replaced by zero. Then we have the theorem:

THEOREM 4. (Duality theorem for complex vector bundles) *If B_1 and B_2 are two complex vector bundles over the same base space X and if $C^{(1)}(t)$ and $C^{(2)}(t)$ are their characteristic polynomials, then the characteristic polynomial of their product bundle $B_1 \boxtimes B_2$ is $C^{(1)}(t)\, C^{(2)}(t)$.*

Using the duality theorem it is easy to prove the following theorem of G. Hirsch and Wu Wen-Tsun,[7] which can be regarded as giving a new definition of the characteristic classes:

---

[7] Hirsch [8]; Wu, unpublished.

THEOREM 5. *Let M be a compact complex manifold of dimension n, and B_{n1}, $\tilde{B}_{n1}$ two of the associate bundles of its tangent bundle, as defined above. Let u_1 be the characteristic class of the bundle $p_{n1}\colon \tilde{B}_{n1} \rightarrow B_{n1}$. Then we have*

$$(10) \qquad u_1{}^n = \sum_{i=1}^{n} (-1)^{i+1} q_{n1}{}^*(C_i) u_1{}^{n-i}.$$

This formula enables us to define the characteristic classes within the framework of homology theory. Unfortunately it does not seem to achieve our purpose of giving a definition applicable to the case when M is an algebraic variety. To obtain still another definition let us notice that the fiber F_{nr} of the bundle B_{nr} is an oriented manifold of topological dimension $r(2n-r-1)$, so that $H^{r(2n-r-1)}(F_{nr}; G)$ is naturally isomorphic to G and the homomorphism (1) can be written

$$(11) \qquad \Psi_{nr}\colon H^m(B_{nr}) \rightarrow H^{m-r(2n-r-1)}(M).$$

Moreover, instead of the characteristic polynomial $C(t)$ we can introduce the dual characteristic polynomial

$$(12) \qquad \bar{C}(t) = \sum_{i=1}^{n} \bar{C}_i t^i, \qquad \bar{C}_0 = 1,$$

defined by the condition

$$(13) \qquad C(t)\bar{C}(t) = 1.$$

Obviously the polynomial $C(t)$ defines $\bar{C}(t)$, and vice versa.

In order to formulate our next theorem, we need some notations. We regard a point of $\tilde{B}_{nr}$ as a sequence of simple multivectors of the form $e_1(x), \cdots, e_1(x) \wedge \cdots \wedge e_r(x) \neq 0$, with the same origin $x \varepsilon M$. Then the sequences of multivectors having the same projection in B_{nr} are exactly the ones obtained from the last sequence by multiplying its multivectors by the non-zero complex numbers $\alpha_1, \cdots, \alpha_r$ respectively. We can therefore regard $\alpha_1, \cdots, \alpha_r$ as the coordinates in the fiber of the bundle $p_{nr}\colon \tilde{B}_{nr} \rightarrow B_{nr}$. The fiber is thus a Cartesian product of r complex centered affine lines, each with the origin deleted. We denote by $v_1, \cdots, v_r$ their characteristic classes in B_{nr}. From them we introduce the cohomology classes $u_1, \cdots, u_r$ according to the equations

$$(14) \qquad v_i = u_1 + \cdots + u_i, \qquad i = 1, \cdots, r.$$

Then we have the theorem:

THEOREM 6. *For $1 \leqq r \leqq n-1$ the following formula holds:*

$$(15) \qquad \Psi_{nr}(u_1{}^{n-2} \cdots u_{r-1}{}^{n-r} u_r{}^{2n-r}) = (-1)^n \bar{C}_{n-r+1}.$$

In terms of homology this formula gives, on account of Theorem 3,

$$(16)\qquad \begin{aligned}(-1)^n \mathfrak{D}_M \bar{C}_{n-r+1} &= (p_{nr})_* \mathfrak{D}_B (u_1^{n-2} \cdots u_{r-1}^{n-r} u_r^{2n-r}) \\ &= (p_{nr})_* \{(\mathfrak{D}_B u_1)^{n-2} \cdots (\mathfrak{D}_B u_{r-1})^{n-r} (\mathfrak{D}_B u_r)^{2n-r}\},\end{aligned}$$

where we write B for B_{nr}. Multiplication in the last expression means intersection of the homology classes.

Now let M be a non-singular algebraic variety in a complex projective space of higher dimension. It is well-known that B_{nr} is a non-singular algebraic variety and that the projection p_{nr} is a rational mapping. As first shown by Weil,[8] each of the classes $\mathfrak{D} u_i$, $i = 1, \cdots, r$, in B_{nr} contains a divisor class which consists of all divisors linearly equivalent to each other. Since the intersection of divisor classes always contains an algebraic cycle and since, under a rational mapping, an algebraic cycle goes into an algebraic cycle, it follows from (16) that $\mathfrak{D}_M \bar{C}_{n-r+1}$ contains an algebraic cycle, for $r = 1, \cdots, n-1$. On the other hand, $\mathfrak{D}_M \bar{C}_1$ is a homology class which contains a divisor class of divisors. We can therefore state the theorem:

THEOREM 7. *Every characteristic homology class on a non-singular algebraic variety contains an algebraic cycle.*

1. On the homology theory of fiber bundles. The homology theory of fiber bundles has been the object of study of many authors. The problem is by nature not a simple one; published accounts of it are either sketchy or need much machinery in algebraic topology. We shall give below a procedure developed by E. H. Spanier and the author[9] which has the advantage of being quite elementary and which will lead to proofs of our Theorems 1, 2, and 3. We begin by discussing some elementary facts on the homology theory of topological spaces.

Let X be a topological space, and A, B, C, D four closed subsets, such that

$$D \subset C \subset A, \qquad D \subset B \subset A.$$

Then the inclusion mapping

$$j\colon (C, D) \subset (A, B)$$

induces a homomorphism of the relative cohomology groups:

$$(17)\qquad j^*\colon H^r(A, B) \to H^r(C, D).$$

---

[8] Weil [16].

[9] Spanier [11].

In these cohomology groups we drop the coefficient group to simplify our notation, whenever there is no danger of confusion. The case that the subset D or both D and B are empty is not excluded. When several inclusion mappings are under consideration, we shall denote them also by i, k, or l, or distinguish them by subscripts. These induced homomorphisms have some simple properties, which have been taken as axioms by Eilenberg and Steenrod in their axiomatic treatment of homology theory.[10] The following axioms will be frequently used in our discussions:

1. **The excision axiom.**[11] If A, B are closed subsets of X, the homomorphism

$$j^*: H^r(A \cup B, B) \to H^r(A, A \cap B) \tag{18}$$

is an onto isomorphism.

2. **The exactness axiom.** Let

$$\delta^*: H^{r-1}(A) \to H^r(X, A) \tag{19}$$

be the coboundary homomorphism. Then the sequence

$$\cdots \to H^{r-1}(A) \xrightarrow{\delta^*} H^r(X, A) \xrightarrow{j^*} H^r(X) \xrightarrow{i^*} H^r(A) \xrightarrow{\delta^*} H^{r+1}(X, A) \to \cdots \tag{20}$$

is exact.

Now let A, B be closed subsets of X, such that $B \subset A \subset X$. Consider the sequence

$$H^r(A, B) \xrightarrow{j^*} H^r(A) \xrightarrow{\delta^*} H^{r+1}(X, A),$$

and define the homomorphism $\Delta^* = \delta^* j^*$. It follows from (20) that the sequence

$$\cdots \to H^r(X, A) \xrightarrow{j^*} H^r(X, B) \xrightarrow{k^*} H^r(A, B) \xrightarrow{\Delta^*} H^{r+1}(X, A) \to \cdots \tag{21}$$

is exact. It is called the *exact sequence of a triple* $B \subset A \subset X$.

---

[10] Eilenberg and Steenrod [5].

[11] This is a strengthened form of the excision axiom and is not true for general homology theory. It is true for homology theories invariant under what Eilenberg and Steenrod called relative homeomorphisms. An example is given by the Cech homology or cohomology theory for the category of compact pairs (Cf. [5], 266, Theorem 5.4). In our applications all the spaces under consideration are finite polyhedra, for which this excision axiom is valid.

LEMMA 1.1. *Let X be a topological space, and A, B, C closed subsets, such that*

$$X = A \cup B, \qquad A \cap B \subset C.$$

Then we have the isomorphism

$$H^r(X, C) \cong H^r(A, A \cap C) \oplus H^r(B, B \cap C). \tag{22}$$

To prove this, we consider the following groups related by homomorphisms, all induced by inclusion mappings:

$$\begin{array}{ccccc} H^r(X, B \cup C) & \xrightarrow{j^*} & H^r(X, C) & \xrightarrow{k^*} & H^r(B \cup C, C) \\ & & \uparrow i^* & \nearrow l^* & \\ & & H^r(X, A \cup C) & & \end{array}$$

The groups of the first row are taken from the exact sequence of the triple $C \subset B \cup C \subset X$, and therefore form an exact sequence. By the excision axiom, l^* is an onto isomorphism. Writing $i^* l^{*-1} = \lambda$, we have $k^* \lambda =$ identity. For $x \in H^r(X, C)$, we find $k^*(x - \lambda k^* x) = 0$, which allows us to put $x - \lambda k^* x = j^* y$, $y \in H^r(X, B \cup C)$. On the other hand, if $x = \lambda z = j^* y$, $z \in H^r(B \cup C, C)$, then $z = k^* \lambda z = k^* j^* y = 0$. It follows that $H^r(X, C)$ is a direct sum of $j^* H^r(X, B \cup C)$ and $\lambda H^r(B \cup C, C)$. λ is clearly an isomorphism (into), so that i^* is an isomorphism. By symmetry between A and B it follows that j^* is an isomorphism. Since, by the excision axiom, the homomorphisms

$$H^r(X, A \cup C) \to H^r(B, B \cap C),$$
$$H^r(X, B \cup C) \to H^r(A, A \cap C),$$

induced by the inclusion mappings are onto isomorphisms, the lemma follows.

By induction this lemma can be put in the following generalized form:

LEMMA 1.1′. *Let X be a topological space, and $A_1, \cdots, A_s$, C closed subsets, such that*

$$X = A_1 \cup \cdots \cup A_s, \quad A_i \cap A_k \subset C, \quad i \neq k;\ i, k = 1, \cdots, s.$$

Then

$$H^r(X, C) \cong \sum_{i=1}^{s} H^r(A_i, A_i \cap C). \tag{22′}$$

the right-hand side being a direct sum of groups.

We now consider a fiber bundle $p\colon B \to X$, whose base space X is a finite connected complex of dimension n and whose fiber F is connected. Denote by X^k the k-dimensional skeleton of X, and put $B_k = p^{-1}(X^k)$. It will turn out that the relative cohomology groups $H^r(B_q, B_{q-1}; G)$ can be interpreted in a simple manner. For simplicity we shall suppose our coefficient system to be simple, an assumption which is fulfilled in all our later applications.

LEMMA 1. 2. *The group* $H^r(B_q, B_{q-1}; G)$ *is isomorphic to the direct sum*

$$\Sigma H^r(p^{-1}(\bar{\sigma}), p^{-1}(\dot{\sigma}); G) \cong C^q(X; H^{r-q}(F; G)), \tag{23}$$

where the summation is over all the q-dimensional cells of X and where the group $C^q(X; H^{r-q}(F; G))$ *is the group of all q-dimensional cochains of X with the coefficient group* $H^{r-q}(F; G)$. *If* λ *denotes this isomorphism of* $H^r(B_q, B_{q-1}; G)$ *onto* $C^q(X; H^{r-q}(F; G))$, *then commutativity holds in the diagram*

$$\tag{24} \begin{array}{ccccc} H^r(B_q, B_{q-1}; G) & \xrightarrow{j^*} & H^r(B_q; G) & \xrightarrow{\delta^*} & H^{r+1}(B_{q+1}, B_q; G) \\ \lambda \downarrow & & & & \downarrow \lambda \\ C^q(X; H^{r-q}(F; G)) & & \xrightarrow{\qquad\delta\qquad} & & C^{q+1}(X; H^{r-q}(F; G)), \end{array}$$

that is, $\delta\lambda = \lambda\delta^* j^*$, *where* δ *is the coboundary operator for the group of cochains of X.*

We denote by $\bar{\sigma}_a{}^q$, $a = 1, \cdots, s$, the closed q-cells of X, by $\dot{\sigma}_a{}^q$ the set-theoretical boundary of $\bar{\sigma}_a{}^q$, and write $\sigma_a{}^q = \bar{\sigma}_a{}^q - \dot{\sigma}_a{}^q$. The latter will also denote the corresponding cell, chain, or cochain, when it is oriented. Putting $A_a = p^{-1}(\bar{\sigma}_a{}^q)$, $C = B_{q-1}$, we get, by applying Lemma 1. 1′ to the space B_q, the direct sum decomposition

$$H^r(B_q, B_{q-1}; G) \cong \sum_{a=1}^{s} H^r(p^{-1}(\bar{\sigma}_a{}^q), p^{-1}(\dot{\sigma}_a{}^q); G). \tag{23′}$$

To describe this isomorphism more explicitly, we put, for every a, $B_{q,a} = p^{-1}(X^q - \sigma_a{}^q)$, and consider the following homomorphisms, all induced by inclusion mappings:

$$\begin{array}{ccc} & H^r(B_q, B_{q,a}; G) & \\ \swarrow i_a{}^* & & \searrow j_a{}^* \\ H^r(p^{-1}(\bar{\sigma}_a{}^q), p^{-1}(\dot{\sigma}_a{}^q); G) & \xleftarrow{k_a{}^*} & H^r(B_q, B_{q-1}; G). \end{array}$$

By tne excision axiom, $i_a{}^*$ is an onto isomorphism, while, by Lemma 1. 1,

j_a^* is an isomorphism. Moreover, we have $i_a^* = k_a^* j_a^*$ or $k_a^*(j_a^* i_a^{*-1}) =$ identity. In the decomposition (23′), $H^r(B_q, B_{q-1}; G)$ is a direct sum of the subgroups $j_a^* i_a^{*-1} H^r(p^{-1}(\bar{\sigma}_a^q), p^{-1}(\dot{\sigma}_a^q); G)$.

Since $p^{-1}(\bar{\sigma}_a^q)$ is homeomorphic to $\bar{\sigma}_a^q \times F$, we have

$$H^r(p^{-1}(\bar{\sigma}_a^q), p^{-1}(\dot{\sigma}_a^q); G) \cong H^{r-q}(F; G) \otimes H^q(\bar{\sigma}_a^q, \dot{\sigma}_a^q; Z),$$

where Z is the additive group of integers. The latter group is isomorphic to the group of cochains $\alpha\sigma_a^q$, $\alpha \varepsilon H^{r-q}(F; G)$. If μ_a denotes this isomorphism, λ is defined componentwise by $\lambda_a = \mu_a k_a^*$.

To prove the commutativity of the diagram (24), it is sufficient to take $x \varepsilon H^r(B_q, B_{q,a}; G)$ and to prove that $\delta\lambda j_a^*(x)$ and $\lambda\delta^* j^* j_a^*(x)$, both $(q+1)$-dimensional cochains of X, have the same value for any $(q+1)$-cell σ_b^{q+1}. For this purpose it is essential to consider the following diagram:

$$\begin{array}{ccccc}
& & H^r(B_q, B_{q,a}; G) & \xrightarrow{\Delta^*} & H^{r+1}(B_{q+1}, B_q; G) \\
& \swarrow i_a^* & \downarrow l_{ab}^* & & \searrow k_b^* \\
H^r(p^{-1}(\bar{\sigma}_a^q), p^{-1}(\dot{\sigma}_a^q); G) & & H^r(p^{-1}(\dot{\sigma}_b^{q+1}), p^{-1}(\dot{\sigma}_b^{q+1} - \sigma_a^q); G) & \xrightarrow{\Delta_1^*} & H^{r+1}(p^{-1}(\bar{\sigma}_b^{q+1}), p^{-1}(\dot{\sigma}_b^{q+1}); G) \\
& \searrow \mu_a & & & \swarrow \mu_b \\
& & C^q(\bar{\sigma}_a^q; H^{r-q}(F; G)) & \xrightarrow{\delta} & C^{q+1}(\bar{\sigma}_b^{q+1}; H^{r-q}(F; G)).
\end{array}$$

In this diagram i_a^*, k_b^*, and l_{ab}^* are induced by inclusion mappings, δ is the coboundary operator of the group of cochains, while $\Delta^* = \delta^* j^* j_a^*$ and Δ_1^* are the coboundary operators of the triples $B_{q,a} \subset B_q \subset B_{q+1}$ and $p^{-1}(\dot{\sigma}_b^{q+1} - \sigma_a^q) \subset p^{-1}(\dot{\sigma}_b^{q+1}) \subset p^{-1}(\bar{\sigma}_b^{q+1})$. In the last notation we adopt the convention that $\dot{\sigma}_b^{q+1} - \sigma_a^q = \dot{\sigma}_b^{q+1}$, if σ_a^q is not a face of σ_b^{q+1}. Since the second triple can be mapped into the first one by the inclusion mapping, we have $k_b^* \Delta^* = \Delta_1^* l_{ab}^*$. Since we are only interested in the values of the cochains for σ_b^{q+1}, we can restrict ourselves to the bundle over $\bar{\sigma}_b^{q+1}$. Then we have

$$\lambda\Delta^* = \mu_b k_b^* \Delta^* = \mu_b \Delta_1^* l_{ab}^*,$$

$$\delta\lambda j_a^* = \delta\mu_a k_a^* j_a^* = \delta\mu_a i_a^*.$$

It suffices to prove that the homomorphisms in the right-hand members of these two equations are identical.

If σ_a^q is not a face of σ_b^{q+1}, both homomorphisms will give zero, because $H^r(p^{-1}(\dot{\sigma}_b^{q+1}), p^{-1}(\dot{\sigma}_b^{q+1} - \sigma_a^q); G) = 0$ and δ obviously gives zero. Suppose now σ_a^q be a face of σ_b^{q+1}. By the excision axiom, the homomorphism

$$i_{ab}^*: H^r(p^{-1}(\dot{\sigma}_b^{q+1}), p^{-1}(\dot{\sigma}_b^{q+1} - \sigma_a^q); G) \to H^r(p^{-1}(\bar{\sigma}_a^q), p^{-1}(\dot{\sigma}_a^q); G)$$

induced by the inclusion mapping is an onto isomorphism. Moreover, $i_a{}^* = i_{ab}{}^* l_{ab}{}^*$. It suffices therefore to prove that $\mu_b \Delta_1{}^* = \delta \mu_a i_{ab}{}^*$. But then all the groups in question refer to the cell $\bar{\sigma}_b{}^{q+1}$ or to the bundle over it, which is homeomorphic to the Cartesian product $\bar{\sigma}_b{}^{q+1} \times F$. The verification of the relation in this case is trivial. This completes the proof of Lemma 1. 2.

Lemma 1. 2 can be briefly described by saying that the relative cohomology group $H^r(B_q, B_{q-1}; G)$ is isomorphic to the group of q-dimensional cochains of X with the coefficient group $H^{r-q}(F; G)$ and that the homomorphism $\delta^* j^*$ becomes then the coboundary operator under this identification. Because of our assumption that the coefficient system is simple, consideration of local coefficients is not necessary.

However, we are interested not in the relative cohomology groups $H^r(B_q, B_{q-1}; G)$, but in the absolute cohomology groups $H^r(B; G) = H^r(B_n; G)$. To derive information about them, we consider successively the groups $H^r(B_q; G)$, $q = 0, 1, \cdots, n$. All these groups are connected by homomorphisms as in the diagram:

$$\begin{array}{ccccc}
H^r(B_n, B_{n-1}) & \xrightarrow{j^*} & H^r(B_n = B) & & \\
 & & \downarrow i^* & & \\
 & & \vdots & & \\
 & & \downarrow i^* & & \\
\cdots \to H^r(B_q, B_{q-1}) & \xrightarrow{j^*} & H^r(B_q) & \xrightarrow{\delta^*} & H^{r+1}(B_{q+1}, B_q) \to \cdots \\
 & & \downarrow i^* & & \\
 & & \vdots & & \\
 & & \downarrow i^* & & \\
\cdots \to H^r(B_r, B_{r-1}) & \xrightarrow{j^*} & H^r(B_r) & \xrightarrow{\delta^*} & H^{r+1}(B_{r+1}, B_r) \to \cdots \\
 & & \downarrow i^* & & \\
 & & \vdots & & \\
 & & \downarrow i^* & & \\
H^r(B_0, B_{-1}) & \xrightarrow{j^*} & H^r(B_0) & \xrightarrow{\delta^*} & H^{r+1}(B_1, B_0). \\
 & & \downarrow & & \\
 & & 0 & &
\end{array} \tag{25}$$

In all these cohomology groups the coefficient group is G, which is omitted for simplicity of notation. Every sequence of groups from the diagram connected by homomorphisms in the cyclic order j^*, i^*, δ^* is exact. We shall denote by $H_0{}^r(B_q)$ that subgroup of $H^r(B_q)$, which is the kernel of δ^*.

First a remark about the homology position of the fiber in the bundle B. We take a vertex $v \in X^0$, and identify F with $p^{-1}(v)$. This gives rise to the inclusion mappings $l_0\colon F \to B_0$ and $l\colon F \to B_r$, and the induced homomorphisms

$$\tag{26} \begin{array}{ccc} & & H^r(B) \\ & \swarrow^{l^*} & \downarrow^{i^*} \\ H^r(F) & \xleftarrow{\;l_0^*\;} & H_0{}^r(B_0). \end{array}$$

The homomorphism $l_0{}^*$ in (26) is an onto isomorphism, for $H_0{}^r(B_0)$ is clearly isomorphic to the group of 0-dimensional cohomology classes of X, with the coefficient group $H^r(F)$, which is isomorphic to $H^r(F)$.

The diagram (25) leads to a homological definition of the primary obstruction, as given by the lemma:

LEMMA 1.3. *Let $r > 1$ be the integer such that $\pi_r(F) \neq 0$ and $\pi_s(F) = 0$ for all $s < r$. Let $H_r(F; Z)$ be the coefficient group G and $\omega \varepsilon H^r(F; G)$ be the cohomology class which assigns to any $z \varepsilon H_r(F; Z)$ the element z itself. There exists $\bar{\omega} \varepsilon H^r(B_r; G)$ such that $(i^*)^r\bar{\omega} = l_0{}^{*-1}\omega$; the element $\lambda\delta^*\bar{\omega}$ is a cocycle and its cohomology class is the negative of the primary obstruction.*

Consider first an element $\phi \varepsilon H_0{}^r(B_0)$. From the relations

$$H^{r+1}(B_s, B_{s-1}) = 0, \qquad 1 < s < r+1,$$

it follows that there exists $\bar{\phi} \varepsilon H^r(B_r)$ such that $(i^*)^r\bar{\phi} = \phi$. By Lemma 1.2 we have

$$\delta(\lambda\delta^*\bar{\phi}) = \lambda\delta^* j^* \delta^* \bar{\phi} = 0,$$

which means that $\lambda\delta^*\bar{\phi}$ is a cocycle. On the other hand, $\bar{\phi}$ is determined up to an additive term j^*y, $y \varepsilon H^r(B_r, B_{r-1})$, so that $\lambda\delta^*\bar{\phi}$ is determined up to an additive term $\lambda\delta^* j^* y = \delta\lambda y$, that is, up to a coboundary. Thus the cohomology class of $\lambda\delta^*\bar{\phi}$ is completely determined.

We also remark that under our assumptions the natural homomorphism

$$\tag{27} f\colon H^r(F; G) \to \operatorname{Hom}(G, G), \qquad G = H_r(F; Z),$$

is an onto isomorphism.[12] The class ω in the statement of the Lemma is therefore well defined. From the above it follows that the same is true of the cohomology class of $\lambda\delta^*\bar{\omega}$. The verification that this is equal to the

[12] Eilenberg-MacLane [4], 808, Theorem 32.1. The theorem quoted was formulated for a star-finite complex and for homology groups with infinite cycles. A similar statement is therefore true for a closure-finite complex and for cohomology groups with infinite cocycles. Our result follows from this theorem and the further fact that $H^{r-1}(F; G) = 0$.

negative of the primary obstruction can be carried out by studying a cross-section over X^r. It is straightforward and we shall omit the details here.

We are now in a position to give a proof of Theorem 1 (Cf. Introduction). Since the primary obstruction is zero, the element $\bar{\omega} \varepsilon H^r(B_r)$ can be so chosen that $\delta^* \bar{\omega} = 0$. Since

$$H^{r+1}(B_s, B_{s-1}) = 0, \qquad s > r+1,$$

there is $\bar{\bar{\omega}} \varepsilon H^r(B)$ such that $(i^*)^{n-r} \bar{\bar{\omega}} = \bar{\omega}$, which implies

$$(i^*)^n \bar{\bar{\omega}} = (i^*)^r \bar{\omega} = l_0^{*-1} \omega.$$

By the definition of our notation for inclusion mappings, we can write i^* for $(i^*)^n$, and the last relation becomes $l_0^* i^* \bar{\bar{\omega}} = l^* \bar{\bar{\omega}} = \omega$. Thus ω belongs to the image of l^*.

To prove that the same is true of any element $\phi \varepsilon H^r(F; G)$, we consider its corresponding endomorphism $f(\phi)$ of G into itself. This endomorphism $f(\phi)$ of the coefficient group induces endomorphisms ϕ_B and ϕ_F of $H^r(B; G)$ and $H^r(F; G)$ respectively. Moreover, it is clear that $\phi_F(\omega) = \phi$. From the commutativity of the diagram

$$\begin{array}{ccc} H^r(B; G) & \xrightarrow{\phi_B} & H^r(B; G) \\ l^* \downarrow & & \downarrow l^* \\ H^r(F; G) & \xrightarrow{\phi_F} & H^r(F; G) \end{array}$$

it follows that ϕ belongs to the image of l^*. This completes the proof of Theorem 1.

The following Lemma describes the operation of "integration over the fiber":

LEMMA 1.4. *Let $H^r(F; Z) \neq 0$, $r > 0$, be the last non-vanishing cohomology group of F, so that $H^s(F; Z) = 0$ for all $s > r$. To any coefficient group G an integration over the fiber*

(1) $$\Psi \colon H^m(B; G) \to H^{m-r}(X; H^r(F; G)),$$

can be defined.

This homomorphism Ψ has been described in the Introduction. To prove its existence we first notice that our assumptions imply

$$H^{m-1}(B_{m-r-2}; G) = H^m(B_{m-r-1}; G) = 0.$$

Let $u \varepsilon H^m(B)$ and $u' = (i^*)^{m-r} u \varepsilon H^m(B_{m-r})$ and consider the cohomology groups

$$\begin{array}{ccccccc} H^{m-1}(B_{m-r-1}, B_{m-r-2}) & \xrightarrow{j^*} & H^{m-1}(B_{m-r-1}) & & & & \\ & & i^* \downarrow & & & & \\ & & 0 & & & & \\ & \xrightarrow{\delta^*} & H^m(B_{m-r}, B_{m-r-1}) & \xrightarrow{j^*} & H^m(B_{m-r}) & \xrightarrow{\delta^*} & H^{m+1}(B_{m-r+1}, B_{m-r}) \\ & & & & i^* \downarrow & & \\ & & & & 0 & & \end{array}$$

where the sequences in the cyclic order j^*, i^*, δ^* are exact. Since $i^*u' = 0$, there is $v \varepsilon H^m(B_{m-r}, B_{m-r-1})$ with $u' = j^*v$. Then $\lambda v \varepsilon C^{m-r}(X; H^r(F; G))$ is a cocycle, for $\delta\lambda v = \lambda\delta^* j^* v = \lambda\delta^* u' = 0$. Moreover, v is defined up to an additive term $\delta^* y$, $y \varepsilon H^{m-1}(B_{m-r-1})$, where $y = j^* z$, $z \varepsilon H^{m-1}(B_{m-r-1}, B_{m-r-2})$. Since $\lambda\delta^* y = \lambda\delta^* j^* z = \delta\lambda z$, λv is defined up to a coboundary. Its cohomology class, which is an element of $H^{m-r}(X; H^r(F; G))$, is therefore completely determined by u. We call it Ψu and thus prove the existence of the homomorphism Ψ.

Concerning the homomorphism Ψ, we have the following useful lemma which follows immediately from its definition:

LEMMA 1. 5. *Let $p\colon B \to X$ and $p'\colon B' \to X'$ be two fiber bundles and $\tilde{f}\colon B \to B'$ be a bundle map which induces a mapping $f\colon X \to X'$ of the base spaces. If Ψ, Ψ' denote integrations over the fiber of the two bundles, then commutativity holds in the diagram:*

$$\begin{array}{ccc} H^m(B) & \xleftarrow{\tilde{f}^*} & H^m(B') \\ \Psi \downarrow & & \downarrow \Psi' \\ H^{m-r}(X; H^r(F)) & \xleftarrow{f^*} & H^{m-r}(X'; H^r(F)), \end{array}$$

that is, $f^*\Psi' = \Psi\tilde{f}^*$.

2. Proof of Theorem 2 (the Generalized Leray-Hirsch Theorem).

To carry out the proof we shall adopt the notations of the preceding section. By double induction on r and q, we proceed to prove the following statements:

a) There exists a homomorphism $\mu_{rq}\colon H_0{}^r(B_q) \to H^r(B)$, such that $(i^*)^{n-q}\mu_{rq}$ is the identity automorphism of $H_0{}^r(B_q)$;

b) $H_0{}^r(B_q)$ is isomorphic to the direct sum of $H_0{}^r(B_{q-1})$ and $H^q(X; H^{r-q}(F))$. Here we make the convention that a cohomology group of negative dimension is vacuous.

We remark that, for $q = n$, b) implies that $H^r(B)$ is isomorphic to the direct sum $\sum_{s=0}^{n} H^s(X; H^{r-s}(F))$. Since the latter is isomorphic to $H^r(X \times F)$ by Künneth's Theorem, our theorem follows from b).

For $q = 0$ we put $\mu_{r0} = \mu l_0^*$. Since l_0^* establishes an isomorphism between $H_0{}^r(B_0)$ and $H^r(F)$, the fact that $l^*\mu = l_0{}^*i^*\mu$ is the identity automorphism of $H^r(F)$ implies that $l_0{}^{*-1}(l_0{}^*i^*\mu)l_0{}^* = i^*\mu_{r0}$ is the identity automorphism of $H_0{}^r(B_0)$. This proves a). Statement b) is obvious, if we define $H_0{}^r(B_{-1})$ to be zero.

Suppose μ_{ts} be defined for $t < r$ and $t = r$, $0 \leqq s \leqq q-1$, fulfilling the conditions a), b). Consider the diagram

$$\begin{array}{ccccccc}
\rightarrow H^{r-1}(B_{q-1}, B_{q-2}) & \xrightarrow{j^*} & H^{r-1}(B_{q-1}) & \xrightarrow{\delta^*} & H^r(B_q, B_{q-1}) & & \\
 & & \downarrow i^* & & \downarrow j^* & & \\
 & & H^{r-1}(B_{q-2}) & & H^r(B_q) & \xrightarrow{\delta^*} & H^{r+1}(B_{q+1}, B_q) \rightarrow \cdots. \\
 & & & & \downarrow i^* & & \\
 & & & & H^r(B_{q-1}) & &
\end{array}$$

We put $\nu = (i^*)^{n-q}\mu_{r,q-1}\colon H_0{}^r(B_{q-1}) \rightarrow H_0{}^r(B_q)$, so that $i^*\nu = $ identity by induction hypothesis. A familiar argument proves that $H^r(B_q)$ is a direct sum of $\nu H_0{}^r(B_{q-1})$ and $j^*H^r(B_q, B_{q-1})$. Moreover, ν is clearly an isomorphism. By exactness the second summand fulfills the isomorphism

$$j^*H^r(B_q, B_{q-1}) \cong H^r(B_q, B_{q-1})/\delta^*H^{r-1}(B_{q-1}).$$

From our induction hypothesis it follows that the group $H^{r-1}(B_{q-1})$ is a direct sum of $(i^*)^{n-q+1}\mu_{r-1,q-2}H_0{}^{r-1}(B_{q-2})$ and $j^*H^{r-1}(B_{q-1}, B_{q-2})$ of which the first summand goes to 0 under δ^*. Therefore we have

$$j^*H^r(B_q, B_{q-1}) \cap H_0{}^r(B_q) \cong K^r(B_q, B_{q-1})/\delta^*j^*H^{r-1}(B_{q-1}, B_{q-2}),$$

where $K^r(B_q, B_{q-1}) \subset H^r(B_q, B_{q-1})$ is the kernel of δ^*j^*. Using the isomorphism λ, we see that this group is isomorphic to $H^q(X; H^{r-q}(F))$. This proves b).

It remains to define μ_{rq} to satisfy a). For an element

$$x = \nu y \,\varepsilon\, \nu H_0{}^r(B_{q-1}),\ y \,\varepsilon\, H_0{}^r(B_{q-1}),$$

we set

$$\mu_{rq}x = \mu_{r,q-1}y = \mu_{r,q-1}i^*x.$$

Then $(i^*)^{n-q}\mu_{rq} = (i^*)^{n-q}\mu_{r,q-1}i^* = \nu i^*$ is the identity automorphism, since ν is an isomorphism of $H_0{}^r(B_{q-1})$ onto $\nu H_0{}^r(B_{q-1})$. To define μ_{rq} for the other summand, we need some preparations.

First we put $B' = X \times B$ and consider B' as a bundle over X with projection p' defined by $p'(x, b) = x$, $x \varepsilon X, b \varepsilon B$. To this bundle the homology theory established in § 1 can be applied, and we shall denote the notions and symbols pertaining to it by dashes. This bundle is, however, a trivial bundle and has very simple properties. In particular, we have, by Künneth's Theorem, the isomorphisms

$$H_0^r(B_q') = H_0^r(X^q \times B) \cong \sum_{s=0}^{q} H^s(X; H^{r-s}(B)),$$

$$H^r(B') = H^r(X \times B) \cong \sum_{s=0}^{n} H^s(X; H^{r-s}(B)).$$

These permit us to define the homomorphisms

$$\mu_{rq}': H_0^r(B_q') \to H^r(B'),$$

such that $(i'^*)^{n-q}\mu_{rq}'$ is the identity.

Next we define the mappings

$$g_q: B_q \to B_q' = X^q \times B$$

by $g_q(b) = (p(b), b)$, $b \varepsilon B_q$, and write $g_n = g$. These mappings induce homomorphisms on the cohomology groups, for which there is commutativity in the diagram

$$\begin{array}{ccccc} H^r(X^q \times B, X^{q-1} \times B) & \xrightarrow{j'^*} & H^r(X^q \times B) & \xrightarrow{\delta'^*} & H^{r+1}(X^{q+1} \times B, X^q \times B) \\ \downarrow g_q^* & & \downarrow g_q^* & & \downarrow g_{q+1}^* \\ H^r(B_q, B_{q-1}) & \xrightarrow{j^*} & H^r(B_q) & \xrightarrow{\delta^*} & H^{r+1}(B_{q+1}, B_q). \end{array}$$

To $y' \varepsilon H^r(X^q \times B, X^{q-1} \times B)$, $y \varepsilon H^r(B_q, B_{q-1})$, we have $\lambda' y' \varepsilon C^q(X; H^{r-q}(B))$, $\lambda y \varepsilon C^q(X; H^{r-q}(F))$, so that we can write

$$\lambda' y' = \sum_i h_i' \sigma_i^q, \qquad h_i' \varepsilon H^{r-q}(B),$$

$$\lambda y = \sum_i h_i \sigma_i^q, \qquad h_i \varepsilon H^{r-q}(F),$$

where σ_i^q are the q-cells of X. It follows from the definition of g_q that

$$g_q^* y' = \lambda^{-1}\{\sum_i l^*(h_i')\sigma_i^q\}.$$

Conversely, because of the existence of the homomorphism μ, we can define the homomorphism

$$\rho_{rq}: H^r(B_q, B_{q-1}) \to H^r(X^q \times B, X^{q-1} \times B)$$

by the equation

$$\rho_{rq}(y) = \lambda'^{-1}\{\sum \mu(h_i)\sigma_i^q\}.$$

It has therefore the property that $g_q{}^*\rho_{rq} = \text{identity}$. Moreover, from the interpretation of δ^*j^* and $\delta'^*j'^*$ as coboundary operators under the isomorphisms λ and λ', we have

$$\rho_{r+1,q+1}\delta^*j^* = \delta'^*j'^*\rho_{rq}.$$

From this it follows that $\delta^*j^*y = 0$ if and only if $\delta'^*j'^*\rho_{rq}y = 0$.

To define μ_{rq} for the summand $j^*H^r(B_q, B_{q-1}) \cap H_0{}^r(B_q)$ of $H_0{}^r(B_q)$, we take $y \in H^r(B_q, B_{q-1})$ with $\delta^*j^*y = 0$. Then $j'^*\rho_{rq}y \in H_0{}^r(B'_q)$ and $\mu_{rq}'j'^*\rho_{rq}y \in H^r(B') = H^r(X \times B)$. Now consider the diagram

$$\begin{array}{ccc} H^r(X \times B) & \xrightarrow{g^*} & H^r(B), \\ \downarrow (i'^*)^{n-q} & & \downarrow (i^*)^{n-q} \\ H^r(X^q \times B) & \xrightarrow{g_q{}^*} & H^r(B_q) \end{array}$$

in which commutativity holds. We define

$$\mu_{rq}j^*y = g^*\mu_{rq}'j'^*\rho_{rq}y.$$

When we modify y by an additive term δ^*j^*z, $z \in H^{r-1}(B_{q-1}, B_{q-2})$, the right-hand side will be modified by an additive term

$$g^*\mu_{rq}'j'^*\rho_{rq}\delta^*j^*z = g^*\mu_{rq}'j'^*\delta'^*j'^*\rho_{r-1,q-1}z = 0.$$

Hence μ_{rq} depends only on j^*y and not on the choice of y. Since

$$(i^*)^{n-q}\mu_{rq}j^*y = (i^*)^{n-q}g^*\mu_{rq}'j'^*\rho_{rq}y = g_q{}^*(i'^*)^{n-q}\mu_{rq}'j'^*\rho_{rq}y$$
$$= g_q{}^*j'^*\rho_{rq}y = j^*g_q{}^*\rho_{rq}y = j^*y,$$

we conclude that the homomorphism μ_{rq} satisfies condition a).

This completes our induction and hence the proof of Theorem 2.

The following corollary follows immediately from the above proof.

COROLLARY 2.1. *Suppose the hypotheses of Theorem 2 be satisfied. Then $H^m(B)$ is isomorphic to the direct sum $\sum_{q=0}^{m} H^q(X; H^{m-q}(F))$. For $u \in H^m(B)$, its image Ψu under the integration over the fiber Ψ is the component of u in the summand $H^{m-r}(X; H^r(F))$, where r is defined by Lemma* 1.4.

Under some further assumptions which will be satisfied in our applications, we can express these relations in a more explicit form, using our homomorphism μ. In fact, from our proof of Theorem 2, we see that $H^r(B)$ is a direct sum of $\mu_{rq}(j^*H^r(B_q, B_{q-1}) \cap H_0{}^r(B_q))$ for $q = 0, 1, \cdots, r$.

Suppose that our coefficient system is a ring R with unit element. We call an element of $Z^q(X; H^{r-q}(F; R))$ a cocycle of the first kind, if it is equal to a finite sum of the form $\sum_i c_i \otimes z_i$, where $c_i \varepsilon Z^q(X; R)$, $z_i \varepsilon H^{r-q}(F; R)$. If this is not the case, we call it a cocycle of the second kind. We now make the assumption that every element of $H^q(X; H^{r-q}(F; R))$ has as representative a cocycle of the first kind. This condition is satisfied when, for instance, R is a field.

To an element $y \varepsilon H^r(B_q, B_{q-1})$, with $\delta^* j^* y = 0$, we can write

$$\lambda y = \sum_i c_i \otimes z_i,$$

where $c_i \varepsilon Z^q(X; R)$, $z_i \varepsilon H^{r-q}(F; R)$. The homomorphism μ_{rq}' can be defined such that

$$\mu_{rq}' j'^* \rho_{rq} y = \sum_i \gamma_i \otimes \mu(z_i),$$

where γ_i is the class of c_i. By definition, $\mu_{rq} j^* y$ is the image of this element under the homomorphism g^*. Now we can decompose g as a product of two mappings: $g = h\Delta$, where $\Delta: B \to B \times B$ is the diagonal map defined by $\Delta(b) = (b, b)$, $b \varepsilon B$, and $h: B \times B \to X \times B$ is defined by $h(b, b') = (p(b), b')$, $b, b' \varepsilon B$. It follows that

$$g^*(\sum_i \gamma_i \otimes \mu(z_i)) = \Delta^* h^*(\sum_i \gamma_i \otimes \mu(z_i)) = \Delta^*(\sum_i p^*(\gamma_i) \otimes \mu(z_i)) = \sum_i p^*(\gamma_i) \cup \mu(z_i).$$

These considerations lead to the following theorem:

COROLLARY 2.2. *Under the hypotheses of Theorem 2 suppose further that, for an integer $m \geqq 0$, every element of $H^q(X; H^{m-q}(F; R))$ has as representative a cocycle of the first kind. Then every element $u \varepsilon H^m(B; R)$ can be written in the form*

$$u = \sum_i \mu(z_i) \cup p^*(\gamma_i), \qquad z_i \varepsilon H^q(F; R), \gamma_i \varepsilon H^{m-q}(X; R),$$

where $\dim z_i + \dim \gamma_i = m$. If r is the integer such that $H^r(F; R)$ is the highest non-vanishing cohomology group of F with the coefficient group R, then

$$\Psi u = \sum_{\dim z_i = r} z_i \otimes \gamma_i.$$

3. The case of manifolds; proof of Theorem 3. Throughout this section we suppose that B, X, F are oriented manifolds, with the convention that these same symbols denote their fundamental homology and cohomology classes. It is easily seen that $\Psi(B) = \pm X$. We supose that the orientations of the manifolds are such that the positive sign holds. Under our assumptions the highest non-vanishing cohomology group of F is $H^s(F;G)$, where s is the dimension of F, and it is naturally isomorphic to G. Integration over the fiber can therefore be considered as a homomorphism

$$\Psi\colon H^m(B;G) \to H^{m-s}(X;G). \tag{28}$$

LEMMA 3.1. *Let the abelian groups G_1 and G_2 be paired to G, and let $v \varepsilon H^q(X;G_1)$, $u \varepsilon H^r(B;G_2)$. Then*

$$\Psi(p^*v \cup u) = v \cup \Psi u. \tag{29}$$

To prove this lemma we denote by $d\colon X \to X \times X$ the diagonal map defined by $d(x) = (x,x)$, $x \varepsilon X$. Similarly, let $\Delta\colon B \to B \times B$ be the diagonal map of B. We define the mappings

$$p'\colon B \times B \to X \times B, \tag{30}$$

$$p''\colon X \times B \to X \times X, \tag{31}$$

by $p'(b,b') = (p(b),b')$, $p''(x,b) = (x,p(b))$, $b, b' \varepsilon B$, $x \varepsilon X$. These give rise to two bundles, both with fibers F. We put $\Gamma = p'\Delta$. Then $\Gamma\colon B \to X \times B$ is a bundle map of the given bundle $p\colon B \to X$ into the bundle (31). By Lemma 1.5, we have commutativity in the diagram $(m = q + r)$:

$$\begin{array}{ccc} & & H^m(B \times B;G) \\ & \swarrow^{\Delta^*} & \uparrow p'^* \\ H^m(B;G) & \xleftarrow{\ \Gamma^*\ } & H^m(X \times B;G) \\ \Psi\downarrow & & \downarrow \Psi'' \\ H^m(X;G) & \xleftarrow{\ d^*\ } & H^m(X \times X;G) \end{array}$$

where Ψ'' is the integration over the fiber of the bundle (31).

Now let u', v' be representative cocycles of the cohomology classes u, v respectively. Then $v' \times u' \varepsilon Z^m(X \times B;G)$, the group of all m-cocycles of $X \times B$. If we denote by the same symbols p', Δ cellular approximations of the respective continuous mappings, we have

$$\Delta^*p'^*(v' \times u') = \Delta^*(p^*v' \times u') = p^*v' \cup u'.$$

This proves that the class containing $v' \times u'$ is mapped by $\Delta^*p'^*$ into $p^*v \cup u$. On the other hand, this class is mapped by Ψ'' into the class containing $v' \times \Psi u'$, which goes to $v \cup \Psi u$ under d^*. From the commutativity of our diagram follows therefore the formula (29).

We are now in a position to prove Theorem 3. As stated in the Theorem,

12

we have $u \varepsilon H^m(B; G)$. Then $\Psi u \varepsilon H^{m-s}(X; G)$ and $\mathfrak{D}_X \Psi u \varepsilon H_{n+s-m}(X; G)$. We take any element $v \varepsilon H^{n+s-m}(X; \text{Char } G)$, and denote by a dot the multiplication of homology and cohomology classes. Then we have

$$v \cdot \mathfrak{D}_X \Psi u = v \cdot (\Psi u \cap X) = (v \cup \Psi u) \cdot X,$$

$$\begin{aligned} v \cdot p_* \mathfrak{D}_B u = (p^* v) \cdot (\mathfrak{D}_B u) &= (p^* v) \cdot (u \cap B) = (p^* v \cup u) \cdot B \\ &= \{\Psi(p^* v \cup u)\} \cdot X. \end{aligned}$$

By Lemma 3. 1, the right-hand sides of the two equations are equal. The same is therefore true of the left-hand sides, with arbitrary v. This proves Theorem 3.

4. On the bundles associated to a complex manifold. We now turn our attention to the study of a compact complex manifold M of (complex) dimension n. Some of the more important bundles associated to M have been described in the Introduction, and we shall follow the notations. This section will be devoted to the proof of a few elementary properties of these bundles.

LEMMA 4. 1. *To $x \varepsilon M$ let $F_{nr} = q_{nr}^{-1}(x)$, $\tilde{F}_{nr} = (q_{nr} p_{nr})^{-1}(x)$ be the fibers of the bundles B_{nr}, $\tilde{B}_{nr}$ respectively. Then $p_{nr}(x): \tilde{F}_{nr} \to F_{rn}$ is a bundle having as fiber the Cartesian product of r complex lines, each with the origin deleted. It gives rise to r characteristic cohomology classes $w_1, \cdots, w_r$ of dimension 2 in the base space F_{nr}, which generate, by ring operations, the cohomology ring of F_{nr} with integer coefficients and which can be so chosen that $w_1^{n-1} \cup \cdots \cup w_r^{n-r}$ is the fundamental cohomology class of F_{nr}.*

Let V_n be the tangent complex vector space at x, with 0 denoting its origin. A point of $\tilde{F}_{nr}$ is then a sequence of simple multivectors of V_n of the form $e_1, e_1 \wedge e_2, \cdots, e_1 \wedge \cdots \wedge e_r \neq 0$, while a point of F_{nr} is a sequence of linear spaces of directions $L_0 \subset L_1 \subset \cdots \subset L_{r-1}$. The projection $p_{nr}(x)$ assigns to the multivector $e_1 \wedge \cdots \wedge e_i$, $i = 1, \cdots, r$, the linear subspace L_{i-1} it determines. Since two simple multivectors determine the same L_{i-1} when and only when they differ from each other by a non-zero complex factor, the fiber of the bundle $p_{nr}(x): \tilde{F}_{nr} \to F_{nr}$ is homeomorphic to a Cartesian product $E_1 \times \cdots \times E_r$ of r complex lines, each with the origin deleted. Moreover, we can suppose E_i to be the space of all $e_1 \wedge \cdots \wedge e_i$ having the same projection L_{i-1}. This bundle is orientable, since the spaces involved are complex manifolds. We denote by w_i the characteristic class corresponding to the factor E_i of the fiber. The classes $w_1, \cdots, w_r$ can be considered to be with integer coefficients.

To prove the assertions on the homology structure of F_{nr} we proceed by induction on r. For $r = 1$, $\tilde{F}_{n1} = V_n - 0$ and F_{n1} is the complex projective space P_{n-1} of dimension $n-1$. In this case we see easily that w_1 is the cohomology class dual to the homology class having as representative a hyperplane of P_{n-1}, so that the Lemma is true. Suppose that the Lemma holds for $r-1$. Let

$$t\colon F_{nr} \to F_{n,r-1} \tag{32}$$

be the mapping which sends the sequence $L_0 \subset L_1 \subset \cdots \subset L_{r-1}$ to $L_0 \subset L_1 \subset \cdots \subset L_{r-2}$. This defines a fiber bundle having as fiber the set of all L_{r-1} through a fixed L_{r-2} in P_{n-1}, which is homeomorphic to the complex projective space P_{n-r} of dimension $n-r$. Denote by $k_{nr}\colon P_{n-r} \to F_{nr}$ the inclusion mapping of a fiber into the bundle. When, in the bundle $\tilde{F}_{nr} \to F_{nr}$, we restrict the fiber to the product $E_1 \times \cdots \times E_{r-1}$, it is induced by the mapping t. Hence the first $r-1$ characteristic classes in F_{nr} are $w_i = t^* w_i'$, $i = 1, \cdots, r-1$, where w_i' are the characteristic classes in $F_{n,r-1}$ of the bundle $\tilde{F}_{n,r-1} \to F_{n,r-1}$. Let w_r be the class corresponding to the fiber E_r. Then $k_{nr}^* w_r = w$ is a generating class of P_{n-r}. We define $\mu(w^j) = w_r^j$, $j = 1, \cdots, n-r-1$, and extend μ by linearity into a homomorphism of the cohomology groups of P_{n-r} into those of F_{nr} (with integer coefficients). Then $k_{nr}^* \mu =$ identity, and the hypotheses of Theorem 2 are satisfied. If Ψ is the integration over the fiber of the bundle (32), we see easily that

$$\Psi(w_1^{n-1} \cup \cdots \cup w_r^{n-r}) = (w_1')^{n-1} \cup \cdots \cup (w_{r-1}')^{n-r+1}.$$

Since the right-hand side of this equation is the fundamental cohomology class of $F_{n,r-1}$ by our induction hypothesis, it follows that $w_1^{n-1} \cup \cdots \cup w_r^{n-r}$ is the fundamental cohomology class of F_{nr}. This completes the proof of the Lemma.

LEMMA 4.2. *The bundle B_{nr} satisfies the hypotheses of Theorem 2.*

As in the proof of Lemma 4.1, the bundle $\tilde{B}_{nr} \to B_{nr}$ has as fiber the Cartesian product of r complex lines, each with the origina deleted. Let $v_1, \cdots, v_r$ be the corresponding characteristic cohomology classes. Since the bundle $p_{nr}(x)\colon \tilde{F}_{nr} \to F_{nr}$ is induced by the inclusion mapping $l_{nr}\colon F_{nr} \to B_{nr}$, we have $l_{nr}^* v_i = w_i$, $i = 1, \cdots, r$. Defining $\mu w_i^{\rho_i} = v_i^{\rho_i}$, $\rho_i = 1, \cdots, n-i$, and extending it by linearity into a homomorphism of the cohomology groups of F_{nr} into those of B_{nr}, we have $l_{nr}^* \mu =$ identity. Hence the hypotheses of Theorem 2 are fulfilled.

5. Proof of the duality theorem. The proof of the duality theorem depends on the consideration of the universal bundles. Let $E_{\nu+N}$ be a complex vector space of dimension $\nu+N$ and let $H(\nu, N)$ be the Grassmann manifold of all ν-dimensional linear spaces through the origin in $E_{\nu+N}$. Over $H(\nu, N)$ as base space we can define a bundle of complex vectors by taking the space $S(\nu, N)$ of all pairs (x, E_ν), where E_ν is a ν-dimensional linear subspace of $E_{\nu+N}$ and $x \in E_\nu$, and defining the projection $P_{\nu N}\colon S(\nu, N) \to H(\nu, N)$ by $P_{\nu N}(x, E_\nu) = E_\nu$. For applications to algebraic geometry it is advantageous to identify all non-zero vectors of $E_{\nu+N}$ which differ from each other by a non-zero complex factor and hence to consider the complex projective space $P_{\nu+N-1}$ of dimension $\nu+N-1$ and the Grassmann variety $G(\nu-1, N)$ of all linear spaces of dimension $\nu-1$ in $P_{\nu+N-1}$. Since $H(\nu, N)$ and $G(\nu-1, N)$ are homeomorphic in a natural way, this does not make any difference for our present purpose, and we shall use $H(\nu, N)$ as the base space of the universal bundle.

We also remark that the characteristic classes described in the Introduction can be defined for any bundle of complex vector spaces having as base space a finite polyhedron. The bundle described in the last section is called universal, because of the following theorem: [13]

Any bundle of complex vector spaces of dimension ν over a finite polyhedron X is induced by a mapping $f\colon X \to H(\nu, N)$, if $\dim X \leqq 2N$. The bundles induced by two such mappings are equivalent if and only if the mappings are homotopic. If Γ_i, $i = 1, \cdots, \nu$, are the characteristic classes of the universal bundle, then those of the induced bundle are $f^*\Gamma_i$.

On the other hand, the characteristic classes of the universal bundle in $H(\nu, N)$ can be described in a simple way in terms of the homology properties of $H(\nu, N)$. The latter have been studied by Ehresmann,[14] whose results can be summarized as follows: We take a sequence of integers

$$0 \leqq a_1 \leqq a_2 \leqq \cdots \leqq a_\nu \leqq N,$$

and, corresponding to such a sequence, a sequence of linear spaces through the origin 0 in $E_{\nu+N}$:

$$0 \subset L_1 \subset L_2 \subset \cdots \subset L_\nu, \tag{33}$$

such that $\dim L_i = a_i + i$, $i = 1, \cdots, \nu$. A Schubert variety, to be denoted by $(a_1 \cdots a_\nu)$, consists of all the E_ν satisfying the conditions

$$\dim (L_i \cap E_\nu) \geqq i, \qquad i = 1, \cdots, \nu. \tag{34}$$

[13] Chern [1], 88-89, Theorems 1 and 2.

[14] Ehresmann [3].

It is of dimension $\sum_{i=1}^{\nu} a_i$. The Schubert varieties have the following properties, which we shall use and which we state here without proof:

1) Every Schubert variety $(a_1 \cdots a_\nu)$ carries a cycle, whose homology class depends only on the symbol and not on the choice of the sequence (33). All these homology classes form a homology base of $H(\nu, N)$. From this it follows that all odd-dimensional Betti numbers of $H(\nu, N)$ are zero and that there are no torsion coefficients.

2) The intersection number $KI((a_1 \cdots a_\nu), (N - a_\nu, \cdots, N - a_1))$ is equal to one. All other intersection numbers of Schubert varieties of complementary dimensions are zero.

3) Let $(a_1 \cdots a_\nu)$ denote also the cohomology class whose value is one for $(a_1 \cdots a_\nu)$ and is zero for all other homology classes. Then Γ_k is the class $(0 \cdots 0\ 1 \cdots 1)$ consisting of $\nu - k$ zeros followed by k ones.

After these preparations we proceed to the proof of Theorem 4.[15] Let $E_{\nu_1+N_1}$ and $E_{\nu_2+N_2}$ be complex vector spaces and let $H(\nu_1, N_1)$, $H(\nu_2, N_2)$ be the corresponding Grassmann manifolds. Let $f_\alpha\colon X \to H(\nu_\alpha, N_\alpha)$, $\alpha = 1, 2$, be the mappings which induce the bundles B_α over X. We denote by $E_{\nu_1+\nu_2+N_1+N_2}$ the vector space spanned by $E_{\nu_1+N_1}$ and $E_{\nu_2+N_2}$, and by $H(\nu_1 + \nu_2, N_1 + N_2)$ the Grassmann manifold of all linear spaces of dimension $\nu_1 + \nu_2$ through the origin in $E_{\nu_1+\nu_2+N_1+N_2}$. An $E_{\nu_1} \subset E_{\nu_1+N_1}$ and an $E_{\nu_2} \subset E_{\nu_2+N_2}$ span an $E_{\nu_1+\nu_2} \subset E_{\nu_1+\nu_2+N_1+N_2}$. This defines a mapping

$$F\colon H(\nu_1, N_1) \times H(\nu_2, N_2) \to H(\nu_1 + \nu_2, N_1 + N_2). \tag{35}$$

Then the bundle $B_1 \boxtimes B_2$ over X is induced by the mapping $F \circ f$, where

$$f\colon X \to H(\nu_1, N_1) \times H(\nu_2, N_2) \tag{36}$$

is defined by $f(x) = (f_1(x), f_2(x))$, $x \in X$. Denote by $\Gamma_1^{(1)}, \cdots, \Gamma_{\nu_1}^{(1)}$; $\Gamma_1^{(2)}, \cdots, \Gamma_{\nu_2}^{(2)}$; $\Gamma_1, \cdots, \Gamma_{\nu_1+\nu_2}$ the characteristic classes of the universal bundles over $H(\nu_1, N_1)$, $H(\nu_2, N_2)$, $H(\nu_1 + \nu_2, N_1 + N_2)$ respectively. We assert that the duality theorem is a consequence of the formula

$$F^*\Gamma_k = \sum_{\substack{0 \leqq i \leqq k \\ k-\nu_2 \leqq i \leqq \nu_1}} \Gamma_i^{(1)} \otimes \Gamma_{k-i}^{(2)}, \qquad k = 1, \cdots, \nu_1 + \nu_2. \tag{37}$$

In fact, we have

$$C_k = f^*F^*\Gamma_k, \qquad C_i^{(1)} = f_1^*\Gamma_i^{(1)}, \qquad C_j^{(2)} = f_2^*\Gamma_j^{(2)},$$

$$C_i^{(1)} \cup C_j^{(2)} = (f_1^*\Gamma_i^{(1)}) \cup (f_2^*\Gamma_j^{(2)}) = f^*(\Gamma_i^{(1)} \otimes \Gamma_j^{(2)}).$$

---

[15] The idea of this proof was indicated by Wu; cf. Wu [17].

Applying the dual homomorphism f^* to (37), we get

$$C_k = \sum_{\substack{0 \leqq i \leqq k \\ k-\nu_2 \leqq i \leqq \nu_1}} C_i^{(1)} \cup C_{k-i}^{(2)}, \tag{38}$$

which is obviously an equivalent formulation of the duality theorem.

To establish (37) it suffices to prove that its two sides, being cohomology classes of $H(\nu_1, N_1) \times H(\nu_2, N_2)$, have the same value for a homology base of dimension k of $H(\nu_1, N_1) \times H(\nu_2, N_2)$. Such a base is provided by the products of Schubert varieties of $H(\nu_1, N_1)$ and $H(\nu_2, N_2)$, whose dimensions have the sum k. Consider first a product $\zeta_1 \times \zeta_2$, of which at least one factor is not of the form $(0 \cdots 0\ 1 \cdots 1)$. The value of the right-hand side of (37) for $\zeta_1 \times \zeta_2$ is then zero, while we have

$$(F^*\Gamma_k) \cdot (\zeta_1 \times \zeta_2) = \Gamma_k \cdot F(\zeta_1 \times \zeta_2) = KI(\eta_k, F(\zeta_1 \times \zeta_2)),$$

where

$$\eta_k = (\underbrace{N_1 + N_2 - 1 \cdots N_1 + N_2 - 1}_{k}\ N_1 + N_2 \cdots N_1 + N_2). \tag{39}$$

To prove that this intersection number is zero, it suffices to assume the two Schubert varieties to be in general position and show that they have no element in common. In fact, from our assumption on $\zeta_1 \times \zeta_2$, each of the elements of $F(\zeta_1 \times \zeta_2)$ passes through a fixed linear space A of dimension $\nu_1 + \nu_2 - k + 1$. On the other hand, a linear space $E_{\nu_1+\nu_2}$ belongs to η_k, if and only if it has a linear space of dimension $\geqq k$ in common with a fixed linear space B of dimension $N_1 + N_2 + k - 1$. We can choose B so that A and B have only the zero vector in common. If there is an $E_{\nu_1+\nu_2}$ through A which has a linear space of dimension $\geqq k$ in common with B, it would follow that $A \cap B$ is of dimension $\geqq 1$. It follows therefore that the intersection number in question is zero.

It remain to take $\zeta_1 = (0 \cdots 1\ \underbrace{1 \cdots 1}_{i})$, $\zeta_2 = (0 \cdots 0\ \underbrace{1 \cdots 1}_{k-i})$, for a fixed i, and to prove that, under this choice,

$$KI(\eta_k, F(\zeta_1 \times \zeta_2)) = 1.$$

Let L_{ν_1-i}, L_{ν_1+1}, M_{ν_2-k+i}, M_{ν_2+1} be linear spaces, with subscripts indicating their dimensions, which are used to define ζ_1, ζ_2, so that

$$L_{\nu_1-i} \subset L_{\nu_1+1} \subset E_{\nu_1+N_1},$$

$$M_{\nu_2-k+i} \subset M_{\nu_2+1} \subset E_{\nu_2+N_2}.$$

By the condition (34) for Schubert varieties the elements $V_{\nu_1} \varepsilon \zeta_1$, $V_{\nu_2} \varepsilon \zeta_2$ are respectively characterized by

$$L_{\nu_1-i} \subset V_{\nu_1} \subset L_{\nu_1+1},$$
$$M_{\nu_2-k+i} \subset V_{\nu_2} \subset M_{\nu_2+1}.$$

An element $V_{\nu_1+\nu_2} \varepsilon F(\zeta_1 \times \zeta_2)$ is spanned by V_{ν_1} and V_{ν_2}. The condition that it belongs also to η_k is

$$\dim (V_{\nu_1+\nu_2} \cap B) \geqq k.$$

Since the L's, M's, and B are supposed to be in general position, B has only the zero vector in common with the space $C_{\nu_1+\nu_2-k}$ of dimension $\nu_1 + \nu_2 - k$ spanned by L_{ν_1-i} and M_{ν_2-k+i}. There exists therefore a space $D_{N_1+N_2+k}$ of dimension $N_1 + N_2 + k$, which contains B and is supplementary to $C_{\nu_1+\nu_2-k}$ in $E_{\nu_1+\nu_2+N_1+N_2}$. The projections of L_{ν_1+1} and M_{ν_2+1} in $D_{N_1+N_2+k}$ from $C_{\nu_1+\nu_2-k}$ are of dimensions $i+1$ and $k-i+1$ respectively, which we denote by L_{i+1}' and M_{k-i+1}'. Their intersections with B, say L_i'' and M_{k-i}'', are then of dimensions i and $k-i$ respectively. For a $V_{\nu_1+\nu_2}$ spanned by $V_{\nu_1} \varepsilon \zeta_1$, $V_{\nu_2} \varepsilon \zeta_2$ to belong to η_k, or, what is the same, to have an intersection of dimension $\geqq k$ with B, it is therefore necessary and sufficient that it contains L_i'' and M_{k-i}''. Such a $V_{\nu_1+\nu_2}$ is uniquely determined, as the space spanned by L_{ν_1-i}, M_{ν_2-k+i}, L_i'', M_{k-i}''. This proves that $F(\zeta_1 \times \zeta_2)$ and η_k have in common exactly one $V_{\nu_1+\nu_2}$.

Using a coordinate system, we can study, in the differentiable manifold $H(\nu_1 + \nu_2, N_1 + N_2)$, the submanifolds η_k and $F(\zeta_1 \times \zeta_2)$ in the neighborhood of their intersection. It is easily seen that the tangent spaces of these submanifolds are skew to each other, so that the points of intersection is to be counted simply. This completes the proof of formula (37) and hence the duality theorem.

6. Proofs of Theorems 5 and 6. Theorem 5 is an easy consequence of the duality theorem. To prove it we put an Hermitian metric on M, so that the structural group of the tangent bundle reduces to the unitary group. To a point $y \varepsilon B_{n1}$, that is, a tangent direction of M, let $T(y)$ be the subspace of dimension $n-1$ of the tangent vector space which is perpendicular to y. The vectors of $T(y)$, for all $y \varepsilon B_{n1}$, form a bundle of complex vector spaces of dimension $n-1$ over B_{n1}. The product of this bundle and the bundle $p_{n1}: \tilde{B}_{n1} \to B_{n1}$ is clearly equivalent to the bundle over B_{n1} induced by the mapping $q_{n1}: B_{n1} \to M$. If we denote by

$$C(t) = \sum_{i=0}^{n} C_i t^i, \qquad C_0 = 1,$$

the characteristic polynomial in M, the characteristic polynomial of the induced bundle is

$$q_{n1}{}^*C(t) = \sum_{i=0}^{n} q_{n1}{}^*(C_i)t^i.$$

On the other hand, the characteristic polynomial of the bundle $p_{n1}\colon \tilde{B}_{n1} \to B_{n1}$ is $1 + u_1 t$, while that of the other factor is of the form

$$D(t) = \sum_{j=0}^{n-1} D_j t^j, \qquad D_0 = 1.$$

By the duality theorem we have therefore, identically in t,

$$\sum_{i=0}^{n} q_{n1}{}^*(C_i)t^i = (\sum_{i=0}^{n-1} D_i t^i)(1 + u_1 t).$$

Equating the corresponding coefficients of t^i, we get

$$q_{n1}{}^*(C_i) = D_i + u_1 D_{i-1}, \qquad i = 1, \cdots, n,$$

where we define $D_{-1} = D_n = 0$. Elimination of the D's gives formula (10). This proves Theorem 5.

In order to prove Theorem 6, we need the following algebraic lemma:

LEMMA 6.1. *Let*

$$C(t) = 1 + C_1 t + \cdots + C_n t^n \tag{40}$$

be a polynomial in t with coefficients in a commutative ring with unit element 1. *To the elements C_i introduce $\bar{C}_k$, $k = 1, 2, \cdots$ as the coefficients of the formal power series*

$$\bar{C}(t) = \sum_{k=0}^{\infty} \bar{C}_k t^k, \qquad \bar{C}_0 = 1, \tag{41}$$

so that

$$C(t)\bar{C}(t) = 1. \tag{42}$$

Let u be an element of the ring satisfying the equation

$$u^n C(-\frac{1}{u}) = u^n(1 - \frac{C_1}{u} + \cdots + (-1)^n \frac{C_n}{u^n}) = 0. \tag{43}$$

Then we have, for any integer $N \geqq 1$,

$$u^{N+n-1} = (-1)^N \bar{C}_N u^{n-1} + \text{lower powers of } u. \tag{44}$$

To prove the lemma we write

$$\begin{aligned} 1 - C(t)(1 + \bar{C}_1 t + \cdots + \bar{C}_{N-1} t^{N-1}) &= C(t)(\bar{C}_N t^N + \cdots) \\ &= t^N D(t) + t^{n+N} P(t), \end{aligned}$$

where $P(t)$ is a power series in t and $D(t)$ a polynomial in t of degree $\leqq n-1$:

$$D(t)=D_0+D_1t+\cdots+D_{n-1}t^{n-1}.$$

Comparing the coefficients of t^N in both sides, we get

$$\bar{C}_N=D_0.$$

We now put $t=-1/u$ in the last identity and multiply the two sides by u^{n+N-1}. This gives

$$u^{n+N-1}-\{u^nC(-\frac{1}{u})\}\{u^{N-1}-\bar{C}_1u^{N-2}+\cdots+(-1)^{N-1}\bar{C}_{N-1}\}$$
$$=(-1)^Nu^{n-1}D(-\frac{1}{u})+(-1)^{n+N}\frac{1}{u}P(-\frac{1}{u}).$$

This is a formal identity in the positive and negative powers of u. Restricting ourselves to the non-negative powers and taking account of the hypothesis of the Lemma, we get

$$u^{n+N-1}=(-1)^Nu^{n-1}D(-\frac{1}{u}).$$

From this the Lemma follows.

We proceed now to the proof of Theorem 6. As in the proof of Theorem 5, we consider the bundle induced over B_{nr} by the mapping $q_{nr}\colon B_{nr}\to M$. Since M has an Hermitian metric, an element $L_0\subset L_1\subset\cdots\subset L_{r-1}$ of B_{nr} determines, and can be determined by, r mutually perpendicular directions $y_1,\cdots,y_r$ with the same origin, such that $y_1,\cdots,y_i$ span L_{i-1}, $i=1,\cdots,r$. The induced bundle is therefore equivalent to the product of a bundle of vector spaces of dimension $n-r$ and the bundles of one-dimensional vector spaces over each of $y_1,\cdots,y_r$. We denote by

(45) $$k(t)=k_0+k_1t+\cdots+k_{n-r}t^{n-r},\qquad k_0=1,$$

and $1+u_it$, $i=1,\cdots,r$, their respective characteristic polynomials. From the duality theorem we have

(46) $$D(t)=\sum_{i=0}^{n}D_it^i=k(t)\prod_{j=1}^{r}(1+u_jt),$$

where

(47) $$D_i=q_{nr}^*(C_i),\qquad i=0,\cdots,n.$$

It is easy to see that the classes u_i are related to the classes v_i by the relations

(14) $$v_i=u_1+\cdots+u_i,\qquad i=1,\cdots,r,$$

as given in the Introduction. We proceed to derive from (46) a formula for the expression under the parenthesis in the left-hand side of (15). In what follows, multiplication of cohomology classes will be understood in the sense of cup product; it is commutative, because the classes concerned are all even-dimensional.

We introduce the polynomial

$$D'(t) = \sum_{i=0}^{n-r+1} D_i' t^i = (1 + u_r t) k(t), \tag{48}$$

and define the formal power series

$$\begin{aligned} \bar{D}'(t) &= \sum_{i=0}^{\infty} \bar{D}_i' t^i, \qquad \bar{D}_0' = 1, \\ \bar{D}(t) &= \sum_{i=0}^{\infty} \bar{D}_i t^i, \qquad \bar{D}_0 = 1, \end{aligned} \tag{49}$$

by means of the relations

$$D(t)\bar{D}(t) = 1, \qquad D'(t)\bar{D}'(t) = 1. \tag{50}$$

Then we have

$$D(t)\bar{D}'(t) = (1 + u_1 t) \cdots (1 + u_{r-1} t).$$

This equation completely determines $\bar{D}'(t)$. It follows by observation that

$$\bar{D}'(t) = \bar{D}(t)(1 + u_1 t) \cdots (1 + u_{r-1} t).$$

From the definition of $D'(t)$ we get

$$u_r^{n-r+1} D'(-\frac{1}{u_r}) = 0.$$

By Lemma 6.1, we have therefore

$$u_r^{2n-r} = (-1)^n \bar{D}'_n u_r^{n-r} + \text{terms in lower powers of } u_r.$$

But

$$\bar{D}'_n = u_1 \cdots u_{r-1} \bar{D}_{n-r+1} + \cdots,$$

where the unwritten terms are of degrees $< r - 1$ in $u_1, \cdots, u_{r-1}$. It follows that

$$u_1^{n-2} \cdots u_{r-1}^{n-r} u_r^{2n-r} = (-1)^n \bar{D}_{n-r+1} u_1^{n-1} \cdots u_{r-1}^{n-r+1} u_r^{n-r} + \cdots.$$

Under the homomorphism Ψ_{nr} the unwritten terms go to 0, while the first term goes to $(-1)^n \bar{C}_{n-r+1}$. This proves Theorem 6.

7. Application to algebraic geometry. We now sketch briefly the application of the above results to the case in which M is a non-singular algebraic variety of dimension n in a complex projective space of higher dimension. For this purpose we introduce another bundle over M. It is the bundle $[\tilde{B}_{nr}]$ obtained from $\tilde{B}_{nr}$ by adjoining to each fiber of $\tilde{B}_{nr}$, which is a Cartesian product of r complex lines with the origins deleted, the origin and a point at infinity. Its fiber is therefore a Cartesian product of r complex projective lines $L_1 \times \cdots \times L_r$. It is acted on, in an intransitive manner, by the general linear group $G(n)$, so that $[\tilde{B}_{nr}]$ is also an associated bundle of the principal bundle over M with $G(n)$ as structural group.

At this point we have to distinguish between the notion of an algebraic variety in the classical sense as one imbedded in the projective complex space and that of an abstract variety in the sense of Weil [16] defined by means of "overlapping neighborhoods." The variety $B_{n,n}^*$, being the principal bundle, is clearly an abstract variety. By the theory of fiber bundles in algebraic geometry, which is entirely analogous to the topological case, it follows that the bundles B_{nr}, $[\tilde{B}_{nr}]$, as associated bundles, are also abstract varieties and that the projections $q_{nr}: B_{nr} \rightarrow M$, $p_{nr}: \tilde{B}_{nr} \rightarrow B_{nr}$ are rational mappings.

On $[\tilde{B}_{nr}]$ there are sub-varieties V_{i0}, $V_{i\infty}$, $i = 1, \cdots, r$, defined by setting to zero or to ∞ the i-th coordinate in the fiber. Utilizing these sub-varieties, we can define the classes v_i in B_{nr} according to the following lemma, which was first given by Weil: [17]

LEMMA 7.1. *To an analytic cross-section f of the bundle $p_{nr}: \tilde{B}_{nr} \rightarrow B_{nr}$ the cycle*

$$(p_{nr})_*(f(B_{nr}) \cdot V_{i\infty} - f(B_{nr}') \cdot V_{i0})$$

is defined up to linear equivalence. Its homology class is $\mathfrak{D} v_i$.

To prove Theorem 7 we need also the following lemma:

LEMMA 7.2. *On a non-singular algebraic variety the intersection class of a finite number of divisor classes contains an algebraic cycle.*

For classical algebraic varieties this follows by induction from the following statement: If Z is an algebraic cycle on M and D is a divisor in M, there exists a divisor D' which is equivalent to D and which intersects Z properly.

---

[16] Weil [15], Chapter VII.

[17] Weil [16].

In fact, let F be a hypersurface in the ambient projective space of M, whose intersection with M consists of D and a divisors D_1 which does not contain a given point of Z.[18] Then D_1 intersetcs Z properly. Let F_1 be a hypersurface, of the same degree as F, which intersects both M and Z properly. Set $D_2 = F_1 \cdot M$. Then the divisor $D' = D_2 - D_1$ is equivalent to D and intersects Z properly.

From these two lemmas the proof of Theorem 7 follows immediately. For $\mathfrak{D} v_1, \cdots, \mathfrak{D} v_r$, and hence $\mathfrak{D} u_1, \cdots, \mathfrak{D} u_r$, contain divisor classes. Therefore $(\mathfrak{D} u_1)^{n-2} \cdots (\mathfrak{D} u_{r-1})^{n-r} (\mathfrak{D} u_r)^{2n-r}$ contains an algebraic cycle. Its projection in M, which is a cycle belonging to the homology class $(-1)^n \bar{C}_{n-r+1}$, is an algebraic cycle. This proves Theorem 7.

To make our proof complete, one should prove Lemma 7.2 for abstract varieties. While this is "undoubtedly" true, no proof of it has been published. Alternately, it would presumably be easy to imbed the fiber bundles B_{nr}, $[\tilde{B}_{nr}]$ into a projective space, and then the theorem would again follow. These are questions of algebraic geometry which cannot be completely discussed without a lengthy introduction. We shall therefore not take them up in this paper.

BIBLIOGRAPHY.

[1] S. S. Chern, "Characteristic classes of Hermitian manifolds," *Annals of Mathematics*, (2), vol. 47 (1946), pp. 85-121.

[2] M. Eger, "Sur les systèmes canoniques d'une variété algébrique à plusieurs dimensions," *Annales Scientifique de l'École Normale Superieure*, (3), vol. 60 (1943), pp. 143-172.

[3] C. Ehresmann, "Sur la topologie de certains espaces homogènes," *Annals of Mathematics*, (2), vol. 35 (1934), pp. 396-443.

[4] S. Eilenberg and S. MacLane, "Group extensions and homology," *ibid.*, (2), vol. 43 (1942), pp. 757-831.

[5] ——— and N. Steenrod, *Foundations of algebraic topology*, Princeton (1952).

[6] W. Gysin, "Zur Homologietheorie der Abbildungen und Faserungen der Mannigfaltigkeiten," *Commentarii Mathematici Helvetici*, vol. 14 (1941), pp. 61-122.

[7] G. Hirsch, "L'anneau de cohomologie d'un espace fibré et les classes caractéristiques," *Comptes rendus de l'Académie des Sciences*, vol. 229 (1949), pp. 1297-1299.

[18] van der Waerden [14], 635.

[8] ———, "Quelques relations entre l'homologie dans les espaces fibrés et les classes caractéristiques relatives à un groupe de structure," *Colloque de Topologie*, Paris (1951), pp. 123-136.

[9] W. V. D. Hodge, "The characteristic classes of algebraic varieties," *Proceedings of the London Mathematical Society*, (3), vol. 1 (1951), pp. 138-151.

[10] J. Leray, "L'homologie d'un espace fibré dont la fibre est connexe," *Journal de Mathematiques*, vol. 29 (1950), pp. 169-213.

[11] E. H. Spanier, "Homology theory of fiber bundles," *Proceedings of the International Congress of Mathematicians*, vol. II (1950), pp. 390-396.

[12] N. Steenrod, *Topology of fibre bundles*, Princeton (1951).

[13] J. A. Todd, "The geometrical invariants of algebraic loci," *Proceedings of the London Mathematical Society*, (2), vol. 43 (1937), pp. 127-138.

[14] B. L. van der Waerden, "Zur algebraischen Geometrie XIV, Schnittpunktszahlen von algebraischen Mannigfaltigkeiten," *Mathematische Annalen*, vol. 115 (1938), pp. 619-642.,

[15] A. Weil, *Foundations of algebraic geometry*, New York (1946).

[16] ———, "Fibre-spaces in algebraic geometry," *Conference on algebraic geometry and algebraic number theory*, Chicago (1949), pp. 55-59.

[17] W. T. Wu, "Les *i*-carrés dans une variété grassmannienne," *Comptes rendus de l'Académie des Sciences*, Paris, vol. 230 (1950), pp. 918-920.

PSEUDO-GROUPES CONTINUS INFINIS

par **SHIING-SHEN CHERN** (CHICAGO)

1. Introduction et exemples. — Les groupes de transformations locales définis par les intégrales générales d'un système d'équations différentielles partielles ont été étudiés pour la première fois, par Sophus Lie (1). Plus tard, Elie Cartan a donné un traitement plus satisfaisant du sujet et a résolu quelques-uns de ses problèmes fondamentaux (2). Mais leurs points de vue étaient entièrement locaux. Nous allons essayer, dans le présent article, de poser les fondements de cette théorie et nous montrerons, en particulier, que c'est une généralisation naturelle et de grande portée de la théorie des groupes de Lie et des variétés complexes.

Une situation qui se présente fréquemment en géométrie différentielle, est la suivante : Soit M une variété analytique réelle de dimension n, avec les voisinages coordonnés U_α, U_β, etc... qui forment un revêtement de M. Supposons que pour chaque U_α on donne n formes linéaires Pfaffiennes indépendantes, celles en U_α étant désignées par θ_α^i(3)($i = 1, \ldots, n$). Nous avons alors, pour le cas où $U_\alpha \cap U_\beta \neq 0$

$$(1) \qquad \theta_\alpha^i = \sum_j g^i_{\alpha\beta,j}(x)\,\theta_\beta^j \qquad x \in U_\alpha \cap U_\beta$$

La matrice

$$(2) \qquad g_{\alpha\beta}(x) = \left(g^i_{\alpha\beta,j}(x)\right)$$

peut être considérée comme un élément du groupe linéaire général GL (n, R) à n variables réelles, et nous supposerons que l'association $x \to g_{\alpha\beta}(x)$ ($x \in U_\alpha \cap U_\beta$) définit une application analytique de $U_\alpha \cap U_\beta$ dans GL (n, R).

Soit maintenant G un sous-groupe de GL (n, R). Un problème fondamental de la théorie des espaces fibrés est le suivant : Peut-on choisir les formes θ_α^ide

(1) [14] Les chiffres entre [ ] se rapportent à la biliographie en fin d'article.

(2) [1], [2], [3], [4], [5].

(3) Sauf indication contraire, nous utiliserons dans tout le présent article, les jeux suivants d'indices : $1 \leqslant i, j, k \leqslant n$; $1 \leqslant \rho, \sigma, \tau \leqslant r$.

telle façon que $g_{\alpha\beta}(x)$ appartienne à G pour toutes les paires d'indices α, β, et tous les $x \in U_\alpha \cap U_\beta$? Si ceci est possible, nous disons, dans la terminologie des espaces fibrés, que le groupe structural de l'espace fibré principal de l'espace fibré tangent à M peut être réduit à G. Pour plus de simplicité, nous disons que M a une structure G et que les ensembles de formes de Pfaff θ^i_α définissent une structure G sur M. Les exemples suivants illustreront la signification de cette notion :

1) G se compose de l'identité seulement. M est alors dit parallélisable.

2) G est le groupe orthogonal 0 (n, R) à n variables réelles. Alors

$$(\theta^1_\alpha)^2 + \ldots + (\theta^n_\alpha)^2 = (\theta^1_\beta)^2 + \ldots + (\theta^n_\beta)^2 \tag{3}$$

est une forme différentielle quadratique définie positive, analytique sur M et définit un métrique de Riemann. Inversement, une métrique de Riemann sur M donne lieu, de façon évidente, à une structure G sur M avec G = 0 (n, R).

3) n = 2 m est pair et G est le groupe linéaire général GL (m, C) à m variables complexes considéré comme un sous-groupe de GL (n, R). La structure G correspondante s'appelle généralement une structure presque complexe et M s'appelle une variété presque complexe.

Alors que la théorie des espaces fibrés s'occupe de l'existence ou de la non-existence de structures G sur M pour un G donné, un des buts de la géométrie différentielle consiste précisément dans l'étude des invariants différentiels d'une structure G particulière et de leurs implications globales. Par exemple, l'existence d'un métrique de Riemann (au moins d'un métrique non analytique) sur M est un simple théorème de la théorie des espaces fibrés, mais la géométrie Riemannienne est précisément l'étude des propriétés d'une métrique Riemannienne dont les aspects les plus intéressants sont les implications globales ou topologiques de ces propriétés.

2. Premiers invariants d'une structure G ; structure intégrable. — Nous allons montrer comment on peut définir les premiers invariants différentiels d'une structure G. Appelons θ_α la matrice à une seule ligne dont les éléments sont $\theta^1_\alpha, \ldots, \theta^n_\alpha$. Les équations (1) peuvent alors s'écrire sous la forme matricielle

$$\theta_\alpha = \theta_\beta \, g_{\alpha\beta}(x) \quad , \quad x \in U_\alpha \cap U_\beta \tag{4}$$

A $x \in U_\alpha \cap U_\beta$ nous identifions les éléments (x, y_α) G $U_\alpha \times G$ et $(x, y_\beta) \in U_\beta \times G$, $y_\alpha, y_\beta \in G$ par la condition $y_\beta = g_{\alpha\beta}(x)\, y_\alpha$. L'espace B_G ainsi obtenu s'appelle l'espace fibré principal avec le groupe structural G et il est localement homéomorphe à un produit cartésien $U_\alpha \times G$. Comme $\theta_\alpha y_\alpha = \theta_\beta y_\beta$, leur expression commune ω est une matrice à une seule ligne de formes de Pfaff définie globalement dans B_G. Par suite, au lieu de rester sur M, nous passons à B_G et nous allons étudier les propriétés de ces formes de Pfaff dans B_G.

Le premier stade est le calcul de la dérivée extérieure $d\omega$. Dans U_α nous utilisons la représentation $\omega = \theta_\alpha y_\alpha$ et nous trouvons :

$$d\omega = -\omega \wedge y_\alpha^{-1} dy_\alpha + d\theta_\alpha y_\alpha \tag{5}$$

où $y_\alpha^{-1}\ dy_\alpha$ est une matrice de formes de Pfaff invariantes à gauche par G. Comme les éléments ω^i, $(1 \leqslant i \leqslant n)$ de ω sont linéairement indépendants, nous pouvons écrire cette équation plus explicitement comme suit :

$$(6) \qquad d\omega^i = -\sum_{\rho k} a^i_{\rho k}\,\omega^k \wedge \pi^\rho + \frac{1}{2}\sum_{jk} c^i_{jk}(b)\,\omega^j \wedge \omega^k$$

où r est la dimension de G, $a^i_{\rho k}$ des constantes et $c^i_{jk}(b)$ sont des fonctions sur $U_\alpha \times G$ soumises à la condition d'être anti-symétriques en j, k, afin d'être entièrement déterminées. Les π^ρ sont des formes de Pfaff invariantes à gauche de G. La formule (6) est locale, n'étant valide que dans $p^{-1}(U_\alpha)$ où p est la projection de B_G sur M. Nous appliquons maintenant aux formes π^ρ la transformation

$$(7) \qquad \pi^\rho \to \pi'^\rho = \pi^\rho + \sum_k b^\rho_k\,\omega^k$$

Selon cette transformation, les équations (6) conservent la même forme, avec de nouveaux coefficients c'^i_{jk} donnés par les équations

$$(8) \qquad c'^i_{jk} = c^i_{jk} + \sum_\rho \left(-a^i_{\rho k}\,b^\rho_j + a^i_{\rho j}\,b^\rho_k\right)$$

Les $n^2(n-1)/2$ expressions $\sum(-a^i_{\rho k}\,b^\rho_j + a^i_{\rho j}\,b^\rho_k)$ sont linéaires en b^ρ_k et à coefficients constants. Supposons que s d'entre elles soient linéairement indépendantes. Si $n^2(n-1)/2 = s$, nous pouvons choisir b^ρ_k de façon que tous les c'^i_{jk} disparaissent.

Si $n^2(n-1)/2 > s$ nous choisissons b^ρ_k de façon à annuler s des c'^i_{jk}, convenablement choisis. Nous supposerons maintenant qu'une telle transformation (7) a été effectuée et nous supprimerons les primes (') des π'^ρ et c'^i_{jk}. Pour que nos conditions restent inchangées, les π^ρ sont déterminées à la transformation (7) près satisfaisant à

$$(9) \qquad \sum_\rho \left(-a^i_{\rho k}\,b^\rho_j + a^i_{\rho j}\,b^\rho_k\right) = 0$$

Il résulte de (8) que les nouveaux coefficients c^i_{jk} sont des fonctions sur B_G définies globalement. Ce sont donc les premiers invariants d'une structure G. Nous disons que la structure G est intégrable si tous ces invariants sont constants.

Nous mentionnerons les exemples suivants de structures G intégrables :

1) M est un groupe Lie connexe ou une sous-variété ouverte de ce groupe et ω^i sont les formes de Pfaff invariantes à gauche. Dans ce cas, G comprend l'identité seulement et les c^i_{jk} sont les constantes de structure du groupe.

2) Considérons une structure presque complexe, comme dans l'exemple 3 du paragraphe 1. Pour étudier ses invariants locaux il est commode d'utiliser des formes de Pfaff à valeurs complexes, c'est-à-dire des formes de Pfaff $\omega = \varphi + i\psi$ où φ, ψ sont analytiques réelles. Nous écrirons aussi $\bar{\omega}$ pour $\varphi - i\psi$. Avec des conventions, supposons $n = 2m$ et supposons une structure GL (m, C), c'est-à-dire une structure presque complexe donnée dans M. Ceci est défini par m formes de Pfaff à valeurs complexes (4) θ^t_α, $(1 \leqslant t \leqslant m)$ dans

(4) Dans tout cet exemple et à l'occasion plus loin, lorsqu'il s'agit d'une structure presque complexe, nous utilisons la série d'indices suivante $1 \leqslant t, u, v \leqslant m$.

chaque U_α, telles que $\theta^t_\alpha, \bar{\theta}^t_\alpha$ soient indépendantes linéairement et que

$$(10) \qquad \theta^t_\alpha = \sum_u g^t_{\alpha\beta,u}(x)\,\theta^u_\beta \;, \quad x \in U_\alpha \cap U_\beta$$

où $(y^t_{\alpha\beta,u}(x)) \in GL(m, C)$. En suivant la méthode générale ci-dessus, nous obtenons dans B_G ($G = GL(m, C)$) m formes de Pfaff à valeurs complexes qui peuvent être représentées localement par

$$(11) \qquad \omega^t = \sum_u y^t_{\alpha u}\,\theta^u_\alpha$$

où $(y^t_{\alpha u}) \in GL(m, C)$. Dans $p^{-1}(U_\alpha) \subset B_G$, leurs dérivées extérieures peuvent s'écrire sous la forme

$$(12) \qquad d\omega^t = \sum_u \pi^t_u \wedge \omega^u + \frac{1}{2}\sum_{u,v} c^t_{uv}\,\bar{\omega}^u \wedge \bar{\omega}^v, \; (c^t_{uv} + c^t_{vu}) = 0$$

où les π^t_u sont déterminés à la transformation près

$$(13) \qquad \pi^t_u \to \pi'^t_u = \pi^t_u + \sum_v b^t_{uv}\,\omega^v \;, \; (b^t_{uv} + b^t_{vu}) = 0$$

Il en résulte que c^t_{uv}, en tant que fonctions sur B_G, sont les premiers invariants d'une structure presque complexe. Cette dernière est intégrable si les c^t_{uv} sont des constantes. On peut vérifier, par un bref calcul, que ceci ne se produit que quand $c^t_{uv} = 0$. Selon un théorème de Frobenius, c'est là une condition nécessaire et suffisante pour que le système $\omega^t = 0$ soit complètement intégrable. Lorsque ces conditions sont remplies, nous pouvons prendre comme coordonnées locales dans M un jeu de m intégrales premières fonctionnellement indépendantes $z^1, \ldots, z^n$. Tout autre jeu d'intégrales premières leur est rattaché par une transformation analytique complexe avec un Jacobien non nul. Ceci montre que la structure presque complexe intégrable sur M définit une structure complexe sur M.

Les invariants c^t_{uv} ont été donnés pour la première fois par Ehresmann-Libermann (5), et sont essentiellement équivalents à un tenseur défini par Eckmann-Frohlicher (6).

Si on revient au cas d'une structure G générale, nous désirons faire remarquer que les constantes a^i_{pk} dans (6) ont une interprétation géométrique simple. En fait, si on considère G comme un groupe linéaire agissant sur un espace vectoriel à n dimensions avec les coordonnées ξ^i conformément aux équations

$$(14) \qquad \xi^i \to \xi'^i = \sum_j y^i_j\,\xi^j \qquad (y^i_j) \in G$$

alors les

$$(15) \qquad X_\rho = \sum_{i,k} a^i_{\rho k}\,\xi^k \frac{\partial}{\partial \xi^i}$$

sont r transformations infinitésimales qui engendrent G.

---

(5) |11|.
(6) |9|.

3. Notion d'intégrale générale d'un système différentiel. — Supposons qu'il existe deux voisinages de coordonnées U et U', avec les coordonnées locales x^i et x'^i respectivement, dans chacun desquels on donne une structure G intégrable dont les premiers invariants ont les mêmes valeurs constantes. Soient θ^i, θ'^i les formes de Pfaff qui définissent ces structures G ; elles sont déterminées à une transformation près de G. Un homéomorphisme de U sur U' défini par les équations

$$(16) \qquad x'^i = x'^i(x^1, \ldots, x^n)$$

est dit admissible s'il satisfait aux équations

$$(17) \qquad \theta'^i\left[x'^1(x^k), \ldots, x'^n(x^k)\right] = \sum_j \theta^j g^i_j(x) \quad (g^i_j(x)) \in b$$

Cette condition reste inchangée si nous remplaçons θ^i ou θ'^i par des formes qui en diffèrent par une transformation de G ; c'est donc une condition sur les structures G. Les équations (17) peuvent être considérées comme un système d'équations aux dérivées partielles sur U × U'. Nous voulons étudier la question de savoir si, en rétrécissant au besoin les voisinages U et U', il admet une « intégrale générale ». Une définition précise de la notion d'intégrale générale et d'autres faits se rapportant à la théorie des systèmes différentiels sur une variété seront données, par conséquent, dans la présente section(7).

Soit X une variété analytique réelle. A chaque point $x \in X$ nous désignons par V (x) l'espace des vecteurs (contravariants) en x et par $V^*(x)$ l'espace des covecteurs. Désignons leurs algèbres extérieures par $\Lambda\ (V(x))$ et $\Lambda\ (V^*(x))$ respectivement. Par système différentiel Σ en X nous entendons une sous-variété $Y \subset X$ et l'association d'un idéal $O(x) \in \Lambda\ (V^*(x))$ à chaque point $x \in Y$ de façon que la condition suivante soit satisfaite : Pour chaque élément homogène $\alpha \in 0(x)$ il existe un voisinage U de x sur Y et une forme différentielle analytique ω, sur U, qui appartient à 0 (y) pour chaque point $y \in U$ et se réduit à α en x. On dit qu'une telle forme différentielle ω appartient à Σ. Les espaces vectoriels $\Lambda\ (V(x))$ et $\Lambda\ (V^*(x))$ étant duaux, un élément homogène $I(x) \in \Lambda\ (V(x))$, $x \in Y$ est dit élément intégral s'il annule tous les éléments de 0 (x). Une sous-variété $Q \subset Y$ s'appelle variété intégrale de Σ si son espace tangent en tout point est un élément intégral. Nous disons que Σ est fermé si la différentielle extérieure de chaque forme différentielle de Σ appartient à Σ. Pour l'étude des variétés intégrales, nous pouvons supposer notre système différentiel fermé.

Donnons-nous sur Y une forme différentielle extérieure analytique décomposable W de degré p, qui ne s'annule jamais. Un problème fondamental dans la théorie locale des systèmes différentiels est la formulation des théorèmes sur les variétés intégrales à p dimensions dont les espaces tangents n'annulent jamais W. On peut formuler une condition suffisante comme suit : Comme W est décomposable, nous écrivons localement $W = \varphi^1 \wedge \ldots \wedge \varphi^p$. Si n désigne la dimension de Y, on choisit n-p autres formes de Pfaff $\varphi^{p+1}, \ldots, \varphi^n$ de façon

(7) Pour un exposé détaillé du sujet nous renvoyons le lecteur à |6| ou à |13|.

que $\varphi^1 \wedge \ldots, \wedge \varphi^n \neq 0$ dans le voisinage de Y considéré. Soit $e_1 \ldots, e_n$ la base duale de $\varphi^1 \ldots, \varphi^n$ et considérons les vecteurs :

$$(18) \qquad V_i = e_i + \sum_{\sigma=p+1}^{n} l_i^\sigma e_\sigma \quad (i = 1, \ldots, p)$$

Désignons par $I_k(x)$, $k = 1, \ldots, p$, l'espace engendré par $v_1 \ldots, v_k$. Soit

$$(19) \qquad O_k(x, l_1^\sigma, \ldots, l_k^\sigma) = 0$$

les conditions pour que $I_k(x)$ soit un élément intégral à k dimensions en x. Il est facile de voir que ces équations sont linéaires en l_k^σ. Nous dirons que Σ est involutif (par rapport à la forme décomposable W) si, pour un certain choix des φ, les équations (19) sont linéairement indépendantes en l_k^σ, après avoir tenu compte des équations

$$(20) \qquad O_1(x, l_1^\sigma) = 0, \ldots, O_{k-1}(x, l_1^\sigma, \ldots, l_{k-1}^\sigma) = 0$$

Un théorème fondamental, dans la théorie locale des systèmes différentiels affirme l'existence d'une variété intégrale à p dimensions satisfaisant à des conditions initiales appropriées et à la condition $W \neq 0$, lorsque le système est involutif. On le démontre par des applications successives du théorème de Cauchy-Kowalewsky. Les variétés intégrales dont l'existence est affirmée par ce théorème d'existence sont dites « générales », cette expression étant utilisée par contraste avec les intégrales dites singulières des équations aux dérivées partielles. Il est important de remarquer que la notion de variété intégrale générale a une signification intrinsèque car la condition pour qu'un système différentiel soit involutif est manifestement indépendante du choix des coordonnées locales.

Une notion qui est étroitement rattachée à la notion d'involution d'un système différentiel est celle du prolongement. Pour définir cette notion, soit $\tilde{M}$ la multiplicité de tous les éléments intégraux à p dimensions non annulés par W. En associant à un élément intégral son origine x, nous obtenons une représentation $\pi : \tilde{M} \to Y$. Dans la représentation duale de π, W devient une forme W' et les formes différentielles appartenant à Σ deviennent des formes de $\tilde{M}$. Ces dernières assignent à chaque point $m \in \tilde{M}$ un ensemble d'éléments de $\wedge (V^*(m))$ qui engendre un idéal. Il n'est pas difficile de voir que ces idéaux définissent un système différentiel Σ' dans $\tilde{M}$. En général, Σ' n'est pas fermé. Nous désignerons par $\tilde{\Sigma}$ le système différentiel fermé correspondant obtenu à partir de Σ' en lui ajoutant les dérivées extérieures de formes différentielles de Σ'. Le système $\tilde{\Sigma}$ dans $\tilde{M}$, joint à la forme W', constitue le prolongement du système donné Σ. Il résulte aisément de nos définitions que le prolongement d'un système involutif est involutif.

Il existe un cas particulier important dans lequel la condition d'involution peut être formulée plus explicitement. C'est le cas où les idéaux O (x), pour x appartenant à un voisinage U de Y, peuvent être engendrés par : 1) h formes Pfaffiennes analytiques $\theta^1, \ldots, \theta^h$; 2) m formes différentielles quadratiques extérieures analytiques

$$(21) \qquad \phi^i = \sum_{\rho=1}^{\nu} \sum_{k=1}^{p} a_{\rho k}^i \pi^\rho \wedge \varphi^k + \sum_{j,k=1}^{p} c_{jk}^i \varphi^j \wedge \varphi^k \quad (1 \leq i \leq m)$$

telles que

$$(22) \qquad W\wedge\theta^1\wedge\ldots\wedge\theta^n\wedge\pi^1\wedge\ldots\wedge\pi^v \neq 0$$

Dans ce cas, soient $t_1^k,\ldots,t_{p-1}^k$, $(1 \leqslant k \leqslant p)$ $p(p-1)$ constantes génériques. Soit σ_1 le nombre de formes linéairement indépendantes parmi $\sum_\rho\sum_k a^i_{\rho k}t_1^k\pi^\rho$, $\sigma_1+\sigma_2$ le nombre de formes linéairement indépendantes parmi

$$\sum_\rho\sum_k a^i_{\rho k}t_1^k\pi^\rho \quad , \quad \sum_\rho\sum_k a^i_{\rho k}t_2^k\pi^\rho$$

et finalement $\sigma_1+\sigma_2+\ldots+\sigma_{p-1}$ le nombre des formes linéairement indépendantes parmi

$$\sum_\rho\sum_k a^i_{\rho k}t_1^k\pi^\rho, \qquad \sum_\rho\sum_k a^i_{\rho k}t_{p-1}^k\pi^\rho$$

Soit q la dimension de la variété des éléments intégraux à p dimensions passant par un point de Y. Nous avons alors l'inégalité

$$(23) \qquad q \leqslant pv-(p-1)\,\sigma_1\ldots,\sigma_{p-1}$$

Notre critère d'involution affirme que le signe « égal » est valable si, et seulement si, le système différentiel donné est involutif. (Il est entendu qu'on suppose le système fermé.) Ce critère peut être appliqué à notre système différentiel (17). Pour ce faire, nous transformons le système en une nouvelle forme. Soit $V = U \times G$ $V' = U' \times G$ et nous introduisons dans V et V' respectivement les formes de Pfaff :

$$(24) \qquad \begin{aligned} \omega^i &= \sum_j y^i_j\,\theta^j \\ \omega'^i &= \sum_j y'^i_j\,\theta'^j \end{aligned}$$

où (y^i_j), (y'^i_j) sont des éléments génériques de G. Le système différentiel (17) est alors équivalent au système suivant :

$$(25) \qquad \omega'^i-\omega^i = 0$$

Sur $V \times V'$, dans notre terminologie, l'idéal associé par le système différentiel à un point $V \times V'$ est engendré par les premiers membres de (25). Pour rendre le système fermé, nous ajoutons à l'idéal en tout point les générateurs $d(\omega'^i-\omega^i)$. Comme les deux structures G sont intégrables avec les mêmes premiers invariants, on peut choisir des formes de Pfaff π^ρ, π'^ρ de sorte que

$$d(\omega'^i-\omega^i) = \sum_{\rho,k} a^i_{\rho k}\,(\pi'^\rho-\pi^\rho)\wedge\omega^k$$

Notre système différentiel est donc du type particulier discuté ci-dessus, avec $W = \omega^1\wedge \ldots\wedge\omega^n$. Pour trouver la condition d'involution, nous introduisons les entiers $q, \sigma_1\ldots,\sigma_{n-1}$ comme précédemment, avec p remplacé par n. Selon le critère ci-dessus, une condition nécessaire et suffisante pour que le système différentiel doit involutif est

$$(26) \qquad q = nv-(n-1)\,\sigma_1\ldots\ \sigma_{n-1}$$

En particulier, nous voyons que cette condition ne dépend que du groupe G. Lorsqu'elle est satisfaite, nous disons que le groupe linéaire G est involutif. On dira qu'une structure G est involutive si G est involutif.

4. Pseudo-groupes définis par une structure intégrable involutive. — Nous allons montrer, dans la présente section, comment une structure intégrable involutive sur une variété définit un pseudo-groupe de transformations et une famille correspondante de systèmes de coordonnées admissibles. Nous rappellerons d'abord la définition d'un pseudo-groupe.

Soit E un espace topologique et F une famille de sous-ensembles ouverts telle que l'union d'un nombre quelconque d'ensembles de F et l'intersection d'un nombre fini d'ensembles de F appartiennent à F. Une famille H d'homéomorphismes, chacun représentant un ensemble de F sur un ensemble de F est dite constituer un pseudo-groupe si les conditions suivantes sont satisfaites :

1) Si $U \in F$, la représentation identité appartient à H. Si $h \in H$ représente U sur V la représentation inverse appartient à H. Si $h, h' \in H$ et si leur produit hh' est défini, alors $hh' \in H$.

2) Soit V une union d'ensembles V_α de F. Pour qu'un homéomorphisme h défini dans V appartienne à H, il est nécessaire et suffisant que sa restriction $h|V_\alpha$ appartienne à H.

Soit maintenant X un espace topologique et soit $\{U_\alpha\}$ un revêtement de X par des sous-ensembles ouverts. Supposons que pour chaque α il y ait un homéomorphisme f_α d'un ensemble V_α de F sur U_α de façon que la condition suivante soit satisfaite : Pour deux indices quelconques α, β avec $U_\alpha \cap U_\beta \neq 0$, il y a une représentation $h_{\alpha\beta} \in H$ de V_α sur V_β avec la propriété que la représentation $g_{\alpha\beta} : f_\alpha^{-1}(U_\alpha \cap U_\beta) \to f^{-1}(U_\alpha \cap U_\beta)$ définie par $f_\alpha(x) = f_\beta(g_{\alpha\beta}(x))$, $x \in f_\alpha^{-1}(U_\alpha \cap U_\beta)$ est la restriction de $h_{\alpha\beta}$ à $f_\alpha^{-1}(U_\alpha \cap U_\beta)$. Si c'est le cas, nous disons que X a une famille de systèmes locaux de coordonnées compatibles avec le pseudo-groupe H.

Supposons que nous avons sur M une structure G intégrale involutive. En utilisant les notations du paragraphe 2, nous avons, en particulier, les équations (6) dans lesquelles les c_{jk}^i sont des constantes, tandis que les $\pi^{\rho'}$ sont déterminés à la transformation (7) près, sous réserve des conditions (9). En outre, le groupe G est supposé maintenant être involutif. Les coefficients $a_{\rho k}^i$, c_{jk}^i sont soumis à certaines relations. Pour les établir, remarquons que $d\pi^\rho$ sont des formes différentielles quadratiques extérieures, de sorte que nous pouvons écrire, localement :

$$(27) \qquad d\pi^\rho = \frac{1}{2} \sum_{\sigma,\tau} \gamma_{\sigma\tau}^\rho \pi^\sigma \wedge \pi^\tau + \sum_{\sigma,i} u_{\sigma i}^\rho \pi^\sigma \wedge \omega^i + \frac{1}{2} \sum_{i,j} v_{ij}^\rho \omega^i \wedge \omega^j$$

où

$$(28) \qquad \gamma_{\sigma\tau}^\rho + \gamma_{\tau\sigma}^\rho = 0 \quad , \quad v_{ij}^\rho + v_{ji}^\rho = 0$$

En utilisant (27) nous trouvons que les conditions $d(d\omega^i) = 0$ peuvent s'écrire explicitement :

$$(29) \qquad \sum_i - a^k_{\rho i} a^i_{\sigma i} + a^k_{\sigma i} a^i_{\rho j} = \sum_\tau a^k_{\tau j} \gamma^\tau_{\sigma\rho}$$

$$(30) \qquad \sum_i \left(-c^k_{im} a^i_{\rho l} + c^k_{il} a^i_{\rho m} + c^i_{lm} a^k_{\rho i}\right) + \sum_\sigma \left(a^k_{\sigma l} u^\sigma_{\rho m} - a^k_{\sigma m} u^\sigma_{\rho l}\right) = 0$$

$$(31) \qquad \sum_i \left(c^k_{ij} c^i_{lm} + c^k_{il} c^i_{mj} + c^k_{im} c^i_{jk}\right) + \sum_\rho \left(a^k_{\rho j} v^\rho_{lm} + a^k_{\rho l} v^\rho_{mj} + a^k_{\rho m} v^\rho_{jl}\right) = 0$$

La compatibilité de ces équations est donc une condition nécessaire sur $a^i_{\rho k}$, c^i_{jk}. Nous remarquerons en passant que si G se compose de l'identité seulement alors (29), (30) sont satisfaites identiquement et (31) se réduit aux identités bien connues de Jacobi.

Les équations (29) ont une interprétation simple. En effet, en termes des transformations infinitésimales X_ρ dans (15) elles sont équivalentes aux équations

$$(32) \qquad (X_\rho, X_\sigma) = \sum_\tau \gamma^\tau_{\rho\sigma} X_\tau$$

Il en résulte que $\gamma^\tau_{\rho\sigma}$ sont des constantes et sont les constantes de structure de G.

Un théorème qui est fondamental énonce que ces résultats ont une réciproque. Pour plus de précision, nous avons le théorème suivant : Supposons que les constantes $a^i_\rho k$, c^i_{jk} soient telles que les équations (29) (30), (31) soient compatibles. Soit (y^i_j) un élément générique de G et que π'^ρ soient r formes de Pfaff linéairement indépendantes invariantes à gauche pour G. Nous pouvons trouver un ensemble connexe ouvert E dans l'espace Euclidien à n dimensions (qui peut être tout l'espace) et, dans E, n formes de Pfaff analytiques linéairement indépendantes φ^i, avec la propriété : Dans E × G il y a des formes de Pfaff.

$$(33) \qquad \pi^\rho = \pi'^\rho + \sum_k d^\rho_k \varphi^k$$

qui, jointes à $\omega^i = \sum_k y^i_k \varphi^k$, satisfont aux équations (6).

Ce théorème est une généralisation de la réciproque du troisième théorème fondamental de Lie pour les groupes de Lie. On le démontre en appliquant le théorème d'existence donné au paragraphe 3. L'ensemble E et les formes φ^k ne sont pas déterminés de façon unique. Mais, dans nos discussions à venir, nous porterons notre attention sur un choix particulier de ces éléments.

Lorsqu'un choix de ce genre a été fait, nous dirons qu'une représentation analytique f d'un sous-ensemble ouvert $U \subset E$ dans un sous-ensemble ouvert $V \subset E$ est admissible si les restrictions φ^i_v des formes φ^i à V sont représentées par la représentation duale de f par les formes

$$(34) \qquad \sum_k y^i_k(x) \varphi^k_U \quad , \quad y^i_k(x) \in G, \; x \in U$$

Si nous prenons comme ensembles de F les sous-ensembles ouverts de E et comme représentations de H les homéomorphismes admissibles, alors tous ces homéomorphismes forment un pseudo-groupe dans le sens défini au début de la présente section.

Supposons que notre variété M ait un revêtement $\{U_\alpha\}$ par des voisinages de coordonnées, par rapport auxquels la structure G intégrale, involutive donnée est définie dans chaque U_α par les formes de Pfaff. θ^i_α. Soit $x \in U_\alpha$. Il résulte du théorème fondamental ci-dessus que nous pouvons trouver un voisinage $V \subset E$ et un homéomorphisme f de V sur un voisinage U, $x \in U \subset U_\alpha$ selon lequel les θ^i_α sont représentés par les formes

$$\sum_j y^i_j(v)\,\varphi^j_V \qquad (y^i_j(v)) \in G\,,\ v \in V$$

Parmi ces voisinages U nous pouvons choisir un revêtement de M, qui, joint aux homéomorphismes correspondants f, définit manifestement des systèmes de coordonnées locales compatibles avec le pseudo-groupe ci-dessus H. De tels systèmes de coordonnées locales seront dits admissibles.

5. Autres exemples. — Pour rendre nos notions plus claires, nous allons donner quelques exemples. Toutes les structures de la présente section seront involutives et intégrables.

1) G se compose de l'identité seulement. Comme dans l'exemple 1, paragraphe 2, une telle structure existe sur la variété d'un groupe de Lie ou sur une de ses sous-variétés ouvertes. Dans un certain sens, ce sont les seules possibles. En fait, définissons la structure par les formes de Pfaff linéairement indépendantes, ω^i. Considérons une courbe paramétrisée $x(t)$ $0 \leq t < 1$, dans M et soit $\omega^i = p^i(t)$ le long de la courbe. La structure est dite complète si $\lim_{t \to 1} x(t)$ existe, chaque fois que

$$\int_0^1 \sqrt{(p^1)^2 + \dots + (p^n)^2}\,dt$$

converge. Nous avons alors le théorème suivant : Soit M une variété connexe et simplement connexe avec une structure G intégrable pour laquelle G se compose de l'identité seulement. Supposons la structure complète. M est alors un groupe de Lie avec ω^i comme formes de Pfaff invariantes à gauche.

Le fait d'être complète est une propriété de la structure. Mais il résulte de la définition que la structure est complète si M est compact. En conséquence, nous voyons que la sphère à sept dimensions S^7 donne un exemple de variété parallélisable sur laquelle n'existe aucune structure intégrable avec G = identité.

2) $G = GL(n, R)$. Une telle structure est toujours intégrable et les équations (6) peuvent s'écrire sous la forme

$$d\omega^i = \sum_k \pi^i_k \wedge \omega^k \tag{35}$$

Dans ce cas, nous prenons pour E l'espace Euclidien à n dimensions et

$$\varphi^i = dx^i \tag{36}$$

où x^i sont des coordonnées dans E. Les systèmes de coordonnées locales admissibles ne sont pas autre chose que les systèmes primitifs, correspondant à la structure analytique donnée sur M.

3) Structure presque complexe sur une variété à nombre pair de dimensions. Ceci a été étudié dans l'exemple 2, paragraphe 2 et nous adopterons la notation qui y a été employée. Nous avons montré que si la structure presque complexe est intégrable des systèmes de coordonnées complexes locales peuvent être introduits dans M.

Il est bien connu que l'existence d'une structure presque complexe sur M est équivalente, topologiquement à celle d'une forme différentielle quadratique extérieure ϕ, partout du rang le plus élevé n. Une telle variété a été appelée symplectique (8) par Ehresmann, si ϕ est fermé ($d\phi = 0$). Parmi les variétés symplectiques se trouvent celles pour lesquelles ϕ est réductible localement à la forme

$$\phi = dx^1 \wedge dx^{m+1} + \ldots + dx^m \wedge dx^{2m} \tag{37}$$

Une telle variété a alors une structure involutive intégrable. Ses systèmes de coordonnées locales admissibles sont ceux pour lesquels ϕ prend la forme (37).

4) M a une dimension impaire $n = 2m+1$ et G est le groupe de toutes les matrices non singulières de la forme

$$\begin{pmatrix} g_{00} & g_{01} \ldots g_{02m} \\ 0 & \\ \vdots & C \\ 0 & \end{pmatrix}, \quad g_{00} > 0 \tag{38}$$

où C est la matrice la plus générale $2m \times 2m$ satisfaisant à la relation $CJ\varepsilon 0 = J$, J étant la matrice :

$$J = \begin{pmatrix} 0 & Im \\ -Im & 0 \end{pmatrix} \qquad Im = \left.\begin{pmatrix} 1 & & 0 \\ & \ddots & \\ 0 & & 1 \end{pmatrix}\right\} m \tag{39}$$

On peut écrire les équations (6) sous la forme

$$\begin{cases} d\omega_0 = \pi_0 \wedge \omega_0 - \sum_k \omega_k \wedge \omega_{k'} \\ d\omega_i = \frac{1}{2}\pi_0 \wedge \omega_i + \pi_i \wedge \omega_0 + \sum_k \pi_{ik} \wedge \omega_k + \sum_{k'} \pi_{ik'} \wedge \omega_{k'} \\ d\omega_{i'} = \frac{1}{2}\pi_0 \wedge \omega_{i'} + \pi_{i'} \wedge \omega_0 + \sum_k \pi_{i'k} \wedge \omega_k + \sum_{k'} \pi_{i'k'} \wedge \omega_{k'} \end{cases} \tag{40}$$

où $i' = m + i$, $k' = m + k$ et

$$\pi_{ik'} + \pi_{ki'} = 0 \;, \quad \pi_{i'k} + \pi_{k'i} = 0 \;, \quad \pi_{ik} + \pi_{i'k'} = 0 \;, \quad 1 \leqslant i, k \leqslant m \tag{41}$$

Comme espace E, nous pouvons prendre l'espace euclidien de dimension $2m + 1$ avec les coordonnées $x_0, \ldots, x_{2m}$ et nous pouvons prendre

$$\begin{aligned} \varphi_0 &= dx_0 + \sum_u x_{u'}\, dx_u \\ \varphi_u &= dx_u \quad \varphi_{u'} = dx_{u'} \end{aligned} \tag{42}$$

---

(8) [11].

Le pseudo-groupe est essentiellement celui des transformations de contacts dans un espace de dimension $m+1$. La structure sur M est définie par une forme de Pfaff θ_0.
définie à un facteur positif près, qui est réductible localement à la forme

$$(43) \qquad \theta_0 = dx_0 + x_{m+1}\, dx_1 + \ldots + x_{2m}\, dx_m$$

Les systèmes de coordonnées locales admissibles sont ceux pour lesquels cette forme Pfaffienne peut être écrite sous la forme (43).

5) M est de dimension 3 et G est le groupe de toutes les matrices de la forme

$$(44) \qquad \begin{pmatrix} u & v & w \\ 0 & 1 & 0 \\ 0 & 0 & 1 \end{pmatrix} \qquad u > 0$$

Les équations (6) deviennent

$$(45) \qquad \begin{cases} d\omega_1 = \pi_1 \wedge \omega_1 \\ d\omega_2 = \pi_2 \wedge \omega_1 \\ d\omega_3 = \pi_3 \wedge \omega_1 - \omega_2 \wedge \omega_3 \end{cases}$$

Comme espace E nous prenons, dans le cas présent, le demi-espace $y > 0$ de l'espace euclidien à trois dimensions (x, y, z). Les équations (45) ont pour solutions

$$(46) \qquad \omega_1 = u\,dx \qquad \omega_2 = v\,dx + \frac{1}{y}\,dy \qquad \omega_3 = w\,dx + \frac{1}{y}\,dz$$

Une représentation admissible d'un voisinage avec les coordonnées x, y, z dans un voisinage ayant les coordonnées X, Y, Z satisfait aux équations différentielles

$$(47) \qquad \begin{aligned} dX &= u\,dx \\ \frac{1}{y}\,dY &= v\,dx + \frac{1}{y}\,dy \\ \frac{1}{y}\,dZ &= w\,dx + \frac{1}{y}\,dz \end{aligned}$$

Les équations finies sont

$$(48) \qquad X = f(x) \qquad Y = y\,g(x) \qquad Z = z\,g(x) + h(x)$$

où f, g, h, sont des fonctions analytiques arbitraires de x.

6. Groupes linéaires involutifs et semi-involutifs ; connexions. — Il résulte des discussions ci-dessus que l'étude des groupes linéaires involutifs joue un rôle important dans la théorie des pseudo-groupes continus infinis. C'est un problème purement algébrique. Nous allons donner cependant quelques considérations géométriques qui devraient aider à clarifier la situation.

Pour continuer les discussions générales du paragraphe 2, nous voyons que les formes π^ρ de B sont déterminées à la transformation (7) près, sous réserve des conditions (9). Ces dernières sont des équations linéaires homogènes en b_k^ρ, à coefficients constants. Supposons que b_{ks}^ρ, $1 \leqslant s \leqslant r_1$ soit un système fondamental de solutions de ce système linéaire de sorte que toute autre solution puisse s'écrire :

$$b_k^\rho = \sum_{s=1}^{r_1} b_{ks}^\rho t^s \tag{49}$$

où t_s sont des paramètres arbitraires. Il s'ensuit que dans B_G nous avons n + r formes de Pfaff ω^i $-\rho$, dont les premières sont entièrement déterminées tandis que les dernières sont déterminées à la transformation près

$$\pi^\rho \to \pi'^\rho = \pi^\rho + \sum_{k,s} b_{ks}^\rho t^s \omega^k \tag{50}$$

Si G_1 désigne le groupe de toutes les matrices de la forme :

$$\begin{pmatrix} I_n & C \\ 0 & I_r \end{pmatrix} \qquad C = \left(\sum_s b_{js}^\rho t^s\right) \tag{51}$$

ceci veut dire exactement que B_G a une structure G_1, G_1 est un groupe abélien linéaire de dimension r_1 agissant sur un espace vectoriel de dimension n + r. Il est complètement déterminé par G et on l'appellera groupe déduit de G.

En répétant ce procédé, nous obtenons une suite de variétés B_G, B_{G_1}, B_{G_2}, ..., ayant des structures correspondant aux groupes G_1, G_2, G_3, ..., respectivement, où G_k est le groupe déduit de G_{k-1}. Naturellement, ces structures ne sont pas nécessairement intégrables, même si la structure G originale sur M est intégrable. Nous dirons que G est semi-involutif si G_k est involutif et différent de l'identité pour un certain $k \geqslant 1$. Puisque, selon notre théorème d'existence sur les équations différentielles, le groupe déduit d'un groupe involutif est involutif, un groupe involutif différent de l'identité est semi-involutif. Une condition nécessaire pour que G soit semi-involutif est que, avec $k \geqslant 1$; aucun G_k ne soit l'identité.

Cette situation est en contraste direct avec le cas où une connexion peut être définie à partir d'une structure G. En fait, la condition G_k = identité, signifie exactement qu'on peut définir des formes de Pfaff linéairement indépendantes sur $B_{G_{k-1}}$ dont le nombre est égal à la dimension de $B_{G_{k-1}}$. Ces formes de Pfaff définissent une connexion dans l'espace fibré $B_{G_{k-1}} \to B_{G_{k-2}}$ (avec la convention $B_{G_0} = B_G$, $B_{G_{-1}} = M$) dont le groupe structurel est G_{k-1}. Comme exemple particulier, nous pouvons prendre G = 0 (n, R). Alors, comme on le montre facilement, G_1 se compose de l'identité seulement et nous définissons une connexion dans le faisceau $B_G \to M$, qui est le parallélisme bien connu de Levi-Civita. Ceci démontre, en particulier, que le groupe orthogonal 0 (n, R) n'est pas involutif.

Par un large emploi de la théorie des représentations des algèbres semi-simples de Lie, E. Cartan a déterminé tous les groupes linéaires complexes semi-simples qui sont irréductibles et dont le groupe déduit n'est pas l'identité.

Sur la base de ce résultat, il a déterminé tous les groupes linéaires qui sont irréductibles et semi-involutifs. En particulier, on montre que les seuls groupes linéaires semi-involutifs, irréductibles, semi-simples, complexes sont le groupe linéaire spécial et le groupe symplectique. La démonstration de ce théorème est longue et demande beaucoup de calculs. Une preuve simplifiée et une extension au cas réel seraient très désirables.

Cartan s'est servi de ce théorème pour déterminer ce qu'il a appelé des pseudo-groupes simples. Dans des développements ultérieurs de la théorie des pseudo-groupes infinis, ces pseudo-groupes simples seraient probablement les premiers objectifs de l'étude. Nous allons par conséquent résumer ses résultats ici. Les conceptions suivantes sont entièrement locales.

Comme d'habitude, un pseudo-groupe est dit simple s'il ne contient aucun pseudo sous-groupe proprement dit distinct de l'identité. On l'appelle imprimitif, s'il transforme entre elles une famille de sous-variétés disjointes k-dimensionnelles, $0 < k < n$, en lesquelles la variété est partagée. S'il n'est pas imprimitif, on dit qu'il est primitif. Un pseudo-groupe simple est nécessairement primitif. En déterminant tous les pseudo-groupes infinis primitifs, Cartan a montré que, dans le domaine complexe, les pseudo-groupes infinis simples sont des quatre espèces suivantes :

A) Le pseudo-groupe de toutes les transformations analytiques complexes sur n variables complexes avec Jacobien non nul.

B) Le pseudo-groupe de toutes les transformations analytiques complexes sur n variables complexes avec Jacobien égal à 1.

C) Le pseudo-groupe de toutes les transformations analytiques complexes sur $n = 2m$ variables complexes $z^1, \ldots, z^{2m}$, laissant la forme différentielle quadratique extérieure invariante

$$dz^1 \wedge dz^{m+1} + \ldots + dz^m \wedge dz^{2m} \tag{52}$$

D) Le pseudo-groupe de toutes les transformations analytiques complexes sur $n = 2m + 1$ variables complexes $z^0, \ldots, z^{2m}$ qui reproduisent la forme de Pfaff

$$dz^0 + z^{m+1} dz^1 + \ldots + z^{2m} dz^m \tag{53}$$

à un facteur non nul près.

Ces pseudo-groupes ont leurs analogues dans les variables réelles. Nous dirons qu'une variété a une structure A si elle a un revêtement par des systèmes de coordonnées locales compatibles avec le pseudo-groupe A ; de même pour les structures B, C, D. Dans le cas réel, nous entendrons par pseudo-groupe D celui de toutes les transformations réelles qui reproduisent la forme (53) par un facteur positif. Ce pseudo-groupe a été considéré dans l'exemple 4, paragraphe 5.

7. Problèmes globaux. — Soit G un groupe linéaire involutif à n variables. Les problèmes globaux immédiats que nous rencontrons sont de savoir si une variété M a une structure G et si elle a une structure G intégrable. Pour $n = 2m$ et $G = G(m, C)$ ceci se ramène aux problèmes sur l'existence de structures presque complexes et complexes sur M, pour lesquelles on a fait beaucoup de travail (9). Nous allons donner dans la présente section un échantillon de ces résultats pour le cas d'une structure D sur une multiplicité de dimension impaire.

Nous rappellerons que le seul invariant local d'une équation Pfaffienne $\omega = 0$ est sa classe, qui est un entier impair $c = 2r + 1$ tel que $2r$ soit le rang de $d\omega$ (mod ω). Sur une multiplicité de dimension impaire $n = 2m + 1$ il existe toujours une forme Pfaffienne non nulle (c'est-à-dire une forme Pfaffienne qui ne s'annule jamais). L'existence d'une structure D est équivalente à l'existence d'une forme Pfaffienne qui est partout de classe $2m + 1$. Une telle propriété a des implications topologiques sur la variété. Il est facile de voir que si M a une structure D, elle doit être orientable.

D'autres conditions nécessaires pour l'existence d'une structure D peuvent être déduites comme suit : Soit E l'espace réel euclidien de dimension $2m + 2N+1$ et 0 un point fixe de E. Soit E' un hyperplan passant par 0 et v un vecteur unité perpendiculaire à E'. E' peut être considéré comme un espace euclidien complexe de dimension complexe $m + N$. Un espace vectoriel complexe ξ de dimension complexe m issu de 0 dans E' détermine un espace vectoriel réel de dimension 2m, qui détermine avec v un espace vectoriel réel de dimension $2m + 1$ issu de 0. Si nous désignons par $G(m, N, C)$ la multiplicité complexe de Grassmann de tous les espaces vectoriels complexes de dimension m issue de 0 et se trouvant dans E' et par $G(2m + 1, 2N, R)$ la multiplicité réelle de Grassmann de tous les espaces vectoriels réels de dimension $2m + 1$ de E ; la construction ci-dessus définit une représentation

$$h : G(m,N,C) \to G(2m+1, 2N, R) \tag{54}$$

Pour N suffisamment grand, soit $\tau : M \to G(2m + 1, 2N, R)$ la représentation tangentielle de M, c'est-à-dire la représentation qui induit l'espace fibré tangent de M. Il résulte de considérations standard sur les espaces fibrés que M n'a une structure D que s'il existe une représentation $\sigma : M \to G(m, N, C)$ telle que τ et $h\sigma$ soient homotopiques. Comme $G(m, N, C)$, en tant que multiplicité complexe de Grassmann, n'a pas de groupe de cohomologie de dimension impaire non nul, nous avons le théorème suivant : Si M a une structure D, toutes les classes caractéristiques de dimension impaire de Stiefel-Whitney doivent être nulles.

Cette condition nécessaire conduit facilement à des multiplicités de dimension impaire sans structure D. En particulier, nous pouvons considérer la variété de Wu Wen-Tsun à cinq dimensions définie comme suit : Soit P_2 le plan projectif complexe et I l'intervalle unité $0 \leqslant t \leqslant 1$. Par un point $z \in P_2$ soit z le point dont les coordonnées homogènes sont les conjuguées complexes de z. Nous prenons le produit cartésien $P_2 \times I$ et nous identifions les points $z \times 0$ et Z et $\bar{z} \times 1$. L'espace résultant est une variété orientable à cinq dimensions, dont la classe de dimension 3 de Stiefel-Whitney ne s'annule pas (10). Par suite, il ne peut pas avoir de structure D.

---

(9) Voir par exemple [8], [11], [12].

(10) [15].

D'autre part, il existe des variétés avec une structure D. Soit μ une variété analytique de dimension $m + 1$. Par un co-rayon de μ, nous entendons la classe de tous les co-vecteurs différant les uns des autres par un facteur positif. La variété de tous les co-rayons de μ est de dimensions $2m + 1$ et a évidemment une structure D. Au lieu de considérer la variété de tous les co-rayons, nous pouvons aussi prendre la variété de tous les rayons, appelée généralement variété tangente de μ. Si nous introduisons un métrique de Riemann sur u, nous voyons que la variété tangente est différentiablement homéomorphe avec la variété de tous les co-rayons, et par suite a une structure D.

Un autre exemple de multiplicités à structure D est donné par une sphère S^{2m+1} de dimension $2m + 1$. En effet, considérons un espace complexe euclidien de dimension $m + 1$ avec les coordonnées $z^0, \ldots, z^m$ et définissons S^{2m+1} comme le lieu d'équation

$$Z^0\bar{Z}^0 + Z^1\bar{Z}^1 + \ldots + Z^m\bar{Z}^m = 0 \tag{55}$$

Alors, la forme de Pfaff

$$\frac{1}{i}\sum_{\alpha=0}^{m}\left(z^\alpha d\bar{z}^\alpha - \bar{z}^\alpha dz^\alpha\right) \tag{56}$$

est réelle et partout de classe $2m + 1$.

Nous désirons faire remarquer que le second exemple n'est pas compris dans le premier. Le théorème suivant m'a été communiqué par le Professeur Spanier : Aucune sphère S de dimension $2m + 1$ n'est la variété tangente d'une multiplicité de dimension $m + 1$.

Ce théorème se démontre comme suit : Supposons que S soit la variété tangente d'une variété μ de dimension $m + 1$. Alors S est un espace fibré en sphères de dimension m sur μ. Si μ est non orientable, la variété tangente de son revêtement orientable à deux feuillets recouvrirait S deux fois. Comme ceci n'est pas possible, μ doit être orientable. Comme le cas $m = 1$ conduit facilement à une contradiction, nous supposons m supérieur à 1. Considérons la suite de Gysin de l'espace fibré : $S \rightarrow \mu$ (11) :

$$\cdots \rightarrow H^{z-1}(S) \xrightarrow{j} H^{r-m-1}(\mu) \xrightarrow{h} H^{z}(\mu) \xrightarrow{p} H^{r}(S) \xrightarrow{j} H^{r-m}(\mu) \xrightarrow{h} H^{r+1}(\mu) \rightarrow \cdots \tag{57}$$

où les groupes sont des groupes de co-homologie avec coefficients entiers, p^* est l'homomorphisme dual induit par p, j est l'intégration sur la fibre et h est la multiplication par la classe caractéristique W^{m+1}. De cette suite exacte, nous pouvons tirer les conclusions suivantes : 1) Pour $r = m - 1$ $H^0(u) \cong H^{m+1}(u)$ et W^{m+1} engendre $H^{m+1}(p)$ ce qui implique que la caractéristique d'Euler-Poincaré $\chi(\mu)$ est égale à ± 1 ; 2) Pour $r < m$, $H^r(\mu) \cong H^r(S) = 0$ 3) ; 3). Pour $r > 1$ $H^r(\mu) \cong H^{\bar{r}+m}(S)$ de sorte que $H^m)\mu) = 0$. Des deux derniers résultats, nous tirons $H^r(\mu) = 0$ $(1 \leqslant r \leqslant m)$. Cette dernière condition implique que $\chi(\mu)$ est égale à 0 ou à 2, ce qui est en contradiction avec 1). Ceci complète la démonstration du théorème.

---

(11) [7].

8. Observations. — Les pseudo-groupes continus infinis considérés dans le présent article ne constituent qu'un cas particulier de ceux qui ont été étudiés par E. Cartan. On peut obtenir d'autres généralités dans deux directions. D'abord, les transformations du pseudo-groupe peuvent être définies comme étant les intégrales générales d'un système d'équations aux dérivées partielles d'ordre supérieur, au lieu d'être du premier ordre, comme les équations (17) dans notre cas. Pour cette généralisation, la théorie des jets de Ehresmann ou la théorie des points infiniment voisins de N. Bourbaki devraient se révéler utiles. Deuxièmement, Cartan a pris en considération ce qu'il a appelé des groupes intransitifs. Ceci est important pour les pseudo-groupes continus infinis, car, avec une notion d'isomorphisme local que nous n'avons pas discutée ici, il existe, contrairement au cas des groupes de Lie, des pseudo-groupes intransitifs qui ne sont isomorphes à aucun pseudo-groupe transitif.

Malgré la nature restreinte de notre classe de pseudo-groupes, il semble qu'il y ait suffisamment de matière à réflexion. Outre le type de problèmes considérés au paragraphe 7, il y a la question de l'étude de familles particulières de fonctions sur M. Si le groupe structural G a un sous-espace invariant, une famille naturelle de fonctions de ce genre se compose de celles dont les vecteurs gradients appartiennent partout aux sous-espaces invariants correspondants. Cette notion contient celle d'une fonction analytique complexe sur une variété presque complexe. Si nous utilisons la notation du paragraphe 2 une fonction f à valeurs complexes sur une variété presque complexe M, est complexe analytique si df est partout une combinaison linéaire des ω^r. Etant donnée l'extrême richesse de la théorie des fonctions à une ou plusieurs variables complexes, nous avons le sentiment qu'il vaut peut-être la peine de considérer cette généralisation d'un peu plus près.

BIBLIOGRAPHIE

|1| CARTAN E. : Sur l'intégration des systèmes d'équations aux différentielles totales, Ann. Ec. Norm. Sup. 18, 241-311 (1901).

|2| CARTAN E. : Sur la structure des groupes infinis de transformations, ibid., 21, 153-206 (1904).

|3| CARTAN E. : Sur la structure des groupes infinis de transformations, ibid., 22, 219-308 (1905).

|4| CARTAN E. : Les sous-groupes des groupes continus de transformations, ibid., 25 57-194 (1908).

|5| CARTAN E. : Les groupes de transformations continus, infinis, simples, ibid., 26, 93-161 (1909).

|6| CARTAN E. : Les systèmes différentiels extérieurs et leurs applications géométriques, Paris (1945).

|7| CHERN S. et SPANIER E. : The homology structure of fiber bundles, Proc. Nat. Acad. Sci., U. S. A., 36, 248-255 (1950).

|8| ECKMANN B. : Complex-analytic manifolds, Proc. Int. Cong. Math. II, 420-427 (1952).

|9| ECKMANN B. et FROHLICHER A. : Sur l'intégrabilité des structures presque complexes, C. R. Acad. Sci. Paris, 232, 2284-2286 (1951).

|10| EHRESMANN C. : Sur la théorie des espaces fibrés, Coll., Top. Alg. C. N. R. S. Paris, (1947), 3-15.

[11] EHRESMANN C. : Sur les variétés presque complexes, Proc. Int. Cong. Math. II, 412-419 (1952).

[12] HOPF H. : Uber komplex-analytische Mannigfaltigkeiten, Rend. Mat. Roma, Série 5, 10, 1-14 (1951).

[13] KAHLER E. : Einführung in die Theorie der Systeme von Differentialgleichungen, Leipzig, (1934).

[14] LIE S. : Die Grundlagen fur die Theorie der unendlichen continuirlichen Transformationsgruppen, Leipziger Berichte, 316-393 (1891) or Gesammelte Abhandlungen, vol. 6, Section 2, 300-364.

[15] WU, WEN-TSUN : Classes caractéristiques et i-carrés d'une variété, C. R. Acad. Paris, 230, 508-511 (1950).

Apart from vol. 28, fasc. 4 (1954)
COMMENTARII MATHEMATICI HELVETICI

On Isothermic Coordinates

By SHIING-SHEN CHERN, PHILIP HARTMAN and AUREL WINTNER

Dedicated to Professor H. Hopf on his 60th birthday

It has recently been shown ([7], pp. 686—687) that if $S: X = X(u, v)$, where $X = (x, y, z)$, is a (small piece of a) surface of class C^n and if $n \geqq 3$, then a local parametrization $X = X(U, V)$ of S can be chosen with the properties that $X(U, V)$ is of class C^n and that the element of arc-length $ds^2 = |dX|^2$ has the conformal normal form

$$ds^2 = \gamma(dU^2 + dV^2) \qquad (\gamma = \gamma(U, V) > 0) . \tag{1}$$

In other words, in dealing with surfaces of class C^n, where $n \geqq 3$, there is no loss of differentiability when (1) is assumed. It remained undecided whether or not the same is true if $n = 2$, an important case which, because of a low degree of differentiability, leads to peculiar difficulties.

It will be shown in this paper that these difficulties can be overcome and that, just as *loc. cit.*, the theorem holds also if the given

$$ds^2 = g_{11}du^2 + 2g_{12}dudv + g_{22}dv^2 \tag{2}$$

is not embedded as $|dX|^2$ on a surface S of class C^2, but has coefficients of class C^1 and possesses a continuous curvature.

The theorem to be proved is as follows:

(*) *On the domain* $u^2 + v^2 < r^2$, *let the coefficients* g_{ik} *of the positive definite metric* (2) *be functions* $g_{ik}(u, v)$ *of class* C^1 *and such that* (2) *possesses a continuous curvature* $K = K(u, v)$. *Then there exist mappings*

$$u = u(U, V),\ v = v(U, V) , \tag{3}$$

of class C^1 *and of non-vanishing Jacobian, which transform* (2) *into the normal form* (1), *and every mapping* (3) *with these properties is of class* C^2, *so that* $\gamma(U, V)$ *is of class* C^1.

It is understood that curvature for a metric (2) with coefficients of class C^1 is meant in the sense of Weyl ([6], pp. 42—44). This can be explained in the notation of H. Cartan ([1], p. 60) as follows:

A Pfaffian form $\omega = Pdu + Qdv$ with continuous coefficients $P(u, v)$, $Q(u, v)$ on a simply connected domain D is called *regular* if there exists a continuous function $f(u, v)$ on D with the property that

$$\int_J \omega = \iint_B f(u, v)\,dudv \tag{4}$$

holds for every domain B bounded by a piecewise smooth Jordan curve J in D. (For example, if P and Q are of class C^1, then ω is regular and $f = Q_u - P_v$.) When ω is regular, the identity (4) will be abbreviated as

$$d\omega = f du \wedge dv\,, \tag{5}$$

and f will be called the density of the Pfaffian form ω, relative to $du \wedge dv$. If $\omega_1 = P_1 du + Q_1 dv$ and $\omega_2 = P_2 du + Q_2 dv$ are two Pfaffian forms, the symbol $\omega_1 \wedge \omega_2$ is understood to be $(P_1Q_2 - P_2Q_1)dudv$ (so that $du \wedge dv = dudv$).

If the coefficients in (2) are of class C^1, then (2) can be written as the sum of the squares of two Pfaffian forms,

$$ds^2 = \omega_1^2 + \omega_2^2\,, \tag{6}$$

each having coefficients of class C^1. Furthermore, there is a unique Pfaffian form ω_{12} with continuous coefficients satisfying

$$d\omega_1 = \omega_{12} \wedge \omega_2\,, \quad d\omega_2 = \omega_1 \wedge \omega_{12}\,. \tag{7}$$

If the coefficients of (2) are of class C^2, then those of ω_1, ω_2 can be chosen of class C^2 and those of ω_{12} become of class C^1. In this case, ω_{12} is regular and Riemann's definition of the curvature $K = K(u, v)$ of (2) is

$$d\omega_{12} = -K\omega_1 \wedge \omega_2\,. \tag{8}$$

If the coefficients of (2) are only of class C^1, then ω_{12} need not be regular. Following Weyl, (2) is said to possess a continuous curvature K if ω_{12} is regular, in which case K is defined by (8).

Remark. It follows from (*) that (1) has a continuous curvature and that the relation corresponding to (8) is

$$\int_J \gamma^{-1}(\gamma_V dU - \gamma_U dV) = \iint_B 2K\gamma\, dU dV\,, \tag{9}$$

where K as a function of (U, V) is $K(u(U, V), v(U, V))$. For consequences of the relation (9), see [7].

Even without the assumption that (2) has a continuous curvature, there exist mappings (3) transforming (2) into the conformal form (1)

302

and such that the mapping functions (3) have first order partial derivatives satisfying a Hölder condition of every index less than 1 (Lichtenstein [5]). Since the mappings transforming the conformal form (1) into another conformal form $ds^2 = \gamma_0(dU_0^2 + dV_0^2)$ are of the form $U + iV = F(Z_0)$, where F is an analytic function of $Z_0 = U_0 + iV_0$, it follows that every mapping (3) taking (2) into the form (1) has the property that the first order partial derivatives of the functions (3) satisfy a Hölder condition. This fact will not be used below.

(*) states that the additional assumption of the existence of a continuous curvature $K = K(u, v)$ implies that the assertion of Hölder continuity for the first order partial derivatives can be improved to the assertion of the existence and continuity of second order partial derivatives. It is known ([3], p. 265) that the assertion of (*) becomes false if the assumption concerning the existence of a continuous curvature is omitted.

The truth of (*) was implied by Weyl but was not proved thus far; cf. [6], pp. 49—50 and [7], p. 685, footnote.

Proof of (*). Let (3) be a mapping of class C^1 transforming (2) into (1). It is sufficient to show that γ in (1) is of class C^1. For, according to [2], p. 222, any mapping of class C^1 which transforms a metric (2) with coefficients of class C^1 into another metric with the same property must be of class C^2. (A simplified proof of this general fact, depending on the methods of this paper, will be given elsewhere. For the case when one of the metrics is of the conformal form (2), as in (*), see [7], pp. 681—682).

Introduce the complex-valued Pfaffian form

$$\varphi = \omega_1 + i\omega_2 . \tag{10}$$

Then (6) becomes

$$ds^2 = \varphi\overline{\varphi} . \tag{11}$$

It also follows from (10) that $\varphi \wedge \overline{\varphi} = -2i\omega_1 \wedge \omega_2$, while (7) and (8) become

$$d\varphi = -i\omega_{12} \wedge \varphi \tag{12}$$

and

$$d\omega_{12} = -\tfrac{1}{2} iK\varphi \wedge \overline{\varphi} \tag{13}$$

The form ω_{12}, with the additional condition that it is real-valued, is uniquely determined by (12).

After the change of parameters (3), it follows from (11) that $\varphi\overline{\varphi}$ becomes

303

the right side of (2). Hence it is readily verified that there exists a continuous (complex-valued) function $\alpha = \alpha(U, V)$ satisfying

$$\alpha\bar{\alpha} = \gamma (\neq 0) \tag{14}$$

and

$$\varphi = \alpha dw, \text{ where } w = U + iV , \tag{15}$$

and $U = U(u, v)$, $V = V(u, v)$ is the mapping inverse to (3).

In the (u, v)-coordinates, the coefficients of φ are of class C^1, so that φ is regular. Since the definition of a regular Pfaffian form shows that regularity is preserved under transformations of class C^1, it follows that $\varphi = \alpha(U, V)(dU + idV)$ is regular. Let the continuous function $\tau = \tau(U, V)$ be defined by

$$d\varphi = \tau\alpha d\bar{w} \wedge dw \tag{16}$$

or, equivalently, by $d\varphi = 2i\tau\alpha dU \wedge dV$; so that τ is $(2i\alpha)^{-1}$ times the density of φ, relative to $dU \wedge dV$.

Since (16) can be written as

$$d\varphi = \tau d\bar{w} \wedge \varphi = (\tau d\bar{w} - \bar{\tau} dw) \wedge \varphi ,$$

where $\tau d\bar{w} - \bar{\tau} dw$ is a purely imaginary form, the remark following (13) shows that

$$- i\,\omega_{12} = \tau d\bar{w} - \bar{\tau} dw . \tag{17}$$

Since regularity is preserved under C^1-mappings, the assumption of (*) that (2) has a continuous curvature (that is, that ω_{12} is regular in (u, v)-coordinates) implies that (17) is a regular form (in dU, dV).

Let the density of the form (17), relative to $dU \wedge dV$, be $-2ik(U, V)$. Since the coefficients of (17) are purely imaginary, k is a real-valued continuous function and

$$d(\tau d\bar{w} - \bar{\tau} dw) = -k d\bar{w} \wedge dw . \tag{18}$$

It has been shown by H. Cartan ([1], pp. 62—63) that a Pfaffian form ω is regular on a simply connected domain D if and only if there exists a sequence of Pfaffian forms $\omega^1, \omega^2, \ldots$ which have smooth coefficients on D and which approximate ω in the sense that, as $n \to \infty$, the coefficients and the densities of ω^n tend to those of ω uniformly on every compact subset of D. [The sufficiency of this condition for the regularity of ω is clear. The necessity is proved by defining ω^n to be the form whose coefficients are the convolutions of the corresponding coefficients of ω

with a smooth, non-negative function $f^n(U, V)$ which vanishes outside the circle $U^2 + V^2 = 1/n^2$ and which satisfies $\iint f^n dU dV = 1$. It follows from Fubini's theorem that the density of ω^n is the convolution of $f^n(U, V)$ with the density of w.]

This theorem of H. Cartan implies that if $\varphi = \alpha dw$ is regular and has the density $2i\tau\alpha$, then the form βdw, where

$$\beta = \log \alpha\ , \tag{19}$$

is regular and has the density $2i\tau$. Thus

$$d(\beta dw) = \tau d\bar{w} \wedge dw\ . \tag{20}$$

Since $\beta + \bar{\beta} = \log \alpha\bar{\alpha} = \log \gamma$, by (14) and (19), the proof of (*) will be complete if it is shown that the real part of β is of class C^1. To this end, let

$$\tfrac{1}{2}\beta = a(U, V) + ib(U, V)\ , \quad \tau = g(U, V) - ih(U, V)\ , \tag{21}$$

where a, b, g, h are real-valued (continuous) functions. Then (20) means that

$$d(adU - bdV) = hdU \wedge dV\ , \quad d(bdU + adV) = gdU \wedge dV \tag{22}$$

and (18) means that

$$d(hdU + gdV) = kdU \wedge dV\ . \tag{23}$$

Thus (*) is contained in the following lemma:

Lemma. *Let $a(U, V)$, $b(U, V)$ be real-valued continuous functions on $U^2 + V^2 < R^2$ with the properties that the Pfaffian forms $adU - bdV$, $bdU + adV$ are regular and that, if $h(U, V)$, $g(U, V)$ are the respective densities relative to $dU \wedge dV$, the form $hdU + gdV$ is regular. Then a, b are of class C^1.*

Proof. If n is a sufficiently large integer, let $f^n(U, V)$ be a non-negative, smooth (say, of class C^2) function satisfying $\iint f^n(U, V) dU dV = 1$ and vanishing outside the circle $U^2 + V^2 = 1/n^2$, and let a^n, b^n, g^n, h^n, k^n, respectively, denote the convolutions of a, b, g, h, k with f^n; for example,

$$a^n(U, V) = \iint a(U + x, V + y) f^n(x, y) dx dy\ .$$

Thus a^n, b^n, g^n, h^n, k^n are smooth (say, of class C^2) functions on $U^2 + V^2 < (R - 1/n)^2$, and tend, as $n \to \infty$, to a, b, g, h, k uniformly on every compact subset of $U^2 + V^2 < R^2$.

According to (22), (23) and H. Cartan's proof of his approximation theorem, outlined above,

$$a_U^n - b_V^n = g^n, \quad b_U^n + a_V^n = -h^n, \quad g_U^n - h_V^n = k^n. \tag{24}$$

These relations show that a^n satisfies Poisson's equation

$$a_{UU}^n + a_{VV}^n = k^n. \tag{25}$$

It follows at once that $a(U, V)$ is of class C^1. In fact, if $G(U, V; x, y)$ is the Green function belonging to the Laplace equation on the circle $D(\varepsilon)$: $U^2 + V^2 < (R - \varepsilon)^2$, then, in this circle, $a^n(U, V)$ is the sum of

$$\iint_{D(\varepsilon)} G(U, V; x, y) k^n(x, y) dx dy$$

and of the harmonic function $p^n(U, V)$ which assumes the same boundary values as $a^n(U, V)$. By the uniformity of the limit processes $a^n \to a$ and $k^n \to k$ on $U^2 + V^2 \leqq (R - \varepsilon)^2$,

$$a(U, V) = p(U, V) + \iint_{D(\varepsilon)} G(U, V; x, y) k(x, y) dx dy, \tag{26}$$

where $p(U, V)$ is the harmonic function which assumes the same boundary values as $a(U, V)$. The C^1-character of $a(U, V)$ in $D(\varepsilon)$ is implied by (26) and the continuity of k. In fact, the logarithmic potential of a continuous density is always of class C^1 (but not necessarily of class C^2).

The fact that $b(U, V)$ is of class C^1 (and has the partial derivatives $b_U = -a_V - h, b_V = a_U - g$) follows from (22), and also from (24). This completes the proof of the Lemma and of (*).

Appendix

In view of that particular case of (*) which concerns the first fundamental form of surfaces $S: X = X(u, v)$ of class C^2, it is natural to raise the question whether or not a surface $S: X = X(u, v)$ of class C^1 always has a parametrization $X = X(U, V)$ of class C^1 in which its first fundamental form has the conformal normal form (1). It turns out that the answer is in the negative.

It is known ([3], p. 262) that there exist positive definite metrics (2) with continuous coefficients for which no mapping (3) of class C^1, with a non-vanishing Jacobian, transforms (2) into (1). What is at stake is

306

to show that there exists a metric which has this property and is, at the same time, the fundamental form of a surface $X = X(u, v)$ of class C^1.

This can be concluded by considering (at $r = 0$) suitable C^1-surfaces S of revolution,

$$S : z = z(x, y) = \int_0^r f(r)\,dr\,, \qquad r = (x^2 + y^2)^{\frac{1}{2}} \geqq 0\,, \tag{27}$$

where, on some interval $0 \leqslant r \leqslant a$, the function $f(r)$ is continuous and such that

$$f(r) \gtreqless 0 \quad \text{according as} \quad r \gtreqless 0\,. \tag{28}$$

A suitable choice of $f(r)$ proves to be $(-\log r)^{-\frac{1}{2}}$ (if $0 < r < 1$), a choice of f made by Lavrentieff ([4], p. 420) for a similar purpose.

Consider first the parametrization of S in terms of polar coordinates (r, θ). This is not an admissible parametrization (at $r = 0$), since the resulting element of arc-length

$$ds^2 = \big(1 + f^2(r)\big)\,dr^2 + r^2 d\theta^2 = r^2\big(r^{-2}(1 + f^2)\,dr^2 + d\theta^2\big) \tag{29}$$

is not positive definite (at $r = 0$). Let $\varrho = \varrho(r)$ be the function

$$\varrho = \varrho(r) = \exp\Big(-\int_r^a r^{-1}\big(1 + f^2(r)\big)^{\frac{1}{2}} dr\Big)\,, \quad (0 < r \leqq a)\,, \tag{30}$$

so that $d\varrho/\varrho = -\,r^{-1}(1 + f^2)^{\frac{1}{2}}\,dr$ and $ds^2 = r^2\varrho^{-2}(d\varrho^2 + \varrho^2 d\theta^2)$. If new parameters are defined by

$$u = \varrho(r)\cos\theta\,, \quad v = \varrho(r)\sin\theta\,, \tag{31}$$

then $du^2 + dv^2 = d\varrho^2 + \varrho^2 d\theta^2$, and so (29) becomes

$$ds^2 = r^2\varrho^{-2}(du^2 + dv^2)\,, \qquad \varrho = \varrho(r)\,. \tag{32}$$

This is a "conformal" form, but the factor $r^2\varrho^{-2}$ may not be continuous and positive at $r = 0$.

Since $\varrho(r)$ is an increasing function of r for $0 \leqslant r \leqslant a$, the transformation $(x, y) \equiv (r\cos\theta, r\sin\theta) \to (u, v)$ is one-to-one. In addition, this transformation is of class C^1 with a non-vanishing Jacobian for $x^2 + y^2 \neq 0$ (and/or $u^2 + v^2 \neq 0$).

Suppose that there exists a mapping

$$U = U(x, y), \quad V = V(x, y) \tag{33}$$

of class C^1 with non-vanishing Jacobian on some circle $x^2 + y^2 < \varepsilon^2$, with the property that the element of arc-length on S has the conformal normal form

$$ds^2 = \gamma(dU^2 + dV^2) \qquad (\gamma = \gamma(U, V)) \tag{34}$$

307

and that, without loss of generality, $U(0,0) = V(0,0) = 0$. Then there exists on some circle $u^2 + v^2 < \delta^2$ a mapping

$$U = U(u, v), \quad V = V(u, v) \tag{35}$$

with the property that (35) is of class C^1, has non-vanishing Jacobian for $u^2 + v^2 \neq 0$ and the inverse of (35) transforms (32) into (34) for $u^2 + v^2 \neq 0$. Hence $U(u, v) + iV(u, v) = W(w)$ is a continuous (single-valued) function of $w = u + iv$ on the circle $|w| < \delta$ and is regular on the punctured circle $0 < |w| < \delta$. Consequently, $W(w)$ is regular on the circle $|w| < \delta$ and $\gamma = r^2 \varrho^{-2} |dW/dw|^{-2}$ for $|w| \neq 0$, where $\varrho = |w|$ and where $r = r(\varrho)$ is the function inverse to (30). Since $W(w)$ ($\not\equiv 0$) is regular at $\varrho = 0$, there exist an integer $m \geqq 0$ and a constant $c > 0$ such that $\varrho |dW/dw| \sim c\varrho^{m+1}$ as $\varrho = |w| \to 0$. Hence,

$$r = r(\varrho) \sim \gamma^{\frac{1}{2}}(0, 0) c \varrho^{m+1} \quad \text{as} \quad \varrho \to 0 . \tag{36}$$

It follows that no neighborhood of the point $(x, y) = (0, 0)$ of S has a C^1-parametrization in which the element of arc-length has the form (34) if the inverse function $r = r(\varrho)$ of (30) fails to satisfy (36) for some constant $\gamma^{\frac{1}{2}}(0, 0)c > 0$ and some integer $m \geqq 0$. This is the case, for example, if

$$f(r) = (-\log r)^{-\frac{1}{2}}, \tag{37}$$

since $\log(1/\varrho) = \text{Const.} + \log r^{-1} + \frac{1}{2} \log \log r^{-1} + o(1)$ as $r \to 0$, by (30) and (37). Hence $C\varrho \sim r/(-\log r)^{\frac{1}{2}}$ as $r \to 0$, where C is a positive constant. Consequently, $r \sim C\varrho(-\log \varrho)^{\frac{1}{2}}$ as $\varrho \to 0$, and so (36) cannot hold.

On the C^1-surface S of Lavrentieff (*loc. cit.*), defined by (27) and (37), the element of arc-length in terms of its Cartesian parameters (x, y) is

$$ds^2 = (1 - x^2/R)dx^2 - 2(xy/R)dxdy + (1 - y^2/R)dy^2, \text{ where } R = r^2 \log r. \tag{38}$$

It is curious that the example given in [3], pp. 269—279 of a (non-embedded) continuous ds^2 which cannot be conformalized is very close to (38), namely,

$$ds^2 = dx^2 + (1 + x/2R)^2 dy^2 ,$$

where $R = r^2 \log r$, as in (38).

It is worth mentioning that the surface S defined by (27) and (37) is strictly convex, since (37) is an increasing function of r.

REFERENCES

[1] *H. Cartan*, Algebraic topology, Mimeographed notes, Harvard University, 1949.

[2] *P. Hartman*, On unsmooth two-dimensional Riemannian metrics, Amer. J. Math., *74* (1952), pp. 215—226.

[3] *P. Hartman* and *A. Wintner*, On the existence of Riemannian manifolds which cannot carry non-constant analytic or harmonic functions in the small, ibid., *75* (1953), pp. 260—276.

[4] *M. Lavrentieff*, Sur une classe de représentations continues, Recueil Mathématique, *42* (1935), pp. 407—424.

[5] *L. Lichtenstein*, Zur Theorie der konformen Abbildung. Konforme Abbildung nicht-analytischer, singularitätenfreier Flächenstücke auf ebene Gebiete, Bulletin International de l'Académie des Sciences de Cracovie, ser. A (1916), pp. 192—217.

[6] *H. Weyl*, Über die Bestimmung einer geschlossen konvexen Fläche durch ihr Linienelement, Viertelj'schr. Naturforsch. Ges. Zürich, *61* (1915), pp. 40—72.

[7] *A. Wintner*, On the rôle of theory of the logarithmic potential in differential geometry, Amer. J. Math., *75* (1953), pp. 679—690.

(Received February 10, 1954.)

DRUCK: ART. INSTITUT ORELL FÜSSLI AG., ZÜRICH

On a Generalization of Kähler Geometry

Shiing-shen Chern

1. Introduction

A Kähler manifold is a complex Hermitian manifold, whose Hermitian metric

(1) $$ds^2 = \sum_{1 \leq \alpha, \beta \leq m} g_{\alpha\beta}(z^1, \ldots, z^m; \bar{z}^1, \ldots, \bar{z}^m)\, dz^\alpha d\bar{z}^\beta \quad (\bar{g}_{\alpha\beta} = g_{\beta\alpha}),$$

has the property that the corresponding exterior differential form

(2) $$\Omega = \sum_{1 \leq \alpha, \beta \leq m} g_{\alpha\beta}\, dz^\alpha \wedge d\bar{z}^\beta$$

is closed. The importance of Kähler manifolds lies in the fact that they include as special cases the non-singular algebraic varieties over the complex field.

So far the most effective tool for the study of the homology properties of compact Kähler manifolds is Hodge's theory of harmonic integrals or harmonic differential forms.† The notion of a harmonic differential form is defined on any orientable Riemann manifold, and can be briefly introduced as follows: The Riemann metric allows us to define the star operator $*$, which transforms a differential form of degree p into one of degree $n-p$, n being the dimension of the manifold. From the operator $*$ and the exterior differentiation operator d we introduce the operators

(3) $$\begin{cases} \delta = (-1)^{np+n+1} * d\, *, \\ \Delta = d\delta + \delta d. \end{cases}$$

If the manifold is compact, as we shall assume from now on, a differential form η is called harmonic, if $\Delta\eta = 0$.

In the case of a complex manifold it will be convenient to consider complex-valued differential forms. The star operator can be extended in an obvious way to such differential forms. For its definition we

† Various accounts of this study are now in existence; cf. [3], [4], [5], [8], [9]. The numbers refer to the Bibliography at the end of this paper.

follow the convention of Weil,† without repeating the details. We only mention that we can define an operator $\bar{*}$ by

$$\bar{*}\eta = *\bar{\eta}. \tag{4}$$

The operator δ is then extended to complex-valued differential forms by the definition

$$\delta = (-1)^{np+n+1}\bar{*}d\bar{*}. \tag{5}$$

By means of this we define Δ by the second equation of (3). For a Kähler manifold we introduce furthermore the operators

$$\begin{cases} L\eta = \Omega \wedge \eta, \\ \Lambda = \bar{*}L\bar{*}. \end{cases} \tag{6}$$

A differential form η on a Kähler manifold is called effective or primitime, if $\Lambda\eta = 0$.

The notion of a primitive harmonic form is a formulation, in terms of cohomology, of the effective cycles of Lefschetz on an algebraic variety.‡ Lefschetz proved that on a complex algebraic variety every cycle is homologous, with respect to rational coefficients, to a linear combination of effective cycles and the intersection cycles, by linear spaces of the ambient projective space, of effective cycles of higher dimension. This result can be expressed in terms of harmonic differential forms by the following decomposition theorem of Hodge:

Every harmonic form ω of degree p on a compact Kähler manifold of (complex) dimension m can be written in a unique way in the form

$$\omega = \Sigma L^k \omega_k, \tag{7}$$

where the summation is extended over the following range of k:

$$\max(0, p-m) \leqq k \leqq q = [\tfrac{1}{2}p],$$

and where ω_k is a primitive harmonic form of degree $p-2k$, completely determined by ω.

The existing proofs of this theorem depend on the establishment of various identities between the operators introduced above. We attempt to give in this paper what seems to be a better understanding of this theorem by generalizing it and proving it in an entirely different way.

It is well known§ that the existence of a positive definite Hermitian metric on a complex manifold allows us to define a connection with the unitary group and that the Kähler property $d\Omega = 0$ is equivalent

† Since we are dealing with real manifolds, our δ operator differs from Weil's in sign.

‡ [6] or [5], p. 182.

§ [1], p. 112.

to the absence of torsion of this connection. Our contention is that the latter condition accounts more for the homology properties of Kähler manifolds than the analytically simpler condition $d\Omega=0$.

Utilizing this idea, we generalize the Kähler property as follows: Let M be a real differentiable manifold of dimension n. Suppose that the structural group of its tangent bundle, which is the general linear group $GL(n, R)$ in n real variables, can be reduced, in the sense of fiber bundles, to a subgroup G of the rotation group $R(n) \subset GL(n, R)$. It will be proved in § 2 that a connection can be defined, with the group G. In general, the torsion tensor of this connection does not vanish. The vanishing of torsion of this connection is then a natural generalization of the Kähler property.

On the other hand, the group G acts on the tangent vector space V of M at a point and also on its dual space V^*. This induces a linear representation of G in the exterior q^{th} power $\Lambda^q(V^*)$ of V^*, which can also be described as the representation of G into the space of all anti-symmetric covariant tensors of order q. If $G=GL(n, R)$, it is well known that this representation is irreducible. However, if G is a proper subgroup of $R(n)$, it is possible that this representation is reducible. When this is the case, suppose W be an invariant subspace of this representation. Since $G \subset R(n)$, there is an inner product defined in $\Lambda^q(V^*)$, and the subspace W' in $\Lambda^q(V^*)$ orthogonal to W is also invariant. The invariance property of W allows us to introduce the notion of a differential form of degree q and type W, as one which assigns to every point $x \in M$ an element of $W(x) \subset \Lambda^q(V^*(x))$. Similarly, we can define an operator P_W on differential forms of degree q, its projection in W. With these preparations we can state our decomposition theorem:

Let M be a compact differentiable manifold of dimension n, which has the following properties: (1) *The structural group of its tangent bundle can be reduced to a subgroup G of the rotation group $R(n)$ in n variables.* (2) *There is a connection with the group G, whose torsion tensor vanishes. Let $W \subset \Lambda^q(V^*)$ be an invariant subspace of $\Lambda^q(V^*)$ under the action of G, and let P_W be the projection of an exterior differential form of degree q into W. Then*

$$P_W \Delta = \Delta P_W. \tag{8}$$

It follows that if $W_1, \ldots, W_k$ are irreducible invariant subspaces of $\Lambda^q(V^*)$ under the action of G and if η is a harmonic form of degree q, then $P_{W_1}\eta, \ldots, P_{W_k}\eta$ are harmonic. Moreover, if η is a form of degree q and type W, then $\Delta\eta$ is also a form of degree q and type W.

When $n=2m$ is even and $G=U(m)$ is the unitary group in m complex variables considered as a subgroup of $GL(n, R)$, our notion of a manifold having a G-connection without torsion includes that of a Kähler manifold. As will be shown in § 4, it also includes a generalization of Kähler manifold studied by A. Lichnerowicz,† namely, an orientable even-dimensional Riemann manifold with the property that there exists an exterior quadratic differential form, everywhere of the highest rank, whose covariant derivative is zero. To derive Hodge's decomposition theorem from ours it remains to solve the following algebraic problem: Let V_1^*, V_2^* be two m-dimensional complex vector spaces and V^* their direct sum. Let $U(m)$ act on V^* such that it acts on V_1^* in the usual way but on V_2^* by the conjugate-complex transformation. This induces a representation of $U(m)$ into the group of linear transformations of $\Lambda^q(V^*)$. Our problem is to decompose this representation into its irreducible parts. It will be shown in § 4 that the summands in (7) correspond to the irreducible parts of the representation. In this sense the Hodge decomposition theorem cannot be further improved.

To illustrate that the scope of our theorem goes beyond Kähler geometry, we consider in § 5 the case that $G=R(s)\times R(n-s)$ $(0<s<n)$ is the direct product of two rotation groups of dimensions s and $n-s$ respectively. As is well known, the reduction of the structural group of the tangent bundle to G is equivalent to the existence of a continuous field of oriented s-dimensional linear spaces over M. The existence of a G-connection without torsion means the existence of such a field with the further property that its linear spaces are parallel with respect to a Riemann metric. Our decomposition theorem shows that the cohomology classes of M can be given a bi-degree. In particular, it follows that the s-dimensional Betti number of such a manifold is $\geqq 1$.‡

2. G-connection in a tangent bundle; torsion tensor

We shall derive in this section the basic notions and formulas for a G-connection in the tangent bundle of a real n-dimensional differentiable manifold M. Since the results are local, M will not be assumed to be compact.

The tangent bundle of M has as structural group the general linear group $GL(n, R)$ in n real variables. We consider $GL(n, R)$ to be the group of all $n\times n$ real non-singular matrices. Let G be a closed sub-

† [7].
‡ I was first informed of this theorem by Dr T. J. Willmore.

group of $GL(n, R)$. By a G-structure in M we mean a covering of M by coordinate neighborhoods U_α and, to each U_α, a set of n linearly independent Pfaffian forms θ^i_α in U_α,† such that, when $U_\alpha \cap U_\beta \neq 0$, we have

(9) $$\theta^i_\alpha = \sum_j g^i_{\alpha\beta,j}(x)\,\theta^j_\beta \quad (x \in U_\alpha \cap U_\beta),$$

where $g_{\alpha\beta}(x) = (g^i_{\alpha\beta,j}(x)) \in G$ and the mapping $U_\alpha \cap U_\beta \to G$ defined by $x \to g_{\alpha\beta}(x)$ is differentiable. Let θ_α and θ_β denote respectively the one-rowed matrices, whose elements are θ^i_α and θ^k_β. Then equation (9) can be abbreviated in the matrix form

(9a) $$\theta_\alpha = \theta_\beta g_{\alpha\beta}(x).$$

The 'coordinate transformations' $g_{\alpha\beta}(x)$ define a principal fiber bundle $p\colon B_G \to M$ with the structural group G. We recall that B_G is the union of the sets $\{U_\alpha \times G\}$, under the identification

(10) $$y_\beta = g_{\alpha\beta}(x)\,y_\alpha,$$

where $\quad (x, y_\alpha) \in U_\alpha \times G, \quad (x, y_\beta) \in U_\beta \times G, \quad x \in U_\alpha \cap U_\beta.$

From (9a) and (10) it follows that the one-rowed matrix of Pfaffian forms

(11) $$\omega = \theta_\alpha y_\alpha = \theta_\beta y_\beta$$

is globally defined in B_G. The elements ω^i in $\omega = (\omega^1, \ldots, \omega^n)$ are clearly linearly independent. By exterior differentiation we find

(12) $$d\omega = -\omega \wedge y_\alpha^{-1} dy_\alpha + d\theta_\alpha y_\alpha,$$

where $y_\alpha^{-1} dy_\alpha$ is a matrix of left-invariant Pfaffian forms of G. If π^ρ $(1 \leqq \rho \leqq r)$ is a set of linearly independent left-invariant Pfaffian forms of G, we can write (12) in the form

(13) $$d\omega^i = -\sum_{\rho,k} a^i_{\rho k}\,\omega^k \wedge \pi^\rho + \tfrac{1}{2} \sum_{j,k} c^i_{jk}(b)\,\omega^j \wedge \omega^k,$$

where $a^i_{\rho k}$ are constants and $c^i_{jk}(b)$ are functions in B_G satisfying the conditions

(14) $$c^i_{jk}(b) + c^i_{kj}(b) = 0 \quad (b \in B_G).$$

The constants $a^i_{\rho k}$ in (13) have a simple geometrical meaning. In fact, we can regard G as acting on an n-dimensional vector space with the coordinates (ξ^i), according to the equations

(15) $$\xi^i \to \xi'^i = \sum_j y^i_j \xi^j \quad ((y^i_j) \in G).$$

† Unless otherwise stated, we agree on the following ranges of indices:
$$1 \leqq i, j, k, l \leqq n, \quad 1 \leqq \rho, \sigma, \tau \leqq r.$$
In this section we use α, β, γ to index the neighborhoods. When we discuss almost complex manifolds of dimension $2m$, we suppose $1 \leqq \alpha, \beta, \gamma \leqq m$.

Then the infinitesimal transformations

$$X_\rho = \sum_{i,k} a^i_{\rho k} \xi^k \frac{\partial}{\partial \xi^i} \tag{16}$$

are linearly independent and generate the group G. It follows that their commutators satisfy equations of the form

$$[X_\rho, X_\sigma] = \sum_\tau \gamma^\tau_{\rho\sigma} X_\tau, \quad \gamma^\tau_{\rho\sigma} = -\gamma^\tau_{\sigma\rho}, \tag{17}$$

where $\gamma^\tau_{\rho\sigma}$ are the constants of structure of G. When the expressions (16) are substituted into (17), we get

$$\sum_i (a^i_{\rho k} a^j_{\sigma i} - a^i_{\sigma k} a^j_{\rho i}) = \sum_\tau \gamma^\tau_{\rho\sigma} a^j_{\tau k}. \tag{18}$$

The equations (12) or (13) are derived with reference to a representation of ω in $p^{-1}(U_\alpha)$ and are therefore local in character. To put this in a different way, the forms π^ρ are not globally defined in B_G. The permissible transformation on π^ρ is given by

$$\pi^\rho \to \pi'^\rho = \pi^\rho + \sum_k b^\rho_k \omega^k. \tag{19}$$

Under this transformation the equations (13) preserve the same form, with new coefficients c'^i_{jk} given by

$$c'^i_{jk} = c^i_{jk} + \sum_\rho (-a^i_{\rho k} b^\rho_j + a^i_{\rho j} b^\rho_k). \tag{20}$$

The $\frac{1}{2}n^2(n-1)$ expressions

$$A^i_{jk} = \sum_\rho (-a^i_{\rho k} b^\rho_j + a^i_{\rho j} b^\rho_k) \tag{21}$$

are linear and homogeneous in b^ρ_k, with constant coefficients. We say that the group G has the property (C), if there are nr linearly independent ones among them, that is, if $A^i_{jk}=0$ implies $b^\rho_k=0$.

If the group G has the property (C), we can define a connection† in the bundle $p'\colon B_{G'} \to M$, where G' is the group of non-homogeneous linear transformations on n variables with G as the homogeneous part and the bundle is obtained from $p\colon B_G \to M$ by enlarging the group from G to G'. To define such a connection it suffices to determine in B_G a set of forms π^ρ satisfying (13). Suppose $A^{i'}_{j'k'}$ be a subset of nr linearly independent expressions among the A^i_{jk}, and suppose $A^{i''}_{j''k''}$ be its complementary set. Then there exists one, and only one, set of forms π^ρ in B_G satisfying the equations (13) together with the conditions $c^{i'}_{j'k'}=0$. These forms, together with ω^i, define a connection in the bundle $p'\colon B_{G'} \to M$. By abus du langage, we call such a connection a G-connection in the tangent bundle.

The bundles $p\colon B_G \to M$ and $p'\colon B_{G'} \to M$ are G'-equivalent in the sense of bundles. From the point of view of connections it is, however,

† [2].

necessary to consider the second bundle. The curvature form of our connection is a tensorial quadratic differential form in M, of type $ad(G')$ and with values in the Lie algebra $L(G')$ of G'. Since the Lie algebra $L(G)$ of G is a subalgebra of $L(G')$, there is a natural projection of $L(G')$ into the quotient space $L(G')/L(G)$. The image of the curvature form under this projection will be called the torsion form or the torsion tensor. If the forms π^ρ in (13) define a G-connection, the vanishing of the torsion form is expressed analytically by the conditions

$$c^{i''}_{j''k''} = 0. \tag{22}$$

We proceed to derive the analytical formulas for the theory of a G-connection without torsion in the tangent bundle. In general we will consider such formulas in B_G. The fact that the G-connection has no torsion simplifies (13) into the form

$$d\omega^i = \sum_{\rho,k} a^i_{\rho k} \pi^\rho \wedge \omega^k. \tag{23}$$

By taking the exterior derivative of (23) and using (18), we get

$$\sum_{\rho,k} a^i_{\rho k} \Pi^\rho \wedge \omega^k = 0, \tag{24}$$

where we put

$$\Pi^\rho = d\pi^\rho + \tfrac{1}{2} \sum_{\sigma,\tau} \gamma^\rho_{\sigma\tau} \pi^\sigma \wedge \pi^\tau. \tag{25}$$

For a fixed value of k we multiply the above equation by

$$\omega^1 \wedge \ldots \wedge \omega^{k-1} \wedge \omega^{k+1} \wedge \ldots \wedge \omega^n,$$

getting

$$\sum_\rho a^i_{\rho k} \Pi^\rho \wedge \omega^1 \wedge \ldots \wedge \omega^n = 0,$$

or

$$\sum_\rho a^i_{\rho k} \Pi^\rho \equiv 0, \quad \operatorname{mod} \omega^j.$$

Since the infinitesimal transformations X_ρ are linearly independent, this implies that

$$\Pi^\rho \equiv 0, \quad \operatorname{mod} \omega^j.$$

It follows that Π^ρ is of the form

$$\Pi^\rho = \sum_j \phi^\rho_j \wedge \omega^j,$$

where ϕ^ρ_j are Pfaffian forms. Substituting these expressions into (24), we get

$$\sum_{\rho,j,k} (a^i_{\rho k} \phi^\rho_j - a^i_{\rho j} \phi^\rho_k) \wedge \omega^j \wedge \omega^k = 0.$$

It follows that

$$\sum_\rho (a^i_{\rho k} \phi^\rho_j - a^i_{\rho j} \phi^\rho_k) \equiv 0, \quad \operatorname{mod} \omega^l.$$

Since G has the property (C), the above equations imply that

$$\phi^\rho_j \equiv 0, \quad \operatorname{mod} \omega^k.$$

In other words, we have

(26) $$\Pi^\rho = \tfrac{1}{2}\textstyle\sum_{j,k} R^\rho_{jk}\,\omega^j \wedge \omega^k, \quad R^\rho_{jk} + R^\rho_{kj} = 0.$$

These are essentially the curvature form of the G-connection.

It will be convenient to introduce the quantities

(27) $$S^i_{jkl} = \textstyle\sum_\rho a^i_{\rho j} R^\rho_{kl}.$$

By substituting (26) into (24) and equating to zero the coefficients of the resulting cubic differential form, we get

(28) $$S^i_{jkl} + S^i_{klj} + S^i_{ljk} = 0.$$

We now consider a differential form of degree q in B_G which belong to the base manifold M, that is, which is the dual image of a differential form in M under the projection p. Such a differential form can be written as

(29) $$\eta = \frac{1}{q!}\textstyle\sum_{i_1,\dots,i_q} P_{i_1\dots i_q}\,\omega^{i_1} \wedge \dots \wedge \omega^{i_q},$$

where $P_{i_1\dots i_q}$ can be supposed to be anti-symmetric in any two of its indices. In order that η belongs to M, it is necessary that $d\eta$ has the same property. By using (23), we see that this implies the relations

(30) $$dP_{i_1\dots i_q} + \textstyle\sum_{s=1}^q \sum_\rho P_{i_1\dots i_{s-1} j\, i_{s+1}\dots i_q}\, a^j_{\rho i_s}\pi^\rho = \sum_l P_{i_1\dots i_q|l}\,\omega^l.$$

Exterior differentiation of this equation gives

$$\begin{aligned}\textstyle\sum_l (dP_{i_1\dots i_q|l} + \sum_{s=1}^q \sum_{\rho,j} P_{i_1\dots i_{s-1} j\, i_{s+1}\dots i_q|l}\, a^j_{\rho i_s}\pi^\rho + \sum_{\rho,j} P_{i_1\dots i_q|j}\, a^j_{\rho l}\pi^\rho) \wedge \omega^l \\ = \tfrac{1}{2}\textstyle\sum_{s=1}^q \sum_{j,k,l} P_{i_1\dots i_{s-1} j\, i_{s+1}\dots i_q} S^j_{i_s kl}\,\omega^k \wedge \omega^l.\end{aligned}$$

This allows us to put

(31) $$\begin{aligned} dP_{i_1\dots i_q|l} + \textstyle\sum_{s=1}^q \sum_{\rho,j} P_{i_1\dots i_{s-1} j\, i_{s+1}\dots i_q|l}\, a^j_{\rho i_s}\pi^\rho \\ + \textstyle\sum_{\rho,j} P_{i_1\dots i_q|j}\, a^j_{\rho l}\pi^\rho = \sum_k P_{i_1\dots i_q|l|k}\,\omega^k.\end{aligned}$$

Substituting this into the last equation and equating to zero the coefficient of $\omega^k \wedge \omega^l$, we get

(32) $$P_{i_1\dots i_q|l|k} - P_{i_1\dots i_q|k|l} = \textstyle\sum_{s=1}^q \sum_j P_{i_1\dots i_{s-1} j\, i_{s+1}\dots i_q} S^j_{i_s kl}.$$

These equations are usually known as the interchange formulas.

An important case of a G-connection is when G is a subgroup of the orthogonal group $O(n)$ in n variables. In this case we shall lower the superscripts of our symbols and use subscripts throughout.

We first remark that *such a group G always has the property* (C). Since the infinitesimal transformations X_ρ leave invariant the quadratic form

$$(\xi^1)^2 + \dots + (\xi^n)^2,$$

we have

(33) $$a_{i\rho k}+a_{k\rho i}=0.$$

Suppose that $$\Sigma_\rho(-a_{i\rho k}b^\rho_j+a_{i\rho j}b^\rho_k)=0.$$

Permuting this equation cyclically in i, j, k, we get

$$\Sigma_\rho(-a_{j\rho i}b^\rho_k+a_{j\rho k}b^\rho_i)=0,$$

$$\Sigma_\rho(-a_{k\rho j}b^\rho_i+a_{k\rho i}b^\rho_j)=0.$$

By subtracting the first equation from the sum of the last two equations, we find

$$\Sigma_\rho a_{j\rho k}b^\rho_i=0.$$

But these equations imply $b^\rho_i=0$. This proves that G has the property (C).

When G is a subgroup of $O(n)$, there are some symmetry properties of S_{ijkl} which will be useful later on. From the second equation of (26) and (33), we have

(34) $$S_{ijkl}=-S_{jikl}=-S_{ijlk}.$$

It is well known that these relations and (28) imply

(35) $$S_{ijkl}=S_{klij}.$$

3. Proof of the decomposition theorem

We are now ready to give a proof of the decomposition theorem as stated in the Introduction. Since $G\subset R(n)$, the G-structure on M defines an orientation on M by the condition $\omega_1\wedge\ldots\wedge\omega_n>0$ and a Riemann metric on M by

(36) $$ds^2=\omega_1^2+\ldots+\omega_n^2.$$

Relative to these the operators in (3) are defined. Our first problem is to compute $\Delta\eta$, with η given by (29).

This is a routine computation, and we shall only give the relevant formulas. First of all we have

(37) $$d\eta=\frac{1}{q!}\Sigma_{i_1,\ldots,i_q,j}P_{i_1\ldots i_q|j}\,\omega_j\wedge\omega_{i_1}\wedge\ldots\wedge\omega_{i_q}.$$

To make the coefficients anti-symmetric, we can write

(38) $$d\eta=\frac{(-1)^q}{(q+1)!}\Sigma_{i_1,\ldots,i_{q+1}}(P_{i_1\ldots i_q|i_{q+1}}-P_{i_{q+1}i_2\ldots i_q|i_1}$$
$$-\ldots-P_{i_1\ldots i_{q-1}i_{q+1}|i_q})\,\omega_{i_1}\wedge\ldots\wedge\omega_{i_{q+1}}.$$

We also have, by definition,

$$(39)\qquad *\eta = \frac{1}{q!\,(n-q)!} \sum_{i_1,\dots i_n} \epsilon_{i_1\dots i_q i_{q+1}\dots i_n} P_{i_1\dots i_q}\,\omega_{i_{q+1}} \wedge \dots \wedge \omega_{i_n},$$

where $\epsilon_{i_1\dots i_n}$ is equal to $+1$ or -1, according as $i_1, \dots, i_n$ form an even or odd permutation of $1, \dots, n$, and is otherwise equal to zero. Using (3), we find

$$(40)\qquad \delta\eta = \frac{(-1)^q}{(q-1)!} \sum_{i_1,\dots,i_{q-1},j} P_{i_1\dots i_{q-1}j|j}\,\omega_{i_1} \wedge \dots \wedge \omega_{i_{q-1}}.$$

Further computation gives

$$(41)\qquad \begin{cases} -(q-1)!\,d\delta\eta = \sum_{i_1,\dots,i_q,j} P_{i_1\dots i_{q-1}j|j|i_q}\,\omega_{i_1} \wedge \dots \wedge \omega_{i_q}, \\ -(q-1)!\,\delta d\eta = \dfrac{1}{q} \sum_{i_1,\dots,i_q,j} P_{i_1\dots i_q|j|j}\,\omega_{i_1} \wedge \dots \wedge \omega_{i_q} \\ \qquad\qquad - \sum_{i_1,\dots,i_q,j} P_{i_1\dots i_{q-1}j|i_q|j}\,\omega_{i_1} \wedge \dots \wedge \omega_{i_q}. \end{cases}$$

By using (32), we get the following fundamental formula:

$$(42)\qquad \begin{aligned} -(q-1)!\,\Delta\eta = {} & \frac{1}{q} \sum_{i_1,\dots,i_q,j} P_{i_1\dots i_q|j|j}\,\omega_{i_1} \wedge \dots \wedge \omega_{i_q} \\ & + \sum_{i_1,\dots,i_q,k,j} P_{i_1\dots i_{q-1}k} S_{kji_qj}\,\omega_{i_1} \wedge \dots \wedge \omega_{i_q} \\ & - (q-1) \sum_{i_1,\dots,i_q,k,j} P_{i_1\dots i_{q-2}kj} S_{ki_{q-1}ji_q}\,\omega_{i_1} \wedge \dots \wedge \omega_{i_q}. \end{aligned}$$

The disadvantage of this formula is that the coefficients are not anti-symmetric in their indices. The following artifice is used to anti-symmetrize the coefficients: Let $\epsilon(i_1 \dots i_q; j_1 \dots j_q)$ denote the number which is equal to $+1$ or -1 according as $j_1, \dots, j_q$ form an even or odd permutation of $i_1, \dots, i_q$, and is otherwise equal to zero. We define

$$(43)\qquad \begin{aligned} & S(i_1 \dots i_q, j_1 \dots j_q;\, k_1 \dots k_q, l_1 \dots l_q) \\ & = \sum \epsilon(i_1 \dots i_q;\, r_1 \dots r_{q-1} g)\, \epsilon(j_1 \dots j_q;\, r_1 \dots r_{q-1} h) \\ & \quad \times \epsilon(k_1 \dots k_q;\, s_1 \dots s_{q-1} u)\, \epsilon(l_1 \dots l_q;\, s_1 \dots s_{q-1} r)\, S_{ghuv} \\ & \qquad\qquad (1 \leqq r, s, g, h, u, v \leqq n), \end{aligned}$$

where all the indices run from 1 to n and the summation is over all the repeated ones. To shorten our notation we write the symbol on the left-hand side also as $S((i)(j);\,(k)(l))$. It is easily seen that these symbols have the following properties:

(1) They are anti-symmetric in any two indices of each of the sets $i_1, \dots, i_q;\ j_1, \dots, j_q;\ k_1, \dots, k_q;\ l_1, \dots, l_q$.

(2) $S((i)(j);\,(k)(l)) = -S((j)(i);\,(k)(l))$.

(3) $S((i)(j);\,(k)(l)) = -S((i)(j);\,(l)(k))$.

(4) $S((i)(j);\,(k)(l)) = S((k)(l);\,(i)(j))$.

From these quantities we define

$$(44)\qquad S((i)(k)) = S(i_1 \dots i_q, k_1 \dots k_q) = \frac{1}{q!}\Sigma_{j_1,\dots j_q} S(i_1 \dots i_q, j_1 \dots j_q; k_1 \dots k_q, j_1 \dots j_q).$$

Then $S((i)(k))$ are anti-symmetric in the indices of each one of the sets $i_1, \dots, i_q$ and $k_1, \dots, k_q$, and

$$(45)\qquad S((i)(k)) = S((k)(i)).$$

It turns out that $S(i_1 \dots i_q; k_1 \dots k_q)$ are the quantities which occur in the expression for $\Delta\eta$. In fact, we find

$$(46)\qquad -\Delta\eta = \frac{1}{q!}\Sigma_{i_1,\dots,i_q,j} P_{i_1 \dots i_q | j | j}\, \omega_{i_1} \wedge \dots \wedge \omega_{i_q} + \frac{1}{(q!\,(n-q)!)^2}\Sigma_{i,k} P_{i_1 \dots i_q} S(i_1 \dots i_q, k_1 \dots k_q)\, \omega_{k_1} \wedge \dots \wedge \omega_{k_q}.$$

The exterior q^{th} power $\Lambda^q(V^*(x))$ of the space of covectors $V^*(x)$ at $x \in M$ of is dimension $N = \binom{n}{q}$ and has as base

$$(47)\qquad \omega_{i_1} \wedge \dots \wedge \omega_{i_q} \quad (1 \leqq i_1 < \dots < i_q \leqq n).$$

In $\Lambda^q(V^*(x))$ an inner product is defined by

$$(48)\qquad (\eta, \eta) = \frac{1}{q!}\Sigma_{i_1,\dots i_q} P^2_{i_1 \dots i_q}.$$

Under the action of G through its linear representation this inner product remains invariant. If W_1 is an invariant subspace of $\Lambda^q(V^*(x))$ under G, its orthogonal space W_2 is also invariant. There exist therefore base vectors $\Phi_1, \dots, \Phi_N \in \Lambda^q(V^*(x))$, which are related to the base (47) by an orthogonal transformation

$$(49)\qquad \omega_{i_1} \wedge \dots \wedge \omega_{i_q} = \Sigma_{\lambda=1}^{N} g_{i_1 \dots i_q, \lambda} \Phi_\lambda,$$

such that $\Phi_1, \dots, \Phi_h$ and $\Phi_{h+1}, \dots, \Phi_N$ span W_1 and W_2 respectively. We can assume $g_{i_1 \dots i_q, \lambda}$ to be defined for all $i_1, \dots, i_q$ and anti-symmetric in any two of its first q indices. Then equations (49) can be solved for Φ_λ, giving

$$(50)\qquad \Phi_\lambda = \frac{1}{q!}\Sigma_{i_1,\dots i_q} g_{i_1 \dots i_q, \lambda}\, \omega_{i_1} \wedge \dots \wedge \omega_{i_q}.$$

From now on till the end of this section we shall agree on the following ranges of indices:

$$(51)\qquad 1 \leqq A, B, C \leqq h, \quad h+1 \leqq \alpha, \beta, \gamma \leqq N.$$

We shall find the condition that W_1 and W_2 are invariant under G. For this purpose we compute the exterior derivative $d\Phi_A$ and find

$$(52)\quad q!\,d\Phi_A = q\sum_{i_1,\ldots,i_q,l,\rho} g_{i_1\ldots i_{q-1}l,A}\, a_{l\rho i_q}\pi_\rho\wedge\omega_{i_1}\wedge\ldots\wedge\omega_{i_q}$$
$$= q\sum_{i_1,\ldots,i_{q-1},l,m,\lambda,\rho} g_{i_1\ldots i_{q-1}l,A}\, g_{i_1\ldots i_{q-1}m,\lambda}\, a_{l\rho m}\pi_\rho\wedge\Phi_\lambda.$$

It follows that the invariance of W_1 under G implies

$$(53)\quad \sum_{i_1,\ldots,i_{q-1},l,m} g_{i_1\ldots i_{q-1}l,A}\, g_{i_1\ldots i_{q-1}m,\alpha}\, a_{l\rho m} = 0$$

or

$$(54)\quad \sum_{i_1,\ldots,i_{q-1},l,m} g_{i_1\ldots i_{q-1}l,A}\, g_{i_1\ldots i_{q-1}m,\alpha}\, S_{lmjk} = 0.$$

As to be expected, this relation is symmetric in A and α.

Our theorem will be established if we prove that the condition that η is of type W_1 implies that $\Delta\eta$ is of type W_1. Suppose therefore that

$$(55)\quad \eta = \sum_A P_A\Phi_A = \frac{1}{q!}\sum_{i_1,\ldots,i_q,A} P_A g_{i_1\ldots i_q,A}\,\omega_{i_1}\wedge\ldots\wedge\omega_{i_q}.$$

By (46) we find

$$(56)\quad -\Delta\eta = \frac{1}{q!}\sum P_{A|j|j}\, g_{i_1\ldots i_q,A}\,\omega_{i_1}\wedge\ldots\wedge\omega_{i_q}$$
$$+\frac{1}{(q!\,(q-1)!)^2}\sum P_A g_{i_1\ldots i_q,A}\, S(i_1\ldots i_q,\, k_1\ldots k_q)\,\omega_{k_1}\wedge\ldots\wedge\omega_{k}$$
$$= \sum P_{A|j|j}\Phi_A + \frac{1}{(q!\,(q-1)!)^2}\sum P_A g_{i_1\ldots i_q,A}\, g_{k_1\ldots k_q,\lambda}\, S(i_1\ldots i_q,\, k_1\ldots k_q)\,\Phi_\lambda.$$

It suffices to prove that

$$(57)\quad \sum_{i_1,\ldots,i_q,k_1,\ldots,k_q} S(i_1\ldots i_q, k_1\ldots k_q)\, g_{i_1\ldots i_q,A}\, g_{k_1\ldots k_q,\alpha} = 0.$$

For this purpose we consider the quantities introduced in (43), and put

$$(58)\quad (q!)^4 R_{\kappa\lambda\mu\nu} = \sum_{i_1,\ldots,l_q} g_{i_1\ldots i_q,\kappa}\, g_{j_1\ldots j_q,\lambda}\, g_{k_1\ldots k_q,\mu}\, g_{l_1\ldots l_q,\nu}$$
$$\times S(i_1\ldots i_q,\, j_1\cdots j_q;\, k_1\ldots k_q,\, l_1\ldots l_q),$$

where the indices of $R_{\kappa\lambda\mu\nu}$ have the ranges

$$(59)\quad 1\leqq \kappa,\lambda,\mu,\nu\leqq N.$$

We also put

$$(60)\quad R_{\kappa\mu} = \sum_\lambda R_{\kappa\lambda\mu\lambda}.$$

Because of similar properties of

$$S(i_1\ldots i_q,\, j_1\cdots j_q;\, k_1\ldots k_q,\, l_1\ldots l_q),$$

$R_{\kappa\lambda\mu\nu}$ has the properties

$$(61)\quad R_{\kappa\lambda\mu\nu} = -R_{\lambda\kappa\mu\nu} = -R_{\kappa\lambda\nu\mu},$$
$$R_{\kappa\lambda\mu\nu} = R_{\mu\nu\kappa\lambda}.$$

From (54) it follows that, on remembering the ranges of indices as agreed upon in (51),

(62) $$R_{A\alpha\mu\nu}=0.$$

From this we find

$$R_{A\alpha}=\Sigma_\lambda R_{A\lambda\alpha\lambda}=\Sigma_B R_{AB\alpha B}+\Sigma_\beta R_{A\beta\alpha\beta}=0.$$

But this is exactly the equation (57) to be proved. Thus the proof of our decomposition theorem is complete.

4. The case of the unitary group

As discussed in the Introduction, our decomposition theorem reduces the proof of (7) to a purely algebraic problem. The latter has been solved in the theory of representations of the unitary group. To be precise, the problem can be formulated as follows:

Let L be a complex vector space of dimension $2m$, which is a direct sum of two complex vector spaces V, $\overline{V}$ of dimension m. Let ω_α, $\overline{\omega}_\alpha$ be base vectors of V, $\overline{V}$ respectively. The equations

(63) $$\begin{cases}\omega_\alpha\to\omega'_\alpha=\Sigma_\beta u_{\alpha\beta}\omega_\beta,\\ \overline{\omega}_\alpha\to\overline{\omega}'_\alpha=\Sigma_\beta \overline{u}_{\alpha\beta}\overline{\omega}_\beta,\end{cases}$$

where $(u_{\alpha\beta})$ is a unitary matrix, define a linear mapping of L, which maps the vector $\Sigma_\alpha(f_\alpha\omega_\alpha+g_\alpha\overline{\omega}_\alpha)$ into the vector $\Sigma_\alpha(f_\alpha\omega'_\alpha+g_\alpha\overline{\omega}'_\alpha)$. The linear mappings so obtained, for all $(u_{\alpha\beta})\in U(m)$, define a representation of $U(m)$. It induces a linear representation of $U(m)$ in the exterior power $\Lambda^r(L)$. Our problem is to decompose this representation into its irreducible parts.

As base vectors of $\Lambda^r(L)$ we can take

(64) $$\omega_{\alpha_1}\wedge\ldots\wedge\omega_{\alpha_p}\wedge\overline{\omega}_{\beta_1}\wedge\ldots\wedge\overline{\omega}_{\beta_q}$$
$$(p+q=r,\ 1\leqq\alpha_1<\ldots<\alpha_p\leqq m,\ 1\leqq\beta_1<\ldots<\beta_q\leqq m).$$

For fixed values of p, q these vectors clearly span an invariant subspace of $\Lambda^r(L)$, to be denoted by $\Lambda^{p,q}(L)$. An element of $\Lambda^{p,q}(L)$ is said to be bi-homogeneous with the bi-degree (p,q). Such an element can be written in the form

(65) $$\Sigma_{\alpha,\beta}P_{\alpha_1\ldots\alpha_p\beta_1\ldots\beta_q}\omega_{\alpha_1}\wedge\ldots\wedge\omega_{\alpha_p}\wedge\overline{\omega}_{\beta_1}\wedge\ldots\wedge\overline{\omega}_{\beta_q},$$

where we can suppose the coefficients to be anti-symmetric in the α's and β's separately. For $p\geqq 1$, $q\geqq 1$, the linear subspace in $\Lambda^{p,q}(L)$, defined by the equation

(66) $$\Sigma_\gamma P_{\alpha_1\ldots\alpha_{p-1}\gamma\beta_1\ldots\beta_{q-1}\gamma}=0,$$

is an invariant subspace.

We wish to remark that the following theorem is true: *For* $p+q\leqq m$ *the representation of* $U(m)$ *in the linear subspace* (66) *of* $\Lambda^{p,q}(L)$ *is irreducible.*

This can be verified by a computation of the character of the representation. In fact, we easily show that in the notation of H. Weyl† this is the representation of signature $(\underbrace{1\ldots 1}_{p}0\ldots 0\underbrace{-1\ldots -1}_{q})$.

Moreover, if we consider the maximal Abelian subgroup of all diagonal matrices

$$\begin{pmatrix} \epsilon_1 & & 0 \\ & \ddots & \\ 0 & & \epsilon_n \end{pmatrix} \tag{67}$$

of $U(m)$ and introduce the integers

$$\begin{cases} l_1 = m, \ldots, l_p = m-p+1, l_{p+1} = m-p-1, \ldots, \\ \qquad l_{m-q} = q, l_{m-q+1} = q-2, \ldots, l_m = -1, \end{cases} \tag{68}$$

this representation has the character

$$\chi = \frac{|\epsilon^{l_1}\ldots\epsilon^{l_m}|}{|\epsilon^{m-1}\ldots\epsilon^0|}, \tag{69}$$

where

$$|\epsilon^{l_1}\ldots\epsilon^{l_m}| = \begin{vmatrix} \epsilon_1^{l_1} & \ldots & \epsilon_1^{l_m} \\ \ldots & \ldots & \ldots \\ \epsilon_m^{l_1} & \ldots & \epsilon_m^{l_m} \end{vmatrix}, \quad |\epsilon^{m-1}\ldots\epsilon^0| = \begin{vmatrix} \epsilon_1^{m-1} & \ldots & \epsilon_1^0 \\ \ldots & \ldots & \ldots \\ \epsilon_m^{m-1} & \ldots & \epsilon_m^0 \end{vmatrix}. \tag{70}$$

As is well known, condition (66) characterizes the primitive elements. Hence all these add to the remark that each summand in (7) corresponds to an irreducible representation of $U(m)$, so that in this sense Hodge's decomposition theorem cannot be improved.

We shall show briefly that our considerations include also as a particular case a generalization of Kähler geometry, which has been studied by A. Lichnerowicz.‡ This is the geometry on a compact even-dimensional manifold on which there are given an exterior quadratic differential form Ω of highest rank and a Riemann metric with the property that the covariant derivative of Ω vanishes. Lichnerowicz proved that if such is the case the Riemann metric can be so modified that we can suppose

$$\begin{cases} ds^2 = \sum_\alpha \theta_\alpha^2 + \sum_{\alpha'} \theta_{\alpha'}^2, \\ \Omega = \sum_\alpha \theta_\alpha \wedge \theta_{\alpha'}, \end{cases} \tag{71}$$

† [10], in particular, pp. 198–201. ‡ [7].

where $\alpha' = \alpha + m$, etc., and where θ_α, $\theta_{\alpha'}$ are linearly independent linear differential forms. By putting

$$\omega_\alpha = \frac{1}{\sqrt{2}}(\theta_\alpha + i\theta_{\alpha'}), \quad \overline{\omega}_\alpha = \frac{1}{\sqrt{2}}(\theta_\alpha - i\theta_{\alpha'}), \tag{72}$$

we can also write

$$\begin{cases} ds^2 = 2\sum_\alpha \omega_\alpha \overline{\omega}_\alpha, \\ \Omega = i\sum_\alpha \omega_\alpha \wedge \overline{\omega}_\alpha. \end{cases} \tag{73}$$

These forms define an almost complex structure on the manifold, and the group of the bundle is reduced to $U(m)$. By following the general discussions in § 2, we see that a connection can be defined in the bundle, with the group $U(m)$. Without going into details, we state that the forms $\omega_{\alpha\beta}$ which define the connection are characterized by the conditions

$$\begin{cases} d\omega_\alpha = \sum_\beta \omega_\beta \wedge \omega_{\beta\alpha} + \Omega_\alpha, \\ \Omega_\alpha = \sum_{\beta,\gamma} (A_{\alpha\beta\gamma}\omega_\beta \wedge \omega_\gamma + B_{\alpha\beta\gamma}\overline{\omega}_\beta \wedge \overline{\omega}_\gamma) \\ \qquad\qquad (A_{\alpha\beta\gamma} + A_{\alpha\gamma\beta} = 0,\ B_{\alpha\beta\gamma} + B_{\alpha\gamma\beta} = 0), \end{cases} \tag{74}$$

and

$$\omega_{\alpha\beta} + \overline{\omega}_{\beta\alpha} = 0. \tag{75}$$

It is possible to express these equations in the real form. For this purpose we write

$$\omega_{\alpha\beta} = \phi_{\alpha\beta} + i\psi_{\alpha\beta}, \tag{76}$$

where $\phi_{\alpha\beta}$, $\psi_{\alpha\beta}$ are real. Conditions (75) are equivalent to the conditions

$$\begin{cases} \phi_{\alpha\beta} + \phi_{\beta\alpha} = 0, \\ \psi_{\alpha\beta} - \psi_{\beta\alpha} = 0, \end{cases} \tag{77}$$

and equations (74) will then take the real form

$$\begin{cases} d\theta_\alpha = \sum_\beta (\theta_\beta \wedge \phi_{\beta\alpha} - \theta_{\beta'} \wedge \psi_{\beta\alpha}) + \Theta_\alpha, \\ d\theta_{\alpha'} = \sum_\beta (\theta_\beta \wedge \psi_{\beta\alpha} + \theta_{\beta'} \wedge \phi_{\beta\alpha}) + \Theta_{\alpha'}, \end{cases} \tag{78}$$

where Θ_α, $\Theta_{\alpha'}$ are defined by

$$\begin{cases} \Omega_\alpha = \frac{1}{\sqrt{2}}(\Theta_\alpha + i\Theta_{\alpha'}), \\ \overline{\Omega}_\alpha = \frac{1}{\sqrt{2}}(\Theta_\alpha - i\Theta_{\alpha'}). \end{cases} \tag{79}$$

On the other hand, the ds^2 in (71) defines a Riemann metric on the manifold. To this structure with the orthogonal group there always exists a connection without torsion, the parallelism of Levi-Civita.

The latter will be defined by the forms $\theta_{AB}=-\theta_{BA}$ $(A,B=1,\ldots,2m)$, satisfying the equations

$$(80)\qquad \begin{cases} d\theta_\alpha=\sum_\beta(\theta_\beta\wedge\theta_{\beta\alpha}+\theta_{\beta'}\wedge\theta_{\beta'\alpha}),\\ d\theta_{\alpha'}=\sum_\beta(\theta_\beta\wedge\theta_{\beta\alpha'}+\theta_{\beta'}\wedge\theta_{\beta'\alpha'}).\end{cases}$$

In terms of this connection we express the condition that the covariant derivative of Ω is zero. This gives

$$(81)\qquad \theta_{\alpha'\beta}=\theta_{\beta'\alpha},\quad \theta_{\alpha'\beta'}=\theta_{\alpha\beta}.$$

Equations (78) are therefore satisfied, if we put

$$(82)\qquad \phi_{\beta\alpha}=\theta_{\beta\alpha},\quad \psi_{\beta\alpha}=-\theta_{\beta'\alpha},\quad \Theta_\alpha=0,\quad \Theta_{\alpha'}=0.$$

Since the forms $\phi_{\beta\alpha}$, $\psi_{\beta\alpha}$, Θ_α, $\Theta_{\alpha'}$ in (78) are completely determined by their symmetry properties (77) and the form of Ω_α, it follows that the condition of Lichnerowicz is equivalent to saying that the group of the tangent bundle can be reduced to $U(m)$ in such a way that the resulting connection has no torsion. Our decomposition theorem applies therefore to this case and gives a decomposition of harmonic forms identical with Hodge's theorem for Kähler manifolds.

As an illustration to derive topological consequences let us prove the following theorem: *If a compact manifold of dimension 2m has a U(m)-connection without torsion, then its odd-dimensional Betti numbers are even.*

On the complex-valued differential forms η of degree r we define the operator

$$(83)\qquad C=\sum_{p+q=r} i^{p-q}P_{p,q},$$

where $P_{p,q}\eta$ is the bihomogeneous component of η of bidegree (p,q). If η is harmonic, $P_{p,q}\eta$ is harmonic, and the same is true of $C\eta$. Moreover, this operator C has the following properties: (1) $C^2=(-1)^r$; (2) C is a real operator, that is, it maps a real form into a real form. If r is odd, C defines a linear mapping on the space of real harmonic forms of degree r, such that $C^2=-1$. It follows that the dimension of this vector space, that is, the r-dimensional Betti number, must be even.

5. Riemann manifolds with a field of parallel oriented linear spaces

We first prove the theorem:

Let $G=R(s)\times R(n-s)\subset R(n)$ be the product of two rotation groups in s and $n-s$ variables respectively, $0<s<n$. The existence, on a manifold of dimension n, of a G-connection without torsion is equivalent to that

of a Riemann metric and a continuous field of oriented s-dimensional linear spaces which are parallel with respect to the Riemann metric.

We adopt in this section the following ranges of indices:

(84) $$1 \leqq \alpha, \beta, \gamma \leqq s, \quad s+1 \leqq a, b, c \leqq n.$$

Following the general method in § 2, we define, in the corresponding principal bundle, a uniquely determined set of forms $\omega_\alpha, \omega_a, \pi_{\alpha\beta} = -\pi_{\beta\alpha}$, $\pi_{ab} = -\pi_{ba}, \omega_{\alpha a}$, such that $\omega_\alpha, \omega_a, \pi_{\beta\alpha}, \pi_{ba}$ are linearly independent and

(85) $$\omega_{\alpha a} = \sum_\beta A_{\alpha a\beta}\omega_\beta + \sum_b B_{\alpha ab}\omega_b,$$

which satisfy the conditions

(86) $$\begin{cases} d\omega_\alpha = \sum_\beta \omega_\beta \wedge \pi_{\beta\alpha} - \sum_a \omega_a \wedge \omega_{\alpha a}, \\ d\omega_a = \sum_\alpha \omega_\alpha \wedge \omega_{\alpha a} + \sum_b \omega_b \wedge \pi_{ba}. \end{cases}$$

The forms $\pi_{\beta\alpha}, \pi_{ba}$ define a G-connection in the bundle. It is without torsion, if and only if

(87) $$\omega_{\alpha a} = 0.$$

Let e_α, e_a be tangent vectors which are dual to the covectors ω_β, ω_b. The s-dimensional linear space spanned by e_α defines a parallel field, if and only if (87) is fulfilled. This proves our theorem.

Suppose from now on that we have a compact manifold with such a G-structure without torsion. A differential form of degree r can be written in the form

(88) $$\eta = \sum_{p+q=r} \sum_{\alpha, a} P_{\alpha_1 \dots \alpha_p a_1 \dots a_q} \omega_{\alpha_1} \wedge \dots \wedge \omega_{\alpha_p} \wedge \omega_{a_1} \wedge \dots \wedge \omega_{a_q},$$

where the coefficients $P_{\alpha_1 \dots \alpha_p a_1 \dots a_q}$ are supposed to be anti-symmetric in the α's and in the a's. For fixed p, q we define the operator

(89) $$P_{p,q}\eta = \sum_{\alpha, a} P_{\alpha_1 \dots \alpha_p a_1 \dots a_q} \omega_{\alpha_1} \wedge \dots \wedge \omega_{\alpha_p} \wedge \omega_{a_1} \wedge \dots \wedge \omega_{a_q}.$$

Then we have

(90) $$\eta = \sum_{p+q=r} P_{p,q}\eta.$$

Each of these summands is said to be bihomogeneous of bidegree p, q. According to our decomposition theorem, $P_{p,q}$ commutes with the operator Δ. It follows that if η is harmonic, then each of the summands in (90) is harmonic. Let $B^{p,q}$ be the number of linearly independent bihomogeneous harmonic forms of bidegree p, q. Then the r-dimensional Betti number of the manifold is given by

(91) $$B^r = \sum_{p+q=r} B^{p,q}.$$

It can be shown that each of the summands in (90) corresponds to an irreducible representation of G. In this sense the decomposition of a harmonic form η given by (90) cannot be improved.

Let

$$\Omega_1 = \omega_1 \wedge \ldots \wedge \omega_s, \tag{92}$$

and let Φ be a bihomogeneous differential form of bidegree $0, q$. We wish to prove the following lemma:

The differential form Φ is harmonic, if and only if $\Omega_1 \wedge \Phi$ is harmonic.

To prove this lemma we remark that the exterior derivative of a bihomogeneous differential form η of bidegree (p, q) is a sum of two bihomogeneous differential forms, of bidegrees $(p+1, q)$ and $(p, q+1)$ respectively. We call them $d_1 \eta$ and $d_2 \eta$, so that

$$d = d_1 + d_2. \tag{93}$$

We write

$$\Phi = \frac{1}{q!} \Sigma_{a_1, \ldots, a_q} P_{a_1 \ldots a_q} \omega_{a_1} \wedge \ldots \wedge \omega_{a_q}, \tag{94}$$

where the coefficients $P_{a_1 \ldots a_q}$ are supposed to be anti-symmetric in their indices. We define an operator $*_2$ by

$$*_2 \Phi = \frac{1}{q!\,(n-s-q)!} \Sigma_{a_1, \ldots, a_{n-s}} \epsilon_{a_1 \ldots a_{n-s}} P_{a_1 \ldots a_q} \omega_{a_{q+1}} \wedge \ldots \wedge \omega_{a_s}. \tag{95}$$

Then we have

$$\begin{cases} * \Phi = \pm \Omega_1 \wedge *_2 \Phi, \\ *(\Omega_1 \wedge \Phi) = \pm *_2 \Phi. \end{cases} \tag{96}$$

It follows immediately from definition that, for a form of type $(0, q)$, the conditions $d_1 \Phi = 0$ and $d_1(*_2 \Phi) = 0$ are equivalent.

To prove our lemma, suppose Φ be harmonic:

$$d\Phi = 0, \quad d * \Phi = 0.$$

From the first equation follow

$$d_1 \Phi = 0, \quad d_1(*_2 \Phi) = 0.$$

From the second equation we get

$$d(\Omega_1 \wedge *_2 \Phi) = 0 \quad \text{or} \quad d_2(*_2 \Phi) = 0.$$

It follows that

$$d(\Omega_1 \wedge \Phi) = 0,$$

$$d * (\Omega_1 \wedge \Phi) = \pm d(*_2 \Phi) = \pm d_1(*_2 \Phi) \pm d_2(*_2 \Phi) = 0.$$

Hence $\Omega_1 \wedge \Phi$ is harmonic.

Conversely, suppose $\Omega_1 \wedge \Phi$ be harmonic:

$$d(\Omega_1 \wedge \Phi) = 0, \quad d * (\Omega_1 \wedge \Phi) = 0.$$

From the first equation we get

$$d_2 \Phi = 0.$$

From the second equation we get

$$d(*_2 \Phi) = 0,$$

which gives $$d_1(*_2 \Phi) = 0, \quad d_1 \Phi = 0.$$

It follows that

$$d\Phi = 0, \quad d * \Phi = \pm d(\Omega_1 \wedge *_2 \Phi) = 0.$$

Therefore Φ is harmonic.

From our lemma we get the following equalities:

$$B^{0,q} = B^{s,q}, \quad B^{p,0} = B^{p,n-s}. \tag{97}$$

In particular, we have

$$B^s \geqq B^{s,0} = B^{0,0} = 1, \tag{98}$$

which is the result stated at the end of the Introduction.

UNIVERSITY OF CHICAGO

REFERENCES

[1] S. CHERN, *Characteristic classes of Hermitian manifolds*, Ann. of Math., 47 (1946), pp. 85–121.

[2] ——, *Topics in differential geometry*, mimeographed notes, Princeton, 1951.

[3] P. R. GARABEDIAN and D. C. SPENCER, *A complex tensor calculus for Kähler manifolds*, Acta Math., 89 (1953), pp. 279–331.

[4] H. GUGGENHEIMER, *Über komplex-analytische Mannigfaltigkeiten mit Kählerscher Metrik*, Comment. Math. Helv., 25 (1951), pp. 257–297.

[5] W. V. D. HODGE, The theory and application of harmonic integrals, Cambridge University Press, 1941.

[6] S. LEFSCHETZ, L'analysis situs et la géometrie algébrique, Paris, Gauthier-Villars, 1950.

[7] A. LICHNEROWICZ, Généralisations de la géométrie kählerienne globale, Colloque de géométrie différentielle, Louvain, 1951, pp. 99–122.

[8] A. WEIL, *Sur la théorie des formes différentielles attachées à une variété analytique complexe*, Comment. Math. Helv., 20 (1947), pp. 110–116.

[9] ——, Theorie der Kählerschen Mannigfaltigkeiten, Göttingen, 1953.

[10] H. WEYL, The classical groups, Princeton, 1939.

ON THE TOTAL CURVATURE OF IMMERSED MANIFOLDS.*

By SHIING-SHEN CHERN [1] and RICHARD K. LASHOF.

Introduction. In the classical theory of surfaces in the ordinary Euclidean space E an important rôle is played by the normal mapping of Gauss: Let M be an oriented surface which has at every point x a well-defined unit normal vector $\nu(x)$. Then the normal mapping $\nu: M \to S_0$ is the mapping of M into the unit sphere S_0 about the origin of E, which sends x to $\nu(x)$.

For differentiably immersed submanifolds in an Euclidean space of higher dimension the following is a generalization of the Gauss mapping: We consider a C^∞-manifold M^n of dimension n, and a C^∞-mapping $x: M^n \to E^{n+N}$ into the Euclidean spase E^{n+N} of dimension $n+N$ $(N \geqq 1)$. M^n, or rather M^n and the mapping x, is called an immersed submanifold, if the induced mapping of the tangent space is univalent everywhere, or, what is the same, if the Jacobian matrix of x is everywhere of rank n. The submanifold M^n is said to be imbedded, if x is one-one; that is, if $x(p) = x(q)$, $p, q \in M^n$, implies that $p = q$. Let B_ν be the bundle of unit normal vectors of $x(M^n)$, so that a point of B_ν is a pair $(p, \nu(p))$, where $\nu(p)$ is a unit normal vector to $x(M^n)$ at $x(p)$. Then B_ν is a bundle of $(N-1)$ dimensional spheres over M^n and is a C^∞-manifold of dimension $n+N-1$. The mapping $\tilde{\nu}: B_\nu \to S_0^{n+N-1}$ of B_ν into the unit sphere S_0^{n+N-1} of E^{n+N} defined by $\tilde{\nu}(p, \nu(p)) = \nu(p)$ is the mapping with which we will be concerned in this paper.

Let dV be the volume element of M^n. There is a differential form $d\sigma_{N-1}$ of degree $N-1$ on B_ν such that its restriction to a fiber is the volume element of the sphere of unit normal vectors at a point $p \in M^n$; then $d\sigma_{N-1} \wedge dV$ is the volume element of B_ν (for detail see Section 2). Let $d\Sigma_{n+N-1}$ be the volume element of S_0^{n+N-1}. The function $G(p, \nu(p)) = G(p, \nu)$ defined by

$$\tilde{\nu}^* d\Sigma_{n+N-1} = G(p, \nu)\, d\sigma_{N-1} \wedge dV, \tag{1}$$

where $\tilde{\nu}^*$ is the dual mapping on differential forms induced by $\tilde{\nu}$, is a function in B_ν. It generalizes the Gauss-Kronecker curvature and we will call

* Received January 4, 1957.

[1] Work done when the first-named author is under partial support from a contract with the National Science Foundation.

it the Lipschitz-Killing curvature at $\nu(p)$. $G(p,\nu)$ has a geometrical interpretation which we will discuss below. It is zero at a point $(p,\nu(p)) \in B_\nu$, if and only if $\tilde{\nu}$ has a critical value at this point. We call the integral

$$K^*(p) = \int |G(p,\nu)|\, d\sigma_{N-1} \geqq 0 \tag{2}$$

over the sphere of unit normal vectors at $x(p)$ the *total curvature of* M^n *at* p, and define as the *total curvature of* M^n itself the integral $\int_{M^n} K^*(p)\,dV$, if it converges.

The main results of this paper are concerned with the conclusions on M^n when its total curvature is "small." They can be stated as the following theorems:

THEOREM 1. *Let* M^n *be a compact oriented* C^∞*-manifold immersed in* E^{n+N}. *Its total curvature satisfies the inequality:*[2]

$$\int_{M^n} K^*(p)\,dV \geqq 2c_{n+N-1} \tag{3}$$

THEOREM 2. *Under the hypothesis of Theorem* 1, *if*

$$\int_{M^n} K^*(p)\,dV < 3c_{n+N-1}, \tag{4}$$

then M^n *is homeomorphic to a sphere of* n*-dimensions.*

THEOREM 3. *Under the same hypothesis, if*

$$\int_{M^n} K^*(p)\,dV = 2c_{n+N-1}, \tag{5}$$

then M^n *belongs to a linear subvariety* E^{n+1} *of dimension* $n+1$, *and is imbedded as a convex hypersurface in* E^{n+1}. *The converse of this is also true.*

These theorems generalize known results of Fenchel, Fary [2], and Milnor [3] for curves and sharpen some results of Milnor and one of us [1]. Theorem 3 can be interpreted as a characterization of convex hypersurfaces among all immersed submanifolds of a given dimension in an Euclidean space of arbitrary dimension. A large part of our paper is devoted to a proof of this theorem.

1. Moving frames. Suppose E^{n+N} be oriented. By a frame $xe_1 \cdots e_{n+N}$ in E^{n+N} we mean a point x and an ordered set of mutually perpendicular

[2] c_{n+N-1} is the area of the unit hypersphere in an Euclidean space of dimension $n+N$.

unit vectors $e_1, \cdots, e_{n+N}$, such that their orientation is coherent with that of E^{n+N}. Unless otherwise stated, we agree on the following ranges of the indices:

(6) $$1 \leqq i, j, k \leqq n, \quad n+1 \leqq r, s, t \leqq n+N, \quad 1 \leqq A, B, C \leqq n+N.$$

Then we have

(7) $$e_A e_B = \delta_{AB},$$

where the left-hand side is the scalar product of vectors. Let $F(n, N)$ be the space of all frames in E^{n+N}, so that $\dim F(n, N) = (n+N)(n+N+1)/2$. In $F(n, N)$ we introduce the linear differential forms ω'_A, ω'_{AB} by the equations

(8) $$de_A = \sum_B \omega'_{AB} e_B, \qquad dx = \sum_A \omega'_A e_A,$$

where

(9) $$\omega'_{AB} + \omega'_{BA} = 0.$$

Their exterior derivatives satisfy the equation of structure:

(10) $$d\omega'_A = \sum_B \omega'_B \wedge \omega'_{BA}, \qquad d\omega'_{AB} = \sum_C \omega'_{AC} \wedge \omega'_{CB}.$$

As explained in the Introduction, we mean by an immersed submanifold in E^{n+N} an abstract C^∞-manifold M^n and a C^∞-mapping $x\colon M^n \to E^{n+N}$, such that the induced mapping x_* on the tangent space is everywhere univalent. Analytically, the mapping can be defined by a vector-valued function $x(p)$, $p \in M^n$. Our assumption implies that the differential $dx(p)$ of $x(p)$, which is a linear differential form in M^n, with value in E^{n+N}, has as values a linear combination of n vectors, $t_1, \cdots, t_n$, and not less. Since x_* is univalent, we can identify the tangent space of M^n at p with the vector space spanned by $t_1, \cdots, t_n$. A linear combination of the latter is called a tangent vector and a vector perpendicular to them is called a normal vector. The immersion of M^n in E^{n+N} gives rise to the following fiber bundles over M^n:

1) The tangent bundle B_T, whose bundle space is the subset of $M^n \times E^{n+N}$, consisting of all points (p, v), such that $p \in M$ and v is a unit tangent vector at $x(p)$.

2) The normal bundle B_ν, whose bundle space is the subset of $M^n \times E^{n+N}$, consisting of all points (p, v), such that $p \in M$ and v is a unit normal vector at $x(p)$.

3) The bundle B, whose bundle space is the subset of $M^n \times F(n, N)$, consisting of $(p, x(p) e_1 \cdots e_n e_{n+1} \cdots e_{n+N}) \in M^n \times F(n, N)$ such that $e_1, \cdots, e_n$

are tangent vectors and $e_{n+1}, \cdots, e_{n+N}$ are normal vectors at $x(p)$. The projection $B \to M^n$ we denote by ψ.

The last bundle B is the space in which most of our computations take place.[3] We define the mappings

$$\psi_T : B \to B_T, \qquad \psi_\nu : B \to B_\nu, \tag{11}$$

by

$$\psi_T(p, x(p)e_1 \cdots e_{n+N}) = (p, e_n), \qquad \psi_\nu(p, x(p)e_1 \cdots e_{n+N}) = (p, e_{n+N}). \tag{12}$$

We also remark that the Whitney sum $B_T \oplus B_\nu$ over M^n is equivalent to the product bundle $M^n \times S^{n+N-1} \to M^n$.

Consider the mappings

$$B \xrightarrow{i} M^n \times F(n,N) \xrightarrow{\lambda} F(n,N), \tag{13}$$

where i is inclusion and λ is the projection into the second factor. Put

$$\omega_A = (\lambda i)^* \omega'_A, \qquad \omega_{AB} = (\lambda i)^* \omega'_{AB}, \tag{14}$$

Then we have, from (9) and (10),

$$\omega_{AB} + \omega_{BA} = 0, \tag{15}$$

$$d\omega_A = \sum_B \omega_B \wedge \omega_{BA}, \qquad d\omega_{AB} = \sum_C \omega_{AC} \wedge \omega_{CB}. \tag{16}$$

From our definition of B it follows that $\omega_r = 0$ and that ω_i are linearly independent. Hence the first equation of (16) gives

$$\sum_i \omega_i \wedge \omega_{ir} = 0.$$

From this it follows that

$$\omega_{ir} = \sum_j A_{rij} \omega_j, \qquad A_{rij} = A_{rji}. \tag{17}$$

2. The total curvature. An essential idea of the method of moving frames is to consider the bundle space B, and to construct differential forms

---

[3] The bundle B can also be defined as follows: Let Q_{n+N} be the group of all orientation-preserving motions in E^{n+N}, and $R_n \times R_N$ the subgroup of Q_{n+N} consisting of all motions which leave fixed the origin 0 of E^{n+N} and a given oriented linear space of dimension n through 0. Then $Q_{n+N} \to Q_{n+N}/R_n \times R_N$ is a bundle whose structure group is $R_n \times R_N$ and whose base space is the space of all elements consisting of a point and an oriented linear space of dimension n through it. The bundle B is induced by the mapping which sends $p \in M$ to the point $x(p)$ and the oriented tangent space at $x(p)$. In particular, the structural group of B is $R_n \times R_N$. Similarly, we can define the bundles B_T and B_ν.

in B, which are inverse images of differential forms in M^n and B_ν, under the mappings ψ and ψ_ν respectively. In particular, the volume element of M^n can be written

$$dV = \omega_1 \wedge \cdots \wedge \omega_n, \tag{18}$$

and the volume element of B_ν is

$$dV \wedge d\sigma_{N-1} = \omega_1 \wedge \cdots \wedge \omega_{n+N,n+1} \wedge \cdots \wedge \omega_{n+N,n+N-1}, \tag{19}$$

$d\sigma_{N-1}$ being equal to the product of the last $N-1$ factors. On the other hand, we have

$$\tilde{\nu}^* d\Sigma_{n+N-1} = \omega_{n+N,1} \wedge \cdots \wedge \omega_{n+N,n} \wedge \omega_{n+N,n+1} \wedge \cdots \wedge \omega_{n+N,n+N-1}.$$

Using (17), we get

$$\tilde{\nu}^* d\Sigma_{n+N-1} = (-1)^n \det(A_{n+N,ij})\omega_1 \wedge \cdots \wedge \omega_n \wedge \omega_{n+N,n+1} \wedge \cdots \wedge \omega_{n+N,n+N-1}.$$

It follows from the definition (1) of the Lipschitz-Killing curvature $G(p,\nu)$ that

$$G(p,\nu) = (-1)^n \det(A_{n+N,ij}). \tag{20}$$

To see how $G(p,\nu)$ depends on ν, we take a local cross-section of M^n in B, described by the functions $\tilde{e}_A(q)$ for q in a neighborhood of p. Then for any frame $e_A(q)$ in B at $x(q)$, we have $e_A = \sum c_{AB}\tilde{e}_B(q)$ and

$$\sum_{i,j} A_{sij}\omega_i\omega_j = \sum_{r,i,j} c_{sr}\tilde{A}_{rij}\tilde{\omega}_i\tilde{\omega}_j,$$

where $\tilde{A}_{rij}$ is the function A_{rij} restricted to the local cross-section. In particular, if for $\nu = e_{n+N}$ we write $\nu = \sum \nu_r \tilde{e}_r$ we have at p:

$$G(p,\nu) = (-1)^n \det(\sum_r \nu_r \tilde{A}_{rij}(p)). \tag{21}$$

Further we get for the scalar product of ν and the vector-valued second differential d^2x on B

$$\nu \cdot d^2x = \sum_{i,r} \nu_r \tilde{\omega}_{ir}\tilde{\omega}_i = \sum_{r,i,j} \nu_r \tilde{A}_{rij}\tilde{\omega}_i\tilde{\omega}_j \tag{22}$$

and hence as forms on B_ν we have:

$$-d\nu dx = \nu \cdot d^2x = \sum_{r,i,j} \nu_r \tilde{A}_{rij}\tilde{\omega}_i\tilde{\omega}_j. \tag{23}$$

Therefore we may interpret $G(p,\nu)$ as generalizing the determinant of the second fundamental form of a surface.

If $N=1$, i.e., if M^n is an immersed oriented hypersurface in E^{n+1}, its orientation (and that of E^{n+1}) defines a unit normal vector $\nu_0(p)$ at $p \in M^n$. Then $G(p, \nu_0(p)) = G(p)$ is called the Gauss-Kronecker curvature of M at p. Any other unit normal vector at p is of the form $\nu(p) = \pm \nu_0(p)$, and

$$G(p, \nu(p)) = G(p, \pm \nu_0(p)) = (\pm 1)^n G(p).$$

It follows that, for n even, $G(p, \nu(p))$ is independent of the orientation of the hypersurface M^n and of the space E^{n+1}. Naturally, $G(p)$ reduces for $n = 2$ to the classical Gaussian curvature.

In the general case $G(p, \nu)$ admits the following geometrical interpretation in terms of the Gauss-Kronecker curvature: Let $L(\nu)$ be the linear space of dimension $n+1$ spanned by the tangent space to $x(M^n)$ at $x(p)$ and the normal vector $\nu(p)$. Then $G(p, \nu)$ is equal to the Gauss-Kronecker curvature at p of the orthogonal projection of $x(M^n)$ into $L(\nu)$.

Since the theorem is local, we take a local cross-section $\tilde{e}_A(q)$ of M^n in B in a neighborhood of p, such that $\nu(p) = \tilde{e}_{n+N}(p)$. We write

$$\tilde{e}_A(p) = (\tilde{e}_A)_0, \qquad x(p) = x_0.$$

If $x'(q)$ denotes the position vector of the projection of $x(q)$ in $L(\nu)$, $x'(q)$ is defined by the equations

$$x'(q) - x(q) = \xi_{n+1}(\tilde{e}_{n+1})_0 + \cdots + \xi_{n+N-1}(\tilde{e}_{n+N-1})_0,$$

$$x'(q) - x_0 \equiv 0, \qquad \operatorname{mod}(\tilde{e}_1)_0, \cdots, (\tilde{e}_n)_0, (\tilde{e}_{n+N})_0.$$

From this it follows that

$$\xi_{n+\lambda} = (x'(q) - x(q)) \cdot (\tilde{e}_{n+\lambda})_0 = (x_0 - x(q)) \cdot (\tilde{e}_{n+\lambda})_0 \quad 1 \leqq \lambda \leqq N-1.$$

If p is fixed and q varies on the manifold M^n, we have

$$dx' = dx + \sum_\lambda d\xi_{n+\lambda}(\tilde{e}_{n+\lambda})_0 = dx - \sum_\lambda (dx \cdot (\tilde{e}_{n+\lambda})_0)(\tilde{e}_{n+\lambda})_0,$$

$$d^2x' = d^2x - \sum_\lambda (d^2x \cdot (\tilde{e}_{n+\lambda})_0)(\tilde{e}_{n+\lambda})_0,$$

so that $(\tilde{e}_{n+N})_0 d^2x' = (\tilde{e}_{n+N})_0 d^2x$. This proves our statement.

3. Proof of Theorem 1. For this proof we will need the following:

THEOREM (Sard)[4] *Let V and W be two C^1-manifolds of the same dimension and f a mapping of class C^1 of V into W. The image $f(E)$ of the set E of critical points of f is a set of measure zero in W.*

[4] Cf. [6], p. 10.

We consider the map $\bar{\nu}: B_\nu \to S_0^{n+N-1}$. Every point of S_0^{n+N-1} is covered at least twice by $\bar{\nu}$. In fact, for a fixed unit vector ν_0, the scalar product $\nu_0 \cdot x(p)$ as a continuous function on M^n has at least one maximum and one minimum, at which $\pm \nu_0 dx(p) = 0$. If, on the other hand, the maximum and minimum points are the same then M^n is contained in a hyperplane perpendicular to ν_0, and every point of M^n has ν_0 as a normal vector.

The set of critical points of $\bar{\nu}$ is the set E of points in B_ν such that $G(p, \nu) = 0$. Hence

$$\int_{M^n} K^*(p)\,dV = \int_{B_\nu} |\,G(p,\nu)\,|\,dV d\sigma_{N-1}$$

is the volume of the image in S_0^{n+N-1} of the set of non-critical points of B_ν. By Sard's theorem and by the above remark that every point of S_0^{n+N-1} is covered at least twice by $\bar{\nu}$, we have immediately

$$\int_{M^n} K^*(p)\,dV \geqq 2c_{n+N-1}.$$

4. Proof of Theorem 2. Suppose that $\int_{M^n} K^*(p)\,dV < 3c_{n+N-1}$; we wish to show that M^n is homeomorphic to a sphere of n dimensions. Our hypothesis implies that there exists a set of positive measure on S_0^{n+N-1} such that if ν_0 is a unit vector in this set, $\nu_0 \cdot x(p)$ has just two critical points. For if not; every point of S_0^{n+N-1}, except for a set of measure zero, would be covered at least three times by ν and hence as in Section 3 we would have

$$\int_{M^n} K^*(p)\,dV \geqq 3c_{n+N-1}.$$

Since, by Sard's theorem, the image of the set of critical points under $\bar{\nu}$ is of measure zero, there is a unit vector ν_0 such that $\nu_0 \cdot x(p)$ has exactly two critical points on M^n and such that ν_0 is the image of non-critical points of B_ν under $\bar{\nu}$. The latter means that $G(p, \nu_0) \neq 0$ at each critical point $p \in M$ of the function $\nu_0 \cdot x(p)$, which is equivalent to saying that $\nu_0 \cdot d^2x$ is a quadratic differential form of determinant not zero. In other words, the function $\nu_0 \cdot x(p)$ on M^n has exactly two non-degenerate critical points. Now a theorem of Reeb [5] (see also [4], p. 401) asserts that if a compact differentiable manifold M has a real-valued differentiable function on it with only two critical points which are non-degenerate, then M is homeomorphic to a sphere. It follows from this that M^n is homeomorphic to a sphere, and our theorem is proved.

5. Proof of Theorem 3. We begin with two lemmas:

LEMMA 1. *Under the hypothesis of Theorem 3, M^n is immersed in a linear subvariety of dimension $n+1$ of E^{n+N}.*

We may assume $N \geqq 2$, since otherwise there is nothing to prove. We will first show that M is contained in a hyperplane of E^{n+N}. In doing this, we show that the hypothesis of Theorem 3 and the assumption that M^n does not lie in a hyperplane of E^{n+N} lead to a contradiction.

Let $(p, \nu_0(p)) \in B_\nu$ be such that $G(p, \nu_0) \neq 0$; such a point exists by Theorem 1. Choose a local cross-section of M^n in B; i. e., the vectors $\tilde{e}_A(q)$, where q belongs to a neighborhood of p, and that $\tilde{e}_{n+N}(p) = \nu_0$. Then any other unit normal vector at p may be written $\nu = \sum_r \nu_r \tilde{e}_r(p)$ and by equation (21), Section 3, we have $G(p, \nu) = (-1)^n \det (\sum_r \nu_r \tilde{A}_{rij}(p))$. Holding p fixed and restricting ourselves to normal vectors $\nu(\theta)$ such that

$$\nu_{n+N} = \cos\theta, \quad \nu_{n+N-1} = \sin\theta, \qquad \nu_r = 0, r \neq n+N-1, n+N,$$

we have $G(p, \nu) = f(\theta)$, where $f(\theta)$ is a polynomial in $\cos\theta$ and $\sin\theta$ and is hence an analytic function of θ. $f(\theta)$ is not identically zero, since $f(0) = G(p, \nu_0) \neq 0$.

Let H_θ be the tangent hyperplane at $x(p)$ perpendicular to $\nu = \nu(\theta)$. Since $x(M^n)$ does not belong to a hyperplane, there exist tangent hyperplanes H_{θ_1}, H_{θ_2}, $\theta_1 < \theta_2$, and points $q_1, q_2 \in M$, such that $x(q_1) \subset H_{\theta_1}$, $x(q_2) \in H_{\theta_2}$, $x(q_1) \notin H_{\theta_2}$, $x(q_2) \notin H_{\theta_1}$. Since $f(\theta)$ does not vanish identically, there is a tangent hyperplane H_{θ_3}, such that $f(\theta_3) \neq 0$ and that $x(q_1)$ and $x(q_2)$ lie on different sides of the tangent hyperplane H_{θ_3}. The condition $f(\theta_3) \neq 0$ implies that the mapping $\tilde{\nu}$ is one-one in a neighborhood W of $(p, \nu(\theta_3))$ of B_ν. We can choose W so small that for $(q', \nu') \in W$, $x(q_1)$ and $x(q_2)$ lie on different sides of the tangent hyperplane perpendicular to ν' at $x(q')$. The function $\nu' \cdot x$ in M^n has at least three point at which $\nu' \cdot dx = 0$; namely, the maximum, the minimum, and the point q'. The last point q' is distinct from the maximum and the minimum, since, by our construction, there are points of $x(M^n)$ on different sides of the tangent hyperplane at q' perpendicular to ν'. It follows that a neighborhood of S_0^{n+N-1} is covered by the image of $\tilde{\nu}$ at least three times. As every point of S_0^{n+N-1} is covered at least twice, we conclude that the total curvature of $x(M^n)$ is strictly greater than $2c_{n+N-1}$. But this is a contradiction.

It follows from this contradiction that $x(M^n)$ belongs to a hyperplane E^{n+N-1} of E^{n+N}. We wish to show that its total curvature in E^{n+N-1} is equal to $2c_{n+N-2}$. We denote by v the unit vector perpendicular to E^{n+N-1} and by S_0^{n+N-2} the unit sphere in E^{n+N-1}. The sphere S_0^{n+N-2} can be imagined as the equator of S^{n+N-1} with v as the north pole. Let B_ν' be the bundle of unit normal vectors of $x(M^n)$ in E^{n+N-1}. Then $B_\nu' \subset B_\nu$ and $\nu(B_\nu') \subset S_0^{n+N-2}$. Denote by $\tilde{\nu}'$ the restriction of $\tilde{\nu}$ to B_ν'. It suffices to prove that, with the exception of a set of measure zero on S_0^{n+N-2}, the points of S_0^{n+N-2} are covered by $\tilde{\nu}'$ exactly twice.

Suppose the contrary be true. There is thus a set A of positive measure on S_0^{n+N-2}, whose points are covered by $\tilde{\nu}'$ more than twice. To any $\mu \in A$ there are distinct points $p_1, \cdots, p_k \in M^n$, $k \geqq 3$, for which $x(p_1), \cdots, x(p_k)$ have normal vectors parallel to μ. All the unit vectors in the great circle spanned by μ and v are then normal to $x(M^n)$ at $x(p_1), \cdots, x(p_k)$. It follows that all the points of S_0^{n+N-1} belonging to great circles spanned by v and the points of A are covered by $\tilde{\nu}$ more than twice. Since A has positive measure on S_0^{n+N-2}, this set has positive measure on S_0^{n+N-1}, which is a contradiction.

Therefore $x(M^n)$ belongs to E^{n+N-1} and has, as a submanifold of E^{n+N-1}, a total curvature equal to $2c_{n+N-2}$. By induction on N we see that $x(M^n)$ must belong to a linear subvariety E^{n+1} of dimension $n+1$ and has in E^{n+1} the total curvature $2c_n$.

LEMMA 2. *Let $x: M^n \to E^{n+1}$ be an immersion of a compact oriented manifold and let $\nu: M^n \to S_0^n$ be the normal map of Gauss. Let $J(p)$ be the Jacobian matrix of ν at p, and let $U_m = \{p \in M^n \mid \operatorname{rank} J(p) = n-m\}$. Then, if U_m contains an open set V, its image under x is generated by m-dimensional planes. Every boundary point of U_m, which is at the same time a limit point of an m-dimensional generating plane, belongs to U_m.*

The fact that the image under x of V is generated by m-dimensional planes is a classical result; we include a proof for completeness. At any interior point p of U_m the assumption on the rank of J implies that we may choose coordinates on M^n in the neighborhood of p, say, $t_1, \cdots, t_n$, such that if ν is the unit normal vector at p, then $\partial\nu/\partial t_\alpha = 0$, and $\partial\nu/\partial t_a$ are linearly independent. Here, and for the remainder of this section, we make the following convention concerning indices:

$$1 \leqq \alpha, \beta, \gamma \leqq m, \qquad m+1 \leqq a, b, c \leqq n, \qquad 1 \leqq i, j, k \leqq n.$$

We have $\nu \cdot \partial x/\partial t_i = 0$. It follows that

$$\nu \cdot \partial^2 x/\partial t_i \partial t_\alpha = 0,\ \partial\nu/\partial t_\alpha \cdot \partial x/\partial t_i + \nu(\partial^2 x/\partial t_i \partial t_\alpha) = 0,\ \partial\nu/\partial t_a \cdot \partial x/\partial t_\alpha = 0.$$

Hence $\partial x/\partial t_\alpha$ are vectors orthogonal to the $n-m+1$ linearly independent vectors ν, $\partial/\partial t_a$. The surfaces $t_a = t_a^0 = \text{const.}$ are therefore m-dimensional planes in E^{n+1}. Since $\partial\nu/\partial t_a = 0$, the tangent hyperplane remains constant along an m-dimensional generating plane.

Consider now the sub-bundle B' of B consisting of all those frames $e_1, \cdots, e_{n+1}$ such that e_a are in the m-dimensional generating planes. Then, as in Section 1, we have

$$dx = \sum_i \omega_i e_i, \quad de_i = \sum_j \omega_{ij} e_j + \omega_{i,n+1} e_{n+1}, \quad de_{n+1} = \sum_j \omega_{n+1,j} e_j,$$

where

$$\omega_{i,n+1} = \sum_j A_{ij}\omega_j, \qquad A_{ij} = A_{ji}.$$

The above assumption on the bundle of frames B' is equivalent to assuming

$$\omega_{a,n+1} = -\omega_{n+1,a} = 0; \tag{24}$$

i. e., that the matrix (A_{ij}) takes the form

$$\begin{pmatrix} 0 & 0 \\ 0 & A_{ab} \end{pmatrix}, \qquad \det(A_{ab}) = D \neq 0,$$

which is an $(n \times n)$-matrix whose elements are zero, except possibly those of the $(n-m)\times(n-m)$ block in the lower right-hand corner. Our proof depends on studying the behavior of D along the m-dimensional generating plane.

From (24) we have

$$0 = d\omega_{a,n+1} = \sum_k \omega_{ak} \wedge \omega_{k,n+1} = \sum_\beta \omega_{a\beta} \wedge \omega_{\beta,n+1} + \sum_a \omega_{aa} \wedge \omega_{a,n+1}.$$

Our assumption implies

$$\omega_{\beta,n+1} = 0, \qquad \omega_{a,n+1} = \sum_b A_{ab}\omega_b,$$

so that $\sum_{a,b} A_{ab}\omega_{aa} \wedge \omega_b = 0$, or $\sum_a A_{ab}\omega_{aa} \wedge \prod_c \omega_c = 0$. Since $\det(A_{ab}) \neq 0$, we get $\omega_{aa} \wedge \prod_c \omega_c = 0$. Hence we can put

$$\omega_{aa} = \sum_b h_{aab}\omega_b. \tag{25}$$

Now we have $\prod_a \omega_{a,n+1} = D \prod_c \omega_c$. Exterior differentiation of this equation gives

$$\sum_a (-1)^{a-m-1}\omega_{m+1,n+1} \wedge \cdots \wedge d\omega_{a,n+1} \wedge \cdots \wedge \omega_{n,n+1}$$
$$= dD \wedge \prod_c \omega_c + D\Big(\sum_a (-1)^{a-m-1}\omega_{m+1} \wedge \cdots \wedge d\omega_a \wedge \cdots \wedge \omega_n\Big).$$

But

$$d\omega_{a,n+1} = \sum_k \omega_{ak} \wedge \omega_{k,n+1} = \sum_b \omega_{ab} \wedge \omega_{b,n+1},$$

$$d\omega_a = \sum_k \omega_k \wedge \omega_{ka} = \sum_{a,b} h_{aab}\omega_a \wedge \omega_b + \sum_b \omega_b \wedge \omega_{ba}.$$

Hence $0 = dD \wedge \prod_c \omega_c + D(\sum_{a,\alpha} h_{a\alpha\alpha}\omega_a \wedge \prod_c \omega_c)$, or

$$dD + D(\sum_{a,\alpha} h_{a\alpha\alpha}\omega_a) \equiv 0, \quad \text{mod } \omega_c. \tag{26}$$

To complete the proof let $p \in M^n$ be a boundary point of U_m, such that $x(p)$ is a limit point of a generating m-dimensional plane L. We choose a neighborhood V of p, in which x is one-one, and we suppose that $x^{-1}(L) \subset V$. Let $\tilde{e}_1(q), \cdots, \tilde{e}_{n+1}(q)$, $q \in V$, be a local cross-section of V into B, such that, for $q \in x^{-1}(L)$, $\tilde{e}_1(q), \cdots, \tilde{e}_m(q)$ span L. Such a cross-section clearly exists. If $\tilde{\omega}_i$, $\tilde{\omega}_j$ are the restrictions of ω_i, ω_{ij} respectively to this cross-section, then $\tilde{\omega}_i$ are linearly independent and we will have

$$\tilde{\omega}_{a\alpha} = \sum_k h_{\alpha a k}\tilde{\omega}_k.$$

The coefficients $\tilde{h}_{\alpha a b}(q)$ are equal to the functions $h_{\alpha a b}(q)$ introduced above, for $q \in x^{-1}(L)$. Let γ be a curve in $x^{-1}(L)$ abutting at p. We have, along γ,

$$dD + D(\sum_{a,\alpha} h_{a\alpha\alpha}\omega_a) = 0.$$

It follows by integration of this differential equation that

$$D(q) = D_0 \exp(-\int \sum h_{a\alpha\alpha}\omega_a),$$

for $q \in \gamma - p_0$, where $D_0 \neq 0$ is the value of D at a fixed point of γ. Since $D(q)$ is a continuous function and since $h_{a\alpha\alpha}$ is bounded, we conclude that $D(p) \neq 0$. This completes the proof of Lemma 2.

We now complete the proof of Theorem 3 as follows:

Let H be the space of hyperplanes of E^{n+1} (with the obvious topology). A tangent hyperplane of $x(M^n)$ is said to be of rank m, if it is tangent to $x(M^n)$ at a point of $x(U_m)$ and at no point of $x(U_l)$, $l < m$. By the argument used in the proof of Theorem 2, a tangent hyperplane of rank zero does not separate the set $x(M^n)$. For otherwise there would be a neighborhood of S_0^n, whose points are covered at least three times by ν, which would be contrary to the assumption that M^n has the minimum total curvature $2c_n$.

We will show that in every neighborhood (in the space H) of a tangent hyperplane π of $x(M^n)$ there is a tangent hyperplane of rank zero. In fact, let W be such a neighborhood. Suppose $x(p)$, $p \in U_m$, be a point of contact of π. Either there is a neighborhood of p in M which belongs completely to U_m or there are points of U_l, $l < m$, in every neighborhood of p. In both cases we can find a point p_1 such that the tangent hyperplane π_1 at $x(p_1)$ belongs to W and such that p_1 has a neighborhood in M^n which belongs completely to U_l, $l \leqq m$. The image under x of this neighborhood of p_1 is generated by l-dimensional planes and the tangent hyperplane to $x(M^n)$ along the generating l-dimensional plane through $x(p_1)$ is π_1. If $x(p_2)$, $p_2 \in M^n$, is a boundary point of this l-plane, p_2 belongs to U_l by Lemma 2 and is not an interior point of U_l. Hence there exists in every neighborhood about p_2 an open set whose points are in U_k, $k < l$, and which contains a point $p_3 \in U_k$ such that the tangent hyperplane at $x(p_3)$ is in W. Continuing this process, we see that W contains a tangent hyperplane of rank zero of $x(M^n)$.

This means that every neighborhood of π in H contains a tangent hyperplane such that $x(M^n)$ lies on one side. It follows that the same is true for π itself. If v_0 is any unit vector in E_0^{n+1}, $v_0 \cdot x(p)$ has a maximum and a minimum on M^n, which must be distinct, since M^n cannot be immersed in an m-dimensional hyperplane. Hence the intersection of all the half-spaces of E^{n+1} bounded by a tangent hyperplane of $x(M^n)$ and containing points of $x(M^n)$ is a closed convex set with a non-empty interior and with $x(M^n)$ on the boundary.

Since the induced homomorphism x_* of tangent spaces is one-one, x is a local homeomorphism of M into the boundary of the convex set. It follows that $x(M^n)$ is both open and closed on the boundary; thus x maps M^n onto the boundary. But the boundary of the convex set is homeomorphic to a sphere S^n and by the above M^n is a covering space of S^n under the map x. Hence, if $n \geqq 2$, x is a homeomorphism. The same is true for $n = 1$, on account of the fact that the total curvature is $2c_1$.

Conversely, let $x(M^n)$ be a convex hypersurface. It is then locally convex. By reversing the orientation of M^n if necessary, we can suppose that $G(p) \geqq 0$. Then $K^*(p) = 2G(p)$, because there are two unit normal vectors at every point. The degree of ν is

$$1/c_n \int_{M^n} G(p)\,dV = 1.$$

Hence the total curvature of $x(M^n)$ is $2c_n$.

6. A further theorem.

THEOREM 4. *Let $x: M^n \to E^{n+1}$ be an immersion of a closed orientable manifold and $\nu: M^n \to S_0^n$ the normal map. Then the following are equivalent:*

1. $\deg \nu = \pm 1$ *and the Gaussian curvature is of constant sign;*
2. *The total curvature is $2c_n$;*
3. *M^n is imbedded as a convex hypersurface.*

It suffices to prove the implications 1) $\Rightarrow$ 2) $\Rightarrow$ 3) $\Rightarrow$ 1). Since 2) $\Rightarrow$ 3) $\Rightarrow$ 1) are contained in Theorem 3, we only have to prove that 1) implies 2).

For this purpose it is sufficient to show that no set V of positive measure on S_0^n is covered more than once by M^n under ν. Suppose the contrary. By reversing the orientation of M^n if necessary, we can suppose that $\deg \nu = +1$ and that the Gaussian curvature is non-negative. By Sard's theorem, there exists a point $y \in V$ such that for any point in $\nu^{-1}(y)$ the curvature is strictly positive. There can only be a finite number of points in $\nu^{-1}(y)$. For otherwise $\nu^{-1}(y)$ will have a limit point p at which ν is not locally one-one, while, on the other hand, the Jacobian of ν at $p \in \nu^{-1}(y)$, being a non-zero multiple of the Gaussian curvature at p, is different from zero. By our assumption on V the number of points in $\nu^{-1}(y)$ is $\geqq 2$. At each point of $\nu^{-1}(y)$ the Jacobian of ν is strictly positive. It follows that $\deg \nu \geqq 2$, which contradicts our assumption. Hence the theorem is proved.

Remark. We would like to conjecture that for $n \geqq 2$ the condition $\deg \nu = 1$ in 1) can be omitted. In other words, it seems likely that a closed orientable hypersurface (of dimension $\geqq 2$) of non-negative Gauss-Kronecker curvature is convex. If the curvature is strictly positive, this follows from Hadamard's principle. On the other hand, it is well-known that this condition is essential for $n = 1$; there are non-convex immersed curves in the plane with non-negative curvature.

UNIVERSITY OF CHICAGO.

REFERENCES.

[1] S. Chern, "La géométrie des sous-variétés d'un espace euclidien à plusieurs dimensions," *L'Enseigement Mathématique*, vol. 40 (1955), pp. 26-46.

[2] I. Fary, "Sur la courbure totale d'une courbe gauche faisant un noeud," *Bull. Soc. Math. de France*, vol. 77 (1949), pp. 128-138.

[3] J. W. Milnor, "On the total curvature of knots," *Annals of Mathematics*, vol. 52 (1950), pp. 248-257.

[4] ———, "On manifolds homeomorphic to the seven-sphere," *ibid.*, vol. 64 (1956), pp. 399-405.

[5] G. Reeb, "Sur certaines propriétés topologiques des variétés feuilletées, *Actual. Sci. et Indus.*, vol. 1183 (1952), pp. 91-154, Paris.

[6] G. de Rham, *Variétés différentiables* (1955), Paris.

ON THE INDEX OF A FIBERED MANIFOLD[1]

S. S. CHERN, F. HIRZEBRUCH, AND J-P. SERRE

Introduction. Let V be a real vector space of dimension r. Let $F(x, y) = \langle x, y \rangle$, $x, y \in V$, be a real-valued symmetric bilinear function. We can find a base e_i, $1 \leqq i \leqq r$, in V, such that

$$(1) \qquad F(x, y) = \sum_{i=1}^{p} x^i y^i - \sum_{i=p+1}^{p+q} x^i y^i$$

where $x = \sum_{i=1}^{r} x^i e_i$ and $y = \sum_{i=1}^{r} y^i e_i$.

The number $p - q$ is called the index of F, to be denoted by $\tau(F)$. It depends only on F. If F is nonsingular (i.e. $p + q = r$), then min (p, q) equals the maximal dimension of the linear subspaces of V contained in the "cone" $F(x, x) = 0$.

Now let M be a compact oriented manifold. The index of M is defined to be zero, if the dimension of M is not a multiple of 4. If M has the dimension $4k$, consider the cohomology group $H^{2k}(M)$ with real coefficients. This is a real vector space, and the equation

$$(2) \qquad \langle x, y \rangle \xi = x \cup y, \qquad x, y \in H^{2k}(M),$$

where ξ is the generator of $H^{4k}(M)$ defined by the given orientation of M, defines a real-valued symmetric bilinear form $\langle x, y \rangle$ over $H^{2k}(M)$. Its index is called the index of M, to be denoted by $\tau(M)$. Reversal of the orientation of M changes the sign of the index. The form $\langle x, y \rangle$ defined by (2) is nonsingular, since, by Poincaré's duality theorem, the equation $x \cup y = 0$ for all $x \in H^{2k}(M)$ implies $y = 0$.

The main purpose of this paper is to prove the theorem:

THEOREM. *Let $E \to B$ be a fiber bundle, with the typical fiber F, such that the following conditions are satisfied:*

(1) *E, B, F are compact connected oriented manifolds;*

(2) *The fundamental group $\pi_1(B)$ acts trivially on the cohomology ring $H^*(F)$ of F.*

Then, if E, B, F are oriented coherently, so that the orientation of E is induced by those of F and B, the index of E is the product of the indices of F and B, that is,

$$\tau(E) = \tau(F)\tau(B).$$

Received by the editors September 7, 1956.

[1] Work done when the first named author was under partial support by the National Science Foundation.

REMARK. We do not know whether condition (2) and the connectedness hypothesis of condition (1) are necessary. For instance, let E be an n-sheeted covering of B (the spaces B and E still being compact oriented manifolds); is it true that $\tau(E)=n\tau(B)$? We know the answer to be positive only when B possesses a differentiable structure: in that case, according to a theorem of one of us, $\tau(B)$ (resp. $\tau(E)$) is equal to the Pontrjagin number $L(B)$ (resp. $L(E)$) and it is clear that $L(E)=n\cdot L(B)$.

1. **Algebraic properties of the index of a matrix.** Let e_i, $1\leqq i\leqq r$, be a base in V. A real-valued symmetric bilinear function $\langle x, y\rangle$ defines a real-valued symmetric matrix $C=(c_{ij})$, $c_{ij}=\langle e_i, e_j\rangle$, $1\leqq i, j\leqq r$, and is determined by it. The index of the bilinear function is equal to the index $\tau(C)$ of C, if we define the latter to be the excess of the number of positive eigenvalues over the number of negative eigenvalues of C, each counted with its proper multiplicity. We have the following properties of the index of a real symmetric matrix:

For a nonsingular $(r\times r)$-matrix T we have

$$\tau(C) = \tau({}^tTCT). \tag{3}$$

Here, as always, we denote by ${}^tT$ the transpose of T. For nonsingular square matrices A, L (with A symmetric) we have

$$\tau\begin{bmatrix} 0 & 0 & L \\ 0 & A & 0 \\ {}^tL & 0 & 0 \end{bmatrix} = \tau\begin{pmatrix} 0 & L \\ {}^tL & 0 \end{pmatrix} + \tau(A) = \tau(A). \tag{4}$$

Here and always we make use of the convention that the index of the empty matrix is zero.

To prove (4) it is enough to show that

$$\tau\begin{pmatrix} 0 & L \\ {}^tL & 0 \end{pmatrix} = 0. \tag{5}$$

In this case, r is even. Put $r=2\mu$. Obviously, the cone $F(x, x)=0$ of the symmetric bilinear function $F(x, y)$ belonging to the matrix

$$\begin{pmatrix} 0 & L \\ {}^tL & 0 \end{pmatrix}$$

contains a linear space of dimension μ. Thus min $(p, q)\geqq\mu$. On the other hand, $p+q=2\mu$. Therefore, $p=q$ and $\tau=0$.

LEMMA 1. *Let C be a real, symmetric, nonsingular matrix of the form*

$$C = \begin{bmatrix} 0 & & L_0 \\ & \cdot & \\ & \cdot & \\ L_m & & * \end{bmatrix}$$

where $L_0, \cdots, L_m$ are square matrices (empty matrices are admitted) and where L_i is the transpose of L_{m-i}. Then

$$\tau(C) = \tau \begin{bmatrix} 0 & & L_0 \\ & \cdot & \\ & \cdot & \\ L_m & & 0 \end{bmatrix} = \begin{cases} 0, & \text{if } m \text{ is odd}, \\ \tau(L_n), & \text{if } m = 2n. \end{cases}$$

PROOF. We put

$$(6) \qquad C_\lambda = \begin{bmatrix} 0 & & L_0 \\ & \cdot & \\ & \cdot & \\ L_m & & \lambda * \end{bmatrix}, \qquad 0 \leqq \lambda \leqq 1.$$

Since $\det (C_\lambda) = \pm \prod_{i=0}^{m} \det (L_i) \neq 0$, the index $\tau(C_\lambda)$ is obviously independent of λ, so that $\tau(C) = \tau(C_1) = \tau(C_0)$. By (4) we have $\tau(C_0) = 0$ resp. $\tau(C_0) = \tau(L_n)$, q.e.d.

LEMMA 2. *Let A and B be two square matrices, which are either both symmetric or both skew-symmetric. Then their tensor product $A \otimes B$ is symmetric, and*

$$(7) \qquad \tau(A \otimes B) = \tau(A)\tau(B) \text{ or } 0,$$

according as both A and B are symmetric or skew-symmetric.

Suppose first that A and B are both symmetric. Let $\alpha_i > 0$, $\alpha_j < 0$, $1 \leqq i \leqq p$, $p+1 \leqq j \leqq p+q$, be the nonzero eigenvalues of A and $\beta_k > 0$, $\beta_l < 0$, $1 \leqq k \leqq p'$, $p'+1 \leqq l \leqq p'+q'$ be the nonzero eigenvalues of B. Then the nonzero eigenvalues of $A \otimes B$ are $\alpha_u \beta_t$, $1 \leqq u \leqq p+q$, $1 \leqq t \leqq p'+q'$. It follows that

$$\tau(A \otimes B) = pp' + qq' - pq' - p'q = \tau(A)\tau(B).$$

Now let A and B be both skew-symmetric. By applying (3) to the matrix $C = A \otimes B$ we can suppose that A and B are both of the form

$$\begin{bmatrix} A_1 & & & 0 \\ & \cdot & & \\ & & \cdot & \\ & & & A_n & \\ 0 & & & 0 \end{bmatrix}$$

where each A_i is a 2×2 block:

$$A_i = \begin{pmatrix} 0 & 1 \\ -1 & 0 \end{pmatrix} = J.$$

Since

$$\tau\left(\begin{pmatrix} 0 & 1 \\ -1 & 0 \end{pmatrix} \otimes \begin{pmatrix} 0 & 1 \\ -1 & 0 \end{pmatrix}\right) = \tau\begin{pmatrix} 0 & J \\ {}^tJ & 0 \end{pmatrix} = 0,$$

we have $\tau(A\otimes B)=0$.

2. **Poincaré rings.** We consider a graded ring A with the following properties:

(1) In the direct sum decomposition

$$A = \sum_{0\leq r<\infty} A^r$$

of A into the subgroups of its homogeneous elements, each A^r is a real vector space of finite dimension. There exists an n with $A^r=0$ for $r>n$ and with $\dim A^n=1$.

(2) If $x\in A^i$, $y\in A^j$ then $xy\in A^{i+j}$ and

$$xy = (-1)^{ij}yx.$$

Let $\xi\neq 0$ be a base element of A^n. Relative to ξ we define a bilinear pairing $\langle x, y\rangle$ of A^r and A^{n-r} into the real field by the equation

$$\langle x, y\rangle\xi = xy, \qquad x\in A^r,\ y\in A^{n-r}.$$

Let i_{n-r} be the linear mapping of A^{n-r} into $(A^r)^*$, the dual vector space of A^r, which assigns to $y\in A^{n-r}$ the linear function $\langle x, y\rangle$ on A^r $(x\in A^r)$.

A graded ring A is called a Poincaré ring if it satisfies (1), (2) and has moreover the following property:

(3) The mapping i_{n-r} is a bijection of A^{n-r} onto $(A^r)^*$.

A consequence of (3) is

$$\dim A^r = \dim A^{n-r}, \qquad 0\leq r\leq n.$$

The cohomology ring of a compact orientable manifold is a Poincaré ring.

A differentiation in a Poincaré ring A is a linear endomorphism $d: A\rightarrow A$, satisfying the following conditions:

(α) $dA^r\subset A^{r+1}$;

(β) $dd=0$;

(γ) $d(xy)=(dx)y+(-1)^r x(dy)$, if $x\in A^r$;

(δ) $dA^{n-1}=0$.

As is well known, such a differentiation defines a derived ring $A' = d^{-1}(0)/dA$. If we put $A'^r = d^{-1}(0) \cap A^r/dA^{r-1}$, we have the direct sum decomposition

$$A' = \sum_{0 \leq r \leq n} A'^r,$$

and A' is a graded ring. It is easy to verify that, if $x' \in A'^i$, $y' \in A'^j$, then $x'y' \in A'^{i+j}$, and

$$x'y' = (-1)^{ij} y'x'.$$

From the property (δ) of d we have $\dim A'^n = 1$. Thus A' satisfies (1) and (2) with the same maximal degree n as A. We denote the residue class of ξ in A'^n by ξ'. Relative to ξ' we have the linear mapping

$$i'_{n-r}: A'^{n-r} \to (A'^r)^*.$$

LEMMA 3. *The derived ring of a Poincaré ring with differentiation is a Poincaré ring, i.e. i'_{n-r} is bijective.*

It remains to prove that A' has the property (3) in the definition of a Poincaré ring. Let $x \in A^r$, $y \in A^{n-r-1}$. By property (δ) of d, we have

$$0 = d(xy) = (dx)y + (-1)^r x(dy).$$

This gives

(8) $$\langle dx, y \rangle = (-1)^{r-1} \langle x, dy \rangle,$$

a relation which is independent of the choice of ξ. This relation is equivalent to saying that the following diagram is commutative:

$$\begin{array}{ccccc}
A^{n-r-1} & \xrightarrow{\quad d \quad} & A^{n-r} & \xrightarrow{\quad d \quad} & A^{n-r+1} \\
\downarrow i_{n-r-1} & & \downarrow i_{n-r} & & \downarrow i_{n-r+1} \\
(A^{r+1})^* & \xrightarrow{(-1)^{r-1}({}^td)} & (A^r)^* & \xrightarrow{(-1)^r({}^td)} & (A^{r-1})^*
\end{array}$$

where $(A^r)^*$ is the dual space of A^r, and ${}^td$ is the dual homomorphism of d. We have the canonical isomorphism

$$(A'^r)^* \cong {}^td^{-1}(0) \cap (A^r)^*/{}^td(A^{r+1})^*.$$

The above diagram shows that i_{n-r} induces an isomorphism, namely i'_{n-r}, of A'^{n-r} onto $(A'^r)^*$. It follows that A'^r and A'^{n-r} are dually paired into the real field relative to the element $\xi' \in A'^n$, which is the residue class of ξ.

In analogy with the index of an oriented manifold we can define the index $\tau_\xi(A)$ of our Poincaré ring A relative to ξ. It is to be zero, if $n \equiv 0$, mod 4. If $n=4k$, $\tau_\xi(A)$ is to be the index of the bilinear function $\langle x, y\rangle$, $x, y \in A^{2k}$. Obviously, $\tau_\xi(A)=\tau_{\xi_1}(A)$, if ξ_1 is a positive multiple of ξ.

LEMMA 4. *In a Poincaré ring A let $\xi \neq 0$ be a base of A^n, and let $\xi' \in A'^n$ be the residue class which contains ξ. Then $\tau_{\xi'}(A')=\tau_\xi(A)$.*

It is only necessary to prove the lemma for the case $n=4k$. Let $Z^{2k}=d^{-1}(0)\cap A^{2k}$, $B^{2k}=dA^{2k-1}$, and let a, b, c be the respective dimensions of A^{2k}, B^{2k}, Z^{2k}. It follows immediately from (8) that each of the two spaces B^{2k} and Z^{2k} is the orthogonal of the other with respect to the symmetric form $\langle x, y\rangle$ of A^{2k}, whence $a = b + c$. We have $B^{2k}\subset Z^{2k}\subset A^{2k}$. If e_i is a base of A^{2k} such that $e_i \in B^{2k}$ for $1 \leq i \leq b$ and $e_i \in Z^{2k}$ for $b+1 \leq i \leq c$, the matrix $(\langle e_i, e_j\rangle)$ has then the form

$$\begin{pmatrix} 0 & 0 & L \\ 0 & Q & * \\ {}^tL & * & * \end{pmatrix},$$

where L and Q are square nonsingular matrices, of orders b and $c-b$ respectively. Its index is $\tau_\xi(A)$, while $\tau(Q)$ is $\tau_{\xi'}(A')$. By Lemma 1, we get therefore $\tau_{\xi'}(A')=\tau_\xi(A)$, as contended.

3. **Proof of the theorem.** It suffices to prove the theorem (see Introduction) for the case dim $E=4k$, which we suppose from now on. We consider the cohomology spectral sequence $E_r^{p,q}$, $2 \leq r \leq \infty$, of the bundle $E \to B$, with the real field as the coefficient field. Let

$$E_r^s = \sum_{p+q=s} E_r^{p,q}, \qquad E_r = \sum_{0 \leq s} E_r^s, \qquad 2 \leq r \leq \infty.$$

Each E_r is a graded ring, satisfying $E_r^s E_r^{s'} \subset E_r^{s+s'}$ and also $E_r^{p,q}E_r^{p',q'} \subset E_r^{p+p',q+q'}$. It has a differentiation d_r, such that E_{r+1} is the derived ring of E_r. In our case d_r is trivial for sufficiently large r and E_∞, or E_r for r sufficiently large, is the graded ring belonging to a certain filtration of the cohomology ring of the manifold E. The term E_2 of the spectral sequence is by hypothesis (2) of our theorem isomorphic to $H^*(B, H^*(F))=H^*(B)\otimes H^*(F)$, such that

$$E_2^{p,q} \cong H^p(B, H^q(F)) \cong H^p(B) \otimes H^q(F).$$

If we identify $E_2^{p,q}$ with $H^p(B)\otimes H^q(F)$ under this isomorphism, the multiplication in E_2 is given by

$$(b \otimes f)(b' \otimes f') = (-1)^{p'q}(b \cup b') \otimes (f \cup f'),$$
$$b \in H^p(B), \quad b' \in H^{p'}(B), \quad f \in H^q(F), \quad f' \in H^{q'}(F).$$

Let $m = \dim F$, so that $\dim B = 4k - m$. Since B and F are manifolds, E_2 is a Poincaré ring with respect to the grading

$$E_2 = \sum_{0 \leq s < \infty} E_2^s \quad (E_2^s = 0 \text{ for } s > 4k,\ E_2^{4k} = E_2^{4k-m,m}).$$

The ring E_2 is isomorphic to the cohomology ring of $B \times F$.

The orientations of B, F define a generator $\xi_2 = \xi_B \otimes \xi_F$ of E_2^{4k}. Here ξ_B (resp. ξ_F) denotes the generator of $H^{4k-m}(B)$ (resp. $H^m(F)$) belonging to the orientation of B (resp. F). We wish to prove that

$$\tau_{\xi_2}(E_2) = \tau(B) \cdot \tau(F).$$

We have

$$E_2^{2k} = E_2^{2k,0} + E_2^{2k-1,1} + \cdots + E_2^{2k-m,m}. \tag{9}$$

Here some of the $E_2^{p,q}$ might vanish, in particular $E_2^{p,q} = 0$ if $p < 0$. Clearly, for $x \in E_2^{2k-q,q}$ and $y \in E_2^{2k-q',q'}$ we have $xy = 0$ unless

$$q + q' = m.$$

By Poincaré duality in B and F, we have

$$\dim E_2^{2k-q,q} = \dim E_2^{2k-m+q,m-q}.$$

Therefore, the symmetric matrix, which defines the bilinear symmetric function over E_2^{2k}, is, when written in blocks relative to the direct sum decomposition (9), of the form

$$\begin{bmatrix} 0 & & L_0 \\ & \cdot & \\ L_m & & 0 \end{bmatrix}$$

where the L_i are nonsingular square matrices, such that L_i is the transpose of L_{m-i}. By Lemma 1 we obtain

$$\tau_{\xi_2}(E_2) = 0 \text{ if } m \text{ is odd}, \qquad \tau_{\xi_2}(E_2) = \tau(L_{m/2}) \text{ if } m \text{ is even}.$$

In the first case the equation $\tau_{\xi_2}(E_2) = \tau(B)\tau(F)$ is proved, since $\tau_{\xi_2}(E_2) = \tau(F) = 0$. In the latter case we have

$$E_2^{2k-m/2,m/2} = H^{2k-m/2}(B) \otimes H^{m/2}(F),$$

and it is clear that up to the sign $(-1)^{m/2}$ the matrix $L_{m/2}$ is the tensor product of the two matrices defining the bilinear forms of B and F. If $m/2$ is odd, both matrices in this tensor product are skew-symmetric, and we have, by Lemma 2, $\tau(L_{m/2})=0$; on the other hand we have $\tau(B)\tau(F)=0$, since dim $F\not\equiv 0$ (mod 4) and thus by definition $\tau(F)=0$. If $m/2$ is even, that is, if $m\equiv 0$ (mod 4), both matrices are symmetric, and Lemma 2 gives: $\tau(L_{m/2})=\tau(B)\tau(F)$. Combining all cases, we get the formula

$$\tau_{\xi_2}(E_2) = \tau(B)\tau(F) \tag{10}$$

in full generality.

The differentiation d_2 of E_2 satisfies the conditions of a differentiation in a Poincaré ring given in §2. In fact, dim $E_\infty^{4k}=1$, since E is a manifold of dimension $4k$. Therefore, dim $E_r^{4k}=1$ for $2\leqq r$. Thus d_2 annihilates E_2^{4k-1}; more generally d_r annihilates E_r^{4k-1}. It follows by Lemma 3 that E_3 is a Poincaré ring. It has d_3 as differentiation and therefore E_4 is a Poincaré ring etc. Finally, E_∞ is a Poincaré ring. By Lemma 4 and (10) we get

$$\tau(B)\tau(F) = \tau_{\xi_2}(E_2) = \tau_{\xi_3}(E_3) = \cdots = \tau_{\xi_\infty}(E_\infty),$$

where ξ_r (resp. ξ_∞) is the image of ξ_2 in E_r (resp. E_∞).

It remains to prove that $\tau_{\xi_\infty}(E_\infty)=\tau(E)$. The cohomology ring $H^*(E)$ is filtered:

$$H^*(E) = D^0 \supset D^1 \supset \cdots \supset D^p \supset D^{p+1} \supset \cdots, \qquad \bigcap D^p = 0,$$

$$D^{p,q} = D^p \cap H^{p+q}(E), \tag{11}$$

$$D^{p,q}\cdot D^{p',q'} \subset D^{p+p',q+q'}.$$

We have the filtration

$$H^r(E) = D^{0,r} \supset D^{1,r-1} \supset \cdots \supset D^{r,0} \supset D^{r+1,-1} = 0$$

and the canonical isomorphism

$$D^{p,q}/D^{p+1,q-1} \cong E_\infty^{p,q}. \tag{12}$$

The ring structure of E_∞ is induced by that of $H^*(E)$ by the canonical homomorphisms $D^{p,q}\to E_\infty^{p,q}$ (see (12) and (11)). Since $E_\infty^{4k}=E_\infty^{4k-m,m}$, (where $m=\dim F$), we have

$$H^{4k}(E) = D^{4k-m,m} \cong E_\infty^{4k-m,m} \tag{13}$$

and

$$D^{4k-i,i} = 0 \qquad \text{for } i < m. \tag{14}$$

Earlier we have chosen a generator $\xi_\infty \in E_\infty^{4k}$. Under the canonical isomorphism (13) ξ_∞ goes over in the generator ξ_E of $H^{4k}(E)$ belonging to the orientation of E generated by the given orientations of B and F in this order.[2] We now consider the bilinear symmetric function $\langle x, y\rangle$ over $H^{2k}(E)$ relative to ξ_E. Choose a direct sum decomposition of $H^{2k}(E)$ in linear subspaces,

$$H^{2k}(E) = V_0 + V_1 + V_2 + \cdots + V_m \tag{15}$$

such that

$$\sum_{j=0}^{q} V_j = D^{2k-q,q} \qquad (0 \leq q \leq m).$$

Here we use that $D^{2k-s,s} = D^{2k-m,m}$ for $s > m$. By (11) and (14) we have

$$\langle x, y\rangle = 0 \qquad \text{for } x \in V_i,\ y \in V_j \text{ and } i + j < m, \tag{16}$$

and moreover by (13)

$$\langle x, y\rangle = \langle \bar{x}, \bar{y}\rangle, \qquad \text{for } x \in V_i,\ y \in V_j \text{ and } i + j = m, \tag{17}$$

where $\bar{x}$ (resp. $\bar{y}$) denotes the image (see (12)) of x (resp. y) in $E_\infty^{2k-i,i}$ (resp. $E_\infty^{2k-j,j}$) and where on the right side of this equation stands the symmetric bilinear form over E_∞^{2k} relative to ξ_∞. Since $\langle \bar{x}, \bar{y}\rangle = 0$ for $\bar{x} \in E_\infty^{2k-q,q}$, $\bar{y} \in E_\infty^{2k-q',q'}$, unless $q + q' = m$, and since E_∞ is a Poincaré algebra, we can conclude

$$\dim E_\infty^{2k-q,q} = \dim E_\infty^{2k-m+q,m-q}. \tag{18}$$

The preceding remarks, in particular (16), (17), (18), imply: The matrix of the symmetric bilinear function over $H^{2k}(E)$ relative to ξ_E can be written in blocks with respect to the direct sum decomposition (15) in the form

$$\begin{bmatrix} 0 & & & L_0 \\ & & L_1 & \\ & \cdot & & \\ & \cdot & & \\ L_m & & & * \end{bmatrix}$$

[2] This is easy to see when E is a trivial bundle, in which case it is almost the definition of the orientation of a product of manifolds. The general case can be reduced to this one by comparing the spectral sequence of E to that of the bundle induced by E on an open cell of the base, the cohomology being taken with compact carriers.

where the L_i are nonsingular square matrices and where L_i is the transpose of L_{m-i}. Moreover,

$$\begin{pmatrix} 0 & & \cdot & L_0 \\ & & \cdot & \\ & \cdot & & \\ L_m & & & 0 \end{pmatrix}$$

is the matrix of the symmetric bilinear function over E_∞^{2k} relative to ξ_∞. By Lemma 1 we have $\tau(E) = \tau_{\xi_\infty}(E_\infty)$. This concludes the proof of our theorem.

University of Chicago,
University of Nancago,
Princeton University and
Universität Bonn

Integral Formulas for Hypersurfaces in Euclidean Space and Their Applications to Uniqueness Theorems

SHIING-SHEN CHERN*

Introduction. Let E be the Euclidean space of dimension $n + 1$. By a hypersurface in E we mean a differentiable manifold M of dimension n and a differentiable mapping $x: M \to E$, whose functional matrix is everywhere of rank n. Since the strength of the differentiability assumptions will not be the issue, we suppose our manifolds and mappings to be of class C^∞.

If M is compact, certain immediate integral formulas are valid. More generally, we will derive integral formulas for the hypersurface x and a second hypersurface $x' : M \to E$. Such formulas are generalizations of well-known formulas in the theory of convex bodies, which express the mixed volumes of two convex bodies as integrals [1].

Given two compact hypersurfaces, a rigidity or uniqueness theorem gives a sufficient condition such that commutativity holds in the diagram

$$\begin{array}{ccc} M & \xrightarrow{x} & E \\ & {\scriptstyle x'}\searrow \quad \swarrow{\scriptstyle T} & \\ & E & \end{array}$$

where T is a motion in E. Of interest in differential geometry are conditions expressed in terms of the relative curvature of the hypersurface. We review its definition as follows:

Suppose M to be oriented. Then to $p \in M$ there is a uniquely determined unit normal vector $\xi(p)$ at $x(p)$. We put

$$\text{(1)} \qquad \mathrm{I} = dx^2, \qquad \mathrm{II} = d\xi\, dx, \qquad \mathrm{III} = d\xi^2,$$

where dx, $d\xi$ are vector-valued linear differential forms in M and multiplication is in the sense of the scalar product in E. These are three quadratic differential forms in M (the "fundamental forms"), of which I is positive definite. The eigenvalues $k_1, \cdots, k_n$ of II relative to I are called the principal curvatures. If the Gauss-Kronecker curvature $K = k_1 \cdots k_n \neq 0$, the reciprocals $1/k_1, \cdots, 1/k_n$ are called the radii of principal curvature; they are the eigenvalues of II

*Work done under partial support from the National Science Foundation.

relative to III, which is also positive definite under the assumption $K \neq 0$. In this case we introduce the lth elementary symmetric function

$$\binom{n}{l} P_l = \sum \frac{1}{k_1} \cdots \frac{1}{k_l}, \quad 1 \leqq l \leqq n. \tag{2}$$

It seems that our integral formulas are more effective for hypersurfaces which are strictly convex, *i.e.*, hypersurfaces for which the Gauss-Kronecker curvature is everywhere > 0. By using an inequality of L. GÅRDING [2], we can immediately derive the uniqueness theorem of ALEXANDROFF-FENCHEL-JESSEN [3, 4], to the effect that two closed strictly convex hypersurfaces differ by a translation if the function P_l (for a fixed l) takes the same value at points with the same normal vector. Actually this proof does not differ essentially from the one given by ALEXANDROFF. Its merit lies perhaps in stating the result in a more general form and in separating the geometrical from the analytical part. One can, however, derive in this way further uniqueness theorems. For instance, it will be proved that a closed strictly convex hypersurface is a hypersphere if, for a fixed l, $2 \leqq l \leqq n$, $P_{l-1}^{\alpha} P_l^{\beta} = \text{const.}$, $\alpha \geqq 0$, $\beta \geqq 0$, $\alpha + \beta > 0$.

In the proofs of more general uniqueness theorems for closed strictly convex (two-dimensional) surfaces the most important tool is the "index method". This does not seem to generalize to higher dimensions, at least not in an obvious way. The need and search for new methods should make the higher dimensional problems more challenging and interesting.

One immediate problem is the following: Is a closed strictly convex hypersurface defined up to a translation if the lth $(1 \leqq l \leqq n-1)$ elementary symmetric function of the principal curvatures is given as a function of the normal vector?

1. Integral formulas. Let M be an oriented differentiable manifold of dimension n, and let $x: M \to E$ be a hypersurface. Let $\xi(p)$, $p \in M$, be the unit normal vector at $x(p)$. We consider the orthonormal frames $e_1, \cdots, e_n$ in the tangent hyperplane at $x(p)$, such that the determinant $(e_1, \cdots, e_n, \xi) = +1$. The space of all $e_1, \cdots, e_n$ can be identified with the principal fiber bundle B of M relative to the Riemannian metric I. We have

$$\begin{aligned} dx &= \omega_1 e_1 + \cdots + \omega_n e_n, \\ d\xi &= \theta_1 e_1 + \cdots + \theta_n e_n, \end{aligned} \tag{3}$$

so that ω_i, θ_i, $1 \leqq i \leqq n$, are linear differential forms in B. Since

$$\xi \, dx = 0, \tag{4}$$

we get, by exterior differentiation,

$$d\xi \wedge dx = 0. \tag{5}$$

The left-hand side in (5) is the exterior product of two vector-valued linear differential forms; vectors are multiplied in the sense of scalar products in E.

In view of (3), equation (5) can be written

$$\sum_i \omega_i \wedge \theta_i = 0. \tag{6}$$

This is the fundamental relation in the theory of hypersurfaces in E.

Since ω_i are linearly independent, we can put, in view of (6),

$$\theta_i = \sum_k \mu_{ik}\omega_k , \qquad \mu_{ik} = \mu_{ki} , \qquad 1 \leqq i, k \leqq n. \tag{7}$$

If $\det (\mu_{ik}) \neq 0$, we introduce the matrix (λ_{ik}) inverse to (μ_{ik}), so that we have

$$\omega_i = \sum_k \lambda_{ik}\theta_k . \tag{8}$$

By (1) and (3) we can also write

$$\begin{aligned} \mathrm{I} &= \sum_i \omega_i^2 , \qquad \mathrm{III} = \sum_i \theta_i^2 , \\ \mathrm{II} &= \sum_i \theta_i\omega_i = \sum_{i,k} \mu_{ik}\omega_i\omega_k = \sum_{i,k} \lambda_{ik}\theta_i\theta_k . \end{aligned} \tag{9}$$

Let

$$\det (\delta_{ik} + \lambda_{ik}y) = \sum_{0 \leq l \leq n} \binom{n}{l} P_l(\lambda)y^l, \tag{10}$$

where y is a parameter. Then $P_l(\lambda)$ is a homogeneous polynomial of degree l in λ_{ik} , and it is easy to see that it is equal to the invariant P_l defined in (2).

We now introduce the differential forms

$$\begin{aligned} A_r &= (x, \xi, \underbrace{d\xi, \cdots , d\xi}_{n-1-r}, \underbrace{dx, \cdots , dx}_{r}), \qquad 0 \leqq r \leqq n-1, \\ C_s &= (x, d\xi, \cdots , d\xi, dx, \cdots , dx), \\ D_s &= (\xi, d\xi, \cdots , d\xi, dx, \cdots , dx), \qquad 0 \leqq s \leqq n. \end{aligned} \tag{11}$$

Each of these expressions is a determinant of order $n + 1$, whose columns are the components of the respective vectors or vector-valued differential forms, with the convention that in the expansion of the determinant the multiplication of differential forms is in the sense of exterior multiplication. A_r is a differential form of degree $n - 1$ in M, and C_s , D_s are differential forms of degree n; the subscripts r, s in each case denote the number of entries dx in these determinants. It is to be observed that A_r , C_s depend on the choice of the origin in the space E, while D_s depends only on the hypersurface $x(M)$.

Let $h = \xi x$ be the support function, the distance from the origin to the tangent hyperplane at $x(p)$. Since $d\xi$ and dx are linear combinations of $e_1 , \cdots , e_n$ only, we have immediately

$$C_s = hD_s . \tag{12}$$

Exterior differentiation gives

$$dA_r = C_r - D_{r+1} = hD_r - D_{r+1}, \qquad 0 \leqq r \leqq n-1.$$

Hence, by Stokes' Theorem, we have, under the assumption that M is compact,

$$\int_M hD_r - D_{r+1} = 0, \qquad 0 \leqq r \leqq n-1. \tag{13}$$

More generally, for a pair of hypersurfaces $x, x' \colon M \to E$ we introduce the differential forms

$$\begin{aligned} A_{rs} &= (x, \ \xi, \underbrace{d\xi, \cdots, d\xi}_{n-1-(r+s)}, \underbrace{dx, \cdots, dx}_{r}, \underbrace{dx', \cdots, dx'}_{s}), \\ A'_{rs} &= (x', \xi, d\xi, \cdots, d\xi, dx, \cdots, dx, dx', \cdots, dx'), \\ B_{rs} &= (x, x', d\xi, \cdots, d\xi, dx, \cdots, dx, dx', \cdots, dx'), \ 0 \leqq r+s \leqq n-1 \\ C_{uv} &= (x, \quad \underbrace{d\xi, \cdots, d\xi}_{n-(u+v)}, \underbrace{dx, \cdots, dx}_{u}, \underbrace{dx', \cdots, dx'}_{v}), \\ C'_{uv} &= (x', d\xi, \cdots, d\xi, dx, \cdots, dx, dx', \cdots, dx'), \\ D_{uv} &= (\xi, d\xi, \cdots, d\xi, dx, \cdots, dx, dx', \cdots, dx'), \qquad 0 \leqq u+v \leqq n. \end{aligned} \tag{14}$$

As above, we have

$$C_{uv} = hD_{uv}, \qquad C'_{uv} = h'D_{uv}, \qquad h' = \xi x'. \tag{15}$$

By introducing the unit normal vector $\xi'(p)$ of the second hypersurface, we could have more general differential forms, but we will not write them down here.

Exterior differentiation gives

$$\begin{aligned} dA_{rs} &= C_{rs} - D_{r+1,s} = hD_{rs} - D_{r+1,s}, \\ dA'_{rs} &= C'_{rs} - D_{r,s+1} = h'D_{rs} - D_{r,s+1}, \\ dB_{rs} &= C_{r,s+1} - C'_{r+1,s} = hD_{r,s+1} - h'D_{r+1,s}, \qquad 0 \leqq r+s \leqq n-1. \end{aligned} \tag{16}$$

From these follow the integral formulas

$$\begin{aligned} &\int hD_{rs} - D_{r+1,s} = 0, \\ &\int h'D_{rs} - D_{r,s+1} = 0, \\ &\int hD_{r,s+1} - h'D_{r+1,s} = 0, \qquad 0 \leqq r+s \leqq n-1, \end{aligned} \tag{17}$$

the integrations being over M, supposed to be compact.

These integral formulas take a particularly simple form when the hypersurfaces are strictly convex. Then the Gauss-Kronecker curvatures are strictly

>0, and the normal mapping $\xi: M \to S_0$, where S_0 is the unit hypersphere in E, is a differentiable homeomorphism with functional determinant everywhere different from zero. We can therefore identify M with S_0 and define the hypersurface by $x: S_0 \to E$. Geometrically $x(\xi)$, $\xi \in S_0$, is the coordinate vector of the point of the hypersurface at which ξ is the normal vector. Similarly, the second hypersurface will be defined by the mapping $x': S_0 \to E$. The functions and differential forms previously defined in M are now in S_0, and the integrals in (17) are over S_0. We can suppose the hypersurfaces so oriented that II is everywhere positive definite.

To apply the formulas (17) we wish to find more explicit expressions for D_{rs}. With two parameters y, y' we have

$$\begin{aligned}
&\sum_{0 \leq r+s \leq n} \frac{n!}{r!s!(n-r-s)!} y^r y'^s D_{rs} \\
&= (-1)^n \sum_{1 \leq i_1, \cdots, i_n \leq n} \epsilon_{i_1 \cdots i_n} (y\omega_{i_1} + y'\omega'_{i_1} + \theta_{i_1}) \wedge \cdots \wedge (y\omega_{i_n} + y'\omega'_{i_n} + \theta_{i_n}) \\
&= (-1)^n n! (y\omega_1 + y'\omega'_1 + \theta_1) \wedge \cdots \wedge (y\omega_n + y'\omega'_n + \theta_n) \\
&= (-1)^n n! \det (y\lambda_{ik} + y'\lambda'_{ik} + \delta_{ik})\, dV,
\end{aligned}$$

where

$$dV = \theta_1 \wedge \cdots \wedge \theta_n$$

is the volume element of S_0. Let

$$\det (y\lambda_{ik} + y'\lambda'_{ik} + \delta_{ik}) = \sum_{0 \leq r+s \leq n} \frac{n!}{r!\, s!\, (n-r-s)!} y^r y'^s P_{rs} \tag{18}$$

so that P_{rs} is a polynomial in λ_{ik}, λ'_{ik}, homogeneous of degrees r and s respectively. In particular, $P_{l0} = P_l(\lambda)$, the latter being defined in (10). The integral formulas (17) can then be written

$$\begin{aligned}
&\int (hP_{rs} - P_{r+1,s})\, dV = 0, \\
&\int (h'P_{rs} - P_{r,s+1})\, dV = 0, \\
&\int (hP_{r,s+1} - h'P_{r+1,s})\, dV = 0, \qquad 0 \leq r+s \leq n-1.
\end{aligned} \tag{19}$$

The third equation in (19) is, for $0 \leq r+s \leq n-2$, a consequence of the first two equations.

An important consequence of the third equation of (19) consists in the formulas

$$\int (hP_{0l} - h'P_{1,l-1})\, dV = 0, \qquad \int (hP_{l-1,1} - h'P_{l0})\, dV = 0, \qquad 1 \leq l \leq n.$$

From these we derive

$$(20)\qquad 2\int h(P_{0l}-P_{l-1,1})\,dV=\int\{h'(P_{1,l-1}-P_{l0})-h(P_{l-1,1}-P_{0l})\}\,dV.$$

Let $F(u_1, \cdots, u_n)$ be a function in n positive variables. We will say that F is of type $l \geqq 2$ if the following conditions are satisfied: (1) $F(P_{10}, \cdots, P_{n0}) = F(P_{01}, \cdots, P_{0n})$ implies $P_{l-1,1} - P_{0l} \geqq 0$; (2) $F(P_{10}, \cdots, P_{n0}) = F(P_{01}, \cdots, P_{0n})$, $P_{l-1,1} - P_{0l} = 0$ if and only if $\lambda'_{ik} = \lambda_{ik}$.

Theorem 1. *Let $F(u_1, \cdots, u_n)$ be a function of type $l \geqq 2$. If two closed strictly convex hypersurfaces have the property that at points with the same unit normal vector the functions $F(P_{10}, \cdots, P_{n0})$ and $F(P_{01}, \cdots, P_{0n})$ have the same value, then the hypersurfaces differ from each other by a translation.*

Proof. It suffices to prove that $\lambda'_{ik} = \lambda_{ik}$ at points with the same unit normal vector. Choose the origin in E such that $h > 0$. Then the integral at the left-hand side of (20) is $\leqq 0$. The same is therefore true of the right-hand side of (20). But the latter is anti-symmetric in the two hypersurfaces, and hence must be zero. It follows that $P_{l-1,1} - P_{0l} = 0$. By the second property of the function F, we get $\lambda'_{ik} = \lambda_{ik}$.

2. The uniqueness theorem of Alexandroff-Fenchel-Jessen.

Theorem 2 (ALEXANDROFF-FENCHEL-JESSEN). *Two closed strictly convex hypersurfaces differ by a translation if the function $P_l(\lambda)$, $1 \leqq l \leqq n$, takes the same value at points with the same unit normal vector.*

For $2 \leqq l \leqq n$ we will prove that the function $F = u_l$ is of type l. Then the theorem for $2 \leqq l \leqq n$ will follow from Theorem 1.

We will need the following inequality of L. GÅRDING [2]:

To the polynomial $P_l(\lambda)$ defined in (10) let $P_l(\lambda^{(1)}, \cdots, \lambda^{(l)})$ be its completely polarized form, so that $P_l(\lambda, \cdots, \lambda) = P_l(\lambda)$. Then, for positive definite symmetric matrices $(\lambda_{ik}^{(1)}), \cdots, (\lambda_{ik}^{(l)})$, the following inequality is valid:

$$(21)\qquad P_l(\lambda^{(1)}, \cdots, \lambda^{(l)}) \geqq P_l(\lambda^{(1)})^{1/l} \cdots P_l(\lambda^{(l)})^{1/l}.$$

The equality sign holds if and only if the l matrices are pairwise proportional.

Suppose now $P_{l0} = P_{0l}$. This condition can be written $P_l(\lambda) = P_l(\lambda')$. By definition $P_{l-1,1} = P_l(\lambda, \cdots, \lambda, \lambda')$. Since (λ_{ik}) and (λ'_{ik}) are positive definite, it follows from (21) that

$$P_l(\lambda, \cdots, \lambda, \lambda') \geqq P_l(\lambda)^{(l-1)/l} P_l(\lambda')^{1/l} = P_l(\lambda'),$$

which implies the first condition for a function of type l. The equality sign holds only if $\lambda'_{ik} = \rho\lambda_{ik}$. Since $P_l(\lambda) = P_l(\lambda')$, we have $\rho = 1$.

The proof of Theorem 2 for $l = 1$ is different, but is actually easier in the sense that an inequality such as (21) will not be needed. The proof has been given on a previous occasion [5]; we present it here in our notation for the sake of completeness.

By definition we have

$$n(n-1)P_{20} = (\sum_i \lambda_{ii})^2 - \sum_{i,k} \lambda_{ik}^2,$$

$$n(n-1)P_{02} = (\sum_i \lambda'_{ii})^2 - \sum_{i,k} \lambda'^2_{ik}, \tag{22}$$

$$2n(n-1)P_{11} = \sum_{i,k} (\lambda_{ii}\lambda'_{kk} + \lambda'_{ii}\lambda_{kk} - 2\lambda_{ik}\lambda'_{ik}).$$

From these it follows that

$$\Lambda \equiv \sum_{i,k} \begin{vmatrix} \lambda'_{ii} - \lambda_{ii} & \lambda'_{ik} - \lambda_{ik} \\ \lambda'_{ki} - \lambda_{ki} & \lambda'_{kk} - \lambda_{kk} \end{vmatrix} = n(n-1)(P_{20} + P_{02} - 2P_{11}).$$

Our integral formulas (19) give

$$\int h(P_{01} - P_{10})\, dV = \int (P_{11} - P_{20})\, dV,$$

$$\int h'(P_{01} - P_{10})\, dV = \int (P_{02} - P_{11})\, dV. \tag{23}$$

From the hypothesis $P_{10} = P_{01}$ we therefore get

$$\int \Lambda\, dV = 0.$$

On the other hand, under the same hypothesis we have

$$\begin{aligned} \Lambda &= 2 \sum_{i<k} \begin{vmatrix} \lambda'_{ii} - \lambda_{ii} & \lambda'_{ik} - \lambda_{ik} \\ \lambda'_{ki} - \lambda_{ki} & \lambda'_{kk} - \lambda_{kk} \end{vmatrix} \\ &= 2 \sum_{i<k} (\lambda'_{ii} - \lambda_{ii})(\lambda'_{kk} - \lambda_{kk}) - 2 \sum_{i<k} (\lambda'_{ik} - \lambda_{ik})^2 \\ &= \sum_{i,k} (\lambda'_{ii} - \lambda_{ii})(\lambda'_{kk} - \lambda_{kk}) - \sum_i (\lambda'_{ii} - \lambda_{ii})^2 - 2 \sum_{i<k} (\lambda'_{ik} - \lambda_{ik})^2 \\ &= -\sum_i (\lambda'_{ii} - \lambda_{ii})^2 - 2 \sum_{i<k} (\lambda'_{ik} - \lambda_{ik})^2. \end{aligned}$$

Therefore

$$\int \{\sum_i (\lambda'_{ii} - \lambda_{ii})^2 + 2 \sum_{i<k} (\lambda'_{ik} - \lambda_{ik})^2\}\, dV = 0,$$

which is possible only when $\lambda'_{ik} = \lambda_{ik}$.

Remark 1. The case $l = n$ of Theorem 2 is called the uniqueness of MINKOWSKI's problem. The theorem for $l = 1, n = 2$ goes back to CHRISTOFFEL and HURWITZ. We refer to BONNESEN & FENCHEL's book for historical facts concerning the problem and for its relation with the fundamental Brunn-Minkowski inequalities in convex bodies.

Remark 2. The uniqueness of the Minkowski problem can be proved without using GÅRDING's inequality as follows: When we replace the matrix $\Delta = (y\lambda_{ik} + y'\lambda'_{ik} + \delta_{ik})$ by ${}^tT\,\Delta\,T$, where T is an arbitrary non-singular matrix; P_{rs} will be multiplied by $(\det T)^2$. Since (λ_{ik}) and (λ'_{ik}) are both positive definite, we can choose T to make $(\lambda'_{ik}) = I$ = unit matrix and (λ_{ik}) = diagonal matrix, while (δ_{ik}) will then be transformed to a general positive definite symmetric matrix. If $\lambda_1, \cdots, \lambda_n$ denote the diagonal elements of (λ_{ik}), we have

$$P_{n-1,1} = \frac{1}{n-1} \sum \lambda_1 \cdots \lambda_{n-1},$$

where the sum denotes the elementary symmetric function of order $n - 1$ in $\lambda_1, \cdots, \lambda_n$. On the other hand, our condition $P_{n0} = P_{0n}$ gives $\lambda_1, \cdots, \lambda_n = 1$. By NEWTON's inequality we get $P_{n-1,1}^{1/(n-1)} \geqq 1$ or $P_{n-1,1} - P_{0n} \geqq 0$. Moreover, $P_{n-1,1} - P_{0n} = 0$ implies $P_{n-1,1}^{1/(n-1)} = 1$, which is possible only when $\lambda_1 = \cdots = \lambda_n = 1$. This proves that the function $F = u_n$ is of type n, and the uniqueness follows from Theorem 1.

Remark 3. GÅRDING's proof of the inequality (21) makes use of general results on hyperbolic polynomials. It may be of interest to observe that for our purpose it would be sufficient to have the inequality for the case that $P_l(\lambda)$ is the lth elementary symmetric function in n ($\geqq l$) positive numbers. It would be interesting to have an elementary proof of this particular case of the inequality.

Remark 4. As is well-known, we can determine our hypersurface by the support function $h(\xi)$, $\xi \in S_0$. Then P_l can be expressed in terms of $h(\xi)$ and its first and second partial derivatives. If $h(\xi)$ is considered to be an unknown function, this relation is, for $l \geqq 2$, a non-linear partial differential equation of the second order in $h(\xi)$. Our uniqueness theorem can be interpreted as asserting that two solutions of this equation on the unit hypersphere S_0 differ from each other by a linear homogeneous combination of the coordinates of the space. In this respect the fact that the cases $l = 1$ (linear case) and $l \geqq 2$ have to be treated differently is of significance.

Remark 5. With integral formulas as the main tool in the proofs the theorems can be immediately extended to hypersurfaces with boundaries. The extension amounts merely to a writing of the formula (19) with a term pertaining to the boundary, and to an interpretation of this term. It is necessary that the boundaries differ by a translation and that corresponding points have the same normal vectors. This condition, together with the same conditions as in the case of hypersurfaces without boundary, is then sufficient.

3. Characterization of the hypersphere. When the second hypersurface is a hypersphere of radius r, Theorem 1 gives

Theorem 3. *Let $F(u_1, \cdots, u_n)$ be a function in n positive variables which has the following properties ($l \geqq 2$): (1) $F(P_1, \cdots, P_n) = F(r, \cdots, r^n) = const.$*

implies that $P_{l-1} - r^{l-1} \geqq 0$; (2) $F(P_1, \cdots, P_n) = F(r, \cdots, r^n)$, *together with* $P_{l-1} = r^{l-1}$, *implies* $\lambda_{ik} = r\delta_{ik}$. *Then a closed strictly convex hypersurface with* $F(P_1, \cdots, P_n) = $ *const. is a hypersphere of radius* r.

As a consequence of this theorem we have the following characterization of the hypersphere:

Theorem 4. *A closed strictly convex hypersurface with* $P_{l-1}^{\alpha} P_l^{\beta} =$ *const,* $2 \leqq l \leqq n$, $\alpha + \beta > 0$, $\alpha \geqq 0$, $\beta \geqq 0$, *is a hypersphere.*

Proof. In the inequality (21) we put

$$\lambda_{ik}^{(1)} = \cdots = \lambda_{ik}^{(l-1)} = \lambda_{ik}, \qquad \lambda_{ik}^{(l)} = r\delta_{ik}.$$

Then we get

$$P_{l-1}^{1/(l-1)} \geqq P_l^{1/l}, \tag{24}$$

where the equality sign holds if and only if $\lambda_{ik} = s\delta_{ik}$.

In proving Theorem 4 we can suppose $\beta > 0$. Let $r > 0$ be defined by

$$P_{l-1}^{\alpha} P_l^{\beta} = r^{\alpha(l-1)+\beta l}, \tag{25}$$

so that

$$P_l = r^{(\alpha/\beta)(l-1)+l} P_{l-1}^{-\alpha/\beta}.$$

From (24) we get

$$P_{l-1} \geqq P_l^{(l-1)/l} = r^{(l-1)\{(\alpha/\beta)[(l-1)/l]+1\}} P_{l-1}^{-(\alpha/\beta)[(l-1)/l]},$$

which gives

$$P_{l-1} \geqq r^{l-1}.$$

Moreover, the equality sign holds here if and only if it holds in (24). Hence $P_{l-1} = r^{l-1}$ implies $\lambda_{ik} = s\delta_{ik}$. But then (25) gives $s = r$. Our theorem therefore follows from Theorem 3.

References

[1] Bonnesen, T. &.Fenchel, W., *Theorie der konvexen Körper*, Berlin 1934.

[2] Gårding, L., An inequality for hyperbolic polynomials, this *Jour. Math. and Mech.* **8** (1959), pp. 957–965.

[3] Alexandroff, A. D., Zur Theorie der gemischten Volumina von konvexen Körpern (Russian), *Recueil Math., Serie nouvelle,* **2** (1937), pp. 947–972, 1205–1238, **3** (1938), pp. 27–46, 227–251.

[4] Fenchel, W. & Jessen, B., Mengenfunktionen und konvexe Körper, *Danske Videns. Selskab., Math.-fysiske Medd.*, **16** (1938), pp. 1–31.

[5] Chern, S., *Topics in Differential Geometry*, Princeton, 1951, pp. 29–30. *Cf.* also Hsiung, C. C., On Differential Geometry of Hypersurfaces in the Large, *Trans. Amer. Math. Soc.*, **81** (1956), pp. 243–252.

University of Chicago
Chicago, Illinois

A Uniqueness Theorem on Closed Convex Hypersurfaces in Euclidean Space

S. S. CHERN, J. HANO & C. C. HSIUNG[1]

The object of this note is to prove the following theorem:

Let Σ, Σ' be two closed, strictly convex, C^2-differentiable hypersurfaces in a Euclidean space of dimension $n + 1$ ($\geqq 3$). Let $f: \Sigma \to \Sigma'$ be a diffeomorphism such that Σ and Σ' have parallel outward normals at $p \in \Sigma$ and $p' = f(p)$ respectively. Denote by $P_l(p)$ (respectively $P_l(p')$) the lth elementary symmetric function of the principal radii of curvature of Σ (resp. Σ') at p (resp. p'). If, for a fixed l, $2 \leqq l \leqq n$, we have

$$(1) \qquad P_{l-1}(p) \leqq P_{l-1}(p'), \qquad P_l(p) \geqq P_l(p')$$

for all points $p \in \Sigma$, then f is a translation.

The interest in this theorem lies in the fact that the conditions (1) involve only inequalities. The proof of the theorem depends on some integral formulas established in a previous paper of one of us[2] and on an algebraic inequality.

Let (λ_{ik}), (λ'_{ik}), $1 \leqq i, k \leqq n$, be positive definite symmetric matrices. We define the polynomials $P_{rs}(\lambda, \lambda')$ by means of the equation

$$(2) \qquad \det(\delta_{ik} + y\lambda_{ik} + y'\lambda'_{ik}) = \sum_{0 \leq r,s \leq n} \frac{n!}{r!\, s!\, (n-r-s)!} P_{rs}(\lambda, \lambda') y^r y'^s.$$

$P_{rs}(\lambda, \lambda')$ are thus homogeneous polynomials of degrees r, s in λ_{ik}, λ'_{ik}, respectively. It is easy to derive from the definition that

$$(3) \qquad rP_{r-1,1}(\lambda, \lambda') = \sum_{i \leq k} \lambda'_{ik} \frac{\partial P_{r0}}{\partial \lambda_{ik}}.$$

Since $P_{r0}(\lambda, \lambda')$ is independent of λ'_{ik}, we shall write $P_{r0}(\lambda, \lambda') = P_r(\lambda)$; then $P_{0r}(\lambda, \lambda') = P_r(\lambda')$.

Lemma. *Let $\alpha = (\alpha_{ik})$, $\lambda = (\lambda_{ik})$ be positive definite symmetric matrices such that, for a fixed l, $2 \leqq l \leqq n$,*

$$(4) \qquad P_{l-1}(\alpha) \leqq P_{l-1}(\lambda), \qquad P_l(\alpha) \geqq P_l(\lambda).$$

[1]The first two authors are partially supported by the National Science Foundation, and the third author is supported by the Air Force Office of Scientific Research.

[2]S. S. Chern, *Integral formulas for hypersurfaces in Euclidean space and their applications to uniqueness theorems, this Journal,* 8 (1959), pp. 947–955. This paper will be quoted as IFH.

Then

(5) $$Q_l(\alpha, \lambda) \equiv P_l(\alpha) + P_l(\lambda) - 2P_{l-1,1}(\alpha, \lambda) \leqq 0.$$

Proof. Let D be the convex domain of all positive definite symmetric matrices of order n. Let $\delta = (\delta_{ik}) \varepsilon D$ be the unit matrix. We have

$$\det(\delta_{ik} + \lambda_{ik}y) = \sum_{0 \leqq r \leqq n} \binom{n}{r} P_r(\lambda) y^r,$$

so that

$$\det(\delta_{ik} + \lambda_{ik}y + t\,\delta_{ik}y) = \sum_{0 \leqq r \leqq n} \binom{n}{r} P_r(\lambda + t\,\delta) y^r.$$

But the left-hand side can also be written

$$\det(\delta_{ik}(1 + ty) + \lambda_{ik}y) = \sum_{0 \leqq r \leqq n} \binom{n}{r} P_r(\lambda) y^r (1 + ty)^{n-r}.$$

Comparing the coefficients of y^l, we get

$$P_l(\lambda + t\,\delta) = P_l(\lambda) + lP_{l-1}(\lambda)t + \sum_{2 \leqq i \leqq l} c_i P_{l-i}(\lambda) t^i, \qquad c_i > 0.$$

Similarly, comparing the coefficients of $y^{l-1}y'$ on both sides of the equation

$$\det(\delta_{ik} + y\alpha_{ik} + y'\,\delta_{ik}) = \sum_{0 \leqq r, s \leqq n} \frac{n!}{r!\,s!\,(n - r - s)!} P_{rs}(\alpha, \delta) y^r y'^s$$
$$= \det((1 + y')\,\delta_{ik} + y\alpha_{ik}) = \sum_{0 \leqq r \leqq n} \frac{n!}{r!\,(n - r)!} P_r(\alpha) y^r (1 + y')^{n-r},$$

we get

(6) $$P_{l-1,1}(\alpha, \delta) = P_{l-1}(\alpha).$$

It follows that

(7) $$P_{l-1,1}(\alpha, \lambda + t\,\delta) = P_{l-1,1}(\alpha, \lambda) + tP_{l-1}(\alpha),$$

and therefore that

$$Q_l(\alpha, \lambda + t\,\delta) = Q_l(\alpha, \lambda) + \{lP_{l-1}(\lambda) - 2P_{l-1}(\alpha)\}t + \sum_{2 \leqq i \leqq l} c_i P_{l-i}(\lambda) t^i, \; c_i > 0.$$

Hence if $t > 0$ and $P_{l-1}(\lambda) \geqq P_{l-1}(\alpha)$, then $Q_l(\alpha, \lambda + t\delta) > Q_l(\alpha, \lambda)$.

The tangent hyperplane at α of the hypersurface $\{\lambda \mid P_l(\lambda) = P_l(\alpha)\}$ has the equation

$$\sum_{i \leqq k} (\mu_{ik} - \alpha_{ik}) \frac{\partial P_l(\alpha)}{\partial \alpha_{ik}} = l(P_{l-1,1}(\alpha, \mu) - P_l(\alpha)) = 0.$$

Since the set $\{\lambda \mid P_l(\lambda) \geqq P_l(\alpha)\}$ is strictly convex, we have $P_{l-1,1}(\alpha, \lambda) -$

$P_l(\alpha) > 0$ if $P_l(\lambda) > P_l(\alpha)$. (This also follows directly from GÅRDING's inequality.)

We now prove that $P_{l-1}(\alpha) \leqq P_{l-1}(\lambda)$, $Q_l(\alpha, \lambda) \geqq 0$ imply $P_{l-1,1}(\alpha, \lambda) \geqq P_l(\alpha)$. Suppose the contrary be true. Then there exists $\lambda_0 \in D$ such that $P_{l-1}(\alpha) \leqq P_{l-1}(\lambda_0)$, $Q_l(\alpha, \lambda_0) \geqq 0$, $P_{l-1,1}(\alpha, \lambda_0) < P_l(\alpha)$. The ray $\lambda_0 + t\delta$, $t > 0$, does not belong to the hyperplane $P_{l-1,1}(\alpha, \lambda) = P_{l-1,1}(\alpha, \lambda_0)$, and hence meets the hyperplane $P_{l-1,1}(\alpha, \lambda) = P_l(\alpha)$ in a point $\lambda_0 + t_0\delta$, $t_0 > 0$. Then we have $P_{l-1,1}(\alpha, \lambda_0 + t_0\delta) = P_l(\alpha)$. From the result of the last paragraph this implies that $P_l(\lambda_0 + t_0\delta) \leqq P_l(\alpha)$. It follows that

$$Q_l(\alpha, \lambda_0 + t_0\, \delta) \leqq 2\{P_l(\alpha) - P_{l-1,1}(\alpha, \lambda_0 + t_0\, \delta)\} = 0.$$

On the other hand, we have, since $P_{l-1}(\alpha) \leqq P_{l-1}(\lambda_0)$,

$$Q_l(\alpha, \lambda_0 + t_0\, \delta) > Q_l(\alpha, \lambda_0) \geqq 0.$$

But this is a contradiction, and our statement is proved.

Suppose therefore that $P_{l-1}(\alpha) \leqq P_{l-1}(\lambda)$, $Q_l(\alpha, \lambda) > 0$. Then

$$P_l(\lambda) - P_{l-1,1}(\alpha, \lambda) > -P_l(\alpha) + P_{l-1,1}(\alpha, \lambda) \geqq 0$$

or

$$P_l(\alpha) \leqq P_{l-1,1}(\alpha, \lambda) < P_l(\lambda).$$

This obviously implies the statement of the lemma.

To prove the theorem we utilize the integral formulas established in IFH. Suppose the first hypersurface to be defined by the mapping $x\colon S_0 \to \Sigma$, where S_0 is the unit hypersphere in the Euclidean space and $x(\xi)$, $\xi \in S_0$, is the point of Σ whose outward unit normal vector is ξ. Similarly, the second hypersurface is defined by a mapping $x'\colon S_0 \to \Sigma'$. From their second fundamental forms we construct the mixed scalar invariants P_{rs}, $0 \leqq r, s \leqq n$ (*cf.* IFH). Then we have the following integral formula:

$$\int_{S_0} (P_{rs} - h'P_{r,s-1})\, dV = 0, \qquad 1 \leqq s \leqq n, \tag{8}$$

where dV is the volume element of S_0 and $h'(\xi)$ is the support function of Σ'. By a proper choice of the origin we can always suppose $h'(\xi) > 0$. From (8) we derive the formula

$$\int_{S_0} \{(P_{l0} + P_{0l} - 2P_{l-1,1}) + (P_{0l} - P_{l0}) + 2h'(P_{l-1,0} - P_{0,l-1})\}\, dV = 0. \tag{9}$$

The hypotheses of the theorem give $P_{l-1,0} \leqq P_{0,l-1}$, $P_{0l} \leqq P_{l0}$. From these and the lemma it follows that $P_{0l} + P_{l0} - 2P_{l-1,1} \leqq 0$. Hence the integrand in (9) is $\leqq 0$, and must vanish identically since it is continuous. In particular, we have $P_{0l} = P_{l0}$ for all $p \in \Sigma$. Our theorem then follows from the theorem of ALEXANDROFF-FENCHEL-JESSEN.

Corollary. *Let Σ be a closed strictly convex hypersurface in Euclidean space. If there is a constant c such that*

$$P_{l-1}^{1/(l-1)}(p) \geqq c \geqq P_l^{1/l}(p), \qquad p \in \Sigma, \tag{10}$$

then Σ is a hypersphere.

Remark. The theorem can obviously be extended to a pair of convex hypersurfaces with boundaries.

University of Chicago
Chicago, Illinois
and
Lehigh University
Bethlehem, Pennsylvania

COMPLEX ANALYTIC MAPPINGS OF RIEMANN SURFACES I.*

Dedicated to Professor Marston Morse

By SHIING-SHEN CHERN.[1]

Introduction. The geometrical nature of the theory of functions in one complex variable is a well-known fact and has been particularly emphasized by L. Ahlfors (cf. Bibliography). It is also the most natural viewpoint, because complex function theory should be regarded as the first chapter of the theory of complex analytic mappings of complex manifolds and the classical study of value distributions is the study of the "size" of the image of a complex analytic mapping. We give in this paper a treatment, from a purely differential-geometric viewpoint, of complex analytic mappings of a Riemann surface (= one-dimensional complex analytic manifold) into a compact Riemann surface. In the case when the first Riemann surface is a compact one with a finite number of points deleted, we derive defect relations which generalize the classical relations of Nevanlinna-Ahlfors. In a subsequent paper we will consider the case when the first Riemann surface is a compact one with a finite number of points and a finite number of disks deleted. The paper is written for differential geometers, so that concepts currently in use in differential geometry are freely used and a minimum of function theory will be required. The explicit models of the Gaussian plane or the unit disk are avoided.

1. Hermitian metric on a Riemann surface. Let M be a Riemann surface. On M suppose there be an Hermitian metric, which is given, in terms of a local coordinate $z = x + iy$, by

$$ds^2 = h^2\, dz d\bar{z}, \tag{1}$$

where h is real and strictly positive. We suppose h to be of class C^∞ in the real local coordinates x, y. With the Hermitian metric it is possible to speak of the unit tangent vectors of M, the totality of which forms a circle bundle B over M. We denote by ψ the projection of B onto M.

* Received May 18, 1959.

[1] Work done under partial support of the National Science Foundation.

To the Hermitian metric there corresponds the associated two-form

$$\Omega = (i/2)h^2\, dz \wedge d\bar{z}. \tag{2}$$

It is a real-valued exterior two-form and is the element of area.

The Hermitian metric defines uniquely a connection in the bundle B, which can be described as follows: Relative to the local coordinate $z = x + iy$, the differentials dx, dy form a base in the cotangent space, and the vectors $\partial/\partial x$, $\partial/\partial y$ form its dual base in the tangent space. (Here $\partial/\partial x$ denotes the vector such that its directional derivative is the partial derivative with respect to x, and the same for $\partial/\partial y$.) As usual we introduce the complex vectors

$$\partial/dz = \tfrac{1}{2}(\partial/\partial x - i\partial/\partial y), \qquad \partial/\partial\bar{z} = \tfrac{1}{2}(\partial/\partial x + i\partial/\partial y). \tag{3}$$

Then the real unit vectors are given by

$$(e^{i\phi}\partial/\partial z + e^{-i\phi}\partial/\partial\bar{z})/h. \tag{4}$$

In $\psi^{-1}(U)$, where U is a neighborhood in which the local coordinate z is valid, z and ϕ can serve as local coordinates. The unit vector (4) defines one and only one complex-valued linear differential form

$$\omega_1 = e^{-i\phi}h\, dz, \tag{5}$$

characterized by the properties that it is of type $(1,0)$ and gives the value one when paired with (4). Let

$$\omega = -d\phi + i(d' - d'')\log h, \tag{6}$$

which is then a real-valued linear differential form in $\psi^{-1}(U)$. It can be verified that ω satisfies the equation

$$d\omega_1 = i\omega \wedge \omega_1. \tag{7}$$

and is the only real-valued linear differential form satisfying (7) and having the property that $\omega \equiv -d\phi$, mod $dz, d\bar{z}$. This characterization of ω has the important implication that it is globally defined in B, independent of the choice of local coordinates. It is said to define a connection in B.

The exterior derivation of ω gives the important formula

$$d\omega = i\tfrac{1}{2}K\omega_1 \wedge \bar{\omega}_1 = K\Omega. \tag{8}$$

The coefficient K is a real-valued function in M, and is the curvature of the Hermitian metric. In terms of a local coordinate z one finds

$$K = -(4/h^2)\partial^2(\log h)/\partial z\partial\bar{z} \tag{9}$$

which is a well-known formula for the Gaussian curvature in isothermal parameters.

The Hermitian metric is called respectively Euclidean, hyperbolic, or elliptic, if the curvature K is constant and is $= 0 < 0$, or > 0. It is well-known that a compact Riemann surface can always be given an Hermitian metric of constant curvature, the sign of the curvature being the same as the sign of its Euler characteristic.

2. Poisson's equation on a compact Riemann surface. Let Ω be a real-valued two-form of class C^1 on a compact Riemann surface M, such that $\int_M \Omega = c > 0$. An equation of the form

$$(1/\pi i)\, d'd''u = (1/c)\Omega \tag{10}$$

is called a Poisson equation. Since the equation can be written

$$(i/2\pi)\, d(d' - d'')u = (1/c)\Omega,$$

it does not have a smooth solution. The following theorem states that the equation has a solution with a logarithmic singularity at a given point of M:

THEOREM 1. *Let a be an arbitrary point on a campact Riemann surface M. Equation (10) has a solution $u(p,a)$, $p \in M$, having the following properties: 1) $u(p,a)$ is of class C^2 in $M - a$; 2) If z_a is a local coordinate at a such that $z_a = 0$ for a, then $u(p,a) - \log|z_a|$ is of class C^2 in a neighborhood of a.*

The function $u(p,a)$, whose existence is asserted in the theorem, is determined up to an additive constant, for the difference of two such functions is a harmonic function which is everywhere regular on M and is therefore a constant. We also remark that condition 2) is independent of the choice of the local coordinate z_a. For if z_a' is another such coordinate, we must have

$$z_a' = z_a f(z_a), \qquad f(0) \neq 0.$$

Then

$$\log|z_a'| = \log|z_a| + \log|f(z_a)|,$$

where $\log|f(z_a)|$ is regular at a.

To prove the theorem let $cv(p,a,b)$ be a harmonic function, which is regular in $M - a - b$, $a, b \in M$, $a \neq b$, and has the singularity $\log|z_a|$ at a and the singularity $-\log|z_b|$ at b, where z_b is a local coordinate at b such that $z_b = 0$ for b. This function $v(p,a,b)$ is defined up to an additive constant.

For $b = a$ we define $v(p,a,a) = \lim_{a \to b} v(p,a,b) = \text{const}$. We shall prove that the function

$$u(p,a) = \int_M v(p,a,b)\Omega_b, \tag{11}$$

obtained by the integration of $v(p,a,b)$ with respect to $b \in M$, fulfills the conditions of our theorem.

In the first place we will show that the integral (11) converges. Suppose $p \neq a$. Let U_p be a neighborhood about p, with the local coordinate z_p and with $a \notin U_p$. We can write

$$\begin{aligned} u(p,a) = \int_{M-U_p} v(p,a,b)\Omega_b + \int_{U_p} \{v(p,a,b) + (1/c)\log|z_p - \zeta|\}\Omega_b \\ - (1/c)\int_{U_p} \log|z_p - \zeta|\,\Omega_b \end{aligned} \tag{12}$$

where ζ is the coordinate of b. The first two integrals in (12) are proper integrals, while the third integral is obviously convergent. Clearly the function $u(p,a)$ has the singularity $\log|z_a|$ at a.

To calculate $d'd''u(p,a)$ it suffices to restrict ourselves to the neighborhood U_p. The first two integrals in (12) can be differentiated under the integral sign and are annihilated by the operator $d'd''$, because the integrands are harmonic functions. In U_p let

$$\Omega_b = iF(\zeta)\,d\zeta \wedge d\bar{\zeta}.$$

Then we have

$$\begin{aligned} (1/c)d'd'' \int_{U_p} -\log|z_p - \zeta|\,\Omega_b &= -(i/c)d'd'' \int_{U_p} \log|z_p - \zeta|\,F(\zeta)\,d\zeta d\bar{\zeta} \\ &= -(i/c)\,dz_p \wedge d\bar{z}_p(\partial^2/\partial z_p \partial \bar{z}_p) \int_{U_p} \log|z_p - \zeta|\,F(\zeta)\,d\zeta d\bar{\zeta}. \end{aligned}$$

By a well-known computation (cf., for instance, I. G. Petrovsky, Lectures on Partial Differential Equations, p. 219) this is equal to $-(1/c)\pi F dz_p \wedge d\bar{z}_p = (\pi i/c)\Omega$. It follows that $d'd''u = (1/c)\pi i\Omega$, so that $u(p,a)$ is a solution of the Poisson equation (10). Similarly, it can be proved that $u(p,a) - \log|z_a|$ is regular at a and satisfies (10).

As an example we give the function $u(p,a)$ in the case that M is the Riemann sphere and Ω is the element of area. We consider the Riemann sphere as the complex projective line P_1, whose points have the homogeneous coordinates $Z = (z_0, z_1)$ and in which there is given an Hermitian scalar product

$$(Z,W) = \overline{(W,Z)} = z_0\bar{w}_0 + z_1\bar{w}_1, \qquad W = (w_0, w_1). \tag{13}$$

We will write $|Z| = +(Z,Z)^{\frac{1}{2}} \geqq 0$. P_1 has an Hermitian metric of constant positive curvature 4, given by

$$ds^2 = [|Z|^2(dZ,dZ) - (Z,dZ)(dZ,Z)]/|Z|^4. \tag{14}$$

With this Hermitian metric, P_1 is also a metric space, and the distance $d(Z,W)$ between the points Z and W is given by the formula

$$\cos d(Z,W) = |(Z,W)|/|Z|\cdot|W|. \tag{15}$$

We now prove that the element of area is $\Omega = id'd'' \log |Z|$. In fact, let

$$Z_0 = Z/|Z|. \tag{16}$$

so that $(Z_0, Z_0) = 1$. Then

$$ds^2 = (dZ_0, dZ_0) = d\zeta_0 d\bar{\zeta}_0 + d\zeta_1 d\bar{\zeta}_1, \qquad Z_0 = (\zeta_0, \zeta_1), \tag{17}$$

and the element of area of P_1 relative to this Hermitian metric is

$$\Omega = \tfrac{1}{2} i (d\zeta_0 \wedge d\bar{\zeta}_0 + d\zeta_1 \wedge d\bar{\zeta}_1). \tag{18}$$

Let

$$\omega_{00} = (dZ_0, Z_0) = d\zeta_0 \bar{\zeta}_0 + d\zeta_1 \bar{\zeta}_1. \tag{19}$$

Then $\Omega = -\tfrac{1}{2} i d\omega_{00}$. On the other hand, we have

$$2|Z|\,d|Z| = (dZ,Z) + (Z,dZ),$$

and

$$\begin{aligned} \omega_{00} = (dZ_0, Z_0) &= (1/|Z|)((1/|Z|)dZ - (Z/|Z|^2)d|Z|, Z) \\ &= (1/2|Z|^2)\{(dZ,Z) - (Z,dZ)\} \\ &= (d' - d'')\log|Z|. \end{aligned} \tag{20}$$

It follows that

$$\Omega = id'd'' \log |Z|. \tag{21}$$

An elementary computation gives $\int_{P_1} \Omega = \pi$, which is therefore the total area of P_1.

Let $A = (a_0, a_1)$, $A\perp = (-\bar{a}_1, \bar{a}_0)$. Since $(Z, A\perp)$ is holomorphic, $\log|(Z, A\perp)|$ is zero under the operator $d'd''$, and we have

$$d'd'' \log(|(Z,A\perp)|/|Z|\cdot|A|) = -d'd'' \log|Z| = i\Omega.$$

It follows that the function

$$u = \log(|(Z,A\perp)|/|Z|\cdot|A|). \tag{22}$$

is a solution of the Poisson equation (10) and has a singularity $\log|z_A|$ at A, where z_A is a local coordinate at A. This is therefore the explicit expression of the function $u(p,a)$ whose existence was asserted by Theorem 1.

Since

$$d(Z,A)+d(Z,A^{\perp})=d(A,A^{\perp})=\pi/2,$$

we can write

$$u=\log\cos d(Z,A^{\perp})=\log\sin d(Z,A). \tag{23}$$

The quantity $\sin d(Z,A)$ is the length of the chord joining Z and A, when P_1 is realized as a sphere of diameter 1 in Euclidean space. This choice was made in the literature by Ahlfors and our discussion gives an intrinsic justification of the choice.

3. The first main theorem. Let D be a compact differentiable oriented domain bounded by a sectionally smooth curve C, and let $f\colon D\to M$ be a differentiable mapping. Suppose $f(\zeta_0)=a$, $\zeta_0\in D-C$, $a\in M$, and suppose that ζ_0 is the only point in a neighborhood of ζ_0 which is mapped into a by f. Then we can define an integer $n(\zeta_0,a)$, to be called the *order of f at* (ζ_0,a), which in a sense measures the number of times a neighborhood of a is covered by a neighborhood of ζ_0 under f. If a is such that $f^{-1}(a)$ is a finite set of points belonging to $D-C$, de define

$$n(a)=\sum_{\zeta\in f^{-1}(a)} n(\zeta,a). \tag{24}$$

The first main theorem expresses the difference of $n(a)$ and the area of $f(D)$ as an integral over the boundary curve C.

The definition of $n(\zeta_0,a)$ is as follows: Since f is a continuous mapping, there are coordinate neighborhoods U, V of ζ_0, a respectively, such that $f(U)\subset V$. Let $\zeta=\rho e^{i\theta}$, $z=re^{i\phi}$ be the local coordinates in U, V respectively, with ζ_0 and a having the coordinates $\zeta=0$, $z=0$. Denote by S_η and Σ_ϵ the circles $\rho=\eta=$ const. and $r=\epsilon=$ const. respectively, where η and ϵ are sufficiently small. There is a mapping g (called a retraction in topology): $V-a\to\Sigma_\epsilon$, which maps a point with the coordinate $z=re^{i\phi}$ into the point with the coordinate $\epsilon e^{i\phi}$. Since both D and M are oriented, the circles S_η, Σ_ϵ have induced orientations, and we use the same symbols to denote their fundamental cycles. Then $gf(S_\eta)$ is a cycle on Σ_ϵ and is homologous to an integral multiple of Σ_ϵ. This multiple, which can be shown to be independent of the various choices we have made, is defined to be the order $n(\zeta_0,a)$.

The order $n(\zeta_0, a)$ can be expressed by an integral formula. In fact, let $u(p, a)$ be a solution of (10) given by Theorem 1. Let

$$\lambda = (i/2\pi)(d' - d'')u, \tag{25}$$

so that $d\lambda = (1/c)\Omega$. Denote by $j: \Sigma_\epsilon \to V - a$ the identity mapping. Then

$$j^*\lambda = -\tfrac{1}{2\pi} d\phi + o(\epsilon),$$

where $o(\epsilon)$ denotes a differential form which tends to zero as $\epsilon \to 0$. It follows that

$$n(\zeta_0, a) = -\lim_{\epsilon\to 0} \int_{gf(S\eta)} j^*\lambda = -\lim_{\epsilon\to 0} \int_{jgf(S\eta)} \lambda.$$

We choose Ω in Theorem 1 to be the element of area of M. The first main theorem then follows by the application of Stokes' Theorem to the integral of $d\lambda$ over D. We state it as follows:

THEOREM 2. *Let D be a compact differentiable, oriented domain bounded by a sectionally smooth curve C, and let f be a differentiable mapping of D into a compact Riemann surface M. If $a \in M$ is such that $f^{-1}(a) \cap C = \emptyset$ and that $f^{-1}(a)$ is a finite set of points, then*

$$n(a) + \int_{f(C)} \lambda = (1/c) \int_{f(D)} \Omega = (1/c)v(D), \tag{26}$$

where $v(D)$ is the area of $f(D)$. In particular, if D has no boundary, then

$$n(a) = (1/c)v(D). \tag{27}$$

Remark. If f is orientation-preserving, then $v(D) \geqq 0$ and we have the following corollary of Theorem 2:

COROLLARY. *Let D be a compact oriented differentiable manifold of two real dimensions. Let $f: D \to M$ be an orientation-preserving differentiable mapping of D into a compact Riemann surface M. The image of D under f covers M completely, provided that the Jacobian of f does not vanish identically.*

4. The Gauss-Bonnet formula and the second main theorem. We now apply Stokes' Theorem to the formula (8). Suppose Δ be a domain in M bounded by a sectionally smooth curve γ. Since (8) is valid in the bundle B, and not in M, we have to "lift" Δ to B in order that Stokes' Theorem be applied. This is done by defining a differentiable field of unit vectors over Δ, such that it has a finite number of singularities in the interior of Δ and that the vectors at the points of γ point to the interior of Δ. According to a known theorem in topology, the sum of indices at the singularities of such a vector field is equal to the Euler characteristic $\chi(\Delta)$ of Δ. From the local expression (6) of ω, the integral of ω along the vectors over a small circle about a singularity has as limit -2π times the index of the singularity, as the radius of the circle tends to zero. It follows by Stokes' Theorem that

$$2\pi\chi(\Delta) + \int_\gamma \omega = \iint_\Delta K\Omega, \tag{28}$$

where the simple integral is over the vector field along γ. There is arbitrariness in the choice of the unit vector field along γ. If γ is smooth, it is natural to choose at each point $p \in \gamma$ the unit interior normal vector to γ at p. Then ω is equal to the element of arc of γ multiplied by its geodesic curvature. If γ is only sectionally smooth, then, corresponding to each corner, we have to add the exterior angle at that corner. Formula (28) is of course the well-known Gauss-Bonnet formula.

Consider now a compact complex analytic domain D bounded by a sectionally smooth curve C and a complex analytic mapping $f: D \to M$. If ζ is a local coordinate about a point $\zeta_0 \in D$, and z a local coordinate about the image point $a = f(\zeta_0)$, ζ_0 is called a stationary point or branch point, if $(dz/d\zeta)_{\zeta_0} = 0$. This condition is obviously independent of the choice of the local coordinates. If f is not a constant mapping, the stationary points are isolated. The Hermitian metric (1) in M induces the Hermitian metric $f^*(ds^2)$ in D, except at the stationary points. We will generalize the Gauss-Bonnet formula (28) to the domain D by taking account of the contributions arising from the isolated stationary points.

Suppose that the local coordinates ζ, z about ζ_0, a respectively are such that they vanish for ζ_0, a. The coordinate z of the image point $f(\zeta)$ is then given by a power series in ζ:

$$z = a_m\zeta^m + a_{m+1}\zeta^{m+1} + \cdots, \qquad a_m \neq 0, m \geqq 1, \tag{29}$$

which is convergent in a neighborhood of $\zeta = 0$. The integer $m - 1$ is called the stationary index; a point of stationary index 0 is a regular point. About ζ_0 draw a small circle S_η of radius η and attach at each of its points the unit outward normal vector, so that it will be an inward normal vector of the complement domain. By (6) the integral of ω over this vector field tends to $-2\pi m$, as $\eta \to 0$.

If there are s stationary points, the Euler characteristic of the domain

D with the stationary points deleted is $\chi(D)-s$. Hence, if no stationary point lies on C, we have the formula

$$2\pi(\chi(D)-s)+\int_C \omega+2\pi\sum_{1\leq i\leq s} m_i=\iint_{f(D)} K\Omega,$$

or

$$2\pi\chi(D)+\int_C \omega+2\pi\sum_{1\leq i\leq s}(m_i-1)=\iint_{f(D)} K\Omega, \tag{30}$$

where m_i-1, $1\leq i\leq s$, are the stationary indices.

If D has no boundary and is an n-leaved covering surface of M, then $\iint_{f(D)} K\Omega=2\pi n\chi(M)$, and (30) becomes

$$\chi(D)+\sum_{1\leq i\leq s}(m_i-1)=n\chi(M). \tag{31}$$

This is a well-known formula of A. Hurwitz.

5. Point at infinity. Let V_1 be a compact Riemann surface, and V the surface obtained from it by removing a finite number of points $v_1,\cdots,v_m$. Such a point v_k, $1\leq k\leq m$, is called a point of infinity of V. Let $\zeta_k=r_ke^{i\theta_k}$ be a local coordinate at v_k, with $\zeta_k(v_k)=0$, which will be fixed from now on. Denote by D_ϵ the domain obtained from V_1 by removing the disks $r_k\leq\epsilon$, so that the boundary of D_ϵ consists of the m circles γ_k defined by $r_k=\epsilon$, and that D_ϵ tends to V as $\epsilon\to 0$. Consider a complex analytic mapping $f\colon V\to M$, where M is a compact Riemann surface, such that f does not send all points of V into one point. Then to every $a\in M$, the set $f^{-1}(a)\cap D_\epsilon$ is a finite set. If in addition $f^{-1}(a)\cap\gamma_k=\emptyset$, Theorem 2 gives

$$n(a,\epsilon)-\sum_{1\leq k\leq m}\int_{\gamma_k} f^*\lambda=(1/c)v(\epsilon), \tag{32}$$

where c is the total area of M, $v(\epsilon)$ is the area of $f(D_\epsilon)$, and $n(a,\epsilon)$ is the number of times that a is covered by $f(D_\epsilon)$. The negative sign before the sum is caused by the fact that γ_k is to be oriented by increasing θ_k.

The integral in (32) can be simplified as follows: f being a complex analytic mapping, the dual mapping f^* commutes with both d' and d'', so that we have

$$f^*\lambda=(i/2\pi)(d'-d'')u(f(\zeta_k)).$$

Dropping the subscript k for the moment, consider the local coordinate $\zeta = re^{i\theta}$. If h is a real-valued differentiable function in r, θ, we have

$$dh = h_r\,dr + h_\theta\,d\theta = h_\zeta\,d\zeta + h_{\bar\zeta}\,d\bar\zeta$$
$$= (h_\zeta e^{i\theta} + h_{\bar\zeta} e^{-i\theta})\,dr + ir\,(h_\zeta e^{i\theta} - h_{\bar\zeta} e^{-i\theta})\,d\theta,$$

so that

$$h_r = h_\zeta e^{i\theta} + h_{\bar\zeta} e^{-i\theta}$$

and

$$(33)\qquad (d' - d'')h = h_\zeta\,d\zeta - h_{\bar\zeta}\,d\bar\zeta = (\cdot\ \cdot\ \cdot)\,dr + irh_r\,d\theta.$$

It follows from (32) that

$$(34)\qquad n(a,\epsilon) + \tfrac{1}{2\pi}\epsilon \sum_k \int_0^{2\pi} (\partial u/\partial\epsilon)\,d\theta_k = (1/c)\,v(\epsilon).$$

This leads us to put

$$(35)\qquad T(\epsilon) = (1/c)\int_\epsilon^{\epsilon_0} v(t)\,dt/t, \qquad N(a,\epsilon) = \int_\epsilon^{\epsilon_0} n(a,t)\,dt/t,$$

where ϵ_0 is a small constant, with $\epsilon < \epsilon_0$. A standard argument will justify the relation

$$\int_0^{2\pi} (\partial u/\partial\epsilon)\,d\theta_k = (\partial/\partial\epsilon)\int_0^{2\pi} u\,d\theta_k,$$

and we get, by integrating the relation (34),

$$N(a,\epsilon) + \tfrac{1}{2\pi}\sum_k \int_0^{2\pi} u(\epsilon e^{i\theta_k}, a)\,d\theta_k \Big|_\epsilon^{\epsilon_0} = T(\epsilon),$$

or

$$(36)\qquad N(a,\epsilon) - \tfrac{1}{2\pi}\sum_k \int_0^{2\pi} u(\epsilon e^{i\theta_k}, a)\,d\theta_k = T(\epsilon) + \text{const.}$$

Since $u(p,a)$ differs by a regular function from $\log|z_a|$ in a neighborhood of a and since M is compact, $u(p,a)$ has an upper bound, and we have the fundamental inequality

$$(37)\qquad N(a,\epsilon) < T(\epsilon) + \text{const.}$$

This shows that the *order function* $T(\epsilon)$ dominates $N(a,\epsilon)$ for all $a \in M$.

We now derive an explicit formula for the second main theorem as applied to the domain D_ϵ. Let U_k be a coordinate neighborhood in V_1 in which the local coordinate ζ_k is valid. The restriction of f gives a mapping $f\colon U_k - v_k \to M$. Let $f^*\Omega = \sigma_k{}^2 r_k dr_k \wedge d\theta_k$, so that $\sigma_k{}^2$ is the ratio of the two elements of area. Since f is complex analytic, we have $\sigma_k{}^2 \geqq 0$ and $\sigma_k = 0$ if and only if the point is a stationary point. To the points of $U_k - v_k$ we attach the radial vectors $\partial/\partial r_k$ and we will compute ω for this family of radial vectors. Since $\partial/\partial r_k = e^{i\theta_k}\partial/\partial\zeta_k + e^{-i\theta_k}\partial/\partial\bar{\zeta}_k$, we get, from (4) and (6),

$$\omega = -d\theta_k + i(d' - d'')\log\sigma_k. \tag{38}$$

If γ_k is free from stationary points, the restriction of ω to γ_k is

$$\omega = -(1 + \epsilon(\partial/\partial\epsilon)\log\sigma_k)\,d\theta_k.$$

Let $n_1(\epsilon)$ be the sum of stationary indices in D_ϵ. Then the second main theorem (30) reduces to the formula

$$\chi(V_1) - m + \tfrac{1}{2\pi}\sum_k \int_0^{2\pi}(1 + \epsilon(\partial/\partial\epsilon)\log\sigma_k)\,d\theta_k + n_1(\epsilon) = \tfrac{1}{2\pi}\iint_{f(D\epsilon)} K\Omega,$$

or

$$\chi(V_1) + (\epsilon/2\pi)\sum_k \int_0^{2\pi}(\partial/\partial\epsilon)\log\sigma_k\,d\theta_k + n_1(\epsilon) = \tfrac{1}{2\pi}\iint_{f(D\epsilon)} K\Omega. \tag{39}$$

In particular, if the Gaussian curvature K is constant, we have

$$\chi(V_1) + (\epsilon/2\pi)\sum_k \int_0^{2\pi}(\partial/\partial\epsilon)\log\sigma_k\,d\theta_k + n_1(\epsilon) = (K/2\pi)v(\epsilon). \tag{40}$$

We put

$$N_1(\epsilon) = \int_\epsilon^{\epsilon_0} n_1(t)\,dt/t, \tag{41}$$

and remark that the relation

$$\int_0^{2\pi}(\partial/\partial\epsilon)\log\sigma_k\,d\theta_k = (\partial/\partial\epsilon)\int_0^{2\pi}\log\sigma_k\,d\theta_k$$

can be justified by a standard argument. Integration of the above equation then gives

$$\chi(V_1)(\log\epsilon_0 - \log\epsilon) + \tfrac{1}{2\pi}\sum_k \int_0^{2\pi}\log\sigma_k\,d\theta_k\Big|_\epsilon^{\epsilon_0} + N_1(\epsilon)$$
$$= (cK/2\pi)T(\epsilon) = \chi(M)T(\epsilon)$$

or

$$-\chi(V_1)\log\epsilon - \tfrac{1}{2\pi}\sum_k \int_0^{2\pi}\log\sigma_k\,d\theta_k + N_1(\epsilon) = \chi(M)T(\epsilon) + \text{const.} \tag{42}$$

This is the integrated form of the second main theorem.

6. Generalization of the defect relation of Nevanlinna-Ahlfors. Let $a_1, \cdots, a_q$ be a finite set of points in M, and let

$$\rho(p) = c_1\Big(\prod_{1\leq i\leq q}\exp u(p, a_i)\Big)^{-2\lambda}, \qquad 0 < \lambda < 1, \tag{43}$$

where c_1 is a constant. Then the integral $\int_M \rho(p)\Omega$, where Ω is the element of area of M, is convergent. For the points where the convergence of the integral should be checked are the points a_i and at a_i the integrand is $1/|z_{a_i}|^{2\lambda}$ times a bounded function. We choose the constant c_1 so that $\int_M \rho(p)\Omega = 1$.

We now integrate the inequality (37) over M with respect to the density $\rho(p)\Omega$. Noticing that $n(p,t)$ is the number of times that the point p is covered by $f(D_t)$, we get

$$\int_\epsilon^{\epsilon_0}(dt/t)\int_{D_t}\rho f^*\Omega < T(\epsilon) + \text{const.}$$

Now

$$\int_{D_t}\rho f^*\Omega > \int_{D_t - D_{\epsilon_0}}\rho f^*\Omega = \int_t^{\epsilon_0} r\,dr \sum_k \int_0^{2\pi}\rho\sigma_k^2\,d\theta_k$$

so that we have

$$\int_\epsilon^{\epsilon_0}(dt/t)\int_t^{\epsilon_0} r\,dr\sum_k\int_0^{2\pi}\rho\sigma_k^2\,d\theta_k < T(\epsilon) + \text{const.} \tag{44}$$

LEMMA. *The inequality*

$$\int_\epsilon^{\epsilon_0}(dt/t)\int_t^{\epsilon_0} e^{\phi(r)} r\,dr < S(\epsilon) \tag{45}$$

implies

$$\phi(\epsilon) \leqq \kappa^2 \log S(\epsilon) - 2\log \epsilon, \qquad \kappa > 1, \tag{46}$$

except for a set of intervals in $(0,\epsilon_0)$ *for which* $\int_\epsilon^{\epsilon_0} d\log\epsilon < \infty$.

Proof. Let $A(\epsilon)$ be a decreasing positive-valued C^1-function and $\Lambda(\epsilon)$ be a positive-valued continuous function, both defined in the interval $0 < \epsilon \leqq \epsilon_0$. Suppose

$$-A'(\epsilon) > A(\epsilon)^\kappa \Lambda(\epsilon), \qquad \kappa > 1.$$

Then

$$\Lambda(\epsilon) < -(1/A^\kappa)dA/d\epsilon = (1/(\kappa-1))\,dA^{1-\kappa}/d\epsilon$$

and

$$\int_\epsilon^{\epsilon_0}\Lambda(\epsilon)\,d\epsilon < (1/(\kappa-1))(A(\epsilon_0)^{1-\kappa} - A(\epsilon)^{1-\kappa}),$$

so that the integral at the left-hand side converges as $\epsilon \to 0$. Hence, with the exception of a set of $0 < \epsilon \leqq \epsilon_0$ for which the integral $\int_0^{\epsilon_0}\Lambda(\epsilon)\,d\epsilon$ converges, we have

$$-A'(\epsilon) \leqq A(\epsilon)^\kappa \Lambda(\epsilon),$$

whence

$$\log(-A') \leqq \kappa \log A(\epsilon) + \log \Lambda(\epsilon).$$

It follows that

$$\phi(\epsilon) + \log \epsilon \leqq \kappa \log \int_\epsilon^{\epsilon_0} e^{\phi(r)} r\, dr + \log \Lambda(\epsilon),$$

$$-\log \epsilon + \log \int_\epsilon^{\epsilon_0} e^{\phi(r)} r\, dr \leqq \kappa \log \int_\epsilon^{\epsilon_0} (dt/t) \int_t^{\epsilon_0} e^{\phi(r)} r\, dr + \log \Lambda(\epsilon),$$

so that

$$\phi(\epsilon) + (1-\kappa)\log \epsilon \leqq \kappa^2 \log S(\epsilon) + (1+\kappa) \log \Lambda(\epsilon).$$

Choose $\Lambda(\epsilon) = 1/\epsilon$; then

$$\phi(\epsilon) \leqq \kappa^2 \log S(\epsilon) - 2 \log \epsilon,$$

with the exception of a subset of $0 < \epsilon \leqq \epsilon_0$ for which $\int_0^{\epsilon_0} d \log \epsilon < \infty$. This completes the proof of the lemma.

Following H. Weyl we use the notation $\|$ to denote that an inequality is not universally valid, but only under the exception stated in the lemma.

It follow from the lemma that

$$(47) \| \qquad \log \sum_k \int_0^{2\pi} \rho \sigma_k^2\, d\theta_k \leqq \kappa^2 \log(T(\epsilon) + C) - 2 \log \epsilon,$$

where C is a constant. By the concavity of the logarithm we have

$$\begin{aligned} \tfrac{1}{2\pi} \sum_k \int_0^{2\pi} \log(\rho\sigma_k^2)\, d\theta_k &\leqq \sum_k \log\{ \tfrac{1}{2\pi} \int_0^{2\pi} \rho\sigma_k^2\, d\theta_k \} \\ (48) \| \qquad &\leqq m \log\{(1/2m\pi) \sum_k \int_0^{2\pi} \rho\sigma_k^2\, d\theta_k\} \\ &\leqq m\kappa^2 \log(T(\epsilon) + C) - 2m \log \epsilon + O(1). \end{aligned}$$

On the other hand, by means of (36), (42), (43), the left-hand side of this inequality can be transformed as follows:

$$\begin{aligned} \tfrac{1}{2\pi} \sum_k \int_0^{2\pi} & \log(\rho\sigma_k^2)\, d\theta_k \\ &= \tfrac{1}{2\pi} \sum_k \int_0^{2\pi} \log \rho\, d\theta_k + (2/2\pi) \sum_k \int_0^{2\pi} \log \sigma_k\, d\theta_k \\ &= \text{const} - (2\lambda/2\pi) \sum_k \sum_{1 \leqq j \leqq q} \int_0^{2\pi} u(\epsilon e^{i\theta_k}, a_j)\, d\theta_k \\ &\qquad - 2\chi(V_1) \log \epsilon + 2N_1(\epsilon) - 2\chi(M) T(\epsilon) \\ &= \text{const} + 2(\lambda q - \chi(M)) T(\epsilon) - 2\lambda \sum_{1 \leqq j \leqq q} N(a_j, \epsilon) \\ &\qquad - 2\chi(V_1) \log \epsilon + 2N_1(\epsilon). \end{aligned}$$

Letting $\lambda \to 1$ and using (48), we get

$$(49)\| \quad (q-\chi(M))T(\epsilon) - \sum_{1\leqq j\leqq q} N(a_j,\epsilon) + N_1(\epsilon) \leqq \tfrac{1}{2}m\kappa^2 \log(T(\epsilon)+C) + \chi(V)\log\epsilon + O(1).$$

For $a \in M$ we define the defect of the mapping f by

$$(50) \qquad \delta(a) = 1 - \overline{\lim}(N(a,\epsilon)/T(\epsilon)).$$

Then $0 \leqq \delta(a) \leqq 1$ and $\delta(a) = 1$, if the point a does not belong to the image $f(V)$.

THEOREM 3. *Let V be a non-compact Riemann surface which is obtained from a compact one by the deletion of a finite number of points. Let M be a compact Riemann surface and let $f\colon V \to M$ be a non-constant complex analytic mapping. Let $a_1, \cdots, a_q \in M$. If $\chi(V) \geqq 0$ or if $\underline{\lim}(-\log\epsilon/T(\epsilon)) = 0$, then*

$$(51) \qquad \sum_{1\leqq j\leqq q} \delta(a_j) \leqq \chi(M).$$

Condition (51) means that $\sum \delta(a_j) \leqq 2$, when M is the Riemann sphere. When M is a complex torus, then (51) implies that all the defects are zero. If M is of genus > 1, (51) is a contradiction, so that such a mapping f does not exist.

The theorem follows immediately from (49), for the latter implies, as $N_1(\epsilon) \geqq 0$,

$$\sum_j \delta(a_j) \leqq \chi(M) - \chi(V)\underline{\lim}(-\log\epsilon/T(\epsilon)).$$

To clarify Theorem 3, we have to study the meaning of the condition

$$\underline{\lim}(-\log\epsilon/T(\epsilon)) = 0.$$

Suppose $\underline{\lim}(-\log\epsilon/T(\epsilon)) \neq 0$. Then $\overline{\lim}(T(\epsilon)/-\log\epsilon) = b < \infty$, and we have $T(\epsilon) = O(-\log\epsilon)$.

THEOREM 4. *Suppose V and M be Riemann surfaces such that M is compact and V is obtained from a compact one V_1 by the deletion of a finite number of points. Let $f\colon V \to M$ be a non-constant complex analytic mapping such that $T(\epsilon) = O(-\log\epsilon)$. Then f can be extended into a complex analytic mapping of V_1 into M.*

Let $v \in V_1$ be a point at infinity of V. It follows from (37) and the condition $T(\epsilon) = O(-\log\epsilon)$ that $n(a,\epsilon)$, $a \in M$, is bounded in a neighborhood of v. Consider first the case that $M = M_0$ is the Riemann sphere. The mapping $f\colon V \to M_0$ defines a meromorphic function in a neighborhood of v with an isolated singularity at v. Since $n(a, \epsilon)$ is bounded, the image of $U - v$ under f will omit a finite number of points of M_0, provided that the neighborhood U is small enough. It follows that v is a removable singularity or a pole, i. e., that the mapping f can be extended to a complex analytic mapping of V_1 into M_0.

In the general case that M is an arbitrary compact Riemann surface let M_0 be the Riemann sphere and let $g\colon M \to M_0$ be a covering of M_0, possibly ramified. The composition $gf\colon V \to M_0$ is a complex analytic mapping for which the condition $T(\epsilon) = O(-\log \epsilon)$ is still fulfilled. Hence gf can be extended to a complex analytic mapping of V_1 into M_0. It follows that, if U is small enough, $f(U - v)$ will belong to a finite number of neighborhoods of M, which could moreover be arbitrarily small. Since this is true for any covering g, this is possible only when $f(U - v)$ belongs to one coordinate neighborhood of M. In terms of local coordinates f defines a bounded holomorphic function. Thus the singularity v is removable, and our proof of Theorem 4 is complete.

It seems reasonable to call a non-constant complex analytic mapping $f\colon V \to M$ transcendental, if $T(\epsilon) \neq O(-\log \epsilon)$. Then we can say that a transcendental mapping f satisfies the defect relation (51).

UNIVERSITY OF CHICAGO.

REFERENCES.

[1] L. V. Ahlfors, "Geometrie der Riemannschen Flächen," *Comptes Rendus du Congrès International des Mathematiciens*, Oslo 1936, pp. 239-248.

[2] ———, "Uber die Anwendung differentialgeometrischer Methoden zur Untersuchung von Uberlagerungsflächen," *Acta Societatis Scientiarum Fennicae*, Nova Series A, Tom II, No. 6, pp. 1-17 (1937).

[3] G. af Hällström, "Uber meromorphe Funktionen mit mehrfach zusammenhängenden Existenzgebieten," *Acta Academiae Aboensis, Mathematica et Physica*, XII, vol. 8 (1939), pp. 1-100.

ANNALS OF MATHEMATICS
Vol. 71, No. 3, May, 1960
Printed in Japan

THE INTEGRATED FORM OF THE FIRST MAIN THEOREM FOR COMPLEX ANALYTIC MAPPINGS IN SEVERAL COMPLEX VARIABLES*

BY SHIING-SHEN CHERN

(Received October 12, 1959)

Introduction

The object of this paper is to make a beginning of the study of complex analytic mappings of the complex euclidean space of m dimensions into the complex projective space of the same dimension in the direction of a generalization of the classical theory of Picard-Borel for the one-dimensional case. The main geometrical conclusion of the Picard-Borel theory can be interpreted as a statement on the "size" of the image set. It is well-known that similar statements are not true in the case of higher dimensions. In fact, there are examples due to Bieberbach and Fatou in which the image set omits open subsets of the complex projective space [1]. A proper approach to the theory should therefore begin with introducing the suitable concepts and draw conclusions on them. In a problem of this nature it is the general dimension of the original set which creates the difference from the classical theory. We take it to be the complex euclidean space. When the situation in this particular case becomes clear, various generalizations are possible.

Let $\zeta_1, \cdots, \zeta_m$ be the coordinates in the complex euclidean space E_m of m dimensions. Let D_r be the solid sphere defined by

$$\zeta_1\bar{\zeta}_1 + \cdots + \zeta_m\bar{\zeta}_m \leqq r^2 , \tag{1}$$

and Σ_r be its boundary. We will exhaust E_m by D_r as $r \to \infty$ and consider various geometrical quantities which are functions of r. Another possibility for the exhaustion of E_m would be by the polycylinders $|\zeta_k| \leqq r$, $1 \leqq k \leqq m$. This seems to be less advantageous, because most of the geometrical quantities will be given as integrals over the boundaries of the domains in question and the boundary of a polycylinder is not a differentiable manifold. Meanwhile, it may be remarked that if we consider E_m to be compactified by adding a hyperplane at infinity, then the exterior of D_r is a tube about the hyperplane. This gives a geometrical justification of our choice of the exhaustion of E_m by D_r.

The space E_m has of course the hermitian metric

$$ds_0^2 = \sum_{1 \leqq k \leqq m} d\zeta_k d\bar{\zeta}_k \tag{2}$$

* Work done under Air Force Contract No. AF49(638)-525.

and the associated two-form

$$(3)\qquad \Omega_0 = \frac{i}{2}\sum_{1\leq k\leq m} d\zeta_k \wedge d\bar{\zeta}_k\,.$$

To the complex projective space P_m of dimension m we give the standard hermitian metric with constant holomorphic curvature and let Ω be its associated two-form. Let $f: E_m \to P_m$ be a complex analytic mapping. We introduce the quantities

$$(4)\qquad v_k(r) = \int_{D_r} f^*\Omega^{m-k} \wedge \Omega_0^k\,, \qquad 0 \leq k \leq m\,,$$

so that $v_m(r)$ is the volume of D_r and $v_0(r)$ is the volume of the image of D_r in P_m. For a fixed $r_0 > 0$ we put

$$(5)\qquad T(r) = \int_{r_0}^{r} \frac{v_0(t)dt}{t^{2m-1}}\,.$$

$T(r)$ is a generalization of the order function of Nevanlinna for $m = 1$, as defined geometrically by Ahlfors and Shimizu.

The main result of our paper is an integrated form of the so-called first main theorem, which we will give below. We state here the following geometrical consequence:

THEOREM. *Let $f: E_m \to P_m$ be a complex analytic mapping, which satisfies the following two asymptotic conditions as $r \to \infty$:*

(1) $T(r) \to \infty$;

(2) $\int_{r_0}^{r} (v_1'(t)dt)/t^{2m} = o(T(r))$.

Then the complement of the set $f(E_m)$ in P_m is of measure zero.

We remark that, for non-trivial mappings, condition (1) is automatically satisfied when $m = 1$. Condition (2) shows the necessity of considering $v_k(r)$ for values of k other than 0 and m.

1. The hypersphere in the real and complex euclidean space

Let $E^n(R)$ be the oriented real euclidean space of dimension n with the coordinates $x_1, \cdots, x_n$. We will study the geometry on the hypersphere Σ_r defined by the equation

$$(6)\qquad x_1^2 + \cdots + x_n^2 = r^2\,.$$

Differentiating this equation we find easily that on Σ_r the exterior differential form

$$(7)\qquad (-1)^{i-1}\frac{r}{x_i}(dx_1 \wedge \cdots \wedge dx_{i-1} \wedge dx_{i+1} \wedge \cdots \wedge dx_n), \quad 1 \leq i \leq n\,,$$

is independent of i. We call it Θ_r and will show that it is the volume element of Σ_r, when Σ_r is oriented so that it is the oriented boundary of its interior (i.e., the set defined by $x_1^2 + \cdots + x_n^2 \leqq r^2$), the latter being coherently oriented with $E^n(R)$.

To prove this, remark that this orientation is the same as the one defined by the property that at every point $x = (x_1, \cdots, x_n)$ of Σ_r the outward normal $\xi = (x_1/r, \cdots, x_n/r)$, followed by the oriented tangent hyperplane of Σ_r at x, defines the orientation of $E^n(R)$. If

$$e_1(x) = (u_{11}(x), \cdots, u_{1n}(x)), \cdots, e_{n-1}(x) = (u_{n-1,1}(x), \cdots, u_{n-1,n}(x))$$

are an ordered set of $n-1$ fields of mutually perpendicular unit tangent vectors to Σ_r at x (defined locally), coherent with the orientation of the tangent hyperplane, then the volume element of Σ_r is

$$dV_r = (dx \cdot e_1) \wedge \cdots \wedge (dx \cdot e_{n-1}) ,$$

where $dx \cdot e_\alpha = \sum_{1 \leqq i \leqq n} dx_i u_{\alpha i}$, $1 \leqq \alpha \leqq n-1$. Since the matrix formed by the components of $\xi, e_1, \cdots, e_{n-1}$ is a proper orthogonal matrix, we find

$$dV_r = \sum_{1 \leqq i \leqq n} (-1)^{i-1} \frac{x_i}{r} dx_1 \wedge \cdots \wedge dx_{i-1} \wedge dx_{i+1} \wedge \cdots \wedge dx_n ,$$

which, by (6) and (7), is equal to Θ_r .

Let $\varepsilon_{i_1 \cdots i_n}$ be the Kronecker index, which is equal to $+1$ or -1, according as $i_1, \cdots, i_n$ form an even or odd permutation of $1, \cdots, n$, and is otherwise equal to zero. Then, if $i, i_2, \cdots, i_n$ are mutually distinct, our result can be written

$$\varepsilon_{i i_2 \cdots i_n} dx_{i_2} \wedge \cdots \wedge dx_{i_n} = \frac{x_i}{r} \Theta_r . \tag{8}$$

This relation is equivalent to the following more symmetrical one:

$$\sum_{1 \leqq i_2, \cdots, i_n \leqq n} \varepsilon_{i i_2 \cdots i_n} dx_{i_2} \wedge \cdots \wedge dx_{i_n} = (n-1)! \frac{x_i}{r} \Theta_r . \tag{9}$$

From (6) we have

$$r dr = x_1 dx_1 + \cdots + x_n dx_n .$$

If follows that

$$dr \wedge \Theta_r = r^{n-1} dr \wedge \Theta_1 = dx_1 \wedge \cdots \wedge dx_n , \tag{10}$$

where Θ_1 is the volume element of the unit hypersphere. This gives an expression for the volume element of $E^n(R)$ in terms of the polar coordinate system, as is well-known.

Consider now a complex euclidean space $E_m(C)$ of complex dimension

m, with the coordinates $\zeta_1, \cdots, \zeta_m$. Let the Greek indices run from 1 to m, and put

(11) $$\zeta_\alpha = x_{2\alpha-1} + ix_{2\alpha} .$$

By convention, $E_m(C)$ will be oriented by the ordered set of coordinates $x_1, \cdots, x_{2m}$. The hypersphere Σ_r is defined by the equation

(12) $$\sum_\alpha \zeta_\alpha \bar{\zeta}_\alpha = r^2 .$$

By differentiation we get

(13) $$\sum_\alpha (\bar{\zeta}_\alpha d\zeta_\alpha + \zeta_\alpha d\bar{\zeta}_\alpha) = 0 .$$

From this it follows that on Σ_r the exterior differential form

$$\frac{(-1)^\alpha}{\bar{\zeta}_\alpha} d\zeta_1 \wedge \cdots \wedge d\zeta_{\alpha-1} \wedge d\zeta_{\alpha+1} \wedge \cdots \wedge d\zeta_m \wedge d\bar{\zeta}_1 \wedge \cdots \wedge d\bar{\zeta}_m$$

is independent of α. It is therefore a multiple $A\Theta_r$ of the volume element Θ_r. The actual value of A can be found by substituting for ζ_α the expressions in (11) and using (8). An elementary calculation gives

(14) $$A = (-1)^{\frac{1}{2}m(m+1)+1} \frac{2^{m-1} i^m}{r} .$$

We can therefore write

(15) $$\begin{aligned} &\sum_{1 \leq \alpha_2, \cdots, \alpha_m \leq m} \varepsilon_{\alpha\alpha_2\cdots\alpha_m} d\bar{\zeta}_1 \wedge \cdots \wedge d\bar{\zeta}_m \wedge d\zeta_{\alpha_2} \wedge \cdots \wedge d\zeta_{\alpha_m} \\ &\qquad = (-1)^{\frac{1}{2}m(m+1)} 2^{m-1} i^m (m-1)! \frac{\bar{\zeta}_\alpha}{r} \Theta_r , \end{aligned}$$

or, if $\alpha, \alpha_2, \cdots, \alpha_m$ are mutually distinct,

(16) $$\begin{aligned} &\varepsilon_{\alpha\alpha_2\cdots\alpha_m} d\bar{\zeta}_1 \wedge \cdots \wedge d\bar{\zeta}_m \wedge d\zeta_{\alpha_2} \wedge \cdots \wedge d\zeta_{\alpha_m} \\ &\qquad = (-1)^{\frac{1}{2}m(m+1)} 2^{m-1} i^m \frac{\bar{\zeta}_\alpha}{r} \Theta_r . \end{aligned}$$

Equally valid are naturally also the formulas obtained from (15), (16) by conjugation.

2. An integral formula

In $E^n(R)$ let

(17) $$Xu = \sum_{1 \leq k \leq n} \xi_k(x_1, \cdots, x_n) \frac{\partial u}{\partial x_k}$$

be a differential operator (or vector field) of class C^1. A main tool in our study is the following integral formula

$$\frac{1}{r^{n-1}}\int_{\Sigma_r}(Xu)\Theta_r = \int_{\Sigma_1}\left\{\frac{\partial}{\partial r}(\mu u) + u\left[(n-1)\frac{\mu}{r} - \sum_k \frac{\partial \xi_k}{\partial x_k}\right]\right\}\Theta_1 . \tag{18}$$

where

$$r\mu = \sum_{1\leq k\leq n} x_k \xi_k , \tag{19}$$

and where the integral on the right is extended over the unit hypersphere Σ_1, the argument in the integral being rx, for $x \in \Sigma_1$.

We remark first that a vector field X on a manifold defines two operators on exterior differential forms: the Lie derivative $\theta(X)$ and the interior product $i(X)$. $\theta(X)$ preserves the degree of an exterior differential form and is a derivation, while $i(X)$ diminishes the degree by 1 and is an anti-derivation. Between them and the exterior differentiation operator d, there is the following formula due to H. Cartan [2]:

$$di(X) + i(X)d = \theta(X) . \tag{20}$$

To prove the formula (18) we need the following lemma:

LEMMA. *Let M be a compact oriented manifold of class C^1 and dimension ν, and let X be a vector field of class C^1 on it. Let Θ be an exterior differential form of degree ν on M. Then, for a C^1-function u on M,*

$$\int_M (Xu)\Theta = -\int_M udi(X)\Theta = -\int_M u\theta(X)\Theta . \tag{21}$$

The equality of the last two integrals follows immediately from (20). To prove the equality of the first two integrals, it suffices to establish the local formula

$$du \wedge i(X)\Theta = (Xu)\Theta . \tag{22}$$

For this purpose let $t_1, \cdots, t_\nu$ be a local coordinate system of M. In terms of the t's let

$$\Theta = a dt_1 \wedge \cdots \wedge dt_\nu ,$$

$$Xu = \sum_{1\leq k\leq \nu} b_k \frac{\partial u}{\partial t_k} .$$

Then we have

$$i(X)\Theta = a\sum_{1\leq i\leq \nu}(-1)^{i-1} b_i dt_1 \wedge \cdots \wedge dt_{i-1} \wedge dt_{i+1} \wedge \cdots \wedge dt_\nu$$

and

$$du \wedge i(X)\Theta = \sum_{1 \le k \le \nu} \frac{\partial u}{\partial t_k} dt_k \wedge i(X)\Theta = (Xu)\Theta \,,$$

which was to be proved.

We now proceed to prove (18). Considering X as a vector field of $E^n(R)$, we decompose it as a sum

$$X = \mu \frac{\partial}{\partial r} + X_1 \,, \tag{23}$$

where X_1 is everywhere tangent to Σ_r. If

$$X_1 = \sum_{1 \le k \le n} \eta_k \frac{\partial}{\partial x_k} \tag{24}$$

the condition for it to be tangent to Σ_r is

$$\sum_k x_k \eta_k = 0 \,. \tag{25}$$

(For the remainder of this section we suppose every small Latin index to run from 1 to n.) Since

$$\frac{\partial}{\partial r} = \frac{1}{r} \sum_k x_k \frac{\partial}{\partial x_k} . \tag{26}$$

equation (23) gives

$$\xi_k = \frac{\mu}{r} x_k + \eta_k \,. \tag{27}$$

Using (25), we get

$$\begin{aligned} \mu &= \frac{1}{r} \sum_k x_k \xi_k \,, \\ \eta_k &= \xi_k - \frac{\mu}{r} x_k \,, \end{aligned} \tag{28}$$

which define the decomposition (23).

It will be necessary to compute $\theta(X_1)\Theta_r$. For this purpose notice that $\theta(X_1)$ is a derivation and that

$$\begin{aligned} \theta(X_1) x_k &= X_1 x_k = \eta_k \,, \\ \theta(X_1) dx_k &= d\eta_k \,. \end{aligned} \tag{29}$$

From (9) we have

$$(n-1)! r \Theta_r = \sum_{i_1, \ldots, i_n} \varepsilon_{i_1 \cdots i_n} x_{i_1} dx_{i_2} \wedge \cdots \wedge dx_{i_n} \,.$$

It follows that

$$\begin{aligned} (n-1)! r \theta(X_1)\Theta_r = &\sum_{i_1, \ldots, i_n} \varepsilon_{i_1 \cdots i_n} \eta_{i_1} dx_{i_2} \wedge \cdots \wedge dx_{i_n} \\ &+ (n-1) \sum_{i_1, \ldots, i_n} \varepsilon_{i_1 \cdots i_n} x_{i_1} d\eta_{i_2} \wedge dx_{i_3} \wedge \cdots \wedge dx_{i_n} \,. \end{aligned}$$

The first sum is zero, by (8) and (25). Substituting for η_i the expressions in (28), we get

$$(n-1)!r\theta(X_1)\Theta_r$$
$$= (n-1)\sum_{i_1,\ldots,i_n} \varepsilon_{i_1\cdots i_n} x_{i_1}\Big(d\xi_{i_2} - \frac{\mu}{r}dx_{i_2}\Big) \wedge dx_{i_3} \wedge \cdots \wedge dx_{i_n}$$
$$= (n-1)\sum_{i_1,\ldots,i_n} \varepsilon_{i_1\cdots i_n} x_{i_1}\Big(\frac{\partial\xi_{i_2}}{\partial x_{i_1}}dx_{i_1} + \frac{\partial\xi_{i_2}}{\partial x_{i_2}}dx_{i_2}\Big) \wedge dx_{i_3} \wedge \cdots \wedge dx_{i_n}$$
$$- (n-1)(n-1)!\mu\Theta_r$$
$$= (n-1)!\sum_{i\neq k}\Big(-x_i x_k\frac{\partial\xi_k}{\partial x_i} + x_i^2\frac{\partial\xi_k}{\partial x_k}\Big)\frac{1}{r}\Theta_r - (n-1)(n-1)!\mu\Theta_r$$
$$= (n-1)!\Big\{-\frac{1}{r}\sum_{i,k} x_i x_k\frac{\partial\xi_k}{\partial x_i} + r\sum_k \frac{\partial\xi_k}{\partial x_k} - (n-1)\mu\Big\}\Theta_r\,.$$

It follows that

$$\theta(X_1)\Theta_r = -\frac{1}{r^2}\sum_{i,k} x_i x_k\frac{\partial\xi_k}{\partial x_i}\Theta_r + \sum_k\frac{\partial\xi_k}{\partial x_k}\Theta_r - (n-1)\frac{\mu}{r}\Theta_r\,. \tag{30}$$

Meanwhile, let us notice the following easily verified formula:

$$\frac{\partial\mu}{\partial r} = \frac{1}{r^2}\sum_{i,k} x_i x_k\frac{\partial\xi_k}{\partial x_i}\,. \tag{31}$$

Using (21), (30), (31), we can transform the integral at the left hand side of (18) as follows:

$$\int_{\Sigma_r}(Xu)\Theta_r = \int_{\Sigma_r}\Big(\mu\frac{\partial u}{\partial r} + X_1 u\Big)\Theta_r = \int_{\Sigma_r}\mu\frac{\partial u}{\partial r}\Theta_r - \int_{\Sigma_r} u\theta(X_1)\Theta_r$$
$$= \int_{\Sigma_r}\frac{\partial}{\partial r}(\mu u)\Theta_r - \int_{\Sigma_r} u\frac{\partial\mu}{\partial r}\Theta_r - \int_{\Sigma_r} u\theta(X_1)\Theta_r$$
$$= \int_{\Sigma_r}\frac{\partial}{\partial r}(\mu u)\Theta_r + \int_{\Sigma_r} u\Big\{(n-1)\frac{\mu}{r} - \sum_k\frac{\partial\xi_k}{\partial x_k}\Big\}\Theta_r\,.$$

From this the formula (18) follows immediately.

We will transform (18) into a formula involving a real operator in the complex euclidean space $E_m(C)$. We follow the notation in § 1, keeping the convention that all Greek indices run from 1 to m. Then the operator X can be written

$$Xu = \sum_\alpha\Big(a_\alpha\frac{\partial u}{\partial\zeta_\alpha} + \bar{a}_\alpha\frac{\partial u}{\partial\bar{\zeta}_\alpha}\Big)\,. \tag{32}$$

Put

$$a_\alpha = b_{2\alpha-1} + ib_{2\alpha}\,,$$

where b_α, $b_{m+\alpha}$ are real-valued functions. In terms of the real coordinates x_α, $x_{m+\alpha}$, we have

$$Xu = \sum_\alpha \left(b_\alpha \frac{\partial u}{\partial x_\alpha} + b_{m+\alpha} \frac{\partial u}{\partial x_{m+\alpha}} \right).$$

A straightforward calculation gives from (18) the formula

$$\begin{aligned} &\frac{1}{r^{2m-1}} \int_{\Sigma_r} (Xu) \Theta_r \\ (33) \qquad &= \int_{\Sigma_1} \frac{\partial}{\partial r} (\mu u) \Theta_1 + \int_{\Sigma_1} u \left\{ (2m-1) \frac{\mu}{r} - \sum_\alpha \left(\frac{\partial a_\alpha}{\partial \zeta_\alpha} + \frac{\partial \bar{a}_\alpha}{\partial \bar{\zeta}_\alpha} \right) \right\} \Theta_1 , \end{aligned}$$

where

$$(34) \qquad \mu = \frac{1}{2r} \sum_\alpha (\zeta_\alpha \bar{a}_\alpha + \bar{\zeta}_\alpha a_\alpha) .$$

Formula (33) will play a fundamental rôle in our subsequent discussions.

3. The first main theorem

We describe the points of the complex projective space P_m of dimension m by their homogeneous coordinate vectors $Z = (z_0, \cdots, z_m)$; two non-zero vectors define the same point of P_m, if and only if they differ by a factor. For convenience we will not distinguish a point from its coordinate vector, with, of course, the understanding that the geometrical properties in question will be invariant when the coordinate vectors are multiplied by non-zero factors. To the vectors Z and $W = (w_0, \cdots, w_m)$ let

$$(35) \qquad (Z, W) = \overline{(W, Z)} = z_0 \bar{w}_0 + \cdots + z_m \bar{w}_m .$$

Let $|Z| = +(Z, Z)^{\frac{1}{2}} \geqq 0$. With this hermitian scalar product (35) a distance $d(Z, W)$ between the points Z and W can be defined by

$$(36) \qquad \cos d(Z, W) = \frac{|(Z, W)|}{|Z| \cdot |W|} .$$

Two points Z and W are called orthogonal, if $(Z, W) = 0$. The space P_m is acted on by the unitary group $U(m+1)$ in $m+1$ variables, and it has the invariant Kähler metric

$$(37) \qquad ds^2 = \frac{1}{|Z|^4} \{ |Z|^2 (dZ, dZ) - (Z, dZ)(dZ, Z) \} .$$

It can be verified that the associated two-form of (37) is $id'd'' \log |Z|$. We put

$$\Omega = \frac{i}{\pi} d'd'' \log |Z| . \tag{38}$$

Then $\int_{P_m} \Omega^m = 1$.

Let A be a fixed point of P_m, with $|A| = 1$. Its polar hyperplane π_A consists of all the points orthogonal to A and is a complex projective space of dimension $m - 1$. For a point Z of $P_m - A$ we can write

$$Z = zA + Y , \tag{39}$$

with $(A, Y) = 0$. The point Y is the point where the line AZ meets π_A and is uniquely determined by Z. Moreover, the quotient $|Y|/|Z| \leq 1$ depends only on the point Z, and not on the choice of its coordinate vector. In fact, it is equal to $\sin d(Z, W)$, as can be immediately verified.

The association of Y to Z defines a complex analytic mapping ψ: $P_m - A \rightarrow \pi_A$. On the other hand, let ρ: $\pi_A \rightarrow P_m$ be the restriction mapping. Let $\Phi = \psi^* \rho^* \Omega$. There is an important relation between Φ and Ω, which we will derive presently.

Let ds_Z^2 be the hermitian metric in $P_m - A$, and let ds_Y^2 be the induced hermitian metric in π_A. From (37) and (39) we find

$$ds_Z^2 = \sin^2 d(Z, A) ds_Y^2 + \varphi\bar{\varphi} , \tag{40}$$

where

$$|Z|^2 \varphi = |Y| dz - \frac{z}{|Y|}(dY, Y) . \tag{41}$$

The form φ is of type (1, 0). It is not completely determined by the point Z: if Z is changed to λZ (λ a scalar), φ changes to $(\lambda/|\lambda|)\varphi$. The relation (40) leads to a corresponding relation on the associated two-forms, which is

$$\Omega = \sin^2 d(Z, A)\Phi + \frac{i}{2\pi}\varphi \wedge \bar{\varphi} . \tag{42}$$

We put

$$u = \log \sin d(Z, A) \leq 0 \tag{43}$$

and

$$\Lambda = \frac{1}{2\pi i}(d' - d'')u \wedge \left(\sum_{0 \leq k \leq m-1} \Phi^k \wedge \Omega^{m-1-k}\right) . \tag{44}$$

Then Λ is a real-valued exterior differential form of degree $2m - 1$ in $P_m - A$.

The first main theorem for complex analytic mappings $f: E_m \rightarrow P_m$ as proved by H. Levine [3] can be stated as follows:

Let A be a point of P_m and D a compact domain in E_m such that $f^{-1}(A) \cap D$ consists only of a finite number of points and that $f^{-1}(A)$ does not meet the boundary ∂D of D. Let $n(D, A)$ be the number of times that A is covered by $f(D)$, counted algebraically, and let $v(D)$ be the volume of $f(D)$. Then

$$n(D, A) - v(D) = \int_{f(\partial D)} \Lambda . \tag{45}$$

The theorem shows the importance of the form Λ and therefore the summands $\Phi^k \wedge \Omega^{m-1-k}$. It is desirable to express this summand in terms of powers of Ω by applying (42), because Ω is independent of the point A, while Φ depends on A. We have

$$\Phi^k \wedge \Omega^{m-1-k} = \frac{1}{\sin^{2k} d(Z, A)} \left(\Omega^{m-1} - k \frac{i}{2\pi} \varphi \wedge \bar{\varphi} \wedge \Omega^{m-2}\right), \tag{46}$$

$$0 \leqq k \leqq m - 1 .$$

The mapping f induces a dual mapping f^* on differential forms. We put

$$f^*\Omega = \frac{i}{\pi} \sum_{\alpha,\beta} a_{\alpha\beta} d\zeta_\alpha \wedge d\bar{\zeta}_\beta , \tag{47}$$

where, by (38),

$$a_{\alpha\beta} = \frac{\partial^2 \log |Z|}{\partial\zeta_\alpha \partial\bar{\zeta}_\beta} . \tag{48}$$

The $a_{\alpha\beta}$ are the elements of an hermitian positive semi-definite matrix. An important consequence of the expressions (48) is that the partial derivatives $\partial a_{\alpha\beta}/\partial\zeta_\gamma$ are symmetric in α, γ. Similarly, denoting by f_A the restriction of f to $E_m - f^{-1}(A)$, we put

$$f_A^*\Phi = \frac{i}{\pi} \sum_{\alpha,\beta} b_{\alpha\beta} d\zeta_\alpha \wedge d\bar{\zeta}_\beta . \tag{49}$$

Then $(b_{\alpha\beta})$ is an hermitian positive semi-definite matrix, also with the property that $\partial b_{\alpha\beta}/\partial\zeta_\gamma$ are symmetric in α, γ.

The differential forms $(f_A^*\Phi)^k \wedge (f^*\Omega)^{m-1-k}$, $1 \leqq k \leqq m - 1$, are of type $(m - 1, m - 1)$ in $E_m - f^{-1}(A)$, and $(f^*\Omega)^{m-1}$ is of type $(m - 1, m - 1)$ in E_m. We can therefore write

$$(50)\quad (f_A^*\Phi)^k \wedge (f^*\Omega)^{m-1-k} = \frac{(-1)^{\frac{1}{2}(m-1)(m-2)} i^{m-1}}{((m-1)!)^2} \times \sum_{\substack{\alpha_1,\cdots,\alpha_m \\ \beta_1,\cdots,\beta_m}} \varepsilon_{\alpha_1\cdots\alpha_m}\varepsilon_{\beta_1\cdots\beta_m} C^{(k)}_{\alpha_1\beta_1} d\zeta_{\alpha_2} \wedge \cdots \wedge d\zeta_{\alpha_m} \wedge d\bar{\zeta}_{\beta_2} \wedge \cdots \wedge d\bar{\zeta}_{\beta_m}$$

$$0 \leqq k \leqq m-1,$$

where

$$(51)\quad \pi^{m-1} C^{(k)}_{\alpha\beta} = \sum_{\substack{\alpha_2,\cdots,\alpha_m \\ \beta_2,\cdots,\beta_m}} \varepsilon_{\alpha\alpha_2\cdots\alpha_m}\varepsilon_{\beta\beta_2\cdots\beta_m} b_{\alpha_2\beta_2}\cdots b_{\alpha_{k+1}\beta_{k+1}} a_{\alpha_{k+2}\beta_{k+2}}\cdots a_{\alpha_m\beta_m}.$$

The fact that the forms in (50) are real is equivalent to the condition that $C^{(k)}_{\alpha\beta}$ are hermitian:

$$(52)\quad C^{(k)}_{\alpha\beta} = \bar{C}^{(k)}_{\beta\alpha}.$$

Observe also that $C^{(0)}_{\alpha\beta}$ are functions in E_m. For later abbreviations we introduce the expressions

$$(53)\quad \Gamma = \sum_{0\leq k\leq m-1} (f_A^*\Phi)^k \wedge (f^*\Omega)^{m-1-k},$$

$$(54)\quad C_{\alpha\beta} = \sum_{0\leq k\leq m-1} C^{(k)}_{\alpha\beta},$$

so that Γ can be expressed in terms of $C_{\alpha\beta}$ by a formula analogous to (50). From these expressions for $C_{\alpha\beta}$ and the symmetry properties mentioned above of the partial derivatives of $a_{\alpha\beta}$ and $b_{\cdot\beta}$ we derive immediately the important relation

$$(55)\quad \sum_\alpha \frac{\partial}{\partial\zeta_\alpha} C_{\alpha\beta} = 0.$$

Under the dual mapping f_A^* the formula (42) gives the relation

$$(56)\quad a_{\alpha\beta} = (\sin^2 d) b_{\alpha\beta} + c_\alpha \bar{c}_\beta,$$

where c_α are functions in ζ_ρ and where we write d for $d(Z, A)$ for simplicity. Solving (56) for $b_{\alpha\beta}$ and substituting the resulting expressions into (51), we get

$$(57)\quad \pi^{m-1} C^{(k)}_{\alpha\beta} = \frac{1}{\sin^{2k} d}(\pi^{m-1} C^{(0)}_{\alpha\beta} - k D_{\alpha\beta}),$$

where

$$(58)\quad D_{\alpha\beta} = \sum_{\substack{\alpha_2,\cdots,\alpha_m \\ \beta_2,\cdots,\beta_m}} \varepsilon_{\alpha\alpha_2\cdots\alpha_m}\varepsilon_{\beta\beta_2\cdots\beta_m} c_{\alpha_2}\bar{c}_{\beta_2} a_{\alpha_3\beta_3}\cdots a_{\alpha_m\beta_m}.$$

For further developments we need the following algebraic lemma:

LEMMA. *Let* $(g_{\alpha\beta})$, $(h_{\alpha\beta})$ *be hermitian positive semi-definite matrices, and let*

$$(59)\quad P^{(k)}_{\alpha\beta} = \sum_{\substack{\alpha_2,\cdots,\alpha_m\\ \beta_2,\cdots,\beta_m}} \varepsilon_{\alpha\alpha_2\cdots\alpha_m}\varepsilon_{\beta\beta_2\cdots\beta_m} g_{\alpha_2\beta_2}\cdots g_{\alpha_{k+1}\beta_{k+1}} h_{\alpha_{k+2}\beta_{k+2}}\cdots h_{\alpha_m\beta_m}\,,$$

$$0 \leqq k \leqq m-1\,.$$

Then the matrix $(P^{(k)}_{\alpha\beta})$ *is hermitian positive semi-definite. Also, for any complex numbers* w_γ, *the matrix with the elements*

$$(60)\qquad Q_{\alpha\beta} = \sum_{\substack{\alpha_2,\cdots,\alpha_m\\ \beta_2,\cdots,\beta_m}} \varepsilon_{\alpha\alpha_2\cdots\alpha_m}\varepsilon_{\beta\beta_2\cdots\beta_m} w_{\alpha_2}\overline{w}_{\beta_2} g_{\alpha_3\beta_3}\cdots g_{\alpha_m\beta_m}$$

is hermitian positive semi-definite.

As a corollary to this lemma, it follows that the matrices $(C^{(k)}_{\alpha\beta})$, $0 \leqq k \leqq m-1$, $(C_{\alpha\beta})$, and $(D_{\alpha\beta})$ are all positive semi-definite.

To prove the first statement of the lemma, we subject $g_{\alpha\beta}$, $h_{\alpha\beta}$ to the cogredient transformations

$$g_{\alpha\beta} = \sum_{\rho,\sigma} u_{\alpha\rho}\overline{u}_{\beta\sigma}g'_{\rho\sigma}, \qquad h_{\alpha\beta} = \sum_{\rho,\sigma} u_{\alpha\rho}\overline{u}_{\beta\sigma}h'_{\rho\sigma}\,,$$

with a non-singular matrix $(u_{\alpha\beta})$. Then $P^{(k)}_{\alpha\beta}$ are transformed according to the equations

$$P^{(k)}_{\alpha\beta} = \sum_{\rho,\sigma} U_{\alpha\rho}\overline{U}_{\beta\sigma}P'^{(k)}_{\rho\sigma}\,.$$

where

$$U_{\alpha\beta} = \frac{1}{((m-1)!)^2}\sum_{\substack{\alpha_2,\cdots,\alpha_m\\ \beta_2,\cdots,\beta_m}} \varepsilon_{\alpha\alpha_2\cdots\alpha_m}\varepsilon_{\beta\beta_2\cdots\beta_m} u_{\alpha_2\beta_2}\cdots u_{\alpha_m\beta_m}$$

and the matrix $(U_{\alpha\beta})$ is also non-singular. Since $(g_{\alpha\beta})$, $(h_{\alpha\beta})$ are both positive semi-definite, we can find a non-singular matrix $(u_{\alpha\beta})$ so that $(g'_{\alpha\beta})$, $(h'_{\alpha\beta})$ are diagonal, with the diagonal elements $\geqq 0$. Then $(P'^{(k)}_{\alpha\beta})$ is also diagonal with non-negative diagonal elements. This implies that $(P^{(k)}_{\alpha\beta})$ is positive semi-definite.

The second statement follows by taking $h_{\alpha\beta} = w_\alpha\overline{w}_\beta$.

Since the trace of an hermitian positive semi-definite matrix is $\geqq 0$, it follows from (57) and the Lemma that

$$\sum_\alpha C^{(k)}_{\alpha\alpha} \leqq \frac{1}{\sin^{2k} d}C^{(0)}\,,$$

and hence that

$$(61)\qquad \sum_\alpha C_{\alpha\alpha} \leqq \left(1 + \frac{1}{\sin^2 d} + \cdots + \frac{1}{\sin^{2m-2} d}\right)C^{(0)}\,,$$

where

$$C^{(0)} = \sum_\alpha C^{(0)}_{\alpha\alpha} . \tag{62}$$

The function $C^{(0)}$ is defined in E_m. It has the following notable geometrical interpretation, as given by the formula

$$(f^*\Omega)^{m-1} \wedge \Omega_0 = \frac{2^{m-1}}{m!} C^{(0)} \Omega_0^m . \tag{63}$$

Since

$$\Omega_0^m = m! r^{2m-1} dr \wedge \Theta_1 ,$$

we have

$$v_1(r) = 2^{m-1} \int_0^r r^{2m-1} dr \int_{\Sigma_1} C^{(0)} \Theta_1 . \tag{64}$$

4. The integrated form of the first main theorem

In the first main theorem (45) we now take D to be the domain D_r defined by (1), so that ∂D is Σ_r. We will then write

$$n(D, A) = n(r, A), \qquad v(D) = v(r) = v_0(r) . \tag{65}$$

We now apply the integral formula (33) to the right-hand side of (45). The integrand can be transformed as follows:

$$\begin{aligned} f^*\Lambda &= \frac{1}{2\pi i}(d' - d'')u \wedge \Gamma = \frac{1}{2\pi i} \sum_\alpha (u_{\zeta_\alpha} d\zeta_\alpha - u_{\bar\zeta_\alpha} d\bar\zeta_\alpha) \wedge \Gamma \\ &= \frac{(-1)^{\frac{1}{2}(m-1)(m-2)} i^{m-2}}{2\pi((m-1)!)^2} \sum_{\substack{\alpha_1, \cdots, \alpha_m \\ \beta_1, \cdots, \beta_m}} \varepsilon_{\alpha_1 \cdots \alpha_m} \varepsilon_{\beta_1 \cdots \beta_m} C_{\alpha_1 \beta_1} u_{\zeta_{\alpha 1}} d\zeta_{\alpha_1} \\ &\qquad \wedge \cdots \wedge d\zeta_{\alpha_m} \wedge d\bar\zeta_{\beta_2} \wedge \cdots \wedge d\bar\zeta_{\beta_m} + \text{complex conjugate} \\ &= \frac{(-1)^{\frac{1}{2}(m-1)(m-2)} i^{m-2}}{2\pi(m-1)!} \sum_{\alpha, \beta_1, \cdots, \beta_m} \varepsilon_{\beta_1 \cdots \beta_m} C_{\alpha\beta_1} u_{\zeta_\alpha} d\zeta_1 \\ &\qquad \wedge \cdots \wedge d\zeta_m \wedge d\bar\zeta_{\beta_2} \wedge \cdots \wedge d\bar\zeta_{\beta_m} + \text{complex conjugate} \\ &= \frac{2^{m-2}}{\pi r} \sum_{\alpha, \beta} (C_{\alpha\beta} \zeta_\beta u_{\zeta_\alpha} + C_{\beta\alpha} \bar\zeta_\beta u_{\bar\zeta_\alpha}) \Theta_r . \end{aligned}$$

During the reduction, use is made of the formulas (50), (53), (54), and the complex conjugate of (15).

The integrand is of the form (32), if we set

$$a_\alpha = \frac{2^{m-2}}{\pi r} \sum_\beta C_{\alpha\beta} \zeta_\beta . \tag{66}$$

Then, by (34), we have

$$\mu = \frac{2^{m-2}}{\pi r^2} \sum_{\alpha,\beta} C_{\alpha\beta} \bar{\zeta}_\alpha \zeta_\beta\,, \tag{67}$$

which is $\geqq 0$. Making use of (55), we get

$$\sum_\alpha \frac{\partial a_\alpha}{\partial \zeta_\alpha} = \frac{2^{m-2}}{\pi r} \sum_\alpha C_{\alpha\alpha} - \frac{1}{2r}\mu\,. \tag{68}$$

The integral formula (33) therefore leads to the relation

$$\begin{aligned} &\frac{1}{r^{2m-1}}\{n(r, A) - v(r)\} \\ &= \int_{\Sigma_1} \frac{\partial}{\partial r}(\mu u)\Theta_1 + \int_{\Sigma_1} u\left(\frac{2m\mu}{r} - \frac{2^{m-1}}{\pi r}\sum_\alpha C_{\alpha\alpha}\right)\Theta_1\,, \end{aligned} \tag{69}$$

where in the integrals on the right-hand side, the argument is rx, with $x \in \Sigma_1$ and $r = \text{const}$. Notice also that the formula (69) is valid only for those r for which Σ_r does not meet $f^{-1}(A)$.

However, as in the case $m = 1$, the form of (69) suggests its integration with respect to r. This requires the examination of the improper integrals

$$\begin{aligned} I(r, A) &= \int_{\Sigma_1} (-u\mu)\Theta_1\,, \\ J(r, A) &= \int_{\Sigma_1} \left(-u\sum_\alpha C_{\alpha\alpha}\right)\Theta_1 \end{aligned} \tag{70}$$

for values of r for which $\Sigma_r \cap f^{-1}(A) \neq \varnothing$. In the first place, the integrands of the integrals in (70) are $\geqq 0$. By (57) and the positive semi-definiteness of $(D_{\alpha\beta})$, we have

$$-u\mu \leqq \frac{2^{m-2}}{\pi r^2}(-\log \sin d)\left(\sum_{0 \leqq k \leqq m-1} \frac{1}{\sin^{2k} d}\right) \sum_{\alpha,\beta} C^{(0)}_{\alpha\beta} \bar{\zeta}_\alpha \zeta_\beta\,.$$

Similarly, by (61), we have

$$(-u)\sum_\alpha C_{\alpha\alpha} \leqq (-\log \sin d)\left(\sum_{0 \leqq k \leqq m-1} \frac{1}{\sin^{2k} d}\right) C^{(0)}\,.$$

If $f^{-1}(A) \cap \Sigma_r$ consists of a finite number of points, as we will suppose, the principal part of the integrand at a singularity is

$$(-\log \sin d)\left(\sum_{0 \leqq k \leqq m-1} \frac{1}{\sin^{2k} d}\right),$$

for the integrals of the functions at the right-hand sides of the inequalities, so that these integrals are convergent. The same is therefore true for the integrals of the functions on the left-hand sides, which proves that $I(r, A)$, $J(r, A)$ are defined for all values of r. Moreover, the integrals of the functions on the right-hand sides over a small domain of radius ε of Σ_1 tend to zero uniformly in r as $\varepsilon \to 0$. From this it follows that $I(r, A)$, $J(r, A)$ are continuous functions in r. In particular, this gives the relation

$$\int_{r_0}^{r} dr \int_{\Sigma_1} \frac{\partial}{\partial r}(-u\mu)\Theta_1 = I(r, A) - I(r_0, A) , \tag{71}$$

for $0 < r_0 \leqq r$. For a fixed $r_0 > 0$ and for $r > r_0$ we define $T(r)$ by (5) and we put

$$N(r, A) = \int_{r_0}^{r} \frac{n(t, A)}{t^{2m-1}} dt . \tag{72}$$

Integrating (69) with respect to r, we then get

$$N(r, A) = \text{const} + T(r) + S(r, A) - I(r, A) - 2m\int_{r_0}^{r} \frac{I(r, A)}{r} dr . \tag{73}$$

where

$$S(r, A) = \frac{2^{m-1}}{\pi}\int_{r_0}^{r} \frac{J(r, A)}{r} dr . \tag{74}$$

Since $I(r, A) \geqq 0$, this gives the inequality

$$N(r, A) < \text{const} + T(r) + S(r, A) . \tag{75}$$

Thus, unlike the classical case $m = 1$, an additional term $S(r, A)$ has to be added to $T(r)$ in order that it majorizes $N(r, A)$ asymptotically. Formula (73) is the integrated form of the first main theorem.

We will now give a proof of the theorem stated in the Introduction. Under the hypotheses suppose the contrary be true. Let $\rho(A)$ be the characteristic function of the set $f(E_m)$, so that

$$\rho(A) = 1, \qquad A \in f(E_m) ,$$
$$\rho(A) = 0, \qquad A \,\overline{\in}\, f(E_m) .$$

Let $dA = \Omega^m$ be the volume element of P_m. The total volume of P_m being 1, we have, by assumption, $\int_{P_m} \rho(A)dA = b < 1$. Clearly we have

$$\int_{P_m} n(t, A)\rho(A)dA = v(t) .$$

Integration of the inequality (75) with respect to $\rho(A)dA$ over P_m gives

$$
\begin{aligned}
T(r) &< \text{const} + bT(r) + \int_{P_m} S(r, A)\rho(A)dA \\
&\leqq \text{const} + bT(r) + \int_{P_m} S(r, A)dA \\
&\leqq \text{const} + bT(r) \\
&\quad + \frac{2^{m-1}}{\pi}\int_{r_0}^{r}\frac{dr}{r}\int_{P_m} dA \int_{\Sigma_1}(-\log\sin d)\Big(\sum_{0\leqq k\leqq m-1}\frac{1}{\sin^{2k} d}\Big)C^{(0)}\Theta_1 .
\end{aligned}
$$

But since both dA and $d(Z, A)$ are invariant under the isometries of P_m, the integral

$$\int_{P_m}(-\log\sin d)\Big(\sum_{0\leqq k\leqq m-1}\frac{1}{\sin^{2k} d}\Big)dA$$

is equal to a constant $(\pi/2^{m-1})h$, independent of Z. It follows that

$$T(r) < \text{const} + bT(r) + h\int_{r_0}^{r}\frac{dr}{r}\int_{\Sigma_1} C^{(0)}\Theta_1 .$$

The last term in this inequality is, by (64), equal to $\frac{h}{2^{m-1}}\int_{r_0}^{r}(v_1'(r)dr/r^{2m})$. By the second hypothesis of the theorem the latter integral is $o(T(r))$. Hence there is a contradiction and the theorem is proved.

UNIVERSITY OF CHICAGO

REFERENCES

1. S. BOCHNER, and W. T. MARTIN, Several Complex Variables, Princeton 1948, p. 45.
2. H. CARTAN, Notions d'algèbre différentielle; application aux groupes de Lie et aux variétés où opère un groupe de Lie, Colloque de Topologie, Bruxelles 1950, pp. 15–27.
3. H. LEVINE, *A theorem on holomorphic mappings into complex projective space*, Ann. of Math. 71 (1960), pp. 529–535.

Extrait de *L'Enseignement Mathématique*, tome VII, 1961

HOLOMORPHIC MAPPINGS OF COMPLEX MANIFOLDS

by Shiing-shen CHERN

1. A complex manifold is, briefly speaking, a connected manifold with local complex coordinates defined up to a holomorphic transformation. Examples of complex manifolds include the number space C_m and the projective space P_m of dimensions m. For $m = 1$ these are known in function theory as the Gaussian plane and the Riemann sphere respectively.

A holomorphic mapping of a complex manifold M of dimension m into another one N of dimension n is a continuous mapping f such that, if $\zeta_1, \ldots, \zeta_m$ are the local coordinates at a point $\zeta \in M$ and $z_1, \ldots, z_n$ are local coordinates at the image point $f(\zeta) \in N$, the mapping is locally defined by the equations

$$z_i = z_i(\zeta_1, \ldots, \zeta_m), \quad l \leqq i \leqq n, \tag{1}$$

where the functions at the right-hand side are holomorphic functions in their arguments. By this definition, a holomorphic mapping $f: C_1 \to P_1$ is precisely a meromorphic function in classical function theory.

The first question that arises is the question of existence. For the condition of a holomorphic mapping is so strong that it is not clear that, for given complex manifolds M, N, a holomorphic mapping $f: M \to N$ should exist which is not a constant (i.e., one that the image $f(M)$ is not a single point of N). In fact, if M, N are compact Riemann surfaces (a Riemann surface is a complex manifold of dimension one), then a non-constant holomorphic mapping $f: M \to N$ exists only when $g(M) \geqq g(N)$, where $g(M)$, $g(N)$ are the genera of M, N respectively. This well-known result can be derived as a consequence of the Riemann-Hurwitz formula (cf. § 2).

A more elementary fact is the result that every holomorphic function on a compact complex manifold is a constant. From this it follows that every holomorphic mapping of P_m into a complex torus of dimension n is a constant, because P_m is simply connected and the complex torus has C_n as its universal covering space.

The above result can be generalized. We recall that an analytic set on a complex manifold M is a set E satisfying the condition that, if $\zeta_0 \in E$, there exist s holomorphic functions $f_1, \ldots, f_s$ in a neighborhood of ζ_0 such that the intersection of E with the neighborhood is defined by the equations $f_1 = \ldots = f_s = 0$. Then the following theorem is known [6, p. 356]: Let $f: M \to N$ be a holomorphic mapping such that M is compact and that every compact analytic set of N consists of a finite number of points. Then f is constant.

2. While these results are of interest, it seems desirable to formulate some problems of general scope on holomorphic mappings. I would consider the following a fundamental one: Given a holomorphic mapping $f: M \to N$. To determine relations between the invariants of the manifolds M, N and the invariants which arise from the mapping f.

A first illustration of this problem is the Riemann-Hurwitz formula on the holomorphic mapping $f: M \to N$ of compact Riemann surfaces. The formula can be written

$$2 - 2g(M) + w = d(2 - 2g(N)), \tag{2}$$

where d is the degree of the mapping and w is the index of ramification, i.e., the sum of the orders of the points of ramification. The genera $g(M)$, $g(N)$ are invariants of M, N themselves, while d, w depend on the mapping.

Another set of relations of this nature consists of the Plücker formulas for an algebraic curve. Let an algebraic curve be defined by a holomorphic mapping $f: M \to P_n$, where M is a compact Riemann surface. Suppose that the curve is non-degenerate, i.e., that the image $f(M)$ does not belong to a subspace of dimension $\leqq n - 1$. To this curve is defined the pth associated curve $f^p: M \to G(n, p)$, $0 \leqq p \leqq n - 1$, formed by the osculating projective spaces of dimension p, where $G(n, p)$ is the Grassmann manifold of all p-dimensional projective spaces in P_n $(G(n, 0) = P_n)$. $f^p(M)$ defines a cycle in $G(n, p)$, which is homologous to a positive integral multiple ν_p of the fundamental two-cycle of $G(n, p)$. The integer $\nu_p \geqq 0$ is called the order of rank p of our algebraic curve. Geometrically it is the number of points of the curve at which the osculating spaces of dimension p meet a fixed generic linear space of dimension $n-p-1$ of P_n. A stationary point of order p is one at which the *pth* associated curve has a tangent with a contact of higher order. The stationary points are isolated and a positive index can be associated to each of them. Let $w_p \geqq 0$ be the sum of indices at the stationary points of rank p. Then Plücker's formulas are

$$(3) \qquad -w_p-\nu_{p-1}+2\nu_p-\nu_{p+1} = 2-2g(M), \quad 0 \leqq p \leqq n-1.$$

Here the right-hand side is an invariant of M, while the left-hand side involves quantities which depend on the mapping.

For non-singular algebraic varieties a much more profound relation between invariants of manifolds and quantities depending on a holomorphic mapping is given by Grothendieck's Riemann-Roch theorem [1]. We will not dwell on a discussion of this theorem. It suffices to say that the theorem contains as a special case the Riemann-Hurwitz formula. Applying the theorem of Grothendieck and the classification of singularities by Thom, I. R. Porteous [5] derived relations between the characteristic classes of non-singular algebraic varieties under the following simple types of mappings: *a)* dilatations; *b)* ramified coverings with singularities of a relatively simple type.

It will be natural to expect that the relations answering our fundamental problem have a bearing on the existence problem of holomorphic mappings. An example is the non-existence theorem of holomorphic mappings between compact Riemann surfaces in § 1 derived as a consequence of the Riemann-Hurwitz formula. But our fundamental problem seems to be wider in scope.

A natural counter-part of the existence problem is the uniqueness problem, namely the determination of a holomorphic mapping by its restriction to a certain subset of the original manifold. Very little seems to be known along this line. As an example I wish to state the following so-called Riemann's theorem [6, p. 343]: Let f: $M \to N$ be a continuous mapping of a complex manifold M into another N. Let E be a nowhere dense analytic set in M. If the restriction of f to $M - E$ is holomorphic, the same is true for f itself.

3. Another important problem on holomorphic mappings is the study of the properties of the image set. If f: $M \to N$ is a holomorphic mapping and M is compact, then $f(M)$ is an analytic set. If $N = P_n$, then a famous theorem of Chow says that $f(M)$ is an algebraic set. (We recall that a subset $E \subset P_n$ is called algebraic, if there exist q polynomials $g_1, \ldots, g_q$ in the $n+1$ homogeneous coordinates of P_n such that E is defined by the equations $g_1 = \ldots = g_q = 0$.)

The case that M, N are of the same dimension has particular properties for the following reasons: 1) M, N are oriented manifolds and f preserves orientation; 2) it will be possible to compare the local degree of the mapping with the global degree. The results so obtained are valid for more general mappings. In fact, the following theorem was proved by S. Sternberg and R. G. Swan [9]: Let M, N be two oriented n-dimensional differentiable manifolds, with M compact and N connected. Let f: $M \to N$ be a differentiable mapping, whose Jacobian $J(f)$ is non-negative. Then either $J(f) \equiv 0$ or N is also compact, f is onto, and f has a positive degree on each component of M on which $J(f) \not\equiv 0$.

In particular, suppose M be connected and compact, and $J(f) \not\equiv 0$. Then N is compact, the degree $d(f)$ of the mapping is positive, and every point $a \in N$ is covered $d(f)$ times when counted with the proper multiplicity. Since N is compact, we can equip it with a riemannian metric, so that the total volume of N is 1. Let $v(M)$ be the volume of the image of M under f, and let $n(a)$ be the local degree of f at a, i.e., the number of times that a is covered by $f(M)$. Then we have

$$d(f) = n(a) = v(M) . \tag{4}$$

These results should be considered as a starting-point of the theory of value distributions in complex function theory, the essential difference being that, in the latter case, M is non-compact.

4. The study of mappings $f: M \to N$ where M is non-compact is radically different from the compact case and much finer analytical considerations will be necessary. The natural idea is to exhaust M by a family of compact domains with boundary, D_t, as $t \to \infty$, and to study the restriction of f to D_t. The asymptotic behavior of the geometrical quantities introduced for the restricted mappings $f \mid D_t$ as $t \to \infty$ will then be the main concern of the problem.

The problem which generalizes (4) to the case of a domain with boundary can be stated as follows: Let M and N be two connected, oriented n-dimensional C^∞-manifolds with M non-compact and N compact. Let $f: M \to N$ be a C^∞-differentiable mapping, whose Jacobian $J(f)$ is $\geqq 0$ and $\not\equiv 0$. Let $a \in N$, and let $f \mid D$ be the restriction of f to a compact domain $D \subset M$, such that the image of the boundary ∂D of D does not contain a. Equip N with a riemannian metric with total volume 1, and denote by $v(D)$ the volume of $f(D) \subset N$. Let $n(a, D)$ be the number of times that the point a is covered by $f(D)$. Our problem is to express the difference $n(a, D) - v(D)$ as an integral over ∂D.

An explicit formula solving this problem, which will then be a generalization of (4), is called the first main theorem. A most convenient way to derive such a formula is by applying the theory of harmonic differential forms on a compact riemannian manifold [7] and proceeds as follows:

We consider the manifold N and denote by Φ its volume element. Let δ_a be the Dirac measure with singularity at a. Then Φ and δ_a are both currents of dimension zero and their difference $\Phi - \delta_a$ is orthogonal to the harmonic form Φ. It follows from the fundamental existence theorem on harmonic integrals on a compact riemannian manifold that the equation

$$\Delta S = \delta_a - \Phi \tag{5}$$

where S is a current of dimension zero and Δ is the Laplacian, has a solution in S and that S is a differential form of degree

n in $N-a$. Moreover, the solution S is defined up to an additive harmonic form. Put

$$\Lambda = \delta S \,, \tag{6}$$

where δ is the codifferential. Then, under the above hypotheses, we have the " first main theorem ":

$$n(a, D) - v(D) = \int_{f(\partial D)} \Lambda \,. \tag{7}$$

In order to derive geometrical consequences from (7), it will be necessary to study the integral at its right-hand side, particularly its asymptotic behavior. Formula (7) contains as a special case the classical first main theorem in the theory of value distributions of meromorphic functions, but is of course much more general in scope. One can say that the reason which accounts more than any other for the properties of value distributions of meromorphic functions is the remarkable behavior of the boundary integral in (7).

I have carried out the study of the boundary integral in (7) for the case that $M = C_n$, $N = P_n$. Let $\zeta_1, \ldots, \zeta_n$ be the coordinates in C_n, and let D_t be the ball defined by

$$\zeta_1 \bar{\zeta}_1 + \ldots + \zeta_n \bar{\zeta}_n \leqq t^2 \,. \tag{8}$$

Let

$$\Omega_0 = \frac{i}{2} (d\zeta_1 \wedge d\bar{\zeta}_1 + \ldots + d\zeta_n \wedge d\bar{\zeta}_n) \,, \tag{9}$$

and let Ω be the fundamental two-form of the elliptic Hermitian metric in P_n such that $\int_{P_n} \Omega^n = 1$. We put

$$v_k(t) = \int_{D_t} f^* \Omega^{n-k} \wedge \Omega_0^k \,, \quad 0 \leqq k \leqq n \,, \tag{10}$$

so that $v_0(t)$ is the volume of $f(D_t)$. By estimating the boundary integral in (7) and applying integral-geometric considerations, the following geometrical result is derived [3]:

Let $f: C_n \to P_n$ be a holomorphic mapping which satisfies the following conditions: 1) The function $T(t) = \int_{t_0}^{t} \frac{v_0(\tau)\, d\tau}{\tau^{2n-1}} \to \infty$;

2) $\int_{t_0}^{t} (v_1' (\tau)\, d\tau)/\tau^{2n} = o\,(T\,(t))$. Then the set $P_n - f\,(C_n)$ is of measure zero.

It is well-known that for an arbitrary holomorphic mapping $f\colon C_n \to P_n$, the set $P_n - f\,(C_n)$ may contain some open subset of P_n, so that the conclusion will certainly not be true without some supplementary condition. On the other hand, it is not necessary to suppose the holomorphy of f, for even in the classical case of value distributions the main results are true for quasi-meromorphic functions. It would be an interesting problem to find the proper restrictions on f for the above conclusion to be true.

5. The fundamental problem posed in the beginning of § 2 has a meaning also for the case of a holomorphic mapping $f\colon M \to N$, where M, N are compact complex manifolds, M being now with boundary. If both M and N are Riemann surfaces, the result so obtained forms a generalization of the Riemann-Hurwitz formula. Such a result is easily derived as a consequence of the Gauss-Bonnet formula. In the particular case when $N = P_1$, this is called the second main theorem of the theory of value distributions of meromorphic functions and constitutes the core of the theory.

By simply writing down the generalized Riemann-Hurwitz formula, one can derive in a purely differential-geometric way the following theorem *): Let $f\colon D \to N$ be a holomorphic mapping, where D is the pointed disk $0 < |\zeta| < 1$ and N is a compact Riemann surface of genus > 1. Then f can be extended as a holomorphic mapping of the whole disk $|\zeta| < 1$ into N.

Similarly, by a combination of the first and second main theorems, one can generalize the defect relations on meromorphic functions to holomorphic mappings $f\colon M \to P_1$, where M is a non-compact Riemann surface such that it can be compactified, as a Riemann surface, by the addition of a finite number of points. In the case that the image Riemann surface N is a complex torus, one derives in this way the result that the defect at every point $a \in N$ is zero. Geometrically the latter means that N is "evenly" covered by the image of M.

---

*) I am indebted to J.-P. Serre for pointing out this conclusion to me.

All these seem to justify the emphasis we have put on our fundamental problem. Unfortunately, for higher dimensions, even when the image manifold is P_n, our knowledge on the problem is still very limited. For a holomorphic mapping $f: M \to P_n$, with M compact, this leads us back to the old theory of projective invariants in algebraic geometry. With recent advances in algebraic geometry, it might be possible and worthwhile to organize the classical results in a better form. The case of non-compact M awaits much further work.

I hope to have pointed out a few guiding ideas on the subject of holomorphic mappings. Only the future can tell whether the topic will lead to results of general mathematical interest. I cannot help to feel, however, that so long as the complex structure remains a subject of investigation, the study of holomorphic mappings should be a logical objective.

In conclusion I wish to say that, while I have discussed the subject from a geometrical viewpoint, there has been an extensive literature to which I am indebted and which it would be impossible to quote in detail. Many of the ideas in geometrical function theory in one variable originated from L. Ahlfors. In the case of high dimensions I should mention in particular the works of H. Schwartz and W. Stoll [8, 10], although they do not seem to have a close contact with the viewpoints envisaged here.

REFERENCES

[1] Borel, A. et J.-P. Serre, Le théorème de Riemann-Roch. *Bull. Soc math. France*, 86,97-136 (1958).

[2] Chern, S. S., Complex analytic mappings of Riemann surfaces, I. *Amer. J. Math.*, 82, 323-337 (1960).

[3] —— The integrated form of the first main theorem for complex analytic mappings in several complex variables. *Annals of Math.*, 71, 536-551 (1960).

[4] —— Some remarks on the quantitative theory of holomorphic mappings, to be submitted to *Comm. Math. Helv.*

[5] Porteous, I. R., *On the Chern classes of algebraic varieties related by a regular algebraic correspondence*, to appear.

[6] Remmert, R., Holomorphe und meromorphe Abbildungen komplexer Räume. *Math. Annalen*, 133, 328-370 (1957).

[7] de Rham, G., *Variétés différentiables.* Paris, 1955.

[8] SCHWARTZ, Marie-Hélène. Formules apparentées à la formule de Gauss-Bonnet pour certaines applications d'une variété à n dimensions dans une autre. *Acta Math.*, 91, 189-244 (1954).
[9] STERNBERG, S. and R. G. SWAN, On maps with non-negative Jacobian. *Mich. Math. J.*, 6, 339-342 (1959).
[10] STOLL, W., Die beiden Hauptsätze der Wertverteilungstheorie bei Funktionen mehrerer komplexer Veränderlichen. Teil I: *Acta Math.*, 90, 1-115 (1953); Teil II: *Acta Math.*, 92, 55-169 (1954).

Department of mathematics
University of California
Berkeley, California.

Pseudo-Riemannian Geometry and the Gauss-Bonnet Formula*)

SHIING-SHEN CHERN

University of California, Berkeley, California, USA

(Received November 9, 1962; presented by L. NACHBIN)

1. INTRODUCTION

Recent investigations in the theory of general relativity have aroused an interest in lorentzian and, more generally, in pseudo-riemannian manifolds. The latter are manifolds with non-degenerate quadratic differential forms defined on them. To a compact oriented pseudo-riemannian manifold the integral in the Gauss-Bonnet formula can be generalized, and it is natural to ask whether its value is still equal to the Euler-Poincaré characteristic of the manifold. The usual proof (see [1], [3]) cannot be extended immediately. It is the main purpose of this paper to prove that the formula, with a suitable modifiction, remains true.

We will develop pseudo-riemannian geometry in more detail than would be necessary for the treatment of the Gauss-Bonnet formula. This is partly motivated by the increasing interest on the subject, but a more pertinent reason is the fact that in the study of pseudo-riemannian manifolds the geometry of pseudo-riemannian vector bundles plays a very essential rôle.

2. PSEUDO-RIEMANNIAN VECTOR BUNDLES

By a manifold we will always mean a connected, paracompact, C^∞-differentiable manifold. Moreover, all our functions and mappings will be understood to be C^∞.

Let M be a manifold of dimension m. Let $\psi: B \to M$ be a bundle of real vector spaces of dimension r over M. This means that there is an open covering $\{U, V, W, \ldots\}$ of M such that $\psi^{-1}(U)$ can be coordinatized, relative to U, by the coordinates (x, y_U), $x \in U$, $y_U \in Y$, where Y is a typical fiber, which is a real vector space of dimension r. Moreover, if $x \in U \cap V$ and if (x, y_V), $y_V \in Y$, are the local coordinates of $\psi^{-1}(x)$ relative to V, then $y_U = \gamma_{UV}\, y_V$, where γ_{UV} is a mapping: $U \cap V \to GL(r; R)$. The mappings γ_{UV}, to be called the transition functions of the bundle relative to the covering $\{U, V, \ldots\}$, satisfy the conditions

$$
(1)\qquad
\begin{aligned}
\gamma_{UU} &= \text{identity},\\
\gamma_{UV}^{-1} &= \gamma_{VU},\\
\gamma_{UV}\,\gamma_{VW}\,\gamma_{WU} &= \text{identity in } U \cap V \cap W \neq \emptyset.
\end{aligned}
$$

*) Work done under *National Science Foundation* grant No. G10700.

We will frequently consider Y to be the vector space of one-columned vectors and GL(r;R) to be the group of all non-singular $(r \times r)$-matrices; the action of GL(r; R) on Y will then be given by matrix multiplication.

The vector bundle B is called pseudo-riemannian, if there exists a non-degenerate symmetric bilinear function H in each fiber $\psi^{-1}(x)$, which varies in a C^∞-way in x. Since M is connected, the signature (p, q) of H, $p + q = r$, is constant. (If H is considered as a quadratic form in $\psi^{-1}(x)$ and reduced to square terms by a choice of basis in $\psi^{-1}(x)$ then p is the number of positive squares and q is the number of negative squares). A pseudo-riemannian bundle is called riemannian if $q = 0$ and lorentzian if $q = 1$.

A manifold with a pseudo-riemannian (respectively riemannian or lorentzian) structure in its tangent bundle is called pseudo-riemannian (respectively riemannian or lorentzian).

As mentioned above, we will consider the transition functions γ_{UV} as non-singular $(r \times r)$-matrices defined in $U \cap V \neq \emptyset$ for any two neighborhoods U, V of our covering. Relative to the same covering, an affine connection is given by an $(r \times r)$-matrix θ_U of linear differential forms in each U such that, in $U \cap V \neq \emptyset$, we have

$$d\,\gamma_{UV} + \theta_U\,\gamma_{UV} = \gamma_{UV}\,\theta_V. \tag{2}$$

It is easy to verify that this relation is compatible, that is, in $U \cap V \cap W \neq \emptyset$, two of these relations imply the third one as a consequence.

From (2) it follows that

$$d\,y_U + \theta_U\,y_U = \gamma_{UV}\,(d\,y_V + \theta_V\,y_V). \tag{3}$$

The vanishing of this common expression is therefore independent of the coordinate neighborhood. If a vector field is defined on a curve C in M (i.e., if a cross-section of the vector bundle B over C is given), it is said to be parallel along C if (3) vanishes.

An affine connection on a pseudo-riemannian vector bundle is called admissible, if the bilinear function H (y, z), $y, z \in \psi^{-1}(x)$, remains unchanged when the vectors y, z are displaced parallelly along curves. We will call H (y, z) the scalar product of y, z and we will find the condition for an affine connection to be admissible. Relative to a coordinate neighborhood U, let y_U, z_U be the column vectors which are the coordinates of y, z. Then the function H (y, z) can ve written in the matrix form

$$H(y, z) = {}^t y_U\,H_U\,z_U, \tag{4}$$

where ${}^t y_U$ denotes the transpose of y_U and H_U is a non-singular symmetric matrix in U. The transformation law for H_U is obviously

$$H_V = {}^t\gamma_{UV}\,H_U\,\gamma_{UV}. \tag{5}$$

Under parallel displacement of the vectors y, z we have

$$d({}^t y_U\,H_U\,z_U) = {}^t y_U\,(-\,{}^t\theta_U\,H_U + d\,H_U - H_U\,\theta_U)\,z_U.$$

Since this is to be zero for arbitrary y, z, the condition for the affine connection to be admissible is

$$d\,H_U - H_U\,\theta_U - {}^t\theta_U\,H_U = 0. \tag{6}$$

It is important now to derive the relations which will follow by exterior differentiation of (2) and (6). Taking the exterior derivative of (2), we get

$$\Theta_V = \gamma_{UV}^{-1} \Theta_U \gamma_{UV}, \tag{7}$$

where

$$\Theta_U = d\theta_U + \theta_U \wedge \theta_V. \tag{8}$$

The last product is the usual product of matrices, with the additional convention that multiplication of exterior differential forms is in the sense of the exterior product. The matrices Θ_U of exterior differential forms of degree two, with the transformation law (7), define the curvature of the affine connection; Θ_U will be called the curvature form (relative to U).

Similarly, exterior differentiation of (6) gives

$$H_U \Theta_U + {}^t\Theta_U H_U = 0, \tag{9}$$

so that the matrix $H_U \Theta_U$ is anti-symmetric. Moreover, it satisfies the transformation law

$$H_V \Theta_V = {}^t\gamma_{UV} H_U \Theta_U \gamma_{UV}. \tag{10}$$

Comparison of (5) and (10) leads to a globally defined form on M in case r is even and the bundle B is oriented. The latter means that the group of the bundle is, instead of GL (r, R), its onnected component of the identity or, what is the same, that the matrices γ_{UV} have positive determinants. From (5) we get

$$\det H_V = (\det \gamma_{UV})^2 (\det H_U).$$

Under the assumption that the bundle is oriented, this implies

$$|\det H_V|^{\frac{1}{2}} = (\det \gamma_{UV}) |\det H_U|^{\frac{1}{2}}, \tag{11}$$

where the square roots are taken to be positive.

We put

$$H_U \Theta_U = (\Theta^{\alpha\beta}, U), \quad \gamma_{UV} = (\gamma^{\beta}_{\alpha, UV}), \; 1 \leqq \alpha, \beta \leqq r. \tag{12}$$

Consider the expression

$$\Sigma \epsilon_{\alpha_1 \ldots \alpha_r} \Theta^{\alpha_1 \alpha_2}{}_{,V} \ldots \Theta^{\alpha_{r-1} \alpha_r}{}_{,V},$$

where $\epsilon_{\alpha_1 \ldots \alpha_r}$ is $+1$ or -1 according as $\alpha_1, \ldots, \alpha_r$ are an even or odd permutation of $1, \ldots, r$, and is otherwise zero, and where the summation is extended over all $\alpha_1, \ldots, \alpha_r$ from 1 to r. By (10) this expression is equal to the corresponding expression with the subscript U, multiplied by det γ_{UV}. It follows that the form

$$\Delta = \frac{(-1)^{\frac{r}{2}}}{2^r \pi^{\frac{r}{2}} \left(\frac{r}{2}\right)! \, |\det H_U|^{\frac{1}{2}}} \Sigma \epsilon_{\alpha_1 \ldots \alpha_r} \Theta^{\alpha_1 \alpha_2}{}_{,U} \ldots \Theta^{\alpha_{r-1} \alpha_r}{}_{,U} \tag{13}$$

is independent of the subscript U and is globally defined in M. It is not hard to verify that $d\Delta = 0$. Hence Δ defines, in the sense of the Rham's theorem, an element of the cohomology group H^{2r} (M:R) of M with real coefficients. We will show that this cohomology class is the Euler characteristic class of the vector bundle B. For riemannian bundles this result is known as the Gauss-Bonnet formula.

If a scalar product H defines a pseudo-riemannian structure of signature (p, q) on a vector bundle B, its negative $-H$ defines a pseudo-riemannian structure of signature (q, p). An affine connection admissible to the one is admissible to the other.

From two vector bundles over the same manifold M their Whitney sum can be constructed. If the bundles are pseudo-riemannian, a pseudo-riemannian structure can be defined on their Whitney sum in an obvious way. The same can be said about admissible affine connections.

3. THE WEIL HOMOMORPHISM

The notion of a pseudo-riemannian vector bundle and its admissible affine connection is a special case of the notion of a connection in a principal fiber bundle with a Lie group. We will establish this relationship and apply a theorem of Weil ([2], [4]) to the effect that the cohomology class determined by Δ is independent of the choice of the connection. For this purpose we recall that a frame of the bundle B is an ordered set of r linearly independent vectors with the same origin x (i.e., in the same fiber $\psi^{-1}(x)$). Relative to a coordinate neighborhood U, a frame can be defined analytically by a nonsingular $(r \times r)$-matrix s_U, whose columns are the components of the r vectors. All the frames of B form the frame bundle, which we will denote by B_F. The projection we will denote by $\psi_F: B_F \longrightarrow M$. The frame bundle has the local coordinates (x, s_U), with the transformation law $s_U = \gamma_{UV}\, s_V$ valid in $U \cap V$.

In $\psi_F^{-1}(U)$ we introduce the matrix of linear differential forms

$$\varphi = s_U^{-1}\, d s_U + s_U^{-1}\, \theta_U\, s_U\,. \tag{14}$$

Equation (2) is then equivalent to the statement that this expression is equal, in $\psi_F^{-1}(U \cap V)$, to the corresponding expression with the subscripts U replaced everywhere by V. In other words, φ is globally defined in B_F. If we put

$$\Phi = d\varphi + \varphi \wedge \varphi, \tag{15}$$

we find

$$\Phi = s_U^{-1}\, \Theta_U\, s_U\,. \tag{16}$$

Thus the matrix Φ, globally defined in B_F, essentially gives the curvature form of the connection.

Similarly, we introduce in B_F the matrix

$$F = {}^t s_U\, H_U\, s_U = {}^t s_V\, H_V\, s_V\,. \tag{17}$$

This matrix has a simple geometrical meaning: The element in its αth row and βth column is the scalar product of the αth and the βth vectors of the frame. The admissibility of the affine connection can be expressed by the condition

$$d F = F \varphi + {}^t\varphi F. \tag{18}$$

Exterior differentiation of (18) gives

$$F \Phi + {}^t \Phi F = 0 \tag{19}$$

If the bundle is oriented and r is even, the expression in (13) can be written

$$(20) \qquad \psi_F^* \Delta = \frac{(-1)^{\frac{r}{2}}}{2^r \pi^{\frac{r}{2}} \left(\frac{r}{2}\right)! \, |\det F|^{\frac{1}{2}}} \Sigma \, \epsilon_{\alpha_1 \ldots \alpha_r} \Phi^{\alpha_1 \alpha_2} \ldots \Phi^{\alpha_{1-r} \alpha_r}, \qquad F \Phi = (\Phi^{\alpha\beta}).$$

The advantage for the consideration of the frame bundle consists in the fact that the quantities are globally defined in it instead of being locally defined in the base manifold.

We now restrict ourselves to frames for which the matrix F is diagonal with the diagonal elements $(\underbrace{1, \ldots, 1}_{p}, \underbrace{-1, \ldots, -1}_{q})$. This matrix we will denote by $F_o(p, q)$ or simply F_o. The frames in question are then characterized analytically by the condition

$$(21) \qquad F_o = {}^t s_U H_U s_U .$$

Their totality constitutes a submanifold B_{FO} of B_F. By reducing the coordinate neighborhoods when necessary, we choose in each U a matrix $s_U(x)$, $x \in U$, of positive determinant and satisfying the condition (21). In the vector bundle B we can use new local coordinates defined by

$$(22) \qquad y'_U = s_U^{-1} y_U .$$

The new transition functions will then be

$$(23) \qquad \gamma'_{UV} = s_U^{-1} \gamma_{UV} s_V ,$$

and they satisfy the condition

$$(24) \qquad {}^t\gamma'_{UV} F_o \gamma'_{UV} = F_o .$$

We wish also to note that the scalar product in terms of the new local coordinates is given by

$$(25) \qquad H(y, z) = {}^t y'_U F_o z'_U .$$

In general, we will denote by O(p,q) the group of all $(r \times r)$-matrices T satisfying the condition

$$(26) \qquad {}^t T F_o T = F_o ,$$

and by SO(p, q) the subgroup of O(p, q) whose matrices have positive determinant. We set O(r, 0) = O(r), SO(r, 0) = SO(r), which are respectively the orthogonal group and the properly orthogonal group in r variables.

Since $\gamma'_{UV} \in O(p, q)$, the pseudo-riemannian structure of B gives a reduction of the structure group to O(p, q). Moreover, if B is oriented, its structure group is reduced to SO(p, q). Conversely, a reduction of the structure group of B to O(p, q) (respectively to SO(p, q)) implies a pseudo-riemannian (respectively an oriented pseudo-riemannian) structure on B. For the scalar product defined by (25) is then independent of the choice of the coordinate neighborhood U.

Suppose that our vector bundle B has an oriented pseudo-riemannian structure. Its structure group is then reduced to SO(p,q), and it has as its principal fiber bundle the sub-manifold B_{FSO} of B_F, which consists of all frames satisfying (21) and the additional condition that $\det s_U > 0$. Given an admissible affine connection, the restriction of φ

to B_{FSO} gives a connection form in the sense of a connection in a principal fiber bundle, while the restriction of Φ gives its curvature form. Equations (18) and (19) can then be written

$$F_0\,\varphi + {}^t\varphi\,F_0 = 0, \tag{27}$$
$$F_0\,\Phi + {}^t\Phi\,F_0 = 0.$$

Weil's homomorphism gives the following result:

For an oriented pseudo-riemannian bundle of vector spaces of even dimension, the cohomology class with real coefficients in M determined by the form Δ is independent of the choice of the admissible affine connection.

4. THE GAUSS-BONNET THEOREM

We will prove the following theorem:

Let B be an oriented pseudo-riemannian bundle of vector spaces of even dimension over a compact manifold M. The cohomology class determined by the form Δ is its Euler characteristic class.

For a riemannian bundle this is the classical Gauss-Bonnet theorem. (The theorem has so far only been formulated for the tangent bundle, but the proof extends in a straightforward way).

Consider therefore the general case that the pseudo-riemannian structure H is of arbitrary signature (p,q). We impose in addition a riemannian structure G on the bundle, so that $G(y,z)$, $y, z \in \psi^{-1}(x)$, is a positive definite symmetric bilinear function, which varies in a C^∞-way with x. For fixed $x \in M$ and fixed $y \in \psi^{-1}(x)$, the eigenvalues of H relative to G are the values λ such that

$$H(y,z) = \lambda\ G(y,z)$$

holds identically in z. The corresponding non-zero y which satisfies this equation is called an eigenvector. It is well-known that there are r real eigenvalues, of which p are positive and q are negative. Let $\psi_1^{-1}(x)$ (respectively $\psi_2^{-1}(x)$) be the subspace of $\psi^{-1}(x)$, which is spanned by the eigenvectors with positive (respectively negative) eigenvalues. Then $\psi^{-1}(x)$ is a direct sum of $\psi_1^{-1}(x)$ and $\psi_2^{-1}(x)$, so that any $y \in \psi^{-1}(x)$ can be written in a unique way as a sum:

$$y = P_1\,y + P_2\,y, \quad P_i\,y \in \psi_i^{-1}(x),\ i = 1, 2. \tag{28}$$

We define the symmetric bilinear functions

$$H_1(y,z) = H(P_1\ y,\ P_1\ z), \tag{29}$$
$$H_2(y,z) = -H(P_2\,y, P_2\,z).$$

Then H_i is positive definite when restricted to $\psi_i^{-1}(x)$, i = 1, 2, and we have

$$H(y,z) = H_1(y,z) - H_2(y,z). \tag{30}$$

Moreover, the function

$$K(y,z) = H_1(u,z) + H_2(y,z) \tag{31}$$

defines a riemannian structure on the bundle B.

We introduce the bundles

$$B_i = \bigcup_{x \in M} \psi_i^{-1}(x), \quad i = 1, 2, \tag{32}$$

whose projections $\psi_i : B_i \to M$ are defined by $\psi_i(\psi_i^{-1}(x)) = x$. Then the given bundle B is a Whitney sum of B_1 and B_2. The function H_i defines a riemannian structure on B_i, from which the given pseudo-riemannian structure H and a new riemannian strucuture K on B are defined by (30) and (31). Take in B_i an affine connection admissible to H_i. Their direct sum is an affine connection in B, which is admissible to both H and K. Let Δ_H and Δ_K be the form Δ in (20) constructed with respect to the pseudo-riemannian strucutures H and K respectively.

To study these two expressions remember that they are constructed from the matrix $F\Phi$, where Φ is the curvature form of the admissible affine connection and the element in the αth row and βth column of F is the scalar product of the αth and βth vectors of the frame. Since the form Δ is in the base manifold M, we can restrict ourselves to frames at x, whose first p vectors are in $\psi_1^{-1}(x)$ and whose last q vectors are in $\psi_2^{-1}(x)$. Then the matrices F relative to H and K are respectively of the forms

$$F_H = \begin{pmatrix} F_1 & 0 \\ 0 & -F_2 \end{pmatrix}, \quad F_K = \begin{pmatrix} F_1 & 0 \\ 0 & F_2 \end{pmatrix},$$

where F_1, F_2 are positive definite symmetric matrices of orders p,q respectively. Moreover, the curvature matrix of the affine connection when restricted to our choice of frames is of the form

$$\Phi = \begin{pmatrix} \Phi_1 & 0 \\ 0 & \Phi_2 \end{pmatrix},$$

where Φ_1, Φ_2 are matrices of exterior quadratic differential forms of orders p, q respectively. From these expressions we immediately conclude that

$$\Delta_H = \Delta_K = 0, \text{ if p or q is odd,}$$

and that

$$\Delta_H = \Delta_K, \text{ if p and q are even.}$$

In both cases we see that the theorem is true in case the affine connection is the one chosen above. Our theorem then follows as a consequence of the theorem at the end of § 3.

Our discussion gives also a new proof, and slight generalization, of a theorem of H. Samelson and T. J. Willmore [5], [6].

If an oriented bundle over a compact manifold has a field of subspaces of odd dimension, its Euler class with real coefficients is zero.

5. PSEUDO-RIEMANNIAN MANIFOLDS

Certain special features arise, when the vector bundle is the tangent bundle of the manifold M, and we will discuss them. In this case the transition function γ_{UV} is the functional matrix of the local coordinates u^i in U with respect to the local coordinates v^k in V, $1 \leqq i, k \leqq m$. We introduce the one-rowed matrices

$$du = (du^1, \ldots, du^m), \quad dv = (dv^1, \ldots, dv^m). \tag{33}$$

Then

(34) $$du\ s_U = dv\ s_V \quad \text{in } U \cap V$$

and this common expression defines, globally in the frame bundle B_F, a one-rowed matrix of linear differential forms, which we will denote by τ. If there is an affine connection in the bundle, we will have, in $\psi_F^{-1}(U)$,

$$d\tau = -\tau \wedge \varphi + du \wedge \Theta_U S_U.$$

The affine connection is said to be without torsion, if the following condition holds:

(35) $$d\tau = -\tau \wedge \varphi.$$

If we put

$$\theta_U = \left(\sum_k \Gamma^j_{ik}\, du^k\right), \quad 1 \leq i, j, k \leq m,$$

then

$$du \wedge \theta_U = \left(\sum_{i,k} \Gamma^j_{ik}\, du^i \wedge du^k\right),$$

and the absence of torsion of the affine connection is equivalent to the analytical condition that Γ^j_{ik} is symmetric in i, k.

The classical argument on the existence and uniqueness of the Levi-Civita connection of a riemannian metric, which we will not repeat here, shows that there is exactly one matrix Θ_U which satisfies the condition (6) and the condition: $du \wedge \Theta_U = 0$. In other words, we have the following result: *Every pseudo-riemannian structure on a manifold has exactly one admissible affine connection without torsion.*

Another special feature for the pseudo-riemannian structure in a tangent bundle is the possibility of introducing the sectional curvature to every two-dimensional subspace of the tangent space. Without introducing more analytical apparatus, we will show how the notion of sectional curvature in the cotangent space can be defined. From here till the end of the paper we will suppose that all small Latin indices have the range from 1 to m. We will restrict ourselves to a coordinate neighborhood U and will compare the quantities defined in it with quantities in $\psi_F^{-1}(U)$. We put

(36) $$s_U = (s_i^j),\ H_U = (h^{ij}),\ F = (f^{ij}),$$

so that, by (17), we have

(37) $$f^{ij} = \sum_{k,l} s_k^i\, s_l^j\, h^{kl}.$$

Let

(38) $$\tau = (\tau^1, \ldots, \tau^m).$$

Then we have, by (34),

(39) $$\tau^i = \sum_j du^j\, s_j^i$$

A bivector in the cotangent space can be written

(40) $$\xi = \sum_{i,j} p_{ij}\, du^i \wedge du^j = \sum_{i,j} q_{ij}\, \tau^i \wedge \tau^j, \quad p_{ij} + p_{ji} = q_{ij} + q_{ji} = 0,$$

where

(41) $$p_{ij} = \sum_{k,l} q_{kl}\, s_i^k\, s_j^l .$$

Introduce the quantities

(42) $$p^{ij} = \sum_{k,l} h^{ik}\, h^{jl}\, p_{kl}, \qquad q^{ij} = \sum_{k,l} f^{ik}\, f^{jl}\, q_{kl} .$$

Then we have

(43) $$q^{ij} = \sum_{a,b} s_a^i\, s_b^j\, p^{ab},$$

and

(44) $$\frac{1}{2} \sum_{i,j} p_{ij}\, p^{ij} = \frac{1}{2} \sum_{i,j} q_{ij}\, q^{ij} .$$

This common expression is, from the right-hand side, independent of U and, from the left-hand side, independent of the frame. Its value is therefore a quantity associated to the bivector ξ; it is the square of the measure of ξ.

Similarly, we write

(45) $$H_U\, \Theta_U = (\Theta^{ij}), \quad F\, \Phi = (\Phi^{ij}),$$

where

(46) $$\Theta^{ij} = \frac{1}{2} \sum_{k,l} R^{ij}_{kl}\, du^k \wedge du^l, \qquad \Phi^{ij} = \frac{1}{2} \sum_{k,l} S^{ij}_{kl}\, \tau^k \wedge \tau^l,$$

and

(47) $$R^{ij}_{kl} = -R^{ij}_{lk} = -R^{ji}_{kl}, \quad S^{ij}_{kl} = -S^{ij}_{lk} = -S^{ji}_{kl} .$$

Then we have, from (16), (17), and (39),

(48) $$\sum_{i,\ldots,l} R^{ij}_{kl}\, p_{ij}\, p^{kl} = \sum_{i,\ldots,l} S^{ij}_{kl}\, q_{ij}\, q^{kl} .$$

This common value is therefore a quantity associated to ξ by the curvature of the connection.

The quotient

(49) $$K = -\sum_{i,\ldots,l} R^{ij}_{kl}\, p_{ij}\, p^{kl} \Big/ 2 \sum_{i,j} p_{ij}\, p^{ij}$$

depends only on the two-dimensional subspace of the cotangent space determined by ξ. It is called its sectional curvature.

Added December 11, 1962. After this paper has gone to press, there appeared a note of ANDRÁ AVEZ: *Formule de Gauss-Bonnet-Chern en métrique de signature quelconque*, Comptes Rendus de l'Acad. des Sciénces (Paris), 255, 2049–2051 (1962). In this note Avez sketched a proof of our main result. His proof has some contact with ours, but is somewhat different".

v. 35 n.º 1, 31 de março de 1963.

BIBLIOGRAPHY

[1] Chern, S., (1945), *Ann. of Math.*, **46**, 674–684.

[2] Chern, S., (1950), *Proc. Int. Cong. Math.*, **II**, 397–411.

[3] Flanders, H., (1953), *Trans. Amer. Math. Soc.*, **75**, 311–326.

[4] Griffiths, P. A., (1962), *Ill. J. of Math.*, **6**, 468–79.

[5] Samelson, H. (1951), *Portug. Math.*, **10**, 129–133.

[6] Willmore, T. J., (1951), *C. R. Acad. Sci. Paris*, 232, 298–299.

Minimal Surfaces in an Euclidean Space of N Dimensions[1]

SHIING-SHEN CHERN

1. Introduction

A classical theorem of S. Bernstein [1] asserts that a minimal surface of class C^2, which can be globally represented in the form $z = f(x, y)$, is a plane. This theorem has been generalized by R. Osserman to the following more geometrical form [9]:

A complete simply connected minimal surface in 3-space whose normal mapping into the unit sphere omits a neighborhood of some point is a plane.

A major tool in the study of such problems is complex function theory. We wish to show that the work of Bernstein-Osserman has a complete generalization to minimal surfaces in euclidean spaces of higher dimensions. The over-riding facts are the following: (1) The Grassmann manifold $SO(n+2)/SO(2) \times SO(n)$, (where $SO(n) = SO(n; R)$), of all oriented planes through a point 0 in an euclidean space of dimension $n+2$ has a complex structure invariant under the action of $SO(n+2)$; (2) with the complex structure on the minimal surface defined by its induced riemannian metric, the Gauss mapping, which assigns to a point p of the minimal surface the oriented plane through 0 parallel to the tangent plane at p, is anti-holomorphic. This makes it possible to apply the theory of holomorphic curves to derive geometrical consequences on minimal surfaces.

2. The complex hyperquadric

The Grassmann manifold mentioned above is an irreducible symmetric hermitian manifold ([7], p. 354) and is complex analytically equivalent to the non-degenerate complex hyperquadric. We propose to describe some of its geometrical properties, to the extent that they will be needed in our problem.

[1] Work done under partial support by National Science Foundation grant No. G-21938 and ONR contract 3656(14).

We write

$$Q_n = SO(n+2)/SO(2) \times SO(n), \tag{1}$$

where the right-hand side stands for the space of right cosets. Then Q_n is a compact homogeneous manifold of real dimension $2n$. To give it a complex structure let $SO(n+2; C)$ be the complex properly orthogonal group, i.e., the group of all $(n+2) \times (n+2)$ matrices A with complex entries such that

$$A \cdot {}^tA = I, \qquad \det A = 1, \tag{2}$$

where ${}^tA$ denotes the transpose of A and I is the unit matrix. Throughout this paper we will agree on the following ranges of indices:

$$1 \leqq \alpha, \beta, \gamma \leqq n+2, \qquad 3 \leqq a, b, c \leqq n+2. \tag{3}$$

Let

$$A = (a_{\alpha\beta}). \tag{4}$$

Introduce the vector

$$Z = (z_1, \ldots, z_{n+2}). \tag{5}$$

Then $SO(n+2; C)$ acts on the vector space V_{n+2} of the vectors Z according to

$$Z \to ZA. \tag{6}$$

It therefore acts on the complex projective space $P_{n+1}(C)$ of dimension $n+1$ obtained from $V_{n+2} - 0$ by identifying the vectors which differ from each other by a non-zero factor. This action is intransitive; in fact, it leaves invariant the hyperquadric

$$z_1^2 + \cdots + z_{n+2}^2 = 0. \tag{7}$$

On the other hand, its induced action on (7) is transitive.

Let N be the subgroup of $SO(n+2; C)$, which leaves invariant the point $(1, i, 0, \ldots, 0)$ of (7). The hyperquadric (7) can then be regarded as a right coset space $SO(n+2; C)/N$. From the inclusion

$$SO(n+2) \subset SO(n+2; C)$$

and the easily checked fact that

$$N \cap SO(n+2) = SO(2) \times SO(n),$$

it follows that (7) is isomorphic to Q_n. Henceforth we will identify these two manifolds.

To find the relation between the complex coordinates z_α on Q_n and its representation by (1), let A be a real properly orthogonal matrix. Then A maps the point $(1, i, 0, \ldots, 0)$ to $(a_{11} + ia_{21}, \ldots, a_{1,n+2} + ia_{2,n+2})$, so that we have

(8) $$z_\alpha = a_{1\alpha} + ia_{2\alpha},$$

where $a_{1\alpha}$, $a_{2\alpha}$ are real and satisfy the relations

(9) $$\sum_\alpha a_{1\alpha}^2 = \sum_\alpha a_{2\alpha}^2 = 1, \qquad \sum_\alpha a_{1\alpha}a_{2\alpha} = 0.$$

The Plücker coordinates of the plane are then given by

(10) $$p_{\alpha\beta} = a_{1\alpha}a_{2\beta} - a_{1\beta}a_{2\alpha} = \frac{i}{2}(z_\alpha\bar{z}_\beta - z_\beta\bar{z}_\alpha).$$

The topology of Q_n is well known ([2], [5], [8]). Q_n is simply connected, has no torsion, and has all its odd-dimensional homology groups equal to zero. Its even-dimensional Betti numbers are given by

(11) $$b^{2h} = 1, \qquad 2h \neq n;\ b^n = 2, \text{ if } n \text{ is even.}$$

For $n \geqq 3$ a generator of the homology group $H_{2n-2}(Q_n)$ is given by the section of Q_n by a hyperplane of $P_{n+1}(C)$. For $n = 2$, Q_2 is complex analytically equivalent to $P_1(C) \times P_1(C)$, the equivalence being established as follows: Q_2 has two families of projective lines, such that: (1) distinct lines of the same family do not meet; (2) lines of different familes meet in exactly one point; (3) through a point of Q_2 there passes one line from each family. Let $q_0 \in Q_2$ be a fixed point, and let L_1, L_2 be the lines through q_0. Define the mapping $Q_2 \to L_1 \times L_2$ by assigning to each point $q \in Q_2$ the points where the lines through q meet L_1, L_2 respectively. This establishes the complex analytic equivalence stated above. It follows that the lines of Q_2 define cycles which are generators of $H_2(Q_2)$.

To study the geometry of Q_n we take the right-invariant linear differential forms of $SO(n + 2)$. These are the entries of the skew-symmetric matrix

(12) $$\varphi = (\varphi_{\alpha\beta}) = dA \cdot A^{-1} = dA \,.\, {}^tA.$$

Explicitly they are given by

(13) $$\varphi_{\alpha\beta} = \sum_\gamma da_{\alpha\gamma}a_{\beta\gamma}.$$

They satisfy the structure equations

(14) $$d\varphi_{\alpha\beta} = \sum_\gamma \varphi_{\alpha\gamma} \wedge \varphi_{\gamma\beta}.$$

From (13) and (8) we get

$$\varphi_{1b} + i\varphi_{2b} = \sum_{\gamma} a_{b\gamma}\, dz_{\gamma} \tag{15}$$

Thus the forms at the left-hand side of (15), to be denoted by θ_b, are the complex-valued linear differential forms, which define the almost complex structure on Q_n. The integrability of this almost complex structure follows from the expression in (15), but can also be verified directly by using the structure equations (14).

On Q_n we now introduce the hermitian structure

$$ds^2 = \tfrac{1}{2}\sum_{a} \theta_a \bar{\theta}_a. \tag{16}$$

The latter is obviously invariant under the action of $SO(n + 2)$. By (15), (8), and making use of the relation

$$\sum_{\alpha} z_\alpha\, dz_\alpha = 0, \tag{17}$$

we find

$$\tfrac{1}{2}\sum \theta_a \bar{\theta}_a = \tfrac{1}{2}\sum dz_\alpha\, d\bar{z}_\alpha - \tfrac{1}{4}(\sum \bar{z}_\alpha\, dz_\alpha)(\sum z_\beta\, d\bar{z}_\beta). \tag{18}$$

In this formula z_α are normalized by the condition

$$\sum_{\alpha} z_\alpha \bar{z}_\alpha = 2, \tag{19}$$

in view of (9).

Now the standard hermitian structure on $P_{n+1}(C)$ with the homogeneous coordinates z- is given by (see [3]):

$$\frac{1}{(\sum z_\alpha \bar{z}_\alpha)^2}\{(\sum z_\alpha \bar{z}_\alpha)(\sum dz_\beta\, d\bar{z}_\beta) - (\sum \bar{z}_\alpha\, dz_\alpha)(\sum z_\beta\, d\bar{z}_\beta)\}. \tag{20}$$

This reduces to (18), when the normalization (19) holds. This means that the hermitian structure (16) on Q_n is induced from the hermitian structure (20) of its ambient space $P_{n+1}(C)$. It is known, and can be easily verified, that the associated two-form of the latter can be written

$$\Omega = \frac{i}{2}\, d'\, d'' \log (\sum_{\alpha} z_\alpha \bar{z}_\alpha). \tag{21}$$

In the open set $z_1 \neq 0$ of P_{n+1} we have

$$\Omega = \frac{i}{2}\, d'\, d'' \log \left(1 + \frac{|z_2|^2 + \cdots + |z_{n+2}|^2}{|z_1|^2}\right). \tag{22}$$

This reduces to the associated two-form of the hermitian structure (16) of Q_n by restriction; the latter is by definition

$$\Omega = \frac{i}{4}\sum_a \theta_a \wedge \bar{\theta}_a.$$

For later applications to minimal surfaces we wish to find other expressions for the associated two-form Ω on Q_n for $n = 1,2$. In the case $n = 1$, the conic Q_1, with the point $(1, -i, 0)$ deleted, has the parametric representation

(23) $$z_1 = 1 + t^2, \quad z_2 = i(1 - t^2), \quad z_3 = 2it,$$

and we get

$$\sum_\alpha z_\alpha \bar{z}_\alpha = 2(1 + t\bar{t})^2.$$

It follows that

(24) $$\Omega = i\, d'\, d'' \log\left(1 + \frac{1}{|t|^2}\right), \quad t \neq 0.$$

Similarly, for $n = 2$ the quadric Q_2 has the parametric representation

(25) $$z_1 = 1 + t\tau, \quad z_2 = i(1 - t\tau), \quad z_3 = -t + \tau, \quad z_4 = -i(t + \tau),$$

and we find

(26) $$\Omega = \frac{i}{2} d'\, d'' \log\left\{\left(1 + \frac{1}{|t|^2}\right)\left(1 + \frac{1}{|\tau|^2}\right)\right\}, \quad t\tau \neq 0.$$

By a holomorphic curve in Q_n we mean a holomorphic mapping $f: M \to Q_n$, where M is a Riemann surface (i.e., a complex manifold of one dimension). The mapping f, followed by the inclusion on Q_n in P_{n+1}, defines a holomorphic curve in P_{n+1}. The latter is called non-degenerate, if it does not lie in a linear space of lower dimension. A classical theorem of E. Borel can now be stated as follows [11]:

Let $g: M \to P_{n+1}$ be a non-degenerate holomorphic curve, where M is the gaussian plane. To $n + 3$ hyperplanes in P_{n+1}, in general position, the image $g(M)$ meets one of them.

Let P^*_{n+1} be the dual projective space of P_{n+1}, i.e., the manifold of all the hyperplanes of P_{n+1}. Under the hypotheses of Borel's theorem let S be the set of all hyperplanes which have non-void intersections with $g(M)$. Then Borel's theorem has the consequence that S is dense in P^*_{n+1}.

3. Geometry of surfaces

Consider now the euclidean space E of dimension $n + 2$ and an oriented surface M immersed in E defined by the C^∞-mapping

$$x: M \to E. \tag{27}$$

To the surface is associated the Gauss mapping

$$f: M \to Q_n, \tag{28}$$

where $f(p)$, $p \in M$, is the oriented plane through a fixed point 0 of E and parallel to the tangent plane to $x(M)$ at $x(p)$.

We consider the bundle B of orthonormal frames $xe_1 \cdots e_{n+2}$ such that $x \in x(M)$ and e_1, e_2 are tangent vectors at $x = x(p)$, $p \in M$. In the bundle B we have

$$\begin{aligned} dx &= \omega_1 e_1 + \omega_2 e_2, \\ de_i &= \sum_\alpha \varphi_{i\alpha} e_\alpha. \end{aligned} \tag{29}$$

In these and in later formulas we will agree on the following range of indices:

$$1 \leqq i, j, k \leqq 2. \tag{30}$$

We consider the following diagram of mappings:

$$\begin{array}{ccc} B & \xrightarrow{F} & SO(n+2) \\ {\scriptstyle\psi}\downarrow & & \downarrow{\scriptstyle\pi} \\ M & \xrightarrow{f} & Q_n = SO(n+2)/SO(2) \times SO(n). \end{array} \tag{31}$$

In this diagram ψ is the projection which sends the frame $xe_1 \cdots e_{n+2}$ to its origin x; F maps the frame to the matrix whose rows are the components of e_α; and π is the natural projection. The diagram (31) is clearly commutative. Our notation is essentially consistent with those of §2; use of $\varphi_{i\alpha}$ in (29) means that we have omitted the dual mapping F^*.

By taking the exterior derivative of the first equation of (29) and considering the terms involving e_a, we get

$$\sum_i \omega_i \wedge \varphi_{ia} = 0.$$

From this it follows that

$$\varphi_{ia} = \sum_k {}_{iak}\omega_k, \tag{32}$$

where

$$l_{iak} = l_{kai}. \tag{33}$$

The quadratic differential forms

$$\Phi_a = \sum_i \omega_i \varphi_{ia} = \sum_{i,k} l_{iak}\omega_i\omega_k \tag{34}$$

are the second fundamental forms of the surface.

The normal vector

$$l_a = \sum_k l_{kak} = l_{1a1} + l_{2a2} \tag{35}$$

is called the mean curvature vector. Its vanishing defines the minimal surfaces. The Gaussian curvature of the induced metric on M is given by

$$K = \sum_a (l_{1a1}l_{2a2} - l_{1a2}^2). \tag{36}$$

The complex structure defined by the induced metric on M is given by the differential form $\omega_1 + i\omega_2$. For a minimal surface we find

$$\varphi_{1a} + i\varphi_{2a} = (l_{1a1} + il_{1a2})(\omega_1 - i\omega_2). \tag{37}$$

This proves the result: *If M is a minimal surface, the Gauss mapping* (28) *is anti-holomorphic.* For $n = 2$ this result was first proved by M. Pinl [10].

4. Generalization of the theorem of Bernstein-Osserman

The theorem we wish to establish is the following:

Let $x: M \to E$ be a complete simply-connected minimal surface immersed in a euclidan space E of dimension $n + 2$ ($\geqq 3$) and not lying in a linear space of lower dimension. Let $f: M \to Q_n \subset P_{n+1}$ be the Gauss mapping. Then the subset of P^*_{n+1} whose sections with Q_n meet $f(M)$ is dense in P^*_{n+1}.

For $n = 2$ write $Q_2 = L_1 \times L_2$ as a product of two projective lines and denote by π_1, π_2 respectively the projections on the two factors. Then either $\pi_1 \circ f(M)$ is dense in L_1 or $\pi_2 \circ f(M)$ is dense in L_2. For $n = 1$ the set $f(M)$ is dense in Q_1.

The theorem can be interpreted as a description of the set $f(M)$ relative to the generating cycles of $H_{2n-2}(Q_n)$. The statements in the second part of the theorem correspond to the fact that, for $n = 1, 2$, these generating cycles are not given by the hyperplane sections of Q_n. The last statement is the Bernstein-Osserman theorem quoted in §1.

First of all, M is non-compact ([4]). By the Riemann mapping theorem, M, with the conformal structure induced from the metric of E, is conformally equivalent to the gaussian plane (parabolic type) or to the unit disk (hyperbolic type). We will prove that, under the assumption that the assertions in the theorem are not true, M is necessarily of parabolic type. The proof is based on the following lemma of Osserman [9]:

Let M be a complete, non-compact, simply-connected riemannian manifold of two dimensions. Let K be its gaussian curvature and Δ the second Beltrami operator. Suppose there exists a real-valued function $u \geqq \varepsilon > 0$ (ε = const.) satisfying the condition

$$\Delta \log u = K. \tag{38}$$

Then M, with the conformal structure defined by its riemannian metric, is parabolic.

Before applying the lemma, we will formulate the condition (38) in a different form. Let x, y be the isothermal parameters of the riemannian metric, so that $z = x + iy$ is a local coordinate of the induced complex structure and that the riemannian metric can be written

$$ds^2 = h^2\, dz\, d\bar{z}, \qquad h > 0.$$

Then

$$\Delta = \frac{1}{h^2}\left(\frac{\partial^2}{\partial x^2} + \frac{\partial^2}{\partial y^2}\right)$$

and

$$2i\, d'\, d'' \log u = (\Delta \log u)\, \frac{i}{2}\, h^2\, dz \wedge d\bar{z},$$

where the last expression after the parentheses is the element of area of M. Hence condition (38) can be written

$$2i\, d'\, d'' \log u = K \cdot \frac{i}{2}\, h^2\, dz \wedge d\bar{z}. \tag{39}$$

Suppose that the assertions in the theorem are not true. In the general case there is a neighborhood in P^*_{n+1} whose hyperplanes do not meet $f(M)$; we can suppose this to be a neighborhood of the hyperplane $z_1 = 0$. Then the function

$$u = \left\{\frac{|z_1|^2}{|z_1|^2 + \cdots + |z_{n+2}|^2}\right\}^{\frac{1}{2}} \tag{40}$$

is $\geqq \varepsilon$ on M, for a certain constant $\varepsilon > 0$. Since f is antiholomorphic, we have

$$d'\, d'' f^* = f^*\, d''\, d'$$

and we find, by (22) and (36),

$$2i\, d'\, d''\, f^* \log u = -2if^*\, d'\, d'' \log u = 2f^*\Omega = K\omega_1 \wedge \omega_2,$$

so that the function $u \circ f$ satisfies the condition (39).

In the case $n = 1$ we suppose that $f(M)$ omits a neighborhood of the point $t = 0$ in the parametric representation (23), and put

$$u = \frac{|t|^2}{1 + |t|^2} \tag{41}$$

For $n = 2$ we can consider t and u as the non-homogeneous coordinates on the projective lines L_1, L_2 respectively. Suppose $\pi_1 \circ f(M)$ omits a neighborhood of $t = 0$ and $\pi_2 \circ f(M)$ omits a neighborhood of $\tau = 0$. Choose

$$u = \left\{\frac{|t|^2|\tau|^2}{(1 + |t|^2)(1 + |\tau|^2)}\right\}^{\frac{1}{2}} \tag{42}$$

Then, in view of (24) and (26), these functions u satisfy the conditions in Osserman's lemma. It follows that in all cases M is of parabolic type.

It is now easy to get a contradiction to our assumption. In the general case it follows from Borel's theorem that the holomorphic curve $f(M)$ is degenerate. Suppose P_{m+1}, $m < n$, be the linear subspace of least dimension in P_{n+1}, which contains $f(M)$. Then P_{m+1} is not a point, because $x(M)$ is not a plane. Borel's Theorem, when applied to $f(M) \subset P_{m+1}$, $0 \leqq m < n$, has the consequence that the set of hyperplanes in P_{m+1}, which meet $f(M)$, is dense in P^*_{m+1} (= space of hyperplanes in P_{m+1}). This in turn implies that the set of hyperplanes in P_{n+1}, which meet $f(M)$ is dense in P^*_{n+1}. But this contradicts our assumption. Similarly, we get contradictions in the other cases. Thus our main theorem is completely proved.

It should be remarked that our proof gives also the following theorem: *Let x: $M \to E$ be a complete simply-connected minimal surface immersed in a euclidean space E of four dimensions. If $\pi_i \circ f(M)$ is not dense in L_i, $i = 1, 2$, then the surface $x(M)$ is a plane.*

As a result of the relation between minimal surfaces and holomorphic curves pointed out in this paper, it seems clear that further works are called for to extend the quantitative results on minimal surfaces in the direction initiated by E. Heinz [6] and developed by E. Hopf, Nitsche, Osserman, etc.

Appendix[2]

The idea mentioned at the end of this paper has been partially carried out, and such an inequality has been found. In fact, following the notation of this paper, let x: $M \to E$ be a simply-connected minimal surface and let $D \subset M$ be a finite piece of geodesic radius s about a point $p_0 \in M$. The image $f(D)$ under the Gauss mapping being a subset of P_{n+1}, let δ

[2] Added November 1, 1963.

be the distance, in the hermitian metric in P_{n+1}, of $f(D)$ from a fixed hyperplane of P_{n+1}. Then we have the following inequality

$$|K(p_0)| \leqq \frac{2(n+1)}{s^2} \cot^2 \delta \cos^2 \delta (1 + 2n \cot^2 \delta) \tag{43}$$

for the Gaussian curvature $K(p_0)$ at p_0. A similar inequality has been obtained by Osserman.

Denote the fixed hyperplane by λ. The homogeneous coordinates in P_{n+1} being $z_1, \ldots, z_{n+2}$, we suppose λ to be defined by $z_1 = 0$. The distance $d(z, \lambda)$ from a point $z \in P_{n+1}$ to λ is then given by the formula

$$\sin d(z, \lambda) = \frac{|z_1|}{\sqrt{|z_1|^2 + \cdots + |z_{n+2}|^2}}. \tag{44}$$

Since the inequality (43) is automatically valid when $\delta = 0$, we will suppose $\delta > 0$, i.e., that $f(D)$ does not meet the hyperplane λ. In $P_{n+1} - \lambda$ let

$$\zeta_\rho = z_\rho / z_1, \qquad 2 \leqq \rho, \sigma \leqq n + 2 \tag{45}$$

The mapping which assigns to $z \in P_{n+1} - \lambda$ the number $\zeta_\rho \in C$ (= complex field) we will denote by ψ_ρ. The formula for $d(z, \lambda)$ can be written

$$\sin^2 d(z, \lambda) = \frac{1}{1 + |\zeta|^2}, \tag{46}$$

where

$$|\zeta|^2 = |\zeta_2|^2 + \cdots + |\zeta_{n+2}|^2. \tag{47}$$

The condition $d(z, \lambda) \geq \delta$ is equivalent to $|\zeta|^2 \leqq R^2$, with $R = \cot \delta$.

With the function

$$u = (1 + |\zeta|^2)^{-\frac{1}{2}} \tag{48}$$

defined in (40), the Gaussian curvature K can be calculated according to the formula

$$-2if^* d' d'' \log u = 2i\, d' d'' f^* \log u = K \omega_1 \wedge \omega_2. \tag{49}$$

By calculation we get

$$2i\, d' d'' \log u = \frac{-i}{(1 + |\zeta|^2)^2} \{ \sum_\rho d\zeta_\rho \wedge d\bar{\zeta}_\rho + \sum_{\rho < \sigma} (\zeta_\rho\, d\zeta_\sigma - \zeta_\sigma\, d\zeta_\rho) \wedge (\bar{\zeta}_\rho\, d\bar{\zeta}_\sigma - \bar{\zeta}_\sigma\, d\bar{\zeta}_\rho \} \tag{50}$$

Formula (49) shows that $(f^*u)\, ds$, where ds is the element of arc on M, has zero Gaussian curvature. With this fact as analytical basis, Osserman showed ([9], p. 230), by making use of the uniformization theorem, that there exists a holomorphic mapping h of M into the complex t-plane, such that

$$h^*|dt| = (f^*u)\, ds. \tag{51}$$

Moreover,[3] if $|\zeta|^2 \leqq R^2$ in D, the image $h(D)$ contains an unbranched disk Δ of radius $r = s/(1 + R^2)^{\frac{1}{2}}$ about $h(p_0)$.

Taking the elements of area of both sides of (51), we get

$$h^*\left(\frac{i}{2}\, dt \wedge d\bar{t}\right) = \frac{1}{1 + |\zeta|^2}\, \omega_1 \wedge \omega_2. \tag{52}$$

We now map Δ into C by the composition $\psi_\rho \circ f \circ h^{-1}$. The latter is an anti-holomorphic mapping, under which ζ_ρ is a holomorphic function of $\bar{t}$. Combining (49), (50), (52), we get the following formula for the Gaussian curvature K:

$$|K| = \frac{2}{(1 + |\zeta|^2)^3}\left\{\sum_\rho \left|\frac{d\zeta_\rho}{d\bar{t}}\right|^2 + \sum_{\rho<\sigma}\left|\zeta_\rho \frac{d\zeta_\sigma}{d\bar{t}} - \zeta_\sigma \frac{d\zeta_\rho}{d\bar{t}}\right|^2\right\}. \tag{53}$$

The image of Δ under $\psi_\rho \circ f \circ h^{-1}$ belongs to a disk of radius R. By supposing the center of Δ to be the point $t = 0$, it follows from Schwarz Lemma that

$$\left|\frac{d\zeta_\rho}{d\bar{t}}\right|_{t=0} \leqq \frac{R}{r}.$$

Since $|\zeta_\rho| \leqq R$, we get, at $t = 0$,

$$\left|\zeta_\rho \frac{d\zeta_\sigma}{d\bar{t}} - \zeta_\sigma \frac{d\zeta_\rho}{d\bar{t}}\right| \leqq \frac{2R^2}{r}.$$

Substituting into (53), we find

$$|K(p_0)| \leqq 2\left\{(n+1)\frac{R^2}{r^2} + \binom{n+1}{2}\frac{4R^4}{r^2}\right\},$$

which gives (43) on simplification.

Allowing $s \to \infty$ and utilizing the fact that a minimal surface with zero Gaussian curvature is a plane, we get the following corollary:

Let $x\colon M \to E$ be a complete, simply-connected minimal surface in an euclidean space E of dimension $n + 2$, such that $x(M)$ is not a plane. Then the set of hyperplanes of P_{n+1} which meet the image $f(M) \subset P_{n+1}$ under the Gauss mapping f is dense in P^*_{n+1}.

Actually this corollary also follows from the proof in §4 of this paper, although the main theorem was stated there under stronger hypotheses.

[3] R. Osserman, An analogue of the Heinz-Hopf inequality, *J. Math. and Mech.*, 8 (1959) p. 383–385.

REFERENCES

[1] S. BERNSTEIN, Sur un théorème de géométrie et ses applications aux équations aux dérivées partielles du type elliptique, *Comm. Inst. Sci. Math. Mech. Univ. Kharkov, 15* (1915–17), p. 38–45.

[2] E. CARTAN, Sur les propriétés topologiques des quadriques complexes, *Publ. math. Univ. Belgrade, 1* (1932), p. 55–74; or *Oeuvres*, Partie I, vol. 2, p. 1227–1246.

[3] S. CHERN, Characteristic classes of Hermitian manifolds, *Annals of Math., 47* (1946), p. 85–121.

[4] S. CHERN and C. C. HSIUNG, On the isometry of compact submanifolds in euclidean space, *Math. Annalen, 149* (1963), p. 278–285.

[5] C. EHRESMANN, Sur la topologie de certains espaces homogènes, *Annals of Math., 35* (1934), p. 396–443.

[6] E. HEINZ, Über die Lösungen der Minimalflächengleichung, *Nachr. Akad. Wiss. Göttingen* 1952, p. 51–56.

[7] S. HELGASON, *Differential Geometry and Symmetric Spaces*, Academic Press, New York, 1962.

[8] W. V. D. HODGE and D. PEDOE, *Methods of Algebraic Geometry*, Vol. 2, Book 4, Cambridge University Press, 1952.

[9] R. OSSERMAN, Proof of a conjecture of Nirenberg, *Comm. on Pure and Applied Math., 12* (1959), p. 229–232.

[10] M. PINL, B-Kugelbilder reeller Minimalflächen in R^4, *Math. Zeits., 59* (1953), p. 290–295.

[11] H. WEYL, *Meromorphic Functions and Analytic Curves*, Princeton University Press, 1943.

On Holomorphic Mappings of Hermitian Manifolds of the Same Dimension*

SHIING-SHEN CHERN

1. **Introduction.** In order to study the holomorphic mappings of one complex manifold into another, it is desirable, and in many cases even necessary, to introduce hermitian metrics in these manifolds. If the two manifolds are of the same dimension, the simplest metrical invariant that arises is the ratio of their volume elements. This has an important bearing on the properties of the mapping itself, and the present paper is devoted to a study of such relations.

More precisely, let $f: M \to N$ be a holomorphic mapping, where M and N are complex manifolds of the same dimension n. Relative to the local coordinates z^i let

$$ds_M^2 = \sum_{i,k} h_{ik} dz^i d\bar{z}^k \tag{1}$$

be a positive definite hermitian metric on M. (*Throughout this paper, all the small Latin indices will run, without exception, from* 1 *to n.*) Similarly, let

$$ds_N^2 = \sum_{j,l} \tilde{h}_{jl} dw^j d\bar{w}^l \tag{1a}$$

be a positive definite hermitian metric on N, with the local coordinates w^j. Under the mapping f, w^j are holomorphic functions of z^i, The function

$$u = \frac{\det(\tilde{h}_{jl})}{\det(h_{ik})} \left| \det\left(\frac{\partial w^r}{\partial z^s}\right)\right|^2 \quad (\geq 0) \tag{2}$$

is independent of the choice of the local coordinates, and is geometrically the ratio of the volume elements. Our main result is a local formula for Δu where Δ is the Laplacian in M (see formula (60)). The formula is particularly simple when $u > 0$; then we have

$$\tfrac{1}{2}\Delta \log u = R - \mathrm{Tr}(f^*(\mathrm{Ric})). \tag{3}$$

---

* Work done under partial support form NSF grant GP-3990 and ONR contract 3656 (14).

At the right-hand side of (3), R denotes the scalar curvature of M and $\mathrm{Tr}(f^*(\mathrm{Ric}))$ is the trace of the inverse image of the Ricci form of N (The Ricci form "Ric" is an hermitian form in the tangent spaces of N and is mapped by the induced linear map into an hermitian form in the tangent spaces of M.)

To draw geometrical consequences it is necessary to impose some conditions on the domain manifold M and the image manifold N. The first property is:

(DO_K). M is exhausted by a sequence of open submanifolds

$$M_1 \subset M_2 \subset M_3 \subset \cdots \subset M \tag{4}$$

whose closures $\bar{M}_\alpha$ are compact, such that: (1) to each $\alpha = 1,2,\cdots$ there is a smooth function $v_\alpha \geqq 0$ defined in M_α, which satisfies the inequality

$$\tfrac{1}{2}\Delta v_\alpha \leqq R + K\exp(v_\alpha/n), \tag{5}$$

where K is a given positive constant; (2) $v_\alpha(p_\beta) \to \infty$, if p_β is a divergent sequence of points in M_α. (An infinite sequence of points $p_\beta \in M_\alpha$ is called divergent, if every compact set of M_α contains only a finite number of them.)

For example, we will see that the unit ball D_1 defined by

$$z_1\bar{z}_1 + \cdots + z_n\bar{z}_n < 1 \tag{6}$$

in the n-dimensional number space C_n with coordinates $(z_1,\cdots,z_n)$ has the property (DO_K), with $K = 2n(n+1)$. It is very likely that all the bounded symmetric domains have this property.

The condition on the image manifold N also involves a positive constant K. To state it, we denote by $H(\xi,\eta)$, where ξ,η are tangent vectors with the same origin in N, the hermitian inner product in N, and by $\mathrm{Ric}(\xi,\eta)$ the Ricci form of the hermitian metric. Then we have

(IM_K). N is said to have the property (IM_K), if

$$\mathrm{Ric}(\xi,\xi) \leqq -(K/n)H(\xi,\xi). \tag{7}$$

In particular, N has the above property, if its hermitian metric is einsteinian and has its scalar curvature bounded above by a negative constant. In general we say therefore that N is *almost einsteinian* if it has the property (IM_K).

If $f: M \to N$ is a holomorphic mapping and M and N have the properties (DO_K) and (IM_K) respectively, then we have an upper bound for the scalar function u (cf. Theorems 2 and 3 below). This includes an n-dimensional version of the Schwarz lemma, which says that if M is the unit n-ball and N is almost einsteinian, the mapping f does not increase volume.

The basic idea of this study is the realization that "negative curvature of the image manifold restricts a holomorphic mapping". Many classical theorems in function theory in one complex variable (such as Schottky's theorem) and on mappings of Riemann surfaces fall under this general theme. However, even if the metric on N does not have the property (IM_K), but is conformally equivalent to one with this property, important geometrical consequences will also follow. This leads in particular to a generalization of the Bloch theorem to n dimensions. (Cf. Theorem 4 below.) In this respect it may be worthwhile to point out that our results are developed for hermitian manifolds, and not merely for kählerian manifolds, because a conformal change of the metric usually destroys the kählerian property. It is not without surprise that the possible lack of the kählerian property does not complicate the problems studied in this paper.

2. **Hermitian geometry.** We will review in this section the basic facts on hermitian manifolds. For details compare [2], [5], [7].

Let M be a complex manifold of dimension n. An hermitian metric in M is given in terms of the local coordinates z^i by the hermitian differential form (1) where h_{ij} are complex valued C^∞ functions satisfying the condition of "hermitian symmetry";

$$h_{ij} = \bar{h}_{ji}, \tag{8}$$

and the further condition that the matrix (h_{ij}) is positive definite.

To the hermitian form (1) is associated the real-valued exterior differential form of degree two and rank $2n$:

$$\Theta = \frac{i}{2} \sum_{j,k} h_{jk} dz^j \wedge d\bar{z}^k, \tag{9}$$

called the associated two-form of ds_M^2. It is so normalized that when $n = 1$ and $z^1 = z = x + iy$ the associated two-form of $dzd\bar{z} = dx^2 + dy^2$ is $dx \wedge dy$.

The hermitian metric (1) is called kählerian if Θ is closed, i.e., if

$$d\Theta = 0. \tag{10}$$

Kähler proved that a necessary and sufficient condition for (10) is the existence of a local "potential", i.e., a real-valued function F, defined locally, such that

$$\Theta = id'd''F/2. \tag{11}$$

The hermitian metric (1) defines an inner product in the tangent spaces of M. A unitary frame is an ordered set of n tangent vectors e_i with the same origin, such that

$$(e_i, e_j) = \delta_{ij}, \tag{12}$$

where the expression at the left-hand side stands for the inner product of e_i, e_j. The dual basis in the cotangent space dual to a unitary frame is called a unitary coframe. A unitary coframe consists therefore of n complex-valued linear differential forms θ_i of type (1,0) such that

$$ds_M^2 = \sum_i \theta_i \bar{\theta}_i. \tag{13}$$

Let B be the bundle of all unitary frames of M. The forms θ_i are forms in B. It is a fundamental theorem in local hermitian geometry that there exists in B a set of complex-valued linear differential forms θ_{ij}, uniquely determined, which satisfy the conditions

$$d\theta = \sum_j \theta_j \wedge \theta_{ji} + \Theta_i, \tag{14}$$

$$\theta_{ij} + \bar{\theta}_{ji} = 0, \tag{15}$$

where Θ_i are of type (2,0). The θ_{ij} are the "connection forms" of the hermitian metric (13).

By the consideration of the exterior derivative of (14), it can be shown that we can put

$$d\theta_{ij} - \sum_k \theta_{ik} \wedge \theta_{kj} = \Theta_{ij}, \tag{16}$$

where Θ_{ij} are of the form

$$\Theta_{ij} = \tfrac{1}{2} \sum_{k,l} R_{ijkl} \theta_k \wedge \bar{\theta}_l, \tag{17}$$

and satisfy the conditions

$$\Theta_{ij} + \bar{\Theta}_{ji} = 0. \tag{18}$$

In terms of the coefficients R_{ijkl} in (17), the conditions (18) are expressed by the symmetry properties

$$R_{ijkl} = \bar{R}_{jilk}. \tag{19}$$

The quantities R_{ijkl}, which in our formulation are functions in B, constitute essentially the curvature tensor of the hermitian metric. From them we form the "Ricci tensor"

$$R_{ij} = \sum_k R_{kkij} = \bar{R}_{ji}, \tag{20}$$

and the scalar curvature

$$R = \sum_i R_{ii}. \tag{21}$$

The latter is a real-valued function in M.

To establish the relationship between our "Ricci tensor" R_{ij} and the Ricci form mentioned in the Introduction, let ξ and η be tangent vectors with the same origin, which are given in terms of a unitary frame e_i by

$$\xi = \sum_i \xi_i e_i, \qquad \eta = \sum_k \eta_k e_k. \tag{22}$$

Then we have

(23) $$\mathrm{Ric}(\xi,\eta) = \sum_{i,k} R_{ik}\xi_i\bar{\eta}_k.$$

In this respect it may be mentioned that the frame e_i being unitary the hermitian inner product of our metric can be expressed by

(24) $$H(\xi,\eta) = \sum \xi_i\bar{\eta}_i.$$

We set

(25) $$\theta = \sum_k \theta_{kk}.$$

By (16) and (17) we have then

(26) $$d\theta = \tfrac{1}{2} \sum_{i,j} R_{ij}\theta_i \wedge \bar{\theta}_j.$$

The form θ, from which R_{ij} is determined by (26), is completely characterized by the conditions

(27) $$d(\theta_1 \wedge \cdots \wedge \theta_n) = -\theta \wedge \theta_1 \wedge \cdots \wedge \theta_n, \quad \theta + \bar{\theta} = 0.$$

The hermitian metric is called einsteinian, if $d\theta$ is a multiple of Θ i.e., if

(28) $$d\theta = -(i/n)R\Theta.$$

We now consider differential operators on real-valued C^∞ functions on M. Let u be such a function. We will either lift u into a function in B or choose a local frame field and restrict ourselves to a neighborhood of M; the formalisms are exactly the same. In fact, we put

(29) $$du = \sum_i (u_i\theta_i + \bar{u}_i\bar{\theta}_i).$$

Taking the exterior derivative of this equation, we get

$$\sum_i \Big(du_i - \sum_k u_k\theta_{ik}\Big) \wedge \theta_i + \sum_i \Big(d\bar{u}_i - \sum_k \bar{u}_k\bar{\theta}_{ik}\Big) \wedge \bar{\theta}_i + \sum_i u_i\Theta_i + \sum_i \bar{u}_i\bar{\Theta}_i = 0.$$

This allows us to put

(30) $$du_i - \sum_k u_k\theta_{ik} = \sum_k u'_{ik}\theta_k + \sum_k u_{ik}\bar{\theta}_k,$$

where

(31) $$\sum_{i,k} u'_{ik}\theta_k \wedge \theta_i + \sum_i u_i\Theta_i = 0.$$

From (29) we get

(32) $$d'u = \sum_i u_i\theta.$$

It follows that

$$d'd''u = -dd'u = \sum_{i,j} u_{i\bar{j}}\theta_i \wedge \bar{\theta}_j. \tag{33}$$

We define the Laplacian of u to be

$$\Delta u = 4 \sum_i u_{i\bar{i}}. \tag{34}$$

The factor in (34) is so chosen that for $n = 1$ and $ds_M^2 = dzd\bar{z}$, we have

$$\Delta u = 4\partial^2 u/\partial z\partial\bar{z} = \partial^2 u/\partial x^2 + \partial^2 u/\partial y^2.$$

If $u > 0$, we find

$$\Delta \log u = \frac{1}{u}\Delta u - \frac{4}{u^2}\sum_i u_i \bar{u}_i, \tag{35}$$

a formula which will be useful later on.

It is worth noting that although kählerian manifolds are the ones with which we will be concerned in applications, the kählerian property is not needed in the establishment of our general formulas.

Before proceeding, we wish to consider the example of the unit ball. Let $z_1, \cdots, z_n$ be the coordinates in C_n, and let $r \geqq 0$ be defined by

$$r^2 = \sum_k z_k \bar{z}_k. \tag{36}$$

The unit ball is the domain D_1 in C_n defined by $r < 1$. In it we introduce the kählerian metric whose associated two-form is the form (11) with

$$F = -\log(1 - r^2). \tag{37}$$

It follows that

$$\Theta = \frac{i}{2}\left\{\frac{1}{1 - r^2}\sum_k dz_k \wedge d\bar{z}_k + \frac{4r^2}{(1 - r^2)^2} d'r \wedge d''r\right\} \tag{38}$$

and

$$ds_M^2 = \frac{1}{1 - r^2}\sum_k dz_k d\bar{z}_k + \frac{4r^2}{(1 - r^2)^2} d'r\, d''r. \tag{39}$$

We wish to calculate the Ricci curvature of this metric and to show in particular that it is einsteinian. For this purpose we use the definition of the Ricci curvature by (26). It suffices to make the calculation relative to a unitary frame field or, what is the same, a unitary coframe field. This amounts to writing

$$ds_M^2 = \sum_k \theta_k \bar{\theta}_k, \tag{40}$$

where

(41) $$\theta_k = (1-r^2)^{-1/2}(dz_k + z_k L(r) d'r),$$

with $L(r)$ given by

(42) $$rL(r) + 2 = 2(1-r^2)^{-1/2}$$

It follows that

$$\bigwedge_k \theta_k = (1-r^2)^{-n/2} \bigwedge_k (dz_k + z_k L d'r) = (1-r^2)^{-(n+1)/2} dz_1 \wedge \cdots \wedge dz_n.$$

Differentiating, we get

$$d\left(\bigwedge_k \theta_k\right) = -(n+1) r(1-r^2)^{-(n+3)/2}(d' - d'')r \wedge dz_1 \wedge \cdots \wedge dz_n.$$

Comparison with (27) gives

(43) $$\theta = (n+1) r(1-r^2)^{-1}(d' - d'')r.$$

Its exterior derivative is found to be

(44) $$d\theta = 2i(n+1)\Theta.$$

By comparing with (28) we see that the metric (39) in the unit ball is einsteinian with its scalar curvature equal to $-4n(n+1)$. For $n=1$ we have

(45) $$ds_M^2 = (1-r^2)^{-2} dz d\bar{z},$$

and the scalar curvarture is equal to -4 as is well known.

3. **Holomorphic mapping of hermitian manifolds.** Our purpose is to study the holomorphic mappings of M into an hermetian manifold N of the same dimension n. The treatment of the last section applies also to N. It is necessary, however, to denote the quantities and differential forms pertaining to N by different symbols, and we list the corresponding ones in the following table:

(46) $$\begin{pmatrix} M & \theta_i & \theta_{ij} & \Theta_i & \Theta_{ij} & R_{ijkl} & R_{ij} & R \\ N & \omega_i & \omega_{ij} & \Omega_i & \Omega_{ij} & S_{ijkl} & S_{ij} & S \end{pmatrix}.$$

Thus all the general formulas in §1 remain valid, when the quantities are replaced by the corresponding ones in the second line of (46).

Let $f: M \to N$ be a holomorphic mapping. The most important scalar invariant which arises from such a situation is the ratio of the volume elements, i.e., the volume element of N by that of M. We denote it by $u \geqq 0$ and our basic analytic tool is a formula for Δu.

The derivation of this formula is local, and we will make use of unitary frame fields in both M and N; our final result will of course be independent of these choices. Under the mapping f let us set

$$\omega_i = \sum_j a_{ij}\theta_j. \tag{47}$$

Then we have

$$u = |\det(a_{ij})|^2. \tag{48}$$

By taking the exterior derivative of (47) and making use of (14) and its corresponding formula for $d\omega_i$, we get

$$\sum_j \left(da_{ij} - \sum_k a_{ik}\theta_{jk} + \sum_k a_{kj}\omega_{ki}\right) \wedge \theta_j - \Omega_i + \sum_j a_{ij}\Theta_j = 0.$$

This allows us to put

$$da_{ij} - \sum_k a_{ik}\theta_{jk} + \sum_k a_{kj}\omega_{ki} = \sum_k a_{ijk}\theta_k \tag{49}$$

where a_{ijk} satisfy the relations

$$\sum_{j,k} a_{ijk}\theta_k \wedge \theta_j - \Omega_i + \sum_j a_{ij}\Theta_j = 0. \tag{50}$$

Similarly, we take the exterior derivative of (49) and make use of the equations (14), (16), (17), and the corresponding equations for the image manifold N. It gives

$$da_{ijk} - \sum_l a_{ilk}\theta_{jl} - \sum_l a_{ijl}\theta_{kl} + \sum_l a_{ljk}\omega_{li} = \sum_l a_{ijkl}\,\theta_l + \sum_l b_{ijkl}\bar\theta_l, \tag{51}$$

where b_{ijkl} satisfy the relation

$$\sum_{k,l} b_{ijkl}\bar\theta_l \wedge \theta_k = -\sum_h a_{ih}\Theta_{jh} + \sum_h a_{hj}\Omega_{hi}. \tag{52}$$

The last equation gives, more explicitly,

$$b_{ijkl} = \tfrac{1}{2}\sum_h a_{ih}R_{jhkl} - \tfrac{1}{2}\sum_{h,r,s} a_{hj}a_{rk}\bar a_{sl}S_{hirs}. \tag{53}$$

To calculate Δu, with u given by (48), we put

$$D = \det(a_{ij}) \tag{54}$$

so that

$$u = D\bar D. \tag{55}$$

It follows from (49) that dD is of the form

$$dD = D(\theta - \omega) + \sum_i D_i\theta_i, \tag{56}$$

where θ is defined by (25) and similarly

$$\omega = \sum_i \omega_{ii}. \tag{57}$$

The u_i defined in (29) are therefore given by

$$u_i = \bar{D}D_i. \tag{58}$$

The D_i are polynomials in a_{ij} and a_{ijk}. In order to evaluate Δu it is necessary (and sufficient) to find u_{ik} as defined in (30). This can be done by the use of (51) and (53). However, as we are only interested in the quantities u_{ik}, i.e., the coefficients of $\bar{\theta}_k$ in du_i, a simpler way is to take directly the exterior derivative of (56). This gives

$$\sum_i \left\{ dD_i - \sum_k D_k\theta_{ik} - D_i(\theta - \omega) \right\} \wedge \theta_i + D(d\theta - d\omega) + \sum_i D_i\Theta_i = 0,$$

from which it follows that

$$u_{ik} = D_i\bar{D}_k + \frac{u}{2}\left(R_{ik} - \sum_{h,j} S_{hj}a_{hi}\bar{a}_{jk}\right). \tag{59}$$

Therefore we have our basic formula

$$\frac{1}{4}\Delta u = \sum_i D_i\bar{D}_i + \frac{u}{2}\left(R - \sum_{h,j,i} S_{hj}a_{hi}\bar{a}_{ji}\right). \tag{60}$$

If $u > 0$, it follows from (35), (58), and (60) that

$$\frac{1}{2}\Delta \log u = R - \sum_{h,j,i} S_{hj}a_{hi}\bar{a}_{ji}. \tag{61}$$

This is precisely the formula (3) given in the Introduction.

To draw geometrical conclusions we start with some definitions: f is said to be *degenerate* at $p \in M$, if u vanishes at p. Geometrically this means that the induced linear map on the tangent spaces is not univalent at p. f is called *totally degenerate* if u vanishes identically. If the common dimension of M and N is 1 and M is connected, a totally degenerate mapping is a constant mapping. From the definition of u we see that f is *volume decreasing* or *volume increasing* according as $u \leqq 1$ or $u \geqq 1$.

Then we have the following theorem:

THEOREM 1. *Let $f: M \to N$ be a holomorphic mapping, where M, N are hermitian manifolds of the same dimension, with M compact and N einsteinian. Let R and S be their scalar curvatures respectively. Then we have:*

(1) *If $R > 0$, $S \leqq 0$, f is totally degenerate.*

(2) *If $R < 0$, $S \geqq 0$, then there is a point of M at which f is degenerate.*

In fact, the condition for N to be einsteinian is

$$S_{ik} = (S/n)\delta_{ik},$$

in which case (60) reduces to

$$\frac{1}{4}\Delta u = \sum_i D_i\bar{D}_i + \frac{u}{2}\left(R - \frac{S}{n}\sum_{i,k}|a_{ik}|^2\right).$$

Since M is compact, u attains its maximum at a point $p_0 \in M$. If u is not identically zero, $u(p_0) > 0$. At p_0 we have

$$(D_i)_{p_0} = 0, \quad (\Delta u)_{p_0} \leqq 0.$$

This is not possible under the conditions in (1). Thus (1) is proved.

Similarly (2) is proved by the consideration of the minimum of u.

4. **A generalization of Schwarz lemma.** The condition (IM_K) for the manifold N to be almost einsteinian can be expsessed analytically by the relation

$$\sum_{i,k} S_{ik}\xi_i\bar{\xi}_k \leqq -\frac{K}{n}\sum_i \xi_i\bar{\xi}_i, \quad all\ \xi_i. \tag{62}$$

By Hadamard's well-known determinant inequality we have

$$\frac{1}{n}\sum_{i,k}|a_{ik}|^2 \geqq |\det(a_{ik})|^{2/n} = u^{1/n}. \tag{63}$$

We now put

$$v = \log u. \tag{64}$$

It follows from (61) that if N satisfies the condition (62) and $u > 0$ we have

$$\tfrac{1}{2}\Delta v \geqq R + K\exp(v/n). \tag{65}$$

THEOREM 2. *Let $f: M \to N$ be a holomorphic mapping, where M and N are hermitian manifolds of the same dimension having the properties* (DO_K) *and* (IM_K) *respectively, with the same positive constant K. Then $v \leqq v_\alpha$.*

To prove this theorem consider the open subset E of M defined by $v > v_\alpha$. In E we have $u > 0$. By (5) and (65) we have

$$\tfrac{1}{2}\Delta(v - v_\alpha) \geqq K(\exp(v/u) - \exp(v_\alpha/u)).$$

Thus $\Delta(v - v_\alpha) > 0$ in E, and the function $v - v$ cannot have a maximum in E. Hence $v - v_\alpha$ must approach its least upper bound on a sequence tending to the boundary of E. This sequence cannot have a limit point p_0 in M_α, for at p_0, $v - v_\alpha > 0$, and p_0 would belong to E and be a maximum for $v - v_\alpha$. It also cannot be divergent, for then $v_\alpha \to \infty$ while v is bounded. It follows that the only possibility is that E is vacuous, so that $v \leqq v_\alpha$.

In applications it is useful to replace a metric by a conformal one. Let ds_N^2 denote the hermitian metric on N. A conformal metric defined in a neighborhood

$V \subset N$ of a point $q_0 \in N$ and given by $ds_N^{*2} = \lambda(q)ds_N^2$, $q \in V$, $\lambda(q) > 0$, is said to be a *supporting metric* at q_0 if: (1) $\lambda(q_0) = 1$; (2) $\lambda(q) \leqq 1$, $q \in V$. Then we have:

THEOREM 2a. *The conclusion in Theorem 2 remains valid if the property* (IM_K) *of N is replaced by the following weaker property: To every point* $q_0 \in N$ *there is an hermitian metric which is a supporting metric at* q_0, *and has the property.*((IM_K).

To prove this we consider as above the open subset $E \subset M_\alpha$ defined by $v > v_\alpha$. Suppose that $v - v_\alpha$ has a maximum at $p_0 \in E$. Let $q_0 = f(p_0)$ and let $ds_N^{*2} = \lambda ds_N^2$ be a supporting metric of ds_N^2 at q_0. Let u' be the quotient of the volume element of N with the metric ds_N^{*2} by that of M. Then $u' = \lambda^n u$ and if $v' = \log u'$, $v' - v_\alpha = v - v_\alpha + d \log \lambda$. Since $\log \lambda(q_0) = 0$ and $\log \lambda(q) \leqq 0$ in a neighborhood of q_0, it follows that p_0 is a maximum of the function $v' - v_\alpha$. As the metric ds_N^{*2} has the property (IM_K), this is not possible according to the proof above. The theorem then follows by the rest of the above argument.

To apply Theorem 2 to a more concrete case, we consider the unit ball $r < 1$ discussed at the end of §2. In it let D_ρ be defined by $r < \rho$, where ρ is a constant satisfying $0 < \rho < 1$. We wish to construct in D_ρ the function $v_\rho \geqq 0$ with the property (DO_K), with $K = 2n(n+1)$, i.e., (1) v_ρ satisfies the inequality

$$\tfrac{1}{2}\Delta v_\rho \leqq -2n(n+1)(1-\exp(v_\rho/n)); \tag{66}$$

(2) $v_\rho(p)$ tends to $+\infty$ as p approaches the boundary of D_ρ.

This function v_ρ will be intimately related to $\log(\rho^2 - r^2)$ and we wish to find the expression for $\Delta \log(\rho^2 - r^2)$. By (33) and (34), this can be read off from the form $d'd'' \log(\rho^2 - r^2)$. We have

$$d'' \log(\rho^2 - r^2) = -(\rho^2 - r^2)^{-1} \sum_k z_k d\bar{z}_k,$$

and

$$-d'd'' \log(\rho^2 - r^2) = -dd'' \log(\rho^2 - r^2) = (\rho^2 - r^2)^{-1} \sum_k dz_k \wedge d\bar{z}_k$$
$$+ 4r^2(\rho^2 - r^2)^{-2} d'r \wedge d''r.$$

From the expressions (39) and (40) for ds_M^2, we see that their associated two-forms are also equal. This gives

$$-d'd'' \log(\rho^2 - r^2) = \frac{1-r^2}{\rho^2 - r^2} \sum_k \theta_k \wedge \bar{\theta}_k + \frac{4(1-\rho^2)r^2}{(\rho^2 - r^2)^2(1-r^2)} d'r \wedge d''r. \tag{67}$$

From (41), (42), and the relation $2rd'r = \sum_k \bar{z}_k dz_k$, we get

$$2rd'r = (1 - r^2) \sum_k \bar{z}_k \theta_k. \tag{68}$$

It follows that

$$(69)\qquad \tfrac{1}{4}\Delta\log(\rho^2 - r^2) = -n\,\frac{1-r^2}{\rho^2 - r^2} - \frac{(1-\rho^2)r^2(1-r^2)}{(\rho^2-r^2)^2}.$$

In particular, this formula is valid for $\rho = 1$. Subtracting them, we get

$$(70)\qquad \begin{aligned}\frac{1}{4(n+1)}\Delta\log\left(\frac{1-r^2}{\rho^2-r^2}\right)^{n+1} &= -n+n\,\frac{1-r^2}{\rho^2-r^2}\left\{1+\frac{1}{n}\frac{(1-\rho^2)r^2}{\rho^2-r^2}\right\}\\ &\leqq -(n+1)+(n+1)\left(\frac{1-r^2}{\rho^2-r^2}\right)^2.\end{aligned}$$

Hence the function

$$v_\rho = \log\left(\frac{1-r^2}{\rho^2-r^2}\right)^{2n}$$

has the property (DO_K) with $K = 4n(n+1)$.

By letting $\rho \to 1$, we derive from Theorem 2 an n-dimensional version of the Schwarz lemma:

THEOREM 3. *Let $f, D_1 \to N$ be a holomorphic mapping where D_1 is the unit n-ball with the standard kähler metric and where N is an n-dimensional hermitian einsteinian manifold with scalar curvature $\leqq -4n(n+1)$. Then f is volume-decreasing.*

For the case that N is kählerian a similar theorem has been given by Dinghas [3].

5. **Bloch's problem.** The above considerations lead naturally to a problem which was in a sense initiated by A. Bloch: Let $f\colon M \to N$ be a holomorphic mapping, where M and N are complex hermitian manifolds of the same dimension. Suppose furthermore that N be simply connected and that the riemannian sectional curvature of N be everywhere $\leqq 0$. An open ball of radius s in N is called *univalent* if there is an open subset $U \subset M$ such that the restriction $f\,|\,U$ is a one-one mapping of U onto it, with a functional determinant everywhere nonzero. Let $b(f)$ be the least upper bound of the radii of univalent balls under f. Given a family $\mathscr{F}$ of mappings f, the Bloch problem is to determine the value of $\inf_{f\in\mathscr{F}} b(f)$.

The classical Bloch theorem concerns the case when M is the unit disk, N is the complex line C, and $\mathscr{F}$ is the family of holomorphic mappings f such that $|f'(0)| = 1$. Bloch's theorem states that $\beta = \inf_{f\in\mathscr{F}} b(f) > 0$ and β is called Bloch's constant. Subsequent works of Ahlfors, Grunsky and R. Robinson have established that $0.433 < \beta < 0.472$ (cf., for instance [1]).

This is a case when N does not have an hermitian metric whose Ricci curvature is negative definite and the above results do not apply. However we have the following theorem:

THEOREM 4. *Let N be a simply connected hermitian manifold of dimension n whose metric ds_N^2 has the following properties:*

(1) *Its riemannian sectional curvature is everywhere $\leqq 0$.*

(2) *There exists a smooth nonincreasing function $\lambda(t) > 1$, $0 < t < c$ (c = a fixed constant), such that for every fixed point $q_0 \in N$, the metric $\lambda(\delta(q_0, q))ds_N^2$, $q \in N$, has the property (IM_K) with $K = 2n(n+1)$, where $\delta(q_0, q)$ is the distance between q_0, q.*

Let D be the unit n-ball with the standard kähler metric and let $\mathcal{F}$ be the family of all holomorphic mappings $f: D \to N$, whose Jacobians are nowhere zero and which have the property that the ratio u of volume elements is equal to 1 at a given point a M. Then $\inf_{f \in \mathcal{F}} b(f) \geqq c$.

To prove this theorem consider a fixed mapping $f \in \mathcal{F}$. We shall show that the hypothesis $b(f) < c$ leads to a contradiction.

To a point $q \in N$ let $\rho(q)$ be the radius of the largest univalent ball about q. Then $0 < \rho(q) \leqq b(f) < c$ and $\rho(q)$ is a continuous function in N. Consider the metric $ds_N^{*2} = \lambda(\rho(q))ds_N^2$ in N. We wish to show that at any point $q_0 \in N$ there is a supporting metric with the property (IM_K). In fact, the open ball U of radius $\rho(q_0)$ must have a singular point m on its boundary. Consider the metric

$$d\sigma_N^2 = \lambda(\delta(m, q))ds_N^2 = \lambda(\delta(m, q))ds_N{}^{*2}/\lambda(\rho(q)) ,$$

which is defined for all q such that $\delta(m, q) < c$, i.e. in some neighborhood V of q_0. By the hypothesis (2), $d\sigma_N^2$ has the property (IM_K). Since $\rho(q) \leqq \delta(m, q)$, $q \in V$, and $\lambda(t)$ is nonincreasing, we have

$$\lambda(\delta(m, q))/\lambda(\rho(q)) \leqq 1 ,$$

and the left-hand side is equal to 1 for $q = q_0$. Therefore $d\sigma_N^2$ is a supporting metric of ds_N^{*2} at q_0. It follows from Theorem 2a that $\lambda^n(\rho(f(p)))u(p) \leqq 1$ for all $p \in D$. This gives the inequality $\lambda(\rho(f(a))) \leqq 1$, when $p = a$, contradicting the assumption $\lambda(t) > 1$. Thus Theorem 4 is proved.

A special case of Theorem 4 is when $n = 1$ and N is the complex line with the coordinate z and the metric $ds_N^2 = dzd\bar{z}$. The condition that the gaussian curvature of the metric $\lambda dzd\bar{z}$ is $\leqq -4$ is

$$\Delta \log \lambda \geqq 4\lambda^2 . \tag{72}$$

If $r \geqq 0$ is defined by $r^2 = z\bar{z}$, a solution of (72) is given by

$$\lambda(r) = \tfrac{1}{2}pAr^{p/2-1}(A^2 - r^p)^{-1} , \tag{73}$$

where p, A are positive constants; in fact, for this function $\lambda(r)$, the equality sign holds in (72). $\lambda(r)$ is a nonincreasing function of r, if

$$r^p \leqq (2-p)A^2/(2+p) , \tag{74}$$

and is strictly decreasing, when the inequality sign holds in (74). We choose

$$A = 2^{-p}(2+p)^{(2+p)/4}(2-p)^{-(2-p)/4}. \tag{75}$$

If $p < 2$ and r_0 is defined by

$$r_0^p = \frac{2-p}{2+p} A^2, \tag{76}$$

then $\lambda(r_0) = 1$. It follows that $\lambda(r) > 1$, $0 < r < r_0$. By Theorem 4 we have

$$\inf_{f \in \mathcal{F}} b(f) \geqq r_0 = \tfrac{1}{4}(4 - p^2)^{1/2}.$$

Letting $p \to 0$, we get $\inf_{f \in \mathcal{F}} b(f) \geqq \frac{1}{2}$. Thus we have the following corollary:

COROLLARY. *Let $f: D \to C$ be a holomorphic mapping of the unit disk D into the complex line C such that $f'(\zeta)$, $\zeta \in D$, is nowhere zero and $|f'(0)| = 1$. Then there is a univalent disk in C of radius $\geqq \frac{1}{2}$.*

Our constant $\frac{1}{2}$ is larger than Bloch's constant. This can be explained by the fact that the assumption that $f'(\zeta)$ is nowhere zero imposes an additional restriction on the family of mappings than Bloch's case. If this assumption is dropped, we have to choose $p \geqq 1$, and the choice $p = 1$ gives the lower bound $3^{1/2}/4$ for Bloch's constant as established by Ahlfors.

Even in the general case it should be possible to drop in Theorem 4 the condition that the Jacobian is nowhere zero. This will require the consideration of hermitian metrics which are allowed to be positive semidefinite on certain divisors, as is customary in the one-dimensional case. We hope to return to this problem on a later occasion.

BIBLIOGRAPHY

1. L. V. Ahlfors, *An extension on Schwarz's lemma*, Trans. Amer. Math. Soc. **43** (1938), 359–364.

2. S. S. Chern, *Characteristic classes of hermitian manifolds*, Ann. of Math. (2) **47** (1946), 85–121.

3. A. Dinghas, "Ein *n*-dimensionales Analogon des Schwarz-Pickschen Flächensatzes für holomorphe Abbildungen der komplexen Einheitskugel in eine Kähler-Mannigfaltigkeit," in *Festschrift zur Gedächtnisfeier für Karl Weierstrass*, 1815–1965, Westdeutscher Verlag, Cologne, 1966, pp. 477–494.

4. H. Grauert and H. Reckziegel, *Hermitesche Metriken und normale Familien holomorpher Abbildungen*, Math. Z. **89** (1965), 108–125.

5. P. Griffiths, *The extension probiem in complex analysis*. II: *embedding with positive normal bundle*, Amer. J. Math. **88** (1966), 366–446.

6. A. Koranyi, *A Schwarz lemma for bounded symmetric domains*, Proc. Amer. Math. Soc. **17** (1966), 210–213.

7. A. Weil, *Introduction à l'étude des variétés kählériennes*, Actualités Sci. Ind. no. 1267, Hermann, Paris, 1958.

UNIVERSITY OF CALIFORNIA
BERKELEY, CALIFORNIA

Extrait de *L'Enseignement mathématique*, T. XV, 1969

SIMPLE PROOFS OF TWO THEOREMS ON MINIMAL SURFACES

Shiing-shen CHERN *)

To the memory of J. Karamata

1. INTRODUCTION

We will give simple proofs of the following uniqueness theorems on minimal surfaces:

THEOREM 1 (Bernstein). *Let* z = f (x, y) *be a minimal surface in euclidean three-space defined for all* x, y. *Then* f (x, y) *is a linear function.*

THEOREM 2. *A closed minimal surface of genus zero on the three-sphere must be totally geodesic and is hence a great sphere.*

Theorem 2 has been proved by Almgren [1] and Calabi [2].

2. PROOF OF THEOREM 1

Let

(1) $$W = \left(1 + f_x^2 + f_y^2\right)^{\frac{1}{2}} \geqq 1 .$$

The proof is based on the identity

(2) $$\Delta \log \left(1 + \frac{1}{W}\right) = K ,$$

where Δ is the Laplacian relative to the induced riemannian metric of the minimal surface M and K is its Gaussian curvature.

Suppose (2) be true. Let ds be the element of arc on M. Introduce the conformal metric

*) Work done under partial support of NSF grant GP 8623.

(3) $$d\sigma = \left(1 + \frac{1}{W}\right) ds\,.$$

If p, q are isothermal coordinates on M, so that

(4) $$ds^2 = \lambda^2 (dp^2 + dq^2)\,,$$

we have

(5) $$K = -\frac{1}{\lambda^2}\left(\frac{\partial^2}{\partial p^2} + \frac{\partial^2}{\partial q^2}\right)\log \lambda\,,$$

$$\Delta = \frac{1}{\lambda^2}\left(\frac{\partial^2}{\partial p^2} + \frac{\partial^2}{\partial q^2}\right).$$

Applying this to the metric $d\sigma$, we find immediately that its gaussian curvature is zero, or that the metric is flat.

On the other hand, it is clear that

(6) $$ds \leqq d\sigma \leqq 2\,ds\,.$$

It follows that the metric $d\sigma$ on M is complete, for it dominates ds and ds is complete. We have therefore on M a complete flat riemannian metric $d\sigma$. By a well-known theorem, M, with the metric $d\sigma$, is isometric to the (ξ, η)-plane with its standard flat metric, i.e.,

(7) $$d\sigma^2 = d\xi^2 + d\eta^2\,.$$

Since $K \leqq 0$, we have, from (2) and (5),

(8) $$\left(\frac{\partial^2}{\partial \xi^2} + \frac{\partial^2}{\partial \eta^2}\right)\log\left(1 + \frac{1}{W}\right) \leqq 0\,.$$

The function $\log\left(1+\frac{1}{W}\right)$, considered as a function in the (ξ, η)-plane, is therefore superharmonic. It is also clearly non-negative. By a well-known theorem on superharmonic functions ([3], p. 130) it must be a constant. Equation (2) then gives $K = 0$, which implies that M is a plane.

The proof of (2) is a standard calculation. It will be proved at the end of § 4 as a special case of a more general formula.

An advantage of this proof is the fact that, unlike many other known proofs, complex function theory is not used.

3. Proof of Theorem 2

Let S^3 be the unit sphere in the euclidean 4-space E^4. By an orthonormal frame in E^4 is meant an ordered set of vectors e_α, $0 \leqq \alpha \leqq 3$, satisfying

$$(9) \qquad (e_\alpha, e_\beta) = \delta_{\alpha\beta}, \quad 0 \leqq \alpha, \beta, \gamma \leqq 3,$$

where the left-hand side is the scalar product of the vectors in question. The space of all orthonormal frames can be identified with the group $SO(4)$. We introduce in $SO(4)$ the Maurer-Cartan forms $\omega_{\alpha\beta}$ according to the equations

$$(10) \qquad de_\alpha = \sum_\beta \omega_{\alpha\beta} e_\beta$$

or

$$(11) \qquad \omega_{\alpha\beta} = (de_\alpha, e_\beta).$$

It follows from (9) that

$$(12) \qquad \omega_{\alpha\beta} + \omega_{\beta\alpha} = 0.$$

Exterior differentiation of (10) gives the Maurer-Cartan structure equations of $SO(4)$, which are

$$(13) \qquad d\omega_{\alpha\beta} = \sum_\gamma \omega_{\alpha\gamma} \wedge \omega_{\gamma\beta}.$$

There is a fibering

$$(14) \qquad SO(4) \to S^3 = SO(4)/SO(3),$$

with the projection defined by sending the frame $e_0\, e_1\, e_2\, e_3$ to the unit vector e_0.

Suppose a smooth surface

$$(15) \qquad M \to S^3$$

be described by the vector e_0. We restrict to frames such that e_3 is the unit normal vector to M at e_0. There are two choices for e_3, any one of which is called an orientation of M. Suppose M be oriented. Then the frames are defined up to a rotation of the vectors e_1, e_2 in the tangent plane. In other words, our restricted family of frames is a circle bundle over M, for which the structure equations (13) are valid.

The condition that e_3 is a normal vector at e_0 implies

$$\omega_{03} = 0 . \tag{16}$$

Taking its exterior derivative and using (13), we get

$$\omega_{01} \wedge \omega_{13} + \omega_{02} \wedge \omega_{23} = 0 .$$

Since M is an immersed surface, we have $\omega_{01} \wedge \omega_{02} \neq 0$ and Cartan's lemma allows us to set

$$\omega_{13} = a\omega_{01} + b\omega_{02} , \qquad \omega_{23} = b\omega_{01} + c\omega_{02} . \tag{17}$$

The condition for a minimal surface is the vanishing of the mean curvature:

$$a + c = 0 . \tag{18}$$

Let

$$\alpha = \omega_{01} + i\omega_{02} , \qquad \beta = \omega_{13} + i\omega_{23} . \tag{19}$$

The structure equations (13) give

$$d\alpha = - i\omega_{12} \wedge \alpha ,$$

$$d\beta = - i\omega_{12} \wedge \beta . \tag{20}$$

Under a rotation of e_1 e_2 both α and β will be multiplied by the same complex number of absolute value 1. It follows that

$$\alpha \wedge \bar{\beta} , \qquad \alpha\bar{\beta} , \tag{21}$$

which are exterior and ordinary two-forms respectively, are globally defined on our oriented surface M.

Suppose from now on that M is a minimal surface. Condition (18) can be written

$$\beta = A\bar{\alpha} , \qquad A = a + ib . \tag{22}$$

In this case the first form in (21) vanishes identically, while

$$\alpha\bar{\beta} = \bar{A}\alpha^2 . \tag{23}$$

Taking the exterior derivative of (22) and using (20), we get

$$dA + 2iA\omega_{12} \equiv 0 , \qquad \text{mod } \bar{\alpha} . \tag{24}$$

The induced riemannian metric on M has an underlying complex structure which makes M into a Riemann surface. We wish to show that the form in (23) is a quadratic differential in the sense of Riemann surfaces. For this purpose let z be a local complex coordinate on M, so that

$$\alpha = \lambda dz\,. \tag{25}$$

Then we have, locally,

$$\alpha\bar{\beta} = (\lambda^2 \bar{A})\,dz^2\,.$$

Exterior differentiation of (25) and use of (20) give

$$d\lambda + i\lambda\omega_{12} \equiv 0\,, \qquad \text{mod}\, dz\,.$$

Combining with (24), we get

$$\frac{\partial}{\partial \bar{z}}(\lambda^2 \bar{A}) = 0$$

i.e., the coefficient of dz^2 in $\alpha\bar{\beta}$ is holomorphic.

Since M is of genus zero, the quadratic differential must vanish. This implies $A = 0$ and that M is a great sphere.

The proof given above is not essentially different from those of Almgren and Calabi. The main idea of using the quadratic differential in surface theory goes back to H. Hopf. The formalism developed in this proof should also be useful in the study of other problems on surfaces in S^3.

4. A formula on non-parametric minimal hypersurfaces in euclidean space

Instead of proving formula (2) we will establish a more general formula for a non-parametric minimal hypersurface in the euclidean $(n+1)$-space E^{n+1}, which seems to have an independent interest.

Suppose x: $M \to E^{n+1}$ be an immersion of an n-dimensional manifold M in E^{n+1}. We consider orthonormal frames $x\, e_1 \ldots e_{n+1}$ in E^{n+1}, such that $x \in M$ and e_{n+1} is the unit normal vector to M at x. We have then

$$\begin{aligned} dx &= \sum_i \omega_i e_i\,, \\ de_i &= \sum_k \omega_{ik} e_k + \omega_{i,n+1} e_{n+1}\,, \qquad 1 \leqq i,j,k,l \leqq n\,, \\ de_{n+1} &= -\sum_i \omega_{i,n+1} e_i\,, \end{aligned} \tag{26}$$

with

$$\omega_{ik} + \omega_{ki} = 0 \tag{27}$$

and

$$\omega_{i,n+1} = \sum_k h_{ik}\,\omega_k\,, \qquad h_{ik} = h_{ki}\,. \tag{28}$$

The quadratic differential form

$$\prod = \sum_i \omega_i\,\omega_{i,n+1} = \sum_{i,k} h_{ik}\,\omega_i\,\omega_k \tag{29}$$

is the second fundamental form of M and the condition for M to be a minimal hypersurface is

$$\sum_i h_{ii} = 0\,. \tag{30}$$

Exterior differentiation of (26) gives the structure equations

$$\begin{aligned} d\omega_i &= \sum_j \omega_j \wedge \omega_{ji}\,, \\ d\omega_{i,n+1} &= \sum_j \omega_{ij} \wedge \omega_{j,n+1}\,, \\ d\omega_{ik} &= \sum_j \omega_{ij} \wedge \omega_{jk} - \omega_{i,n+1} \wedge \omega_{k,n+1}\,. \end{aligned} \tag{31}$$

The ω_{ik} are connection forms of the riemannian metric induced on M. If we define its curvature by the equation

$$d\omega_{ik} = \sum_j \omega_{ij} \wedge \omega_{jk} - \frac{1}{2}\sum_{j,l} R_{ikjl}\,\omega_j \wedge \omega_l\,, \tag{32}$$

where R_{ikjl} satisfy the symmetry relations

$$\begin{aligned} R_{ikjl} &= -R_{kijl} = -R_{iklj}\,, \\ R_{ikjl} &+ R_{ijlk} + R_{ilkj} = 0\,, \end{aligned} \tag{33}$$

the R_{ikjl} in this case of a hypersurface are expressible in terms of the h_{ik} by

$$R_{ikjl} = h_{ij}\,h_{kl} - h_{il}\,h_{jk}\,. \tag{34}$$

Taking the exterior derivative of (28) and using the second equation of (31), we get

$$\sum_k \left(dh_{ik} + \sum_j h_{jk}\,\omega_{ji} + \sum_j h_{ij}\,\omega_{jk}\right) \wedge \omega_k = 0\,.$$

This allows us to put

$$dh_{ik} + \sum_j h_{jk}\,\omega_{ji} + \sum_j h_{ij}\,\omega_{jk} = \sum_j h_{ikj}\,\omega_j\,, \tag{35}$$

where h_{ikj} is symmetric in any two of the indices i, k, j. It follows that *for a minimal hypersurface the contraction of* h_{ikj} *with respect to any two indices is zero*. The left-hand side of (35) is the covariant differential of h_{ik}.

Let u be a real-valued smooth function on M. Then we have

$$du = \sum_i u_i\,\omega_i\,, \tag{36}$$

$$Du_i = du_i + \sum_j u_j\,\omega_{ji} = \sum_j u_{ij}\,\omega_j\,, \quad u_{ij} = u_{ji}\,, \tag{37}$$

where Du_i is the covariant differential of the gradient vector u_i. The square of the gradient of u and the Laplacian of u are respectively defined by

$$(\text{grad}\ u)^2 = \sum_i u_i^2\,, \tag{38}$$

$$\Delta u = \sum_i u_{ii}\,. \tag{39}$$

If $\varphi(u)$ is a smooth function of u, we have

$$d\,\varphi(u) = \varphi'(u)\,d\,u\,,$$

$$D(\varphi'(u)\,u_i) = \sum_k (\varphi'(u)\,u_{ik} + \varphi''(u)\,u_i\,u_k)\,\omega_k\,,$$

so that

$$\Delta\varphi(u) = \varphi'(u)\,\Delta u + \varphi''(u)\,(\text{grad}\ u)^2\,. \tag{40}$$

From now on suppose M be a minimal hypersurface, so that the condition (30) is fulfilled. The Ricci curvature is given by

$$R_{ij} = \sum_k R_{ikjk} = -\sum_k h_{ik}\,h_{jk}\,, \tag{41}$$

which is negative semi-definite. The scalar curvature is

$$R = -\sum_{i,k} h_{ik}^2 \leqq 0\,. \tag{42}$$

For $n = 2$ we have $R = 2K$, K being the gaussian curvature.

Now let $a_1, \ldots, a_{n+1}$ be a fixed orthonormal frame in E^{n+1}. We can write

$$x = \sum_i x_i\,a_i + z a_{n+1}\,, \quad 1 \leqq i, k \leqq n\,, \tag{43}$$

and a non-parametric hypersurface will be defined by the equation

(44) $$z = z(x_1, \ldots, x_n).$$

Let

(45) $$(a_A, e_B) = v_{AB}, \qquad 1 \leqq A, B \leqq n+1,$$

where the left-hand side stands for the scalar product of the vectors in question and (v_{AB}) is a properly orthogonal matrix. In particular, $v_{A,n+1}$ are the components of the unit normal vector e_{n+1} with respect to the fixed frame a_A. If we put

(46) $$p_i = \frac{\partial z}{\partial x_i}, \qquad W = \left(1 + \sum_i p_i^2\right)^{\frac{1}{2}} \geqq 1,$$

we have

(47) $$v_{i,n+1} = \frac{p_i}{W}, \qquad v_{n+1,n+1} = -\frac{1}{W}.$$

For simplicity we will write $v = v_{n+1,n+1}$. We wish to establish the formula

(48) $$\Delta v = Rv.$$

In fact, we have, by (45) and (26),

$$dv = dv_{n+1,n+1} = (a_{n+1}, de_{n+1}) = -\sum_{i,k} v_{n+1,i}\, h_{ik}\, \omega_k,$$

and, by (37),

$$D\Big(-\sum_i v_{n+1,i}\, h_{ik}\Big) = -v \sum_{i,j} h_{ik}\, h_{ij}\, \omega_j - \sum_{i,j} v_{n+1,i}\, h_{ikj}\, \omega_j.$$

Formula (48) then follows from the definition of the Laplacian.

Formula (48) has the interesting consequence that on a minimal hypersurface the corresponding equation (48), with v as an unknown function, has a negative solution. In general, I do not know whether on a complete simply-connected non-compact riemannian manifold with negative semi-definite Ricci curvature the equation (48) has a negative solution other than constants; in the latter case we will have $R = 0$. If the answer to this question is no, it will give a proof of the n-dimensional Bernstein conjecture.

Formula (2) now follows as an easy consequence. Suppose therefore $n = 2$. In this case we have, for a minimal surface,

(49) $$\sum_i h_{ij}\, h_{ik} = -\frac{1}{2} R\delta_{jk} = -K\delta_{jk},$$

so that

$$(50) \qquad (\text{grad}\, v)^2 = -K(1-v^2)\,.$$

Formula (2) then follows immediately from (40).

BIBLIOGRAPHY

[1] ALMGREN, F. J. Jr. Some interior regularity theorems for minimal surfaces and an extension of Bernstein's theorem. *Ann. of Math.*, 85 (1966), 277-292.

[2] CALABI, E. Minimal immersions of surfaces in euclidean spheres. *J. of Diff. Geom.*, 1 (1967), 111-125.

[3] PROTTER, M. H. and WEINBERGER, H. *Maximum principles in differential equations*, Prentice-Hall, 1967.

(Reçu le 10 septembre 1968).

S. S. Chern
Dep. of Math.
University of California
Berkeley, Ca. 94720.

Intrinsic norms on a complex manifold

S. S. CHERN, HAROLD I. LEVINE AND LOUIS NIRENBERG*

1. Introduction

We propose to define in this paper certain norms (or more precisely, semi-norms) on the homology groups of a complex manifold. They will be invariants of the complex structure and do not increase under holomorphic mappings. Their definitions depend on the bounded plurisubharmonic functions on the manifold and are modelled after the notion of harmonic length introduced by H. Landau and R. Osserman [8] for Riemann surfaces (= one-dimensional complex manifolds). It is possible to extend the definition to certain families of chains. In particular we get in this way an intrinsic pseudo-metric on the manifold which is closely related to that of Caratheodory [3] and to one recently introduced by S. Kobayashi [7].

To define one of these semi-norms (possibly the most significant one among those to be introduced below), let M be a complex manifold of complex dimension n. Let

$$d^c = i(\bar{\partial} - \partial)\,, \tag{1}$$

so that d^c is a differential operator of degree one on smooth complex-valued exterior differential forms and maps a real form into a real form. Let $\mathcal{F}$ be the family of plurisubharmonic functions u of class C^2 on M satisfying the condition $0 < u < 1$. To a homology class γ of M with real coefficients we set

$$\begin{aligned} N\{\gamma\} &= \sup_{u\in\mathcal{F}} \inf_{T\in\gamma} | T[d^c u \wedge (dd^c u)^{k-1}] | \\ &\qquad \text{if } \dim\gamma = 2k-1\,, \\ N\{\gamma\} &= \sup_{u\in\mathcal{F}} \inf_{T\in\gamma} | T[du \wedge d^c u \wedge (dd^c u)^{k-1}] | \\ &\qquad \text{if } \dim\gamma = 2k\,, \end{aligned} \tag{2}$$

* The first author was partially supported by NSF grant GP-8623. The second author was partially supported by NSF grant GP-6761. The third author was partially supported by Air Force Office of Scientific Research, Grant AF-49(638)-1719.

where T runs over all currents (in the sense of de Rham [10]) of γ. Our main theorem (cf. § 3) asserts that $N\{\gamma\}$ is always finite. The following properties are easily verified.

$$N\{a\gamma\} = |a| N\{\gamma\}, \qquad a \in R; \tag{3}$$

$$N\{\gamma_1 + \gamma_2\} \leqq N\{\gamma_1\} + N\{\gamma_2\}, \qquad \dim \gamma_1 = \dim \gamma_2. \tag{4}$$

Furthermore, under a holomorphic mapping, $N\{\gamma\}$ is non-increasing. These properties make it a useful tool in the study of holomorphic mappings.

Unfortunately in the case when M is compact, the family $\mathcal{F}$ will consist only of constants and the seminorm will be identically zero. We will show, however, that our definition can be refined to give a meaningful invariant which, in the case of compact Riemann surfaces, is "equivalent" to the extremal length of Ahlfors-Beurling [2]. As is well known, the latter, together with the classical topological invariants, gives a complete system of conformal invariants to compact Riemann surfaces, in the sense of the following theorem of Accola [1]. Let $f: M \to M'$ be a diffeomorphism between two compact Riemann surfaces under which corresponding homology classes of curves have the same extremal length. Then f is a conformal equivalence.

2. A lemma on bounded plurisubharmonic functions in C_n

Note: Just the corollary of this paragraph is used in the proof of our Theorem 1. The lemma itself is used only in Remark 3 following Theorem 2.

Let C_n be the complex number space of dimension n with the coordinates z^k, $1 \leqq k \leqq n$. We denote its volume element by

$$dV = \left(\frac{i}{2}\right)^n \bigwedge_k dz^k \wedge d\bar{z}^k. \tag{5}$$

For a real-valued smooth function v defined in an open subset of C_n its partial derivatives will be denoted by

$$v_j = \frac{\partial v}{\partial z^j}, \quad v_{\bar{k}} = \frac{\partial v}{\partial \bar{z}^k}, \quad v_{j\bar{k}} = \frac{\partial^2 v}{\partial z^j \partial \bar{z}^k}, \text{ etc.} \quad 1 \leqq j, k \leqq n. \tag{6}$$

We recall that such a function is called *plurisubharmonic* if the hermitian matrix $(v_{j\bar{k}})$ is positive semi-definite; v is called *pluriharmonic* if $(v_{j\bar{k}}) = 0$ or $dd^c v = 0$.

LEMMA. *Let v be a plurisubharmonic negative-valued function of class C^2 in a polydisc*

$$\Delta: |z^i| < r_i\,, \qquad 1 \leqq i \leqq n\,,$$

in C_n. Let Δ_1 be a compact subpolydisc

$$|z^i| \leqq \rho_i < r_i\,, \qquad 1 \leqq i \leqq n\,.$$

Then there is a constant A independent of v, depending only on the numbers ρ_i, r_i, $1 \leqq i \leqq n$, such that for the integral of any $r \times r$ minor of $(-v_{j\bar{k}}/v)$ over Δ_1 we have the estimate

$$(7) \qquad \int_{\Delta_1} \text{abs.} \begin{vmatrix} v_{i_1\bar{k}_1} & \cdots & v_{i_1\bar{k}_r} \\ & \cdots & \\ v_{i_r\bar{k}_1} & \cdots & v_{i_r\bar{k}_r} \end{vmatrix} \frac{1}{|v|^r} dV \leqq A\,.$$

We prove the lemma by induction on r. For $r = 1$ we prove a stronger form of (7), namely,

$$(7') \qquad \int_{\Delta_1} \left(\left| \frac{v_{i\bar{k}}}{v} \right| + \frac{1}{2} \left| \frac{v_k}{v} \right|^2 \right) dV \leqq A\,, \qquad 1 \leqq i, k \leqq n\,.$$

Since $(v_{i\bar{k}})$ is a positive semi-definite matrix, we have $2\,|v_{i\bar{k}}| \leqq v_{i\bar{i}} + v_{k\bar{k}}$. Hence it suffices to prove (7′) for $i = k$. Let $\zeta \geqq 0$ be a C^∞ function with support in Δ and equal to one in Δ_1. Then, by Green's theorem we have

$$\int \left(\frac{v_{k\bar{k}}}{-v} + \left| \frac{v_k}{v} \right|^2 \right) \zeta^2 dV = \int \left(\frac{v_k}{-v} \right)_{\bar{k}} \zeta^2 dV = 2 \int \frac{v_k}{v} \zeta \zeta_{\bar{k}} dV$$

$$\leqq \int \left(\frac{1}{2} \left| \frac{v_k}{v} \right|^2 \zeta^2 + 2\,|\zeta_{\bar{k}}|^2 \right) dV\,,$$

the integrations being over Δ. It follows that

$$\int \left(\frac{v_{k\bar{k}}}{-v} + \frac{1}{2} \left| \frac{v_k}{v} \right|^2 \right) \zeta^2 dV \leqq 2 \int |\zeta_{\bar{k}}|^2\, dV = A,$$

which yields (7′).

To proceed by induction we suppose the truth of (7) for $r-1$. Since v is plurisubharmonic, the absolute value of a general $r \times r$ minor of $(v_{i\bar{k}})$ is less than or equal to the maximum of the principal minors of order r. It therefore suffices to prove (7) for a principal minor, and we can restrict ourselves to the case $i_1 = k_1 = 1, \cdots, i_r = k_r = r$. We set

$$\Psi = 2^{-2r} \left(\frac{i}{2} \right)^{n-r} \wedge_{k>r} dz^k \wedge d\bar{z}^k$$

and choose ζ as before. Then we have

$$I_r = \int_{\Delta_1} \begin{vmatrix} v_{1\bar{1}} & \cdots & v_{1\bar{r}} \\ & \cdots & \\ v_{r\bar{1}} & \cdots & v_{r\bar{r}} \end{vmatrix} \frac{dV}{(-v)^r} = \int_{\Delta_1} \left(\frac{dd^c v}{-v}\right)^r \wedge \psi \leqq \int_{\Delta} \zeta\left(\frac{dd^c v}{-v}\right)^r \wedge \psi$$

$$= -r\int_{\Delta} \frac{dv \wedge d^c v \wedge (dd^c v)^{r-1}}{(-v)^{r+1}} \zeta\psi - \int_{\Delta} \frac{d\zeta \wedge d^c v \wedge (dd^c v)^{r-1}}{(-v)^r} \wedge \psi ,$$

where the last equality follows from Green's theorem. Since

$$d\zeta \wedge d^c v - dv \wedge d^c\zeta = i(\partial + \bar{\partial})\zeta \wedge (\bar{\partial} - \partial)v - i(\partial + \bar{\partial})v \wedge (\bar{\partial} - \partial)\zeta$$
$$= 2i(\bar{\partial}\zeta \wedge \bar{\partial}v - \partial\zeta \wedge \partial v) ,$$

it contains no term of type (1,1). It follows that

$$\int_{\Delta} \frac{d\zeta \wedge d^c v \wedge (dd^c v)^{r-1}}{(-v)^r} \wedge \psi = \int_{\Delta} \frac{dv \wedge d^c\zeta \wedge (dd^c v)^{r-1}}{(-v)^r} \wedge \psi ,$$

and by Green's theorem this is equal to

$$-\frac{1}{r-1}\int_{\Delta} \frac{dd^c\zeta \wedge (dd^c v)^{r-1}}{(-v)^{r-1}} \wedge \psi .$$

On the other hand, one sees easily that $dv \wedge d^c v \wedge (dd^c v)^{r-1} \wedge \psi$ is a non-negative multiple of dV. Hence we get

$$I_r \leqq \frac{1}{r-1}\int_{\Delta} dd^c\zeta \wedge \left(\frac{dd^c v}{-v}\right)^{r-1} \wedge \psi ,$$

and the desired inequality (7) follows from induction hypothesis.

COROLLARY. *Let the polydiscs Δ, Δ_1 be defined as in the lemma. Let u be a C^2-plurisubharmonic function in Δ with $0 < u < 1$. Then there is a constant B independent of u such that for any $r \times r$ minor of $(u_{i\bar{k}})$ we have the estimate*

$$\int_{\Delta_1} \left[\text{abs.} \begin{vmatrix} u_{i_1\bar{k}_1} & \cdots & u_{i_1\bar{k}_r} \\ & \cdots & \\ u_{i_r\bar{k}_1} & \cdots & u_{i_r\bar{k}_r} \end{vmatrix} + \sum_k |u_k|^2 \right] dV \leqq B . \tag{8}$$

To deduce this corollary from the lemma we set $v = u - 1$. Then $0 < -v < 1$ and

$$\text{abs.} \begin{vmatrix} u_{i_1\bar{k}_1} & \cdots & u_{i_1\bar{k}_r} \\ & \cdots & \\ u_{i_r\bar{k}_1} & \cdots & u_{i_r\bar{k}_r} \end{vmatrix} \leqq \frac{1}{(-v)^r} \text{abs.} \begin{vmatrix} v_{i_1\bar{k}_1} & \cdots & v_{i_1\bar{k}_r} \\ & \cdots & \\ v_{i_r\bar{k}_1} & \cdots & v_{i_r\bar{k}_r} \end{vmatrix} .$$

Thus the corollary follows from the Lemma and (7′).

For $r = 1$ the corollary was proved by P. Lelong [9].

The lemma proved above has a real analogue whose proof is *similar. Let v be a negative convex function of class C^2 in a domain $\mathfrak{D}$ in $\mathbf{R}^n$*, i.e., *a function whose hession matrix is positive semidefinite,*

$$\sum_{i,j} v_{x_i x_j}\xi_i\xi_j \geqq 0 .$$

For any subdomain K with compact closure in $\mathfrak{D}$, there is a constant A, independent of v, such that

$$\int_K \left\{ \left| \frac{v_{x_k}}{v} \right|^2 + \frac{1}{|v|^r} \left| \text{any } r \times r \text{ minor of } (v_{x_i x_j}) \right| \right\} dV \leqq A .$$

3. Semi-norms and their properties

Using the definition (2) we shall prove the theorem.

THEOREM 1. *Let M be a complex manifold and γ a homology class with real coefficients. Then $N\{\gamma\}$ is finite.*

To prove the theorem, let T be a closed current belonging to γ. By a theorem of de Rham [10, §. 15], there exist operators RT, AT whose supports belong to an arbitrarily small neighborhood of the support of T such that

$$RT = T + bAT + AbT , \tag{9}$$

where b is the boundary operator of currents. The operator R is a regularizing operator, constructed by convolution with a smooth kernel, which is given by

$$RT[\varphi] = \int_M \varphi \wedge \psi , \tag{10}$$

where ψ is a closed C^∞-form with support in a neighborhood of the support of T. Since $bT = 0$, there exists in every homology class a regular current and it suffices to show that $\int_M \varphi \wedge \psi$ has a finite upper bound independent of u, where

$$\varphi = d^c u \wedge (dd^c u)^{k-1} , \qquad \dim \gamma = 2k - 1 ,$$
$$\varphi = du \wedge d^c u \wedge (dd^c u)^{k-1} , \qquad \dim \gamma = 2k .$$

Consider first the case $\dim \gamma = 2k - 1$. By Green's theorem we have

$$\int_M \varphi \wedge \psi = - \int_M u(dd^c u)^{k-1} \wedge d^c\psi .$$

Since ψ is a fixed C^∞-form, it follows from our Corollary in § 2 that this integral is bounded in absolute value by a constant independent of $u \in \mathcal{F}$.

If dim $\gamma = 2k$, we have, since $d\psi = 0$,

$$\int_M \varphi \wedge \psi = - \int_M u(dd^c u)^k \wedge \psi \, .$$

The existence of an upper bound for the absolute value of this integral again follows from our Corollary.

The following theorem is an immediate consequence of our definition.

THEOREM 2. *Let $H_l\,(M, R)$ be the l-dimensional homology group of M with real coefficients. Then $N\{\gamma\}$, $\gamma \in H_l(M, R)$, defines a semi-norm on the real vector space $H_l(M, R)$,* i.e.,

$$\text{(11)} \qquad \begin{aligned} &N\{a\gamma\} = |\,a\,|\, N\{\gamma\}\, , && a \in R\, ; \\ &N\{\gamma_1 + \gamma_2\} \leqq N\{\gamma_1\} + N\{\gamma_2\}\, , && \gamma_1, \gamma_2 \in H_l(M, R)\, . \end{aligned}$$

Moreover, under a holomorphic mapping $f: M \to P$ we have

$$\text{(12)} \qquad N_M\{\gamma\} \geqq N_P\{f_*\gamma\}\, ,$$

where f_* is the induced homomorphism on the homology classes, and the semi-norms are taken in M and P respectively.

Remark 1. If the closed currents T_1, T_2 of γ are such that $T_1[\varphi] \neq T_2[\varphi]$, there exists a real number t with

$$\{tT_1 + (1 - t)T_2\}[\varphi] = 0\, ,$$

while the current $tT_1 + (1 - t)T_2$ still belongs to γ. Therefore in the definition (2) we need only consider functions u for which φ is closed. These functions u satisfy respectively the differential equations

$$\text{(13)} \qquad (dd^c u)^k = 0\, , \qquad \dim \gamma = 2k - 1\, ,$$

$$\text{(13a)} \qquad du \wedge (dd^c u)^k = 0\, , \qquad \dim \gamma = 2k\, .$$

In § 5 we will give examples for which $N\{\gamma\}$ is a norm i.e., $N\{\gamma\} > 0$ when $\gamma \neq 0$.

Remark 2. The family, $\mathcal{F}$, of plurisubharmonic functions on M with values between 0 and 1 can be used to define another semi-norm which assigns to a homology class γ the number

$$N'\{\gamma\} = \sup_{u_i \in \mathcal{F}} \inf_{T \in \gamma} |\, T[d^c u_0 \wedge dd^c u_1 \wedge \cdots \wedge dd^c u_k]\,|\, , \qquad \text{if } \dim \gamma = 2k + 1$$

$$N'\{\gamma\} = \sup_{u_i \in \mathcal{F}} \inf_{T \in \gamma} | T[du_0 \wedge d^c u_1 \wedge dd^c u_2 \wedge \cdots \wedge dd^c u_k] | , \qquad \text{if } \dim \gamma = 2k .$$

The proof of finiteness of $N'(\gamma)$ requires a slight modification of the preceding arguments. The analogue of the corollary needed here is

LEMMA. *Let the polydiscs Δ, Δ_1 be defined as in lemma of § 2. Let $u_1, u_2, \cdots, u_r$ be C^2 plurisubharmonic functions in Δ with $0 < u_i < 1$. Then there is constant C independent of the u_i, such that if $J = (j_1, \cdots, j_r)$ and $K = (k_1, \cdots, k_r)$ $1 \leqq j_1 < \cdots < j_r \leqq n$ and $1 \leqq k_1 < \cdots < k_r \leqq n$,*

$$\int_{\Delta_1} | U_{J\bar{K}} | \, dV \leqq C ,$$

where $U_{J\bar{K}}$ is the coefficient of $dz_{j_1} \wedge d\bar{z}_{k_1} \wedge \cdots \wedge dz_{j_r} \wedge d\bar{z}_{k_r}$ in $dd^c u_1 \wedge \cdots \wedge dd^c u_r$, and dV is the element of volume in C_n. To prove this we note first that the matrix with $(J, K)^{\text{th}}$ entry, $U_{J\bar{K}}$ is positive semi-definite (by induction on r), and so we need only consider the case $J = K = (1, 2, \cdots, r)$. Then using the notation and technique of the lemma of § 2, we have

$$\begin{aligned}
\int_{\Delta_1} | U_{J\bar{J}} | \, dV &= \int_{\Delta_1} dd^c u_1 \wedge \cdots \wedge dd^c u_r \wedge \psi \\
&\leqq \int_{\Delta} \zeta \cdot dd^c u_1 \wedge \cdots \wedge dd^c u_r \wedge \psi \\
&= -\int_{\Delta} d\zeta \wedge d^c u_1 \wedge dd^c u_2 \wedge \cdots dd^c u_r \wedge \psi \\
&= -\int_{\Delta} du_1 \wedge dd^c u_2 \wedge \cdots \wedge dd^c u_r \wedge d^c \zeta \wedge \psi \\
&= \int_{\Delta} u_1 \cdot dd^c u_2 \wedge \cdots \wedge dd^c u_r \wedge dd^c \zeta \wedge \psi .
\end{aligned}$$

Theorem 2 and Remark 1 following it are true for N', and $N' \geqq N$. The equations replacing (13) and (13a) are their multilinear versions.

(13′) $\qquad dd^c u_0 \wedge \cdots \wedge dd^c u_k = 0 , \qquad$ for N' on $H_{2k+1}(M, R)$.

(13′a) $\qquad du_0 \wedge dd^c u_1 \wedge \cdots \wedge dd^c u_k = 0 , \qquad$ for N' on $H_{2k}(M, R)$.

Another, possibly larger, semi-norm results if we change the definition of N' by allowing u_0 to be any C^2 function with $0 < u_0 < 1$, but still requiring $u_1, \cdots, u_k$ to be in $\mathcal{F}$. The other norms which we introduce may also be modified in a similar manner with the aid of $k + 1$ functions in place of one.

Remark 3. Another seminorm can be defined by the consideration of a different family of functions. Let $\mathcal{F}_1$ be the family of negative C^2-functions, defined locally up to a multiplicative positive constant, which are plurisubharmonic. For such a function v the forms

$$\frac{dv}{v}, \qquad \frac{d^c v}{v}, \qquad \frac{dd^c v}{v} \tag{14}$$

are well defined on M. With the aid of the functions of $\mathcal{F}_1$ we define, to a homology class γ,

$$\begin{aligned} N_1\{\gamma\} &= \sup_{v\in\mathcal{F}_1} \inf_{T\in\gamma} \left| T\left[\frac{d^c v \wedge (dd^c v)^{k-1}}{(-v)^k}\right]\right|, \\ &\qquad\qquad\qquad \text{if } \dim\gamma = 2k-1, \\ N_1\{\gamma\} &= \sup_{v\in\mathcal{F}_1} \inf_{T\in\gamma} \left| T\left[\frac{dv \wedge d^c v \wedge (dd^c v)^{k-1}}{(-v)^{k+1}}\right]\right|, \\ &\qquad\qquad\qquad \text{if } \dim\gamma = 2k. \end{aligned} \tag{15}$$

By applying the Lemma in § 2 it can be proved that $N_1\{\gamma\}$ is always finite, and is hence a semi-norm in the homology vector space $H_l(M, R)$ ($l = 2k - 1$ or $2k$). Unfortunately we know no example for which $N_1\{\gamma\}$ is not zero. In particular, if $\dim\gamma = 1$, then necessarily $N_1\{\gamma\} = 0$.

It may be observed that in the proof of Theorem 1, only the property of local boundedness of the functions u is utilized. We will therefore introduce wider families of functions and thereby generalize the semi-norms introduced above. Let $\mathfrak{U} = \{U_i\}$ be a locally finite open covering of M. We denote by $\mathcal{F}(\mathfrak{U})$ the family of plurisubharmonic C^2-functions u_i: $U_i \to R$ defined in each member of the covering which satisfy the following conditions:

(1) the oscillation of u_i in U_i is less than one;

(2) $du_i = du_j$ in $U_i \cap U_j \neq \emptyset$.

The latter defines a closed real one-form in M. Similarly, $d^c u_i$ and $dd^c u_i$ are also well defined in M. Without ambiguity, we can denote them without the indices. Analogous to (2) we define

$$\begin{aligned} N\{\gamma, \mathfrak{U}\} &= \sup_{u\in\mathcal{F}(\mathfrak{U})} \inf_{T\in\gamma} | T[d^c u \wedge (dd^c u)^{k-1}] | \\ &\qquad\qquad\qquad \text{if } \dim\gamma = 2k-1, \\ N\{\gamma, \mathfrak{U}\} &= \sup_{u\in\mathcal{F}(\mathfrak{U})} \inf_{T\in\gamma} | T[du \wedge d^c u \wedge (dd^c u)^{k-1}] | \\ &\qquad\qquad\qquad \text{if } \dim\gamma = 2k. \end{aligned} \tag{16}$$

Let $\pi\colon \tilde{M} \to M$ be the universal covering manifold of M and let $\tilde{U}$ be a fundamental domain on $\tilde{M}$. We denote by $\mathcal{F}(\tilde{U})$ the family of plurisubharmonic C^2-functions on $\tilde{M}$ such that their oscillation in $\tilde{U}$ is less than one and their differentials are well defined on M. We define

$$(17)\quad \begin{aligned} N\{\gamma, \tilde{U}\} &= \sup_{u \in \mathcal{F}(\tilde{U})} \inf_{T \in \gamma} | T[d^c u \wedge (dd^c u)^{k-1}] | , && \text{if } \dim \gamma = 2k-1 , \\ N\{\gamma, \tilde{U}\} &= \sup_{u \in \mathcal{F}(\tilde{U})} \inf_{T \in \gamma} | T[du \wedge d^c u \wedge (dd^c u)^{k-1}] | , && \text{if } \dim \gamma = 2k . \end{aligned}$$

We will suppose of the fundamental domain $\tilde{U}$ that each of its points is in the interior of the union of $\tilde{U}$ and a finite number of its translates by deck transformations.

By a partition of unity the proof of Theorem 1 also gives the following theorem.

THEOREM 3. *The* $N\{\gamma, \mathfrak{U}\}$ *and* $N\{\gamma, \tilde{U}\}$ *defined in* (16) *and* (17) *are finite and define seminorms in the homology vector spaces* $H_l(M, R)$. *Between them and* $N\{\gamma\}$ *there are the inequalities*

$$(18)\quad N\{\gamma\} \leqq N\{\gamma, \mathfrak{U}\} , \qquad N\{\gamma\} \leqq N\{\gamma, \tilde{U}\} .$$

Addendum. For non-compact complex manifolds M and P and f a holomorphic map from M to P, we do not in general have an inequality analogous to (12) for the semi-norms, $N\{\cdot, \mathfrak{U}\}$ and $N\{\cdot, \tilde{U}\}$. However, let $\tilde{U}$ and $\tilde{V}$ be fundamental domains for the universal covering spaces $\tilde{M}$ and $\tilde{P}$ and let $\tilde{f}$ cover f. Then if $\tilde{f}(\tilde{U})$ is covered by a finite number T of deck-transforms of $\tilde{V}$, we have

$$N_P\{f_*(\gamma), \tilde{V}\} \leqq T^k N_M\{\gamma, \tilde{U}\} , \qquad \text{for } \dim \gamma = 2k-1 \text{ or } 2k-2 .$$

In particular, if M is compact we always have such inequalities. Similarly, let $\mathfrak{U}$ be a *simple* open covering of M, that is a locally finite open covering of M each member of which is simply connected, and let $\mathfrak{V}$ be an arbitrary, locally finite open covering of P. If for any element $U_i \in \mathfrak{U}$, $f(U_i)$ is covered by S or fewer elements of $\mathfrak{V}$, we again have

$$N_P\{f_*(\gamma), \mathfrak{V}\} \leqq S^k N_M\{\gamma, \mathfrak{U}\} , \qquad \text{for } \dim \gamma = 2k-1 \text{ or } 2k-2 .$$

Of course if M is compact and $\mathfrak{U}$ is simple, we have such inequalities.

4. Comparision of semi-norms.

Very little is known about the relations between the different semi-norms. If we apply the above Addendum to the case that $M = P$ is compact and f is the identity map, we find that the equivalence class of $N\{\cdot, \tilde{U}\}$ and the equivalence class of $N\{\cdot, \mathfrak{U}\}$ are independent of the choice of fundamental domain $\tilde{U}$ and simple covering U respectively. Both of these facts are implied by

THEOREM 4. *Let M be a compact complex manifold without boundary. Let $\mathfrak{U} = \{U_i\}$ be a finite open simple covering of M, and let $\tilde{U}$ be a fundamental domain in the universal covering manifold of M. Then the seminorms $N\{\gamma, \mathfrak{U}\}$ and $N\{\gamma, \tilde{U}\}$ are equivalent,* i.e., *there is a constant $C > 0$, independent of γ, such that*

$$C^{-1}N\{\gamma, \mathfrak{U}\} \leqq N\{\gamma, \tilde{U}\} \leqq CN\{\gamma, \mathfrak{U}\}\,. \tag{19}$$

It follows that for any two finite simple open coverings $\mathfrak{U}$ and $\mathfrak{V}$ the semi-norms $N\{\gamma, \mathfrak{U}\}$ and $N\{\gamma, \mathfrak{V}\}$ are equivalent.

In fact, if a point of U_i is lifted to a point of $\tilde{U}$, the lifting of U_i to $\tilde{M}$ is uniquely determined. It follows that if $u \in \mathcal{F}(\tilde{U})$, then the oscillation of u on each of the U_i is bounded by some constant c. Hence $c^{-1}u \in \mathcal{F}(\mathfrak{U})$, from which the second inequality of (19) follows, with $C = c^k$ or c^{k+1} according as $\dim \gamma = 2k - 1$ or $2k$. In a similar way it is easily seen that if $u \in \mathcal{F}(\mathfrak{U})$ then $c^{-1} u \in \mathcal{F}(\tilde{U})$ for some constant c and the first of (19) follows.

If M is a compact Riemann surface and γ is a one-dimensional homology class, the *extremal length* $\lambda(\gamma)$ of Ahlfors-Beurling is defined by

$$\lambda^{\frac{1}{2}}(\gamma) = \sup_\rho \frac{\inf_{C\in\gamma} \int_C \rho\,|dz|}{\left(\iint_M \rho^2 dxdy\right)^{\frac{1}{2}}} \tag{20}$$

where $\rho \geqq 0$ ranges over all lower semicontinuous densities which are not identically zero [2].

THEOREM 5. *Let M be a compact Riemann surface without boundary and $\mathfrak{U} = \{U_i\}$ a finite open simple covering of M. Then the semi-norms $N\{\gamma, \mathfrak{U}\}$ and $\lambda^{\frac{1}{2}}(\gamma)$ defined over the one-dimensional homology group $H_1(M, R)$ are equivalent,* i.e., *there is a constant $C > 0$, independent of $\gamma \in H_1(M, R)$, such that*

$$(21)\qquad C^{-1}\lambda^{\frac{1}{2}}(\gamma) \leqq N(\gamma, \mathfrak{U}) \leqq C\lambda^{\frac{1}{2}}(\gamma)\,.$$

We proceed to prove this theorem. For $u \in \mathcal{F}(\mathfrak{U})$ we set $\rho = |u_z|$. Let C be any closed curve belonging to the homology class γ. Since $d^c u = 2\,\mathrm{Im}\,(u_z d_z)$, we have

$$\int_C |d^c u| \leqq 2\int_C |u_z|\,|d_z|\,,$$

and hence

$$\inf_{C\in\gamma} \int_C |d^c u| \leqq 2\lambda^{\frac{1}{2}}(\gamma)\Big\{\iint_M |u_z|^2\,dxdy\Big\}^{\frac{1}{2}}\,.$$

On U_i the oscillation of u is less than 1. It follows that on any compact subset K of U_i we can find a uniform bound for

$$\iint_K |u_z|^2\,dxdy$$

for all harmonic functions u. Hence there is a constant C_1 depending only on $\mathfrak{U}$ such that

$$\iint_M |u_z|^2\,dxdy \leqq C_1^2\,.$$

Thus

$$\inf_{C\in\gamma} \int_C |d^c u| \leqq 2C_1\lambda^{\frac{1}{2}}(\gamma)\,.$$

Since this holds for all $u \in \mathcal{F}(\mathfrak{U})$, the last inequality of (21) follows.

To prove the first inequality of (21) we make use of a theorem of Accola [1] which says that there is a harmonic one-form σ on M representing the homology class γ such that

$$(22)\qquad \|\sigma\|^2 = \lambda(\gamma)\,,$$

where $\|\sigma\|$ is the L_2-norm of σ. In U_i we write $\sigma = du$, where u is a harmonic function defined up to an additive constant. Then we have

$$\lambda(\gamma) = \|\sigma\|^2 = 2\iint_M |u_z|^2\,dxdy\,.$$

On the other hand, by standard results on harmonic functions we have

$$|\,\mathrm{osc}\ u \text{ in } U_i\,| \leqq \mathrm{const}\Big(\iint_M |u_z|^2\,dxdy\Big)^{\frac{1}{2}}\,.$$

Thus we may suppose u be so chosen that on each U_i,

$$0 < u < C_2\lambda^{\frac{1}{2}}(\gamma) = a \ ,$$

say, where C_2 is a constant. It follows that $u/a \in \mathcal{F}(\mathfrak{U})$. Since $d^c u$ is closed, we have, for a current $T \in \gamma$,

$$T\left(\frac{d^c u}{a}\right) = \frac{1}{a}\int_M d^c u \wedge \sigma = \frac{1}{a}\int_M \sigma^* \wedge \sigma$$
$$= \frac{1}{a}\,||\,\sigma\,||^2 = \frac{1}{C_2}\,\lambda^{\frac{1}{2}}(\gamma)\ .$$

Hence $\lambda^{\frac{1}{2}}(\gamma) \leqq C_2 N\{\gamma, \mathfrak{U}\}$, and the first inequality of (21) is proved.

Remark. From the comparison with the extremal length it seems natural to extend our definition to a family of chains of a fixed dimension. We could also take the integrals of the absolute values of the corresponding differential forms. For instance, let G be a family of chains of dimension $2k - 1$. We define

$$\hat{N}\{G\} = \sup_{u \in \mathcal{F}} \inf_{g \in G} \int_g |\, d^c u \wedge (dd^c u)^{k-1}\,| \ . \tag{23}$$

Unfortunately we are unable to prove that $\hat{N}\{G\}$ is finite, except the following case. *If $k = 1$ and G contains all curves homotopically equivalent to a closed curve, then $\hat{N}\{G\}$ is finite.*

5. Some examples

In C_n with the coordinates z^k, $1 \leqq k \leqq n$, we set

$$r = |\,z\,| = (\textstyle\sum_k |\,z^k\,|^2)^{\frac{1}{2}}\ . \tag{24}$$

We consider the annulus A_n defined by $1 < r < a$. The homology group $H_{2n-1}(A_n, Z)$ is free cyclic and we denote by γ its generator defined by the natural orientation of C_n.

By definition we find

$$\begin{aligned} d^c \log r &= \frac{-i}{2r^2} \textstyle\sum_k (\bar{z}^k dz^k - z^k d\bar{z}^k)\ , \\ dd^c \log r &= \frac{i}{r^4}\{r^2 \textstyle\sum_k dz^k \wedge d\bar{z}^k - (\sum_k \bar{z}^k dz^k) \wedge (\sum_k z^k d\bar{z}^k)\}\ . \end{aligned} \tag{25}$$

The differential form in the last expression is a real-valued two-form of type (1,1). It remains unchanged when z^k are multiplied by the same factor. This means that if we denote by

$$\psi\colon C_n - \{0\} \to P_{n-1}$$

the identification of the space of lines through 0 in C_n with the complex projective space P_{n-1} of dimension $n-1$, $dd^c \log r$ can be regarded as a form in P_{n-1}. The function $u = \log r/\log a$ satisfies in A_n the condition $0 < u < 1$. Since P_{n-1} is of real dimension $2n-2$, we have

$$(dd^c u)^n = 0 . \tag{26}$$

It follows that the integral

$$\int d^c u \wedge (dd^c u)^{n-1} = \frac{1}{(\log a)^n} \int d^c \log r \wedge (dd^c \log r)^{n-1} \tag{27}$$

over a cycle of the homology class γ depends only on γ. It is an easy computation that over the unit sphere in C_n the form

$$d^c \log r (dd^c \log r)^{n-1}$$

is equal to $(n-1)!\, 2^{n-1}$ times its volume element. Using the value of the volume of the unit sphere in C_n, we find that the integral (27) is equal to $(2\pi/\log a)^n$. By definition we have

$$N\{\gamma\} \geqq \left(\frac{2\pi}{\log a}\right)^n > 0 . \tag{28}$$

Thus $N\{\gamma\}$ is a norm on $H_{2n-1}(A_n, R)$.

Since the norm is non-increasing under a holomorphic mapping, we derive from this the theorem: *Let $f: A_n \to A_n$ be a holomorphic mapping. Then $f_* \gamma = \pm \gamma$ or 0, f_* being the induced homomorphism on homology.* This generalizes a theorem of M. Schiffer [12] and H. Huber [6] for $n = 1$.

We do not know the exact value of $N\{\gamma\}$. In the case $n=1$ Landau and Osserman [8] showed that the equality sign holds in (28). Let $\tilde{U}$ be the domain: $0 \leqq \arg z < 2\pi$. We wish to show that for $n = 1$ we have

$$N\{\gamma\} = N\{\gamma, \tilde{U}\} = \frac{2\pi}{\log a} . \tag{29}$$

In fact, let u be a harmonic function in $\mathbf{F}(\tilde{U})$. Imagining the ring slit on the positive x-axis between 1 and a, the function u is well defined in $\tilde{U}$. It suffices to prove the inequality

$$\left| \int_{|z|=r} d^c u \right| = \left| \int_0^{2\pi} r u_r d\theta \right| \leqq \frac{2\pi}{\log a} ,$$

where $z = re^{i\theta}$. Let S denote the operator of averaging with respect to angle. Since u_θ is periodic, it is easily seen that $v(r) = Su$

is harmonic. Thus $v = c \log r$, with $c \leqq 1/\log a$, since $0 < v < 1$. It follows that

$$|rv_r| \leqq \frac{1}{\log a},$$

which is the inequality to be proved.

Our next example is concerned with the torus $M = S^1 \times S^1$ and with γ the homology class of the torus itself. For any current $T \in \gamma$ we have then

$$T[\varphi] = \int_M \varphi ,$$

where φ is a C^∞ two-form. We consider M to be covered by the (x, y)-plane. Let $\tilde{U}$ be the fundamental domain consisting of the open square

$$0 < x < 1, 0 < y < 1$$

and the segments $y = 0$, $0 \leqq x < 1$ and $x = 0$, $0 \leqq y < 1$. We shall prove that $N\{\gamma, \tilde{U}\} = 1$.

For this purpose let $u \in \mathcal{F}(\tilde{U})$. Since du is well defined on M and M is compact, we have

$$\int_M dd^c u = 0 .$$

Since u is plurisubharmonic, this implies $dd^c u = 0$, i.e., u is harmonic. Its derivatives with respect to x and y are single-valued harmonic functions on M and are therefore constants. Thus u is a linear function and we may take

$$u = az + \overline{a}\overline{z} . \tag{30}$$

For $T \in \gamma$ we have

$$T[du \wedge d^c u] = \int_{\tilde{U}} du \wedge d^c u = 4 \iint_{\tilde{U}} |u_z|^2 \, dxdy = 4 \, |a|^2 .$$

Now the values of u at the corners of $\tilde{U}$ are 0, $a + \overline{a}$, $i(a - \overline{a})$, $a(1 + i) + \overline{a}(1 - i)$. The fact that the oscillation of u in $\tilde{U}$ is at most one means that

$$\begin{aligned} &2\,|\operatorname{Re} a| < 1 , && 2\,|\operatorname{Im} a| < 1 , \\ &2\,|\operatorname{Re} a(1 + i)| < 1 , && 2\,|\operatorname{Re} a(1 - i)| < 1 . \end{aligned} \tag{30a}$$

These imply $|a| < (1/2)$. Consequently we have

$$N\{\gamma, \tilde{U}\} = \sup_u 4\,|a|^2 = 1 .$$

6. Intrinsic pseudo-metrics

In § 4 we remarked about the possibility of defining the semi-norm of a family of chains of a fixed dimension. The simplest case is the family γ of *curves* (rather than chains) having two given points $z, \zeta \in M$ as boundary and containing all curves homotopic to a given one in the family. To indicate that the notions so introduced will be pseudo-distances on M we will change our notation and repeat the definitions as follows:

$$\rho_\gamma(z, \zeta) = \sup_{u \in \mathcal{F}} \inf_{T \in \gamma} |T[d^c u]| , \tag{31}$$

$$\rho_\gamma(z, \zeta; \mathfrak{U}) = \sup_{u \in \mathcal{F}(\mathfrak{U})} \inf_{T \in \gamma} |T[d^c u]| , \tag{32}$$

$$\rho_\gamma(z, \zeta; \tilde{U}) = \sup_{u \in \mathcal{F}(\tilde{U})} \inf_{T \in \gamma} |T[d^c u]| . \tag{33}$$

We shall omit the subscript γ if the family consists of *all* (smooth) curves joining z and ζ and we shall denote $\rho_\gamma(z, \zeta)$ by $\rho_c(z, \zeta)$ if the family consists of all *chains* joining z and ζ. Clearly $\rho \leqq \rho_\gamma$ for the three definitions, with equality if M is simply connected. The definitions (32) and (33) refer respectively to a locally finite open covering $\mathfrak{U}$ of M and a fundamental domain $\tilde{U}$ in the universal covering manifold $\tilde{M}$ of M. In all three formulas T denotes a curve with $z - \zeta$ as boundary. As in the general case it suffices to restrict ourselves to functions u which satisfy the additional condition $dd^c u = 0$, i.e., which are pluriharmonic. From the definitions we have

$$\rho(z, \zeta) \leqq \rho(z, \zeta; \mathfrak{U}) ; \qquad \rho(z, \zeta) \leqq \rho(z, \zeta; \tilde{U}) . \tag{34}$$

Remark. If M is compact, then $\rho(z, \zeta) = 0$. However, the quantities in (32), (33) need not be zero. Consider for example the torus discussed in the end of the last section with $\tilde{U}$ defined as before. The harmonic functions in $\mathcal{F}(\tilde{U})$ are given by (30), where a satisfies the inequalities (30a). For z, ζ on the torus we have

$$T[d^c u] = \int_z^\zeta d^c u = 2 \operatorname{Im}[a(\tilde{\zeta} - \tilde{z})] , \tag{35}$$

where $\tilde{z}$ and $\tilde{\zeta}$ are points in the plane, the covering surface, lying over z and ζ respectively. N. Kerzman has found that in this case $\rho(z, \zeta; \tilde{U})$ is equal to the maximum of the horizontal and vertical distances from $\tilde{\zeta} - \tilde{z}$ to the sides of a period square containing $\tilde{\zeta} - \tilde{z}$.

THEOREM 6. *The quantities $\rho(z, \zeta)$, $\rho(z, \zeta; \mathfrak{U})$, $\rho(z, \zeta; \tilde{U})$ are* *pseudo-distances*, i.e., *they are finite and satisfy the triangle inequalities.*

The finiteness follows from the fact that the only functions u which enter into consideration are bounded pluriharmonic functions, so that their first partial derivatives are uniformly bounded on compact sets. The last statement follows immediately from definition.

We wish to compare our pseudo-metric with those of Caratheodory [3] and S. Kobayashi [7]. We recall their definitions as follows. Let D be the unit disk $|\tau| < 1$, whose hyperbolic distance we denote by $h(\tau_1, \tau_2)$, $\tau_1, \tau_2 \in D$. Then the Caratheodory pseudo-distance is defined by

$$c(z, \zeta) = \sup h\big(f(z), f(\zeta)\big) , \tag{36}$$

as f runs over the family of all holomorphic mappings $f: M \to D$.

To define the Kobayashi pseudo-distance let

$$f_i: D \to M , \qquad 1 \leqq i \leqq k ,$$

be holomorphic mappings which satisfy the conditions

$$z \in f_1(D) , \qquad \zeta \in f_k(D) ,$$
$$f_i(D) \cap f_{i+1}(D) \neq \emptyset , \qquad 1 \leqq i \leqq k-1 .$$

We choose $z_0 = z, z_1, \cdots, z_{k-1}, z_k = \zeta$, such that

$$z_i \in f_i(D) \cap f_{i+1}(D) , \qquad 1 \leqq i \leqq k-1 .$$

Let $a_i, b_i \in D$ be points satisfying

$$z_{i-1} = f_i(a_i), z_i = f_i(b_i) , \qquad 1 \leqq i \leqq k .$$

Then the Kobayashi pseudo-distance is defined by

$$d(z, \zeta) = \inf \textstyle\sum_{1 \leqq i \leqq k} h(a_i, b_i) , \tag{37}$$

where the infimum is taken with respect to all the choices made.

Kobayashi proved that $c(z, \zeta) \leqq d(z, \zeta)$. We will establish the theorem.

THEOREM 7[1]. *Between the pseudo-distances the following inequalities are valid*

$$c(z, \zeta) \leqq \frac{\pi}{2} \rho_c(z, \zeta) \leqq \frac{\pi}{2} \rho(z, \zeta) \leqq d(z, \zeta) . \tag{38}$$

[1] In our original proof we showed $c \leqq (2/\pi)\rho \leqq d$. Kerzman observed that our argument could be used to prove the more general result (38).

The first inequality becomes an equality if M is simply connected.

By the conformal mapping

$$\tau = \frac{i - \exp(\pi i w)}{i + \exp(\pi i w)}, \tag{39}$$

we map the unit disk D onto the infinite strip $S: 0 < u < 1$, $w = u + iv$. Under (39) the real axis of D corresponds to the line $\frac{1}{2} + iv$ and the origin of D to the point $w = \frac{1}{2}$. S has a hyperbolic metric induced from that of D by the mapping (39), which is given by

$$ds^2 = \frac{d\tau d\bar{\tau}}{(1 - \tau\bar{\tau})^2} = \frac{\pi^2 dw d\bar{w}}{2(1 - \cos 2\pi u)}. \tag{40}$$

Thus the hyperbolic distance on S between the points $\frac{1}{2}$ and $\frac{1}{2} + vi$ is

$$h_S\left(\frac{1}{2}, \frac{1}{2} + vi\right) = \frac{\pi}{2} |v|. \tag{41}$$

Since S admits the group of hyperbolic motions, we can normalize the holomorphic mappings $M \to S$ such that the image of z is the point $\frac{1}{2}$ and the image of ζ lies on the line $\operatorname{Re} w = \frac{1}{2}$. Hence the Caratheodory pseudo-distance can be redefined as follows.

$$c(z, \zeta) = \frac{\pi}{2} \sup_g |\operatorname{Im} g(\zeta)|, \tag{42}$$

where g runs over all holomorphic mappings $g: M \to S$ such that $g(z) = \frac{1}{2}$ and $\operatorname{Re} g(\zeta) = \frac{1}{2}$. If σ denotes the segment joining $g(z)$ to $g(\zeta)$, we have

$$\operatorname{Im} g(\zeta) = \int_\sigma dv = \int_\sigma d^c u = \int_z^\zeta d^c u.$$

where $u = \operatorname{Re} g$ is a pluriharmonic function which belongs to the family $\mathcal{F}$. Since $T[d^c u] = \operatorname{Im} g(\zeta)$ for any chain T this proves that

$$c(z, \zeta) \leqq \frac{\pi}{2} \rho_c(z, \zeta).$$

To prove equality consider a pluriharmonic function u in $\mathcal{F}$. If among all *chains* T, $\inf_T |T[d^c u]| \neq 0$, then the integral of $d^c u$ around any closed path vanishes. But then the function

$$v(z) = \int_{z_0}^z d^c u$$

is well defined independent of the path of integration and is therefore a conjugate pluriharmonic function of u. Hence $w = u + iv$ defines a holomorphic mapping $w: M \rightarrow S$. By (40) we have

$$\frac{\pi}{2} |dv| \leqq ds .$$

Since $dv = d^c u$, it follows that

$$\frac{\pi}{2} \rho_c(z, \zeta) \leqq c(z, \zeta) ,$$

and hence

$$c(z, \zeta) = \frac{\pi}{2} \rho_c(z, \zeta) .$$

If M is simply connected then to every pluriharmonic function u in $\mathcal{F}$ there exists a conjugate pluriharmonic function v defined up to an additive constant. It follows that for any chain T bounded by ζ and z, $T[d^c u]$ equals $v(\zeta) - v(z)$ and is therefore independent of the chain T. Consequently

$$\rho_c(z, \zeta) = \rho(z, \zeta) .$$

To prove the last inequality in (38) we use the fact that ρ is non-increasing under a holomorphic mapping. We will also follow the above notation in the definition of the Kobayashi pseudo-distance. Let l_i be the straight segment in D joining a_i to b_i and let $f_i(l_i) = L_i$, $1 \leqq i \leqq k$. Let u be a pluriharmonic function on M, with $0 < u < 1$. Then we have

$$\frac{\pi}{2} \left| \int_{L_i} d^c u \right| \leqq \frac{\pi}{2} \rho_0(a_i, b_i) = h(a_i, b_i) , \qquad 1 \leqq i \leqq k ,$$

where ρ_0 is our metric in D. The last equality follows from what we just proved, as D is simply connected. It follows that

$$\frac{\pi}{2} \left| \sum_i \int_{L_i} d^c u \right| \leqq \sum_i h(a_i, b_i) , \qquad 1 \leqq i \leqq k .$$

Now the right-hand side may be chosen as close to $d(z, \zeta)$ as we like, while the left-hand side is not smaller than $(\pi/2)(\rho(z, \zeta))$. This implies the desired inequality.

Remark. We do not know when $(\pi/2)\rho(z, \zeta) = d(z, \zeta)$, nor how $(\pi/2)\rho(z, \zeta; \tilde{U})$ compares with $d(z, \zeta)$. Using chains one may also introduce pseudo-metrics $\rho_c(z, \zeta; \mathfrak{U})$ and $\rho_c(z, \zeta; \tilde{U})$; but on a compact manifold these are zezo.

7. Remarks on the differential equations

The differential equations (13) and (13a) are, in general, overdetermined systems of non-linear differential equations. For $k=1$, equation (13),

$$dd^c u = 0\,,$$

asserts that u is pluriharmonic, while in general, equation (13) means that the rank of the hessian $u_{i\bar{j}}$ is less than k. Almost nothing is known about the solvability of these equations. The case $k = n$ reduces to a single equation which is the complex analogue of the Monge-Ampère equation

$$\det\{u_{i\bar{j}}\} = 0\,;$$

it is non-linear degenerate elliptic in view of our requirement that the matrix $u_{i\bar{j}}$ be positive semi-definite.

It would be interesting to formulate boundary value problems for these equations. We remark that the equation $(dd^c u)^n = 0$ also arises as the Euler equation for a stationary point of the functional

$$I(u) = \int_M du \wedge d^c u \wedge (dd^c u)^{n-1} \tag{43}$$

under, perhaps, some boundary conditions. Consider for example the class B of C^2 plurisubharmonic functions which are required to equal one on some components of the (smooth) boundary of a compact manifold M, and zero on the others. If v is a member of B, let γ denote the $(2n-1)$-dimensional homology class of the level hypersurfaces $v =$ constant. Then we observe that for $T \in \gamma$, if v satisfies $(dd^c v)^n = 0$,

$$\int_T dv \wedge (dd^c v)^{n-1} = \int_M dv \wedge d^c v \wedge (dd^c v)^{n-1} = I(v)\,. \tag{44}$$

It is not difficult to verify that the functional I is convex and one is therefore tempted to conjecture that

$$N\{\gamma\} = \inf_{v \in B} I(v)\,. \tag{45}$$

If this is the case then $N\{\gamma\}$, which is defined as the supremum of a functional would also be characterized as the infimum of another, a situation that often arises in, so called, dual variational problems in the calculus of variations. The problem of minimizing $I(v)$ seems an interesting one. In the case of the annulus $1 < |z| < a$, the function $v_0 = \log|z|/\log a$ is indeed the minimizing function, since the convex functional I is stationary at v_0.

The differential equation (13) has a real analogue, which is

$$\text{rank}\left(\frac{\partial^2 u}{\partial x^i \partial x^j}\right) \leqq k , \qquad 1 \leqq i, j \leqq n , \tag{46}$$

where $u(x^1, \cdots, x^n)$ is a real-valued C^2-function in the real variables $x^1, \cdots, x^n$. Equation (46) and its generalizations have been studied in connection with some geometrical problems (cf. [4], [5], [11]). In fact, if $u = u(x^1, \cdots, x^n)$ is considered as the equation of a non-parametric hypersurface in the euclidean $(n+1)$-space E^{n+1}, the left-hand side of (46) is called the index of relative nullity, being the rank of its second fundamental form. Hartman and Nirenberg [5, p. 912] proved that for $n = 2$, $k = 1$ (in which case condition (46) means that the surface has zero gaussian curvature) the surface is a cylinder if $u(x^1, x^2)$ is defined for all $(x^1, x^2) \in R^2$. For higher dimensions a similar result is not true, as shown by an example of Sacksteder [11]. In this respect we wish to refer to a general theorem of Hartman [4] concerned with sufficient conditions for an isometrically immersed submanifold in an euclidean space to be cylindrical.

UNIVERSITY OF CALIFORNIA, BERKELEY
BRANDEIS UNIVERSITY
COURANT INSTITUTE, NEW YORK UNIVERSITY.

REFERENCES

[1] R. D. M. ACCOLA, *Differentials and extremal length on Riemann surfaces.* Proc. Nat. Acad. Sci. USA **46** (1960), 540-543.

[2] L. V. AHLFORS and L. SARIO, Riemann Surfaces, Princeton University Press, Princeton, 1960.

[3] C. CARATHEODORY, *Über eine spezielle Metrik, die in der Theorie der analytischen Funktionen auftritt.* Atti. Pont. Acad. Sci. Nuovo Lincei **80** (1927), 135-141.

[4] P. HARTMAN, *On isometric immersions in euclidean space of manifolds with non-negative sectional curvatures.* Trans. Amer. Math. Soc. **115** (1965), 94-109.

[5] ——— and L. NIRENBERG, *On spherical image maps whose jacobians do not change sign.* Amer. J. Math. **81** (1959), 901-920.

[6] H. HUBER, *Über analytische Abbildungen von Ringgebieten in Ringgebiete.* Compos. Math. **9** (1951), 161-168.

[7] S. KOBAYASHI, *Invariant distances on complex manifolds and holomorphic mappings.* J. Math. Soc. Japan **19** (1967), 460-480.

[8] H. J. Landau and R. Osserman, *On analytic mappings of Riemann surfaces.* J. Anal. Math. **7** (1959-60), 249-279.

[9] P. Lelong, *Sur les dérivées d'une fonction plurisousharmonique.* C. R. Acad. Sci. Paris **238** (1954), 2276-2278.

[10] G. de Rham, Variétés Différentiables, Actualités Sci. et Ind. No. 1222, Hermann, Paris, 1955.

[11] R. Sacksteder, *On hypersurfaces with no negative sectional curvatures.* Amer. J. Math. **82** (1960), 609-630.

[12] M. Schiffer, *On the modulus of doubly connected domains.* Quart. J. Math. **17** (1946), 197-213.

(Received September 9, 1968)

Reprint from
Functional Analysis and Related Fields
Edited by Felix E. Browder

Springer-Verlag Berlin · Heidelberg · New York 1970
Printed in Germany · Not in trade

Minimal Submanifolds of a Sphere with Second Fundamental Form of Constant Length

By S. S. Chern*, M. do Carmo**, and S. Kobayashi***
University of California, Berkeley, and IMPA, Rio de Janeiro

1. Introduction

Let M be an n-dimensional manifold which is minimally immersed in a unit sphere S^{n+p} of dimension $n+p$. Let h be the second fundamental form of this immersion; it is a certain symmetric bilinear mapping $T_x \times T_x \to T_x^\perp$ for $x \in M$, where T_x is the tangent space of M at x and $T_x^\perp$ is the normal space to M at x. We denote by S the square of the length of h. It is known that if M is moreover compact, then

$$\int_M \left(\left(2-\frac{1}{p}\right)S-n\right)S \cdot *1 \geqq 0,$$

where $*1$ denotes the volume element of M, (Simons [3]; a slightly more general formula will be proved in §§ 2 and 3). It follows that if $S \leqq n \Big/ \left(2-\frac{1}{p}\right)$ everywhere on M, then either

1) $S=0$ (i.e., M is totally geodesic)

or

2) $S = n \Big/ \left(2-\frac{1}{p}\right)$.

The purpose of the present paper is to determine all minimal submanifolds M of S^{n+p} satisfying $S = n \Big/ \left(2-\frac{1}{p}\right)$. The proof depends upon the results presented by the first named author in his lectures on minimal submanifolds in Berkeley in the Winter of 1968, in which an exposition of the work of Simons [3] was made by the use of moving frames. To describe our result, we begin with examples of minimal submanifolds.

In general, let $S^q(r)$ denote a q-dimensional sphere in $\boldsymbol{R}^{q+1}$ with radius r. Let m and n be positive integers such that $m<n$ and let

* Work done under partial support by NSF Grant GP-6974;
** Work done under partial support by NSF Grant GP-6974 and Guggenheim Foundation;
*** Work done under partial support by NSF Grant GP-8008.

$M_{m,n-m} = S^m\left(\sqrt{\frac{m}{n}}\right) \times S^{n-m}\left(\sqrt{\frac{n-m}{n}}\right)$. We imbed $M_{m,n-m}$ into $S^{n+1} = S^{n+1}(1)$ as follows. Let (u, v) be a point of $M_{m,n-m}$ where u (resp. v) is a vector in $\boldsymbol{R}^{m+1}$ (resp. $\boldsymbol{R}^{n-m+1}$) of length $\sqrt{\frac{m}{n}}\left(\text{resp.} \sqrt{\frac{n-m}{n}}\right)$. We can consider (u, v) as a unit vector in $\boldsymbol{R}^{n+2} = \boldsymbol{R}^{m+1} \times \boldsymbol{R}^{n-m+1}$. It will be shown that $M_{m,n-m}$ is a minimal submanifold of S^{n+1} satisfying $S = n$.

We shall now define the Veronese surface. Let (x, y, z) be the natural coordinate system in $\boldsymbol{R}^3$ and $(u^1, u^2, u^3, u^4, u^5)$ the natural coordinate system in $\boldsymbol{R}^5$. We consider the mapping defined by

$$u^1 = \frac{1}{\sqrt{3}} yz, \quad u^2 = \frac{1}{\sqrt{3}} zx, \quad u^3 = \frac{1}{\sqrt{3}} xy, \quad u^4 = \frac{1}{2\sqrt{3}} (x^2 - y^2),$$
$$u = \frac{1}{6} (x^2 + y^2 - 2z^2).$$

This defines an isometric immersion of $S^2(\sqrt{3})$ into $S^4 = S^4(1)$. Two points (x, y, z) and $(-x, -y, -z)$ of $S^2(\sqrt{3})$ are mapped into the same point of S^4, and this mapping defines an imbedding of the real projective plane into S^4. This real projective plane imbedded in S^4 will be called the *Veronese surface*. It will be shown that the Veronese surface is a minimal submanifold of S^4 satisfying $S = 4/3$.

Main theorem. *The Veronese surface in S^4 and the submanifolds $M_{m,n-m}$ in S^{n+1} are the only compact minimal submanifolds of dimension n in S^{n+p} satisfying $S = n\Big/\left(2 - \frac{1}{p}\right)$.*

The proof is by local argument and the corresponding local result also holds: An *n-dimensional* minimal submanifold of S^{n+p} satisfying $S = n\Big/\left(2 - \frac{1}{p}\right)$ is locally $M_{m,n-m}$ or the Veronese surface.

For $p = 1$, the theorem was proved independently by B. LAWSON [2].

It should be pointed out that, by the equation of GAUSS, $S = n(n-1) - R$ for an n-dimensional minimal submanifold M of S^{n+p}, where R is the scalar curvature of M. In particular, S is an intrinsic invariant when M is minimal.

2. Local formulas for a minimal submanifold

In this section we shall compute the Laplacian of the second fundamental form of a minimal submanifold of a symmetric space.

Let M be an n-dimensional manifold immersed in an $(n+p)$-dimensional riemannian manifold N. We choose a local field of orthonormal frames $e_1, \ldots, e_{n+p}$ in N such that, restricted to M, the vectors $e_1, \ldots, e_n$ are tangent to M (and, consequently, the remaining vectors $e_{n+1}, \ldots, e_{n+p}$

are normal to M). We shall make use of the following convention on the ranges of indices:

$$1 \leqq A, B, C, \ldots, \leqq n+p; \quad 1 \leqq i, j, k, \ldots, \leqq n;$$
$$n+1 \leqq \alpha, \beta, \gamma, \ldots, \leqq n+p,$$

and we shall agree that repeated indices are summed over the respective ranges. With respect to the frame field of N chosen above, let $\omega^1, \ldots, \omega^{n+p}$ be the field of dual frames. Then the structure equations of N are given by

$$d\omega^A = -\sum \omega^A_B \wedge \omega^B, \quad \omega^A_B + \omega^B_A = 0, \tag{2.1}$$

$$d\omega^A_B = -\sum \omega^A_C \wedge \omega^C_B + \Phi^A_B, \quad \Phi^A_B = \tfrac{1}{2} \sum K^A{}_{BCD}\, \omega^C \wedge \omega^D,$$
$$K^A{}_{BCD} + K^A{}_{BDC} = 0. \tag{2.2}$$

We restrict these forms to M. Then

$$\omega^\alpha = 0. \tag{2.3}$$

Since $0 = d\omega^\alpha = -\sum \omega^\alpha_i \wedge \omega^i$, by CARTAN's lemma we may write

$$\omega^\alpha_i = \sum h^\alpha_{ij}\, \omega^j, \quad h^\alpha_{ij} = h^\alpha_{ji}. \tag{2.4}$$

From these formulas, we obtain

$$d\omega^i = -\sum \omega^i_j \wedge \omega^j, \quad \omega^i_j + \omega^j_i = 0, \tag{2.5}$$

$$d\omega^i_j = -\sum \omega^i_k \wedge \omega^k_j + \Omega^i_j, \quad \Omega^i_j = \tfrac{1}{2} \sum R^i{}_{jkl}\, \omega^k \wedge \omega^l, \tag{2.6}$$

$$R^i{}_{jkl} = K^i{}_{jkl} + \sum_\alpha (h^\alpha_{ik} h^\alpha_{jl} - h^\alpha_{il} h^\alpha_{jk}), \tag{2.7}$$

$$d\omega^\alpha_\beta = -\sum \omega^\alpha_\gamma \wedge \omega^\gamma_\beta + \Omega^\alpha_\beta, \quad \Omega^\alpha_\beta = \tfrac{1}{2} \sum R^\alpha{}_{\beta kl}\, \omega^k \wedge \omega^l, \tag{2.8}$$

$$R^\alpha{}_{\beta kl} = K^\alpha{}_{\beta kl} + \sum_i (h^\alpha_{ik} h^\beta_{il} - h^\alpha_{il} h^\beta_{jk}). \tag{2.9}$$

The riemannian connection of M is defined by (ω^i_j). The form (ω^α_β) defines a connection in the normal bundle of M. We call $\sum h^\alpha_{ij}\, \omega^i \omega^j e_\alpha$ the *second fundamental form* of the immersed manifold M. Sometimes we shall denote the second fundamental form by its components h^α_{ij}. We call $\sum_\alpha \frac{1}{n} \left(\sum_i h^\alpha_{ii} \right) e_\alpha$ the *mean curvature normal* or the *mean curvature vector*. An immersion is said to be *minimal* if its mean curvature normal vanishes identically, i.e., if $\sum_i h^\alpha_{ii} = 0$ for all α.

We take exterior differentiation of (2.4) and define h^α_{ijk} by

$$\sum h^\alpha_{ijk}\, \omega^k = d h^\alpha_{ij} - \sum h^\alpha_{il}\, \omega^l_j - \sum h^\alpha_{lj}\, \omega^l_i + \sum h^\beta_{ij}\, \omega^\alpha_\beta. \tag{2.10}$$

Then

$$\sum (h^\alpha_{ijk} + \tfrac{1}{2} K^\alpha_{ijk})\, \omega^j \wedge \omega^k = 0, \tag{2.11}$$

$$h^\alpha_{ijk} - h^\alpha_{ikj} = K^\alpha{}_{ikj} = -K^\alpha{}_{ijk}. \tag{2.12}$$

We take exterior differentiation of (2.10) and define h^α_{ijkl} by

$$\sum h^\alpha_{ijkl}\, \omega^l = d h^\alpha_{ijk} - \sum h^\alpha_{ljk}\, \omega^l_i - \sum h^\alpha_{ilk}\, \omega^l_j - \sum h^\alpha_{ijl}\, \omega^l_k + \sum h^\beta_{ijk}\, \omega^\alpha_\beta. \tag{2.13}$$

Then

$$\sum (h^\alpha_{ijkl} - \tfrac{1}{2} \sum h^\alpha_{im} R^m{}_{jkl} - \tfrac{1}{2} \sum h^\alpha_{mj} R^m{}_{ikl} + \tfrac{1}{2} \sum h^\beta_{ij} R^\alpha{}_{\beta kl})\, \omega^k \wedge \omega^l = 0, \tag{2.14}$$

$$h^\alpha_{ijkl} - h^\alpha_{ijlk} = \sum h^\alpha_{im} R^m{}_{jkl} + \sum h^\alpha_{mj} R^m{}_{ikl} - \sum h^\beta_{ij} R^\alpha{}_{\beta kl}. \tag{2.15}$$

We stated earlier that (ω^i_j) defines a connection in the tangent bundle $T = T(M)$ [and, hence, a connection in the cotangent bundle $T^* = T^*(M)$ also] and that (ω^α_β) defines a connection in the normal bundle $T^\perp = T^\perp(M)$. Consequently, we have covariant differentiation which maps a section of $T^\perp \otimes T^* \otimes \ldots \otimes T^*$, ($T^*$: k times), into a section of $T^\perp \otimes T^* \otimes \ldots \otimes T^* \otimes T^*$, ($T^*$: $k+1$ times). The second fundamental form h^α_{ij} is a section of the vector bundle $T^\perp \otimes T^* \otimes T^*$, and h^α_{ijk} is the covariant derivative of h^α_{ij}. Similarly, h^α_{ijkl} is the covariant derivative of h^α_{ijk}.

Similarly, we may consider $K^\alpha{}_{ijk}$ as a section of the bundle $T^\perp \otimes T^* \otimes T^* \otimes T^*$. Its covariant derivative $K^\alpha{}_{ijkl}$ is defined by

$$\begin{aligned} \sum K^\alpha{}_{ijkl}\, \omega^l = d K^\alpha{}_{ijk} - \sum K^\alpha{}_{mjk}\, \omega^m_i - \sum K^\alpha{}_{imk}\, \omega^m_j \\ - \sum K^\alpha{}_{ijm}\, \omega^m_k + \sum K^\beta{}_{ijk}\, \omega^\alpha_\beta. \end{aligned} \tag{2.16}$$

This covariant derivative of $K^\alpha{}_{ijkl}$ must be distinguished from the covariant derivative of $K^A{}_{BCD}$ as a curvature tensor of N, which will be denoted by $K^A{}_{BCD;E}$. Restricted to M, $K^\alpha{}_{ijk;l}$ is given by

$$\begin{aligned} K^\alpha{}_{ijk;l} = K^\alpha{}_{ijkl} - \sum K^\alpha{}_{\beta jk} h^\beta_{il} - \sum K^\alpha{}_{i\beta k} h^\beta_{jl} - \sum K^\alpha{}_{ij\beta} h^\beta_{kl} \\ + \sum K^m{}_{ijk} h^\alpha_{ml}. \end{aligned} \tag{2.17}$$

In this section, *we shall assume that N is locally symmetric, i.e.,* $K^A{}_{BCD;E} = 0$.

The Laplacian Δh^α_{ij} of the second fundamental form h^α_{ij} is defined by

$$\Delta h^\alpha_{ij} = \sum_k h^\alpha_{ijkk}. \tag{2.18}$$

From (2.12) we obtain

$$\Delta h^\alpha_{ij} = \sum_k h^\alpha_{ikjk} - \sum_k K^\alpha{}_{ijkk} = \sum_k h^\alpha_{kijk} - \sum_k K^\alpha{}_{ijkk}. \tag{2.19}$$

From (2.15) we obtain

$$h^\alpha_{kijk} = h^\alpha_{kikj} + \sum h^\alpha_{km} R^m{}_{ijk} + \sum h^\alpha_{mi} R^m{}_{kjk} - \sum h^\beta_{ki} R^\alpha{}_{\beta jk}. \tag{2.20}$$

Replace h_{kikj} in (2.20) by $h_{kkij} - K^{\alpha}{}_{kikj}$ [see (2.12)] and then substitue the right hand side of (2.20) into h^{α}_{kijk} of (2.19). Then

$$\begin{aligned} \Delta h^{\alpha}_{ij} &= \sum_k (h^{\alpha}_{kkij} - K^{\alpha}{}_{kikj} - K^{\alpha}{}_{ijkk}) \\ &\quad + \sum_k \Big(\sum_m h^{\alpha}_{km} R^m{}_{ijk} + \sum_m h^{\alpha}_{mi} R^m{}_{kjk} - \sum_\beta h^{\beta}_{ki} R^{\alpha}{}_{\beta jk} \Big). \end{aligned} \tag{2.21}$$

From (2.7), (2.9), (2.17) and (2.21) we obtain

$$\begin{aligned} \Delta h^{\alpha}_{ij} &= \sum_k h^{\alpha}_{kkij} + \sum_{\beta,k} (-K^{\alpha}{}_{ij\beta} h^{\beta}_{kk} + 2K^{\alpha}{}_{\beta ki} h^{\beta}_{jk} - K^{\alpha}{}_{k\beta k} h^{\beta}_{ij} \\ &\quad + 2K^{\alpha}{}_{\beta kj} h^{\beta}_{ki}) + \sum_{m,k} (K^m{}_{kik} h^{\alpha}_{mj} + K^m{}_{kjk} h^{\alpha}_{mi} \\ &\quad + 2K^m{}_{ijk} h^{\alpha}_{mk}) + \sum_{\beta,m,k} (h^{\alpha}_{mi} h^{\beta}_{mj} h^{\beta}_{kk} + 2h^{\alpha}_{km} h^{\beta}_{ki} h^{\beta}_{mj} \\ &\quad - h^{\alpha}_{km} h^{\beta}_{km} h^{\beta}_{ij} - h^{\alpha}_{mi} h^{\beta}_{mk} h^{\beta}_{kj} - h^{\alpha}_{mj} h^{\beta}_{ki} h^{\beta}_{mk}). \end{aligned} \tag{2.22}$$

Now, *we assume that M is minimal in N so that* $\sum h^{\beta}_{kk} = 0$ *for all* β. Then, from (2.22) we obtain

$$\begin{aligned} \sum h^{\alpha}_{ij} \cdot \Delta h^{\alpha}_{ij} &= \sum_{\alpha,\beta,i,j,k} (4K^{\alpha}{}_{\beta ki} h^{\beta}_{jk} h^{\alpha}_{ij} - K^{\alpha}{}_{k\beta k} h^{\alpha}_{ij} h^{\beta}_{ij}) \\ &\quad + \sum_{\alpha,m,i,j,k} (2K^m{}_{kik} h^{\alpha}_{mj} h^{\alpha}_{ij} + 2K^m{}_{ijk} h^{\alpha}_{mk} h^{\alpha}_{ij}) \\ &\quad - \sum_{\alpha,\beta,i,j,k,l} (h^{\alpha}_{ik} h^{\beta}_{jk} - h^{\alpha}_{jk} h^{\beta}_{ik}) (h^{\alpha}_{il} h^{\beta}_{jl} - h^{\alpha}_{jl} h^{\beta}_{il}) \\ &\quad - \sum_{\alpha,\beta,i,j,k,l} h^{\alpha}_{ij} h^{\alpha}_{kl} h^{\beta}_{ij} h^{\beta}_{kl}. \end{aligned} \tag{2.23}$$

3. Minimal submanifolds of a space of constant curvature

Throughout this section we shall *assume that the ambient space N is a space of constant curvature c*. Then

$$K^A{}_{BCD} = c(\delta_{AC}\delta_{BD} - \delta_{AD}\delta_{BC}).$$

Then (2.23) reduces to

$$\begin{aligned} \sum h^{\alpha}_{ij} \cdot \Delta h^{\alpha}_{ij} &= -\sum (h^{\alpha}_{ik} h^{\beta}_{kj} - h^{\beta}_{ik} h^{\alpha}_{kj}) (h^{\alpha}_{il} h^{\beta}_{lj} - h^{\beta}_{il} h^{\alpha}_{lj}) \\ &\quad - \sum h^{\alpha}_{ij} h^{\alpha}_{lk} h^{\beta}_{ij} h^{\beta}_{kl} + nc \sum (h^{\alpha}_{ij})^2. \end{aligned} \tag{3.1}$$

For each α, let H_α denote the symmetric matrix (h^{α}_{ij}), and set

$$S_{\alpha\beta} = \sum_{i,j} h^{\alpha}_{ij} h^{\beta}_{ij}. \tag{3.2}$$

Then the $(p \times p)$-matrix $(S_{\alpha\beta})$ is symmetric and can be assumed to be diagonal for a suitable choice of $e_{n+1}, \ldots, e_{n+p}$. We set

$$S_\alpha = S_{\alpha\alpha}. \tag{3.3}$$

We denote the square of the length of the second fundamental form by S, i.e.,

$$S = \sum h_{ij}^{\alpha} h_{ij}^{\alpha} = \sum_{\alpha} S_{\alpha}. \tag{3.4}$$

In general, for a matrix $A = (a_{ij})$ we denote by $N(A)$ the square of the norm of A, i.e.,

$$N(A) = \text{trace } A \cdot {}^t A = \sum (a_{ij})^2.$$

Clearly, $N(A) = N(T^{-1} A T)$ for any orthogonal matrix T. Now, (3.1) may be rewritten as follows:

$$\sum h_{ij}^{\alpha} \cdot \Delta h_{ij}^{\alpha} = -\sum_{\alpha,\beta} N(H_{\alpha} H_{\beta} - H_{\beta} H_{\alpha}) - \sum_{\alpha} S + ncS. \tag{3.5}$$

We need the following algebraic lemma.

Lemma 1. *Let A and B be symmetric $(n \times n)$-matrices. Then*

$$N(AB - BA) \leqq 2N(A) \cdot N(B),$$

and the equality holds for nonzero matrices A and B if and only if A and B can be transformed simultaneously by an orthogonal matrix into scalar multiples of $\tilde{A}$ and $\tilde{B}$ respectively, where

$$\tilde{A} = \left(\begin{array}{cc|c} 0 & 1 & \\ 1 & 0 & 0 \\ \hline 0 & & 0 \end{array}\right); \quad \tilde{B} = \left(\begin{array}{cc|c} 1 & 0 & \\ 0 & -1 & 0 \\ \hline 0 & & 0 \end{array}\right).$$

Moreover, if A_1, A_2 and A_3 are $(n \times n)$-symmetric matrices and if

$$N(A_{\alpha} A_{\beta} - A_{\beta} A_{\alpha}) = 2N(A_{\alpha}) \cdot N(A_{\beta}) \qquad 1 \leqq \alpha, \beta \leqq 3,$$

then at least one of the matrices A_{α} must be zero.

Proof. We may assume that B is diagonal and we denote by $b_1, \ldots, b_n$ the diagonal entries in B. By a simple calculation we obtain

$$N(AB - BA) = \sum_{i \neq k} a_{ik}^2 \cdot (b_i - b_k)^2,$$

where $A = (a_{ij})$. Since $(b_i - b_k)^2 \leqq 2(b_i^2 + b_k^2)$, we obtain

$$\begin{aligned} N(AB - BA) &= \sum_{i \neq k} a_{ik}^2 (b_i - b_k)^2 \leqq 2 \sum_{i \neq k} a_{ik}^2 (b_i^2 + b_k^2) \\ &\leqq 2 \Big(\sum_{i,k} a_{ik}^2\Big)\Big(\sum_i b_i^2\Big) = 2N(A) \cdot N(B). \end{aligned} \tag{3.6}$$

Now, assume that A and B are nonzero matrices and that the equality holds. Then the equality must holds everywhere in (3.6). From the second equality in (3.6), it follows that

$$a_{11} = \ldots = a_{nn} = 0,$$

and that

$$b_i + b_k = 0 \quad \text{if} \quad a_{ik} \neq 0.$$

Without loss of generality, we may assume that $a_{12} \neq 0$. Then $b_1 = -b_2$. From the third equality, we now obtain

$$b_3 = \ldots = b_n = 0.$$

Since $B \neq 0$, we must have $b_1 = -b_2 \neq 0$ and we conclude that $a_{ik} = 0$ for $(i, k) \neq (1, 2)$. To prove the last statement, let A_1, A_2, A_3 be all nonzero symmetric matrices. From the second statement we have just proved, we see that one of these matrices can be transformed to a scalar multiple of $\tilde{A}$ as well as to a scalar multiple of $\tilde{B}$ by orthogonal matrices. But this is impossible since $\tilde{A}$ and $\tilde{B}$ are not orthogonally equivalent. q.e.d.

Applying Lemma 1 to (3.5), we obtain

$$\begin{aligned} -\sum h^\alpha_{ji} \cdot \Delta h^\alpha_{ij} &\leqq 2 \sum_{\alpha \neq \beta} N(H_\alpha) \cdot N(H_\beta) + \sum_\alpha S^2_\alpha - ncS \\ &= 2 \sum_{\alpha \neq \beta} S_\alpha S_\beta + \sum_\alpha S^2_\alpha - ncS \\ &= \Big(\sum_\alpha S_\alpha\Big)^2 + 2 \sum_{\alpha < \beta} S_\alpha S_\beta - ncS \\ &= p\sigma_1^2 + p(p-1)\sigma_2 - ncS, \end{aligned} \tag{3.7}$$

where

$$p\sigma_1 = \sum_\alpha S_\alpha = S, \quad \frac{p(p-1)}{2} \sigma_2 = \sum_{\alpha<\beta} S_\alpha S_\beta. \tag{3.8}$$

It can be easily seen that

$$p^2(p-1)(\sigma_1^2 - \sigma_2) = \sum_{\alpha<\beta} (S_\alpha - S_\beta)^2 \geqq 0, \tag{3.9}$$

and therefore

$$\begin{aligned} -\sum h^\alpha_{ij} \cdot \Delta h^\alpha_{ij} &\leqq p^2 \sigma_1^2 + p(p-1)\sigma_2 - ncS \\ &= (2p^2 - p)\sigma_1^2 - p(p-1)(\sigma_1^2 - \sigma_2) - ncS \\ &\leqq p(2p-1)\sigma_1^2 - ncS \\ &= \Big(2 - \frac{1}{p}\Big) S^2 - ncS. \end{aligned} \tag{3.10}$$

Theorem 1. *Let M be an n-dimensional compact oriented manifold which is minimally immersed in an $(n+p)$-dimensional space of constant curvature c. Then*

$$\int_M \Big[\Big(2 - \frac{1}{p}\Big) S - nc\Big] S *1 \geqq 0, \tag{3.11}$$

where $*1$ denotes the volume element of M.

Proof. This follows from (3.10) and the following lemma.

Lemma 2. *If M is an n-dimensional oriented compact manifold immersed in an $(n+p)$-dimensional riemannian manifold N, then*

$$\int_M \left(\sum h_{ij}^\alpha \cdot \Delta h_{ij}^\alpha\right) *1 = -\int_M \sum (h_{ijk}^\alpha)^2 *1 \leqq 0.$$

Proof of Lemma 2. We have

$$\tfrac{1}{2}\Delta\left(\sum (h_{ij}^\alpha)^2\right) = \sum (h_{ijk}^\alpha)^2 + \sum h_{ij}^\alpha \cdot \Delta h_{ij}^\alpha. \tag{3.12}$$

Integrating (3.12) over M and applying Green's theorem to the left hand side, we see that the integral of the left hand side and hence that of the right hand side also vanish. q.e.d.

Corollary. *Let M be a compact manifold minimally immersed in a space N of constant curvature c. If M is not totally geodesic and if $S \leqq nc\Big/\left(2-\frac{1}{p}\right)$ everywhere on M, then $S = nc\Big/\left(2-\frac{1}{p}\right)$.*

Assume that $S = \sum (h_{ij}^\alpha)^2$ is a constant. Whether M is compact or not, (3.12) implies

$$0 = \sum (h_{ijk}^\alpha)^2 + \sum h_{ij}^\alpha \cdot \Delta h_{ij}^\alpha.$$

This combined with (3.10) yields

$$\left[\left(2-\frac{1}{p}\right)S - nc\right] S \geqq \sum (h_{ijk}^\alpha)^2.$$

We may therefore conclude that if $S = nc\Big/\left(2-\frac{1}{p}\right)$, then $h_{ijk}^\alpha = 0$, i.e., the second fundamental form h_{ij}^α is parallel.

4. Minimal submanifolds of a unit sphere with $S = n\Big/\left(2-\frac{1}{p}\right)$.

Throughout this section, we shall assume that N is a space of constant curvature 1, that M is not totally geodesic and that

$$S = \sum (h_{ij}^\alpha)^2 = n\Big/\left(2-\frac{1}{p}\right).$$

At the end of § 3 we proved that $h_{ijk}^\alpha = 0$. Then $\Delta h_{ij}^\alpha = 0$, and the terms at the both ends of (3.10) vanish. It follows that all inequalities in (3.7), (3.9) and (3.10) are actually equalities. In deriving (3.7) from (3.5), we made use of the inequality $N(H_\alpha H_\beta - H_\beta H_\alpha) \leqq 2N(H_\alpha)\cdot N(H_\beta)$. Hence,

$$N(H_\alpha H_\beta - H_\beta H_\alpha) = 2N(H_\alpha)\cdot N(H_\beta) \qquad \alpha \neq \beta. \tag{4.1}$$

From (3.9) we obtain

$$p(p-1)(\sigma_1^2 - \sigma_2) = 0. \tag{4.2}$$

From (4.1) and Lemma 1, we conclude that at most two of the matrices H_α are nonzero, in which case they can be assumed to be scalar multiples of $\tilde{A}$ and $\tilde{B}$ in Lemma 1. We now consider the cases $p=1$ and $p\geqq 2$ separately.

Case $p=1$. *We set*

$$h_{ij}=h_{ij}^{n+1}.$$

We choose our frame field in such a way that

$$h_{ij}=0 \quad \text{for} \quad i\neq j, \tag{4.3}$$

and we set

$$h_i=h_{ii}.$$

Lemma 3. *After a suitable renumbering of the basis elements* $e_1, \ldots, e_n$, *we have*

(a) $h_1=\ldots=h_m=\lambda=$ *constant*,
$h_{m+1}=\ldots=h_n=\mu=$ *constant*, $(1<m<n)$,
$\lambda\mu=-1$,

(b) $\omega_j^i=0$ *for* $1\leqq i\leqq m$ *and* $m+1\leqq j\leqq n$.

Proof. Since $h_{ijk}=0$, setting $i=j$ in (2.10) and noting (4.3) we obtain

$$0=dh_i-2\sum h_{il}\,\omega_i^l=dh_i, \tag{4.4}$$

which shows that h_i is a constant. Since $h_{ijk}=0$ and $dh_{ij}=0$, (2.10) implies

$$0=\sum h_{il}\,\omega_j^l+\sum h_{lj}\,\omega_i^l=(h_i-h_j)\,\omega_j^i,$$

which shows that $\omega_j^i=0$ whenever $h_i\neq h_j$. Thus, if $h_i\neq h_j$, then

$$0=d\omega_j^i=-\sum\omega_k^i\wedge\omega_j^k-\omega_{n+1}^i\wedge\omega_j^{n+1}+\omega^i\wedge\omega^j.$$

The first sum of the equation above is zero, because $\omega_k^i\neq 0$ and $\omega_j^k\neq 0$ would imply $h_i=h_k=h_j$, contradicting the hypothesis. Hence,

$$\begin{aligned}0&=-\omega_{n+1}^i\wedge\omega_j^{n+1}+\omega^i\wedge\omega^j\\&=\sum h_{ik}\,h_{jl}\,\omega^k\wedge\omega^l+\omega^i\wedge\omega^j\\&=(h_i\,h_j+1)\,\omega^i\wedge\omega^j.\end{aligned}$$

This shows that if $h_i\neq h_j$, then $h_i\,h_j=-1$. Set $\lambda=h_1$. By renumbering the indices of $e_1,\ldots,e_n$, let $\lambda=h_1=\ldots=h_m$ and $\lambda\neq h_j$ for $j\geqq m+1$. Since $\sum h_i=0$ and M is not totally geodesic, not all $h_1,\ldots,h_n$ are equal to λ. Since $h_1\,h_j=-1$ for $j\geqq m+1$, we obtain $h_{m+1}=\ldots=h_n=-\dfrac{1}{\lambda}$. We set $\mu=-\dfrac{1}{\lambda}$. q.e.d.

5*

From (b) of Lemma 3, it follows that the two distributions defined by $\omega^1 = \ldots = \omega^m = 0$ and $\omega^{m+1} = \ldots = \omega^n = 0$ are both integrable and give a local decomposition of M. Then every point of M has a neighborhood U which is a riemannian product $V_1 \times V_2$ with $\dim V_1 = m$ and $\dim V_2 = n - m$. The curvatures of V_1 and V_2 are given by [see (2.7)]

$$R^i{}_{jkl} = (1+\lambda^2)(\delta_{ik}\delta_{jl} - \delta_{il}\delta_{jk}) \quad \text{for} \quad 1 \leqq i, j, k, l \leqq m; \tag{4.5}$$

$$R^i{}_{jkl} = (1+\mu^2)(\delta_{ik}\delta_{jl} - \delta_{il}\delta_{jk}) \quad \text{for} \quad m+1 \leqq i, j, k, l \leqq n. \tag{4.6}$$

If $m \geqq 2$ (resp. $n - m \geqq 2$), then V_1 (resp. V_2) is a space of constant curvature $1 + \lambda^2$ (resp. $1 + \mu^2$). If $m = 1$ (resp. $n - m = 1$), then V_1 (resp. V_2) is a curve and hence is also a space of constant curvature.

The minimality of the immersion implies

$$0 = \sum h_i = m\lambda + (n-m)\mu.$$

On the other hand, the assumption $S = n \Big/ \left(2 - \frac{1}{p}\right) = n$ implies

$$n = S = \sum h_i^2 = n\lambda^2 + (n-m)\mu^2.$$

These two relations together with $\lambda\mu = -1$ imply

$$\lambda = \sqrt{(n-m)/m}, \qquad \mu = -\sqrt{m/(n-m)}$$

or

$$\lambda = -\sqrt{(n-m)/m}, \qquad \mu = \sqrt{m/(n-m)}.$$

Replacing e_{n+1} by $-e_{n+1}$ if necessary, we may assume that $\lambda = \sqrt{(n-m)/m}$ and $\mu = -\sqrt{m/(n-m)}$. In summary, we have

Theorem 2. *Let M be a minimal hypersurface immersed in an $(n+1)$-dimensional space N of constant curvature 1 satisfying $S = n$. Then M is locally a riemannian direct product $M \supset U = V_1 \times V_2$ of spaces V_1 and V_2 of constant curvature, $\dim V_1 = m \geqq 1$ and $\dim V_2 = n - m \geqq 1$. With respect to an adapted frame field, the connection form (ω_B^A) of N, restricted to M, is given by*

$$\left(\begin{array}{ccc|ccc|c}
\omega_1^1 & \cdots & \omega_m^1 & & & & \lambda\omega^1 \\
\vdots & \vdots & \vdots & & 0 & & \vdots \\
\omega_1^m & \cdots & \omega_m^m & & & & \lambda\omega^n \\
\hline
 & & & \omega_{m+1}^{m+1} & \cdots & \omega_n^{m+1} & \mu\omega^{m+1} \\
 & 0 & & \vdots & \vdots & \vdots & \vdots \\
 & & & \omega_{m+1}^n & \cdots & \omega_n^n & \mu\omega^n \\
\hline
-\lambda\omega^1 & \cdots & -\lambda\omega^m & -\mu\omega^{m+1} & \cdots & -\mu\omega^n & 0
\end{array}\right) \tag{4.7}$$

where $\lambda = \sqrt{(n-m)/m}$ and $\mu = -\sqrt{n/(n-m)}$.

We consider now the submanifold $M_{m,\,n-m}$ of S^{n+1} defined in § 1 and shall prove that the connection form of S^{n+1}, restricted to $M_{m,\,n-m}$, is given by (4.7). Let $f_0, f_1, \ldots, f_m$ be an orthonormal frame field for $\boldsymbol{R}^{m+1}$ such that f_0 is normal to $S^m\left(\sqrt{\frac{m}{n}}\right)$ and let $\varphi^0, \varphi^1, \ldots, \varphi^{m+1}$ be the dual frame field. Similarly, for $S^{n-m}\left(\sqrt{\frac{n-m}{n}}\right)$ in $\boldsymbol{R}^{n-m+1}$, we choose an orthonormal frame field $f_{m+1}, \ldots, f_{n+1}$ such that f_{n+1} is normal to $S^{n-m}\left(\sqrt{\frac{n-m}{n}}\right)$ and let $\varphi^{m+1}, \ldots, \varphi^{n+1}$ be the dual frame field. Let $(\varphi^A_B)_{A,B=0,1,\cdots,n+1}$ be the connection form for $\boldsymbol{R}^{n+2}$ with respect to the dual frame field $(\varphi^A)_{A=0,1,\cdots,n+1}$. These forms, restricted to $M_{m,n-m}$, satisfy

$$\varphi^0 = \varphi^{n+1} = 0,$$

$$\varphi^0_i = -\varphi^i_0 = -\sqrt{\frac{n}{m}}\,\varphi^i \qquad i = 1, \ldots, m,$$

$$\varphi^j_{n+1} = -\varphi^{n+1}_j = \sqrt{\frac{n}{n-m}}\,\varphi^j, \quad j = m+1, \ldots, n,$$

$$\varphi^A_B = -\varphi^B_A = 0 \quad \text{for} \quad A = 0, 1, \ldots, m \quad \text{and} \quad B = m+1, \ldots, n+1.$$

The image of the imbedding $M_{m,n-m} \to \boldsymbol{R}^{n+2}$ lies in the unit sphere S^{n+1}. We take a new frame field $e_0, \ldots, e_{n+1}$ for $\boldsymbol{R}^{n+2}$ as follows:

$$e_0 = \sqrt{\frac{m}{n}}\,f_0 + \sqrt{\frac{n-m}{n}}\,f_{n+1},$$

$$e_i = f_i, \quad i = 1, \ldots, n,$$

$$e_{n+1} = \sqrt{\frac{n-m}{n}}\,f_0 - \sqrt{\frac{m}{n}}\,f_{n+1}.$$

Then e_0 is normal to S^{n+1} and e_{n+1} is normal to $M_{m,n-m}$. Let $\omega^0, \ldots, \omega^{n+1}$ be the dual frame field. Then

$$\omega^0 = \sqrt{\frac{m}{n}}\,\varphi^0 + \sqrt{\frac{n-m}{n}}\,\varphi^{n+1},$$

$$\omega^i = \varphi^i, \quad i = 1, \ldots, n,$$

$$\omega^{n+1} = \sqrt{\frac{n-m}{n}}\,\varphi^0 - \sqrt{\frac{m}{n}}\,\varphi^{n+1}.$$

The connection form $(\omega^A_B)_{A,B=0,1,\cdots,n}$ for $\boldsymbol{R}^{n+2}$ with respect to the dual frame field (ω_A) is then given by

$$\omega^0_j = -\omega^j_0 = \sqrt{\frac{m}{n}}\,\varphi^0_j + \sqrt{\frac{n-m}{n}}\,\varphi^{n+1}_j \quad \text{for} \quad j = 1, \ldots, n,$$

$$\omega^0_{n+1} = -\omega^{n+1}_0 = -\varphi^0_{n+1},$$

$$\omega^i_j = \varphi^i_j \quad \text{for} \quad i, j = 1, \ldots, n,$$

$$\omega^i_{n+1} = -\omega^{n+1}_i = \sqrt{\frac{n-m}{n}}\,\varphi^i_0 - \sqrt{\frac{m}{n}}\,\varphi^i_{n+1}.$$

We restrict these forms to $M_{m,\,n-m}$. Then by a straightforward calculation we can verify easily that the connection form $(\omega^A_B)_{A,\,B=1,\cdots,\,n+1}$ of S^{n+1}, restricted to $M_{m,\,n-m}$ coincides with the form in (4.7). We may therefore conclude that a minimal hypersurface of S^{n+1} satisfying $S=n$ coincides locally with $M_{m,\,n-m}$. If it is compact, then it coincides with $M_{m,\,n-m}$.

Case $p \geqq 2$. In this case, (4.2) implies

$$\sigma_1^2 = \sigma_2.$$

We know that at most two of H_α, $\alpha = n+1, \dots, n+p$, are different from zero. Assume that only one of them, say H_α, is different from zero. Then we have $\sigma_1 = \frac{1}{p} S_\alpha$ and $\sigma_2 = 0$, in contradiction to $\sigma_1^2 = \sigma_2$. We may therefore assume that

$$H_{n+1} = \lambda \tilde{A}, \quad H_{n+2} = \mu \tilde{B}, \quad \lambda, \mu \neq 0,$$
$$H_\alpha = 0 \quad \text{for} \quad \alpha \geqq n+3,$$

where $\tilde{A}$ and $\tilde{B}$ are defined in Lemma 1. In other words,

$$\omega_1^{n+1} = \lambda\omega^2, \quad \omega_2^{n+1} = \lambda\omega^1, \quad \omega_i^{n+1} = 0 \quad \text{for} \quad i = 3, \dots, n,$$
$$\omega_1^{n+2} = \mu\omega^1, \quad \omega_2^{n+2} = -\mu\omega^2, \; \omega_i^{n+2} = 0 \quad \text{for} \quad i = 3, \dots, n,$$
$$\omega_i^\alpha = 0 \quad \text{for} \quad \alpha = n+3, \dots, n+p \quad \text{and} \quad i = 1, \dots, n.$$

Since $h^\alpha_{ijk} = 0$, we have [see (2.10)]

$$d h^\alpha_{ij} = \sum h^\alpha_{ik}\, \omega^k_j + \sum h^\alpha_{kj}\, \omega^k_i - \sum h^\beta_{ij}\, \omega^\alpha_\beta. \tag{4.8}$$

Setting $\alpha = n+1$, $i = 1$ and $j = 2$, we see that $d\lambda = d h^{n+1}_{12} = 0$, i.e., λ is constant. Setting $\alpha = n+1$, $i = 1$ and $j \geqq 3$, we see that

$$\omega^2_j = 0 \quad \text{for} \quad j \geqq 3. \tag{4.9}$$

Setting $\alpha = n+1$, $i = 2$ and $j \geqq 3$, we see that

$$\omega^1_j = 0 \quad \text{for} \quad j \geqq 3. \tag{4.10}$$

Similarly, setting $\alpha = n+2$, and $i = j = 1$, we see that μ is a constant. From (4.8), (4.9) and (4.10), it follows that if $j \geqq 3$, then

$$0 = d\omega^1_j = -\sum \omega^1_k \wedge \omega^k_j + \omega^1 \wedge \omega^j = \omega^1 \wedge \omega^j.$$

Since $\omega^1, \dots, \omega^n$ are orthonormal, $\omega^1 \wedge \omega^j = 0$ implies $\omega^j = 0$ for $j \geqq 3$. This shows that $\dim M = 2$. From

$$p\,\sigma_1 = 2(\lambda^2 + \mu^2) \quad \text{and} \quad p(p-1)\,\sigma^2 = 8\lambda^2\mu^2,$$

we obtain

$$p^2(p-1)(\sigma_1^2-\sigma_2) = 4[(p-1)\lambda^4 - 2\lambda^2\mu^2 + (p-1)\mu^4].$$

Since the left hand side is zero by (4.2), the discriminant of the right hand side must be non-negative, i.e.,

$$1-(p-1)^2 \geqq 0.$$

Since $p \geqq 2$, p must be 2. Hence, dim $N=4$. From $0=\lambda^4-2\lambda^2\mu^2+\mu^4$, it follows that $\lambda^2=\mu^2$. Since $\frac{4}{3}=S=4\lambda^2$, we have

$$\lambda^2=\mu^2=\tfrac{1}{3}. \tag{4.11}$$

Replacing e_3 by $-e_3$ and e_4 by $-e_4$ if necessary, we may assume that $-\lambda=\mu=\sqrt{1/3}$.

Setting $\alpha=3$ and $i=j=1$, we obtain

$$\omega_4^3 = \frac{2\lambda}{\mu}\,\omega_1^2 = -2\omega_1^2. \tag{4.12}$$

The curvature of M is given by

$$\Omega_2^1 = \omega^1\wedge\omega^2 + \omega_1^3\wedge\omega_2^3 + \omega_1^4\wedge\omega_2^4 = (1-\lambda^2-\mu^2)\,\omega^1\wedge\omega^2 = \tfrac{1}{3}\omega^1\wedge\omega^2. \tag{4.13}$$

In summary, we have

Theorem 3. *Let M be an n-dimensional manifold immersed minimally in an $(n+p)$-dimensional space N of constant curvature 1 satisfying $S = n\Big/\left(2-\frac{1}{p}\right)$. If $p\geqq 2$, then $n=p=2$. With respect to a an adapted dual orthonormal frame field $\omega^1, \omega^2, \omega^3, \omega^4$, the connection form (ω_B^A) of N, restricted to M, is given by*

$$\begin{pmatrix} 0 & \omega_2^1 & \mu\omega^2 & -\mu\omega^1 \\ \omega_1^2 & 0 & \mu\omega^1 & \mu\omega^2 \\ \lambda\omega^2 & \lambda\omega^1 & 0 & 2\omega_2^1 \\ -\lambda\omega^1 & \lambda\omega^2 & 2\omega_1^2 & 0 \end{pmatrix}, \qquad -\lambda=\mu=\sqrt{\frac{1}{3}} \tag{4.14}$$

We consider now the Veronese surface defined in § 1. We shall compute its structure equations by group theoretic means. Let

$$u^1=\frac{1}{\sqrt{3}}\,yz, \quad u^2=\frac{1}{\sqrt{3}}\,zx, \quad u^3=\frac{1}{\sqrt{3}}\,xy, \quad u^4=\frac{1}{2\sqrt{3}}\,(x^2-y^2),$$

$$u^5=\frac{1}{6}\,(x^2+y^2-2z^2)$$

be as in § 1. These equations define an immersion of $S^2(\sqrt{3})$ into S^4 and induce an action of $SO(3)$ on S^4 so that the immersion $S^2(\sqrt{3})\to S^4$

is equivariant. In other words, we obtain a representation of $SO(3)$ into $SO(5)$. This induces a representation of the Lie algebra $so(3)$ into the Lie algebra $so(5)$. By a straightforward, simple calculation, we see that this representation maps a matrix of the form

$$\begin{pmatrix} 0 & -\alpha & -\gamma \\ \alpha & 0 & -\beta \\ \gamma & \beta & 0 \end{pmatrix} \in so(3)$$

into a matrix of the form

$$\begin{pmatrix} 0 & 0 & \gamma & -\beta & \sqrt{3}\beta \\ -\alpha & 3 & \beta & \gamma & \sqrt{3}\gamma \\ -\gamma & -\beta & 0 & 2\alpha & 0 \\ \beta & -\gamma & -2\alpha & 0 & 0 \\ -\sqrt{3}\beta & -\sqrt{3}\gamma & 0 & 0 & 0 \end{pmatrix} \in so(5).$$

Let (ω_B^A) be the Maurer-Cartan form for $SO(5)$ and set

$$\omega^i = \omega_5^i \qquad i = 1, \ldots, 4.$$

Then the restriction of (ω_B^A) to the image of $SO(3)$ in $SO(5)$ is given by

$$\begin{pmatrix} 0 & \omega_2^1 & \mu\omega^2 & -\mu\omega^1 & \omega^1 \\ \omega_1^2 & 0 & \mu\omega^1 & \mu\omega^2 & \omega^2 \\ \lambda\omega^2 & \lambda\omega^1 & 0 & 2\omega_2^1 & 0 \\ -\lambda\omega^1 & \lambda\omega^2 & 2\omega_1^2 & 0 & 0 \\ -\omega^1 & -\omega^2 & 0 & 0 & 0 \end{pmatrix}, \qquad -\lambda = \mu = \sqrt{\frac{1}{3}}.$$

Comparing this with (4.14), we may conclude that a minimal surface in S^4 satisfying $S = \frac{4}{3}$ coincides locally with the Veronese surface. If it is compact, it coincides with the Veronese surface. This completes the proof of the main theorem.

5. Related examples

Example 1.

$$S^m \times S^q \to S^{m+q+mq}$$

Let

$$S^m = \{(x_0, x_1, \ldots, x_m) \in \boldsymbol{R}^{m+1};\ \sum x_i^2 = 1\},$$
$$S^q = \{(y_0, y_1, \ldots, y_q) \in R^{q+1};\ \sum y_i^2 = 1\},$$
$$S^{m+q+mq} = \{(u_{ij});\ i = 0, 1, \ldots, m;\ j = 0, 1, \ldots, q;\ \sum u_{ij}{}^2 = 1\}.$$

Then the mapping $S^m \times S^q \to S^{m+q+mq}$ defined by $u_{ij} = x_i y_j$ is an isometric immersion. Two points (x, y) and $(-x, -y)$ of $S^m \times S^q$ are mapped into the same point. We have

$$R = \text{scalar curvature of } S^m \times S^q = m(m-1) + q(q-1),$$
$$S = (m+q)(m+q-1) - R = 2mq.$$

On the other hand, if we denote $m+q$ by n and the codimension mq by p, then

$$n\Big/\left(2-\frac{1}{p}\right) = mq(m+q)/(2mq-1).$$

Consider the case $m=q=1$. Then $S = 2 = n\Big/\left(2-\frac{1}{p}\right)$, and this minimal immersion satisfies the assumption in our main theorem. We shall show that this immersion of $S^1 \times S^1 \to S^3$ is a double covering of the immersion of $M_{1,1} = S^1\left(\frac{1}{\sqrt{2}}\right) \times S^1\left(\frac{1}{\sqrt{2}}\right)$ into S^3 defined in § 1. The immersion

$$(x_0, x_1, y_0, y_1) \in S^1 \times S^1 \to (v_0, v_1 \cdot w_0, w_1) \in S^1\left(\frac{1}{\sqrt{2}}\right) \times S^1\left(\frac{1}{\sqrt{2}}\right) \in S^3$$

defined by

$$v_0 = \frac{1}{\sqrt{2}}(x_0 y_0 - x_1 y_1), \qquad v_1 = \frac{1}{\sqrt{2}}(x_1 y_0 + x_0 y_1),$$
$$w_0 = \frac{1}{\sqrt{2}}(x_0 y_0 + x_1 y_1), \qquad w_1 = \frac{1}{\sqrt{2}}(x_1 y_0 - x_0 y_1),$$

differs from the immersion

$$(x_0, x_1, y_0, y_1) \in S^1 \times S^1 \to (x_0 y_0, x_0 y_1, x_1 y_0, x_1 y_1) \in S^3$$

by a rigid motion of S^3.

Example 2. $S^n\left(\sqrt{\frac{2(n+1)}{n}}\right) \to S^{n+p}$ with $p = \frac{1}{2}(n-1)(n+2)$.

Let

$$S^n\left(\sqrt{\frac{2(n+1)}{n}}\right) = \left\{(x_0, x_1, \ldots, x_n) \in \mathbf{R}^{n+1}; \sum x_i^2 = \frac{2(n+1)}{n}\right\}.$$

Let E be the space of $(n+1) \times (n+1)$ symmetric matrices (u_{ij}), $(i, j = 1, \ldots, n)$, such that $\sum u_{ii} = 0$; it is a vector space of dimension $\frac{1}{2}n(n+3)$. We define a norm in E by $\|(u_{ij})\|^2 = \sum u_{ij}{}^2$. Let S^{n+p} with $p = \frac{1}{2}(n-1)(n+2)$ be the unit hypersphere in E. The mapping of $S^n\left(\sqrt{\frac{2(n+1)}{n}}\right)$ into S^{n+p} defined by

$$u_{ij} = \frac{1}{2}\sqrt{\frac{n}{n+1}}\left(x_i x_j - \frac{2}{n}\delta_{ij}\right)$$

is an isometric minimal immersion. (Actually, this gives an imbedding of the real projective space of n-dimension into S^{n+p}.) We have

$$R = \text{scalar curvature of } S^n\left(\sqrt{\frac{2(n+1)}{n}}\right) = n^2(n-1)/2(n+1),$$

$$S = n(n-1) - R = n(n-1)(n+2)/2(n+1).$$

On the other hand,

$$n\Big/\left(2-\frac{1}{p}\right) = n(n-1)(n+2)/2(n^2+n-3).$$

For $n=2$, we recover the Veronese surface.

Example 3. $S^2(\sqrt{6}) \to S^6$.

Making use of harmonic polynomials of degree 3, we consider the following minimal immersion of $S^2(\sqrt{6})$ into S^3 defined by

$$u_0 = \frac{\sqrt{6}}{72} z(-3x^2 - 3y^2 + 2z^2), \qquad u_1 = \frac{1}{24} x(-x^2 - y^2 + 4z^2),$$

$$u_2 = \frac{\sqrt{10}}{24} z(x^2 - y^2), \qquad u_3 = \frac{\sqrt{15}}{72} x(x^2 - 3y^2),$$

$$u_4 = \frac{1}{24} y(-x^2 - y^2 + 4z^2), \qquad u_5 = \frac{\sqrt{10}}{12} xyz,$$

$$u_6 = \frac{\sqrt{15}}{72} y(3x^2 - y^2).$$

We have

$$R = \text{scalar curvature of } S^2(\sqrt{6}) = \tfrac{1}{3},$$

$$S = 2 - R = \tfrac{5}{3},$$

whereas

$$n\Big/\left(2-\frac{1}{p}\right) = \frac{8}{7}.$$

For each positive integer k, the space of harmonic polynomials of degree k in variables x, y, z is a vector space of dimension $2k+1$. Introducing an inner product in this vector space in a well known manner, we get a minimal immersion of a 2-sphere into the hypersphere S^{2k} in a natural manner. The case $k=2$ gives the Veronese surface. The case $k=3$ was described above. Generally, for every positive integer k, we have an isometric minimal immersion

$$S^2\left(\sqrt{\frac{k(k+1)}{2}}\right) \to S^{2k},$$

for which $S = 2 - \frac{4}{k(k+1)}$. For a systematic study of minimal immersions of 2-spheres obtained in this manner, see BORUVKA [1] and a forthcoming paper of DO CARMO and WALLACH.

Example 4. $S^{m_1}\left(\sqrt{\frac{m_1}{n}}\right)\times\cdots\times S^{m_k}\left(\sqrt{\frac{m_k}{n}}\right)\to S^{n+k-1},\ n=\sum_{i=1}^{k} m_i.$

We can generalize the construction of $M_{m,\,n-m}$ as follows. Let $m_1,\ldots,m_k$ be positive integers and $n=m_1+\cdots+m_k$. Let x_i be a point of $S^{m_i}\left(\sqrt{\frac{m_i}{n}}\right)$, i.e., a vector of length $\sqrt{\frac{m_i}{n}}$ in $\boldsymbol{R}^{m_i+1}$. Then $(x_1,\ldots,x_k)$ is a unit vector in $\mathbb{R}^{n+k}$. This defines a minimal immersion of $M_{m_1,\ldots,m_k}=\prod S^{m_i}\left(\sqrt{\frac{m_i}{n}}\right)$ into S^{n+k-1}. We have

$$R = \text{scalar curvature of } M_{m_1,\ldots,m_k}=(n-k)\,n,$$
$$S=n(n-1)-R=(k-1)\,n,$$
$$n\Big/\left(2-\frac{1}{p}\right)=(k-1)\,n/(2k-3).$$

6. Some questions

The above discussions seem to show the interest of the study of compact minimal submanifolds on the sphere with $S=$ constant. With fixed n and p the question naturally arises as to the possible values for S. We proved in the above that S does not take values in the open interval $\left(0,\ n\Big/\left(2-\frac{1}{p}\right)\right)$. It is plausible that the set of values for S is discrete, at least for S not arbitrarily large. If this is the case, an estimate of the value for S next to $n\Big/\left(2-\frac{1}{p}\right)$ should be of interest. This problem can be restricted by imposing further conditions on M, such as M be topologically or metrically a sphere.

Another natural question is that of uniqueness. At least for compact minimal hypersurfaces (codim 1) it seems likely that the values of S should determine the hypersurface up to a rigid motion in the ambient sphere S^{n+1}.

References

1. BORUVKA, O.: Sur les surfaces representées par les fonctions sphériques de première espèce. J. Math. pure et appl. **12**, 337—383 (1933).
2. LAWSON, B.: Local rigidity theorems for minimal hypersurfaces. Ann. of Math. **89**, 187—197 (1969).
3. SIMONS, J.: Minimal varieties in riemannian manifolds. Ann. of Math. **88**, 62—105 (1968).

Some Formulas Related to Complex Transgression

RAOUL BOTT and SHIING S. CHERN[1]

1. Introduction

Let X be a complex manifold of complex dimension n and $\pi: E \to X$ a holomorphic vector bundle whose fiber dimension is also n. On E we introduce a positive definite hermitian norm N and denote by $B^*(E) = \{e \in E \mid 0 < N(e)\}$ the subset of non-zero vectors of E. Let $c_n(E)$ be the Chern form of the hermitian structure (to be described again below). In [1] we proved the theorem:

There exists a real-valued form ρ of type $(n-1, n-1)$ on $B^*(E)$ such that

(1) $$\pi^* c_n(E) = \frac{d d^c}{4\pi} \rho.$$

In this paper we wish to give a proof of (1) based on an explicit construction. We will follow and extend a formalism developed in the real case by H. Flanders [3]. It is possible that this formalism will be useful for later purposes.

2. Multilinear Algebra Over a Complex Vector Space

Let V be a complex vector space of dimension n. Let $\bar{V}$ be its conjugate space, i.e., another copy of V with the complex structure defined by $-i$. The identity mapping $j: V \to \bar{V}$ will be denoted by $j(x) = \bar{x}, x \in V$, and this mapping is antilinear:

(2) $$\overline{\alpha x + \beta y} = \bar{\alpha}\bar{x} + \bar{\beta}\bar{y}, \quad x, y \in V, \quad \alpha, \beta \in C.$$

We will denote j^{-1} also by a bar, so that $\bar{\bar{x}} = x$.

The exterior algebra $\Lambda(V \oplus \bar{V})$ has a bigrading:

(3) $$\Lambda(V \oplus \bar{V}) = \sum \Lambda^{pq}(V \oplus \bar{V}).$$

---

[1] Work done under partial support of the National Science Foundation.

In fact, if $e_1, \ldots, e_n$ form a basis of V, then $\bar{e}_1, \ldots, \bar{e}_n$ form a basis of $\bar{V}$ and an element of $\Lambda^{pq}(V \oplus \bar{V})$ is a sum of terms of the form

$$\alpha e_{i_1} \wedge \cdots \wedge e_{i_p} \wedge \bar{e}_{j_1} \wedge \cdots \wedge \bar{e}_{j_q}, \qquad \alpha \in C.$$

The bar operation acts on $\Lambda(V \oplus \bar{V})$ and is an anti-linear map of $\Lambda(V \oplus \bar{V})$ onto itself. An element $\xi \in \Lambda(V \oplus \bar{V})$ is called *real* if $\bar{\xi} = \xi$. The exterior algebra $\Lambda(V_R)$ of the underlying real vector space V_R of V can be identified with the real subalgebra of real elements of $\Lambda(V \oplus \bar{V})$. In particular, V_R itself can be identified with the real vector space of the real elements of $V \oplus \bar{V}$.

The space $\Lambda^{nn}(V \oplus \bar{V})$ is one-dimensional, and it has a generator $i^n e_1 \wedge \bar{e}_1 \wedge \cdots \wedge e_n \wedge \bar{e}_n$ defined up to a positive factor. It therefore makes sense to talk about the sign of a non-zero real element of $\Lambda^{nn}(V \oplus \bar{V})$.

If W is a second vector space, we have a direct sum decomposition

(4) $$\Lambda(V \oplus \bar{V}) \otimes \Lambda(W \oplus \bar{W}) = \sum \Lambda^{pq}_{rs}$$

where each summand has four degrees, the superscripts being the bidegrees relative to the first factor and the subscripts those to the second factor. If

$$\xi, \xi' \in \Lambda(V \oplus \bar{V}), \qquad \eta, \eta' \in \Lambda(W \oplus \bar{W}),$$

we define a multiplication by

(5) $$(\xi \otimes \eta) \wedge (\xi' \otimes \eta') = (\xi \wedge \xi') \otimes (\eta \wedge \eta').$$

By linearity this is extended to a multiplication in $\Lambda(V \oplus \bar{V}) \otimes (W \oplus \bar{W})$. Having several kinds of products, we will, in later formulas, frequently drop all multiplication signs whenever there is no danger of confusion. We observe that, if

(6) $$\xi \in \Lambda^{pq}_{rs}, \qquad \xi' \in \Lambda^{p'q'}_{r's'},$$

then

(7) $$\xi \xi' = (-1)^k \xi' \xi$$

where

(7a) $$k = (p+q)(p'+q') + (r+s)(r'+s').$$

3. Affine Connection

Let X be a complex manifold of complex dimension m and $\pi: E \to X$ be a holomorphic vector bundle of fiber dimension n. At a point $x \in X$ let E_x be the fiber of E and T_x^* the cotangent space of X. We consider the C^∞-bundle

(8) $$\bigcup_{x\in X} \Lambda(E_x \oplus \bar{E}_x) \otimes \Lambda(T_x^* \oplus \bar{T}_x^*) \to X$$

and the bundles

(9) $$\bigcup_{x\in X} \Lambda_{rs,x}^{pq} \to X,$$

whose fibers are respectively the summands in the decompositions of the fibers of (8) according to (4) (with V and W replaced by E_x and T_x^* respectively). Let $\tilde{\Lambda}_{rs}^{pq}$ be the space of C^∞ sections of the bundle (9). Then $\tilde{\Lambda}_{00}^{00}$, which we will denote by $A^0(X)$ or simply by A^0, is the ring of C^∞ complex-valued functions on X and $\tilde{\Lambda}_{rs}^{pq}$ is a module over A^0. More generally, $\tilde{\Lambda}_{rs}^{00}$ is the space of C^∞-forms of type (r,s) and exterior differentiation in X defines the mappings

(10) $$d': \tilde{\Lambda}_{rs}^{00} \to \tilde{\Lambda}_{r+1,s}^{00}; \quad d'': \tilde{\Lambda}_{rs}^{00} \to \tilde{\Lambda}_{r,s+1}^{00}.$$

An affine connection in the bundle E is a structure which allows us to extend the operators in (10) to the more general spaces $\tilde{\Lambda}_{rs}^{pq}$.

An affine connection in E is defined to be the operators

(11) $$d': \tilde{\Lambda}_{00}^{10} \to \tilde{\Lambda}_{10}^{10}; \quad d'': \tilde{\Lambda}_{00}^{10} \to \tilde{\Lambda}_{01}^{10},$$

(12) $$d = d' + d'',$$

which have the properties:

(13) $$\begin{aligned} &d(v+w) = dv + dw, \\ &d(fv) = df \otimes v + f\,dv, \quad v,w \in \tilde{\Lambda}_{00}^{10}, \quad f \in A^0. \end{aligned}$$

An element $v \in \tilde{\Lambda}_{00}^{10}$ (i.e., a vector field) is called holomorphic if the corresponding mapping $v: X \to E$ is a holomorphic mapping. The connection d is said to be of type (1,0), if $d''v = 0$ for holomorphic v.

We have the following theorem:

Theorem. *If d is an affine connection in E, there exists a unique collection of operators*

(14) $$d': \tilde{\Lambda}_{rs}^{pq} \to \tilde{\Lambda}_{r+1,s}^{pq}; \quad d'': \tilde{\Lambda}_{rs}^{pq} \to \tilde{\Lambda}_{r,s+1}^{pq},$$

(15) $$d = d' + d'',$$

which have the following properties:

α) $$d(\xi + \eta) = d\xi + d\eta, \qquad \xi, \eta \in \tilde{\Lambda}_{rs}^{pq},$$

β) $$d(\xi\zeta) = d\xi\,\zeta + (-1)^{r+s} \xi\, d\zeta, \qquad \zeta \in \tilde{\Lambda}_{r's'}^{p'q'},$$

γ) *d commutes with the bar operation and coincides with the given affine connection on $\tilde{\Lambda}_{00}^{10}$ and with the exterior differentiation on $\tilde{\Lambda}_{rs}^{00}$.*

The proof of this theorem parallels that of Theorem 7.1 in [3] and is straightforward. We will omit it here.

The local properties of an affine connection arise most readily from the consideration of a frame field. By this we mean elements $e_i \in \tilde{\Lambda}^{10}_{00}$, $1 \leq i \leq n$, defined locally in a neighborhood U of X, such that $e_1, \ldots, e_n \neq 0$. (Notice that by our convention the latter is the exterior product.) Then we can write

$$de_i = \sum_j \omega_i^j e_j, \quad 1 \leq i, j \leq n, \tag{16}$$

where ω_i^j are complex-valued one-forms in U. Differentiating (16) according to our Theorem, we get

$$d^2 e_i = \sum_j \Omega_i^j e_j, \tag{17}$$

where

$$\Omega_i^j = d\omega_i^j - \sum_k \omega_i^k \wedge \omega_k^j, \quad 1 \leq i, j, k \leq n. \tag{18}$$

The matrices

$$\omega = (\omega_i^j), \quad \Omega = (\Omega_i^j) \tag{19}$$

are called the connection and curvature matrices respectively (relative to the frame field). It can be verified (cf. [2]) that under a change of the frame field Ω goes into a similar matrix. Therefore the $2n$-form $\left(\frac{\sqrt{-1}}{2\pi}\right)^n \det \Omega$ is globally defined in X; it is called the n-th Chern form of the affine connection.

Since E is a holomorphic vector bundle, the bundle space E is a complex manifold and the projection $\pi: E \to X$ is a holomorphic mapping. Over E we have the induced holomorphic bundle $\pi^* E \to E$ and the above results apply. We denote by (x, v), $v \in E_x$, a point of E and by $\tau^*_{(x,v)}$ the cotangent space to E at (x, v). Analogous to (8) and (9) consider the bundles

$$\bigcup_{(x,v)\in E} \Lambda(E_x \oplus \bar{E}_x) \otimes \Lambda(\tau^*_{(x,v)} \oplus \bar{\tau}^*_{(x,v)}) \to E \tag{20}$$

and

$$\bigcup_{(x,v)\in E} \Lambda^{pq}_{rs,(x,v)} \to E. \tag{21}$$

Let $\tilde{M}^{pq}_{rs}$ be the space of C^∞ sections of the bundle (21). The projection π induces the mapping

$$\pi^*: \tilde{\Lambda}^{pq}_{rs} \to \tilde{M}^{pq}_{rs}, \tag{22}$$

which permits us to identify $\tilde{\Lambda}^{pq}_{rs}$ with a subset of $\tilde{M}^{pq}_{rs}$.

The affine connection in E induces an affine connection in π^*E and gives rise to the operators

$$d': \tilde{M}^{pq}_{rs} \to \tilde{M}^{pq}_{r+1,s}; \quad d'': \tilde{M}^{pq}_{rs} \to \tilde{M}^{pq}_{r,s+1}, \tag{23}$$

with $d = d' + d''$.

By assigning v to $(x,v) \in E$, $v \in E_x$, we can consider v as an element of $\tilde{M}^{10}_{00}$. Similarly, we have $\bar{v} \in \tilde{M}^{01}_{00}$. If the connection is of type (1,0), we have $dv \in \tilde{M}^{10}_{10}$ and $d\bar{v} \in \tilde{M}^{01}_{01}$. Then

$$\alpha = dv\, d\bar{v} = \bar{\alpha} \tag{24}$$

is real and belong to $\tilde{M}^{11}_{11}$; α will play an important rôle later on.

4. Hermitian Structure

Suppose an hermitian structure be given in our holomorphic vector bundle E, i.e., a scalar product $\langle v, w \rangle_x$, $x \in X$, $v, w \in E_x$, which is positive definite hermitian and is C^∞ in x. Let e_i, $1 \leqq i \leqq n$, be a frame field and let

$$h_{ij} = \langle e_i, e_j \rangle, \quad 1 \leqq i, j \leqq n. \tag{25}$$

In order to simplify the formulas which follow, we introduce the matrices

$$H = (h_{ij}), \quad H^{-1} = (h^{ij}), \tag{26}$$

so that

$$\sum_i h_{ij} h^{jk} = \delta_i^k, \quad 1 \leqq i, j, k \leqq n. \tag{27}$$

The hermitian structure defines uniquely an admissible connection of type (1,0), whose connection matrix relative to our frame field is given by (cf. [2, p. 45])

$$\omega = d'H \cdot H^{-1}. \tag{28}$$

It follows from (18) that the curvature matrix is

$$\Omega = -d'd''H \cdot H^{-1} + d'H \cdot H^{-1} \wedge d''H \cdot H^{-1}. \tag{29}$$

Thus the elements of the matrix Ω are two-forms of type (1,1) and the n-th Chern form is a form of type (n,n).

Algebraically the curvature is perhaps best described by the element

$$K = \bar{K} = \sum_{1 \leqq i,j,k \leqq n} \bar{e}_i e_j h^{ik} \Omega^j_k \in \tilde{\Lambda}^{11}_{11} \tag{30}$$

which is independent of the choice of the frame field e_i and is globally defined in X. It can be verified that the Bianchi identity is equivalent to the relation

(31) $$dK=0.$$

5. The Recursive Formulas

We restrict ourselves to the submanifold $B^*(E)$, i.e., the bundle E with the zero section omitted. We introduce the elements

(32)
$$\begin{aligned} y_k &= -\bar{y}_k = v\bar{v}\alpha^{k-1}K^{n-k}\in \tilde{M}^{nn}_{n-1,n-1}, \\ t_k &= \bar{t}_k = (v\,d\bar{v}+\bar{v}\,dv)\alpha^{k-1}K^{n-k}\in \tilde{M}^{nn}_{n,n-1}\oplus \tilde{M}^{nn}_{n-1,n}, && 1\leqq k\leqq n, \\ w_k &= \bar{w}_k = \alpha^k K^{n-k}\in \tilde{M}^{nn}_{nn}, && 0\leqq k\leqq n \end{aligned}$$

and

(33)
$$\left.\begin{aligned} & t'_k=|v|^{-2k}t_k, \quad y'_k=|v|^{-2k}y_k, \quad w'_k=|v|^{-2k}w_k, \\ & t''_k=t'_k+2k\,y'_k(d'-d'')\log|v|, \end{aligned}\right\} 1\leqq k\leqq n,$$
$$w''_k=w'_k+k\,t'_k\,d\log|v|, \qquad 0\leqq k\leqq n,$$

where $|v|$ is the norm of v. We wish to establish the following recursive formulas

(34) $$(d'-d'')y'_k=-t''_k-\frac{k-1}{n-k+1}t''_{k-1},$$

(35) $$\frac{1}{2}dt'_k=w''_k+\frac{k}{n-k+1}w''_{k-1}.$$

The method of proof is to expand the quantities introduced above in terms of $v,\bar{v}$. The calculations being local, we choose a frame field $e_1,\dots,e_n$ such that $e_n=v$ and such that the matrices in (26) take the forms

(36) $$H=\begin{pmatrix} * & 0 \\ \underbrace{0}_{n-1} & |v|^2 \end{pmatrix}, \quad H^{-1}=\begin{pmatrix} * & 0 \\ \underbrace{0}_{n-1} & |v|^{-2} \end{pmatrix}.$$

By (28) we have then

(37) $$\omega^n_n=d'|v|^2(|v|^{-2})=2d'\log|v|,$$

and this differential form will be denoted for simplicity by θ. Equation (16) gives

(38) $$dv = de_n = \sum_{1 \leqq i \leqq n} \omega_n^i e_i = \beta + v\theta,$$

where β stands for the terms not involving v. It follows that

$$\alpha = dv\,d\bar{v} = \alpha_1 + v\theta\bar{\beta} + \bar{v}\bar{\theta}\beta + v\bar{v}\theta\bar{\theta}.$$

By induction we have the formula

(39) $$\alpha^{k-1} = \alpha_1^{k-1} + (k-1)\alpha_1^{k-2}(v\theta\bar{\beta} + \bar{v}\bar{\theta}\beta) + (k-1)^2 \alpha_1^{k-2} v\bar{v}\theta\bar{\theta}_1$$

where α_1 does not involve $v, \bar{v}$.

Next we consider the curvature matrix $\Omega = (\Omega_i^j)$. It will be convenient to introduce the two-forms

(40) $$\Omega^{ij} = \sum_k h^{ik} \Omega_k^j, \quad 1 \leqq i, j, k \leqq n,$$

so that

(40a) $$\Omega_k^j = \sum_i h_{ki} \Omega^{ij},$$

and, because of the special form of our matrix H, we have

(40b) $$\Omega_n^j = |v|^2 \Omega^{nj}.$$

By setting

(41) $$\varphi = \Omega^{nn}, \quad \gamma = \sum_{1 \leqq \lambda \leqq n-1} \Omega^{n\lambda} e_\lambda,$$

we get, from (17),

(42) $$d^2 v = |v|^2 (\gamma + v\varphi).$$

Observe that the matrix $H^{-1}\Omega$ is skew-hermitian ([2], formula (5.56)), so that we have

(43) $$\varphi + \bar{\varphi} = 0.$$

With these new notations we write the element K in (30) as

(44) $$K = \sum_{i,j} \bar{e}_i e_j \Omega^{ij} = K_1 + v\bar{\gamma} + \bar{v}\gamma - v\bar{v}\varphi,$$

where K_1 does not involve $v, \bar{v}$. By induction we have

(45) $$\begin{aligned} K^{n-k} = K_1^{n-k} &+ (n-k)(v\bar{\gamma} + \bar{v}\gamma) K_1^{n-k-1} \\ &+ (n-k) v\bar{v} K_1^{n-k-2} \{-K_1\varphi + (n-k-1)\gamma\bar{\gamma}\}. \end{aligned}$$

From (39) and (45) we get expansions for t_k and w_k, which are

$$t_k = k(\bar{\theta}-\theta)y_k + (n-k)v\bar{v}(\beta\bar{\gamma}-\bar{\beta}\gamma)\alpha^{k-1}K^{n-k-1}, \tag{46}$$

$$\begin{aligned}\frac{1}{n-k}w_k = & \, v\bar{v}\alpha^k K^{n-k-2}\{-K\varphi+(n-k-1)\gamma\bar{\gamma}\} \\ & + kv\bar{v}(-\theta\bar{\beta}\gamma+\bar{\theta}\beta\bar{\gamma})\alpha^{k-1}K^{n-k-1} \\ & + \frac{k^2}{(n-k)}v\bar{v}\theta\bar{\theta}\alpha^{k-1}K^{n-k}.\end{aligned} \tag{47}$$

In verifying these formulas care should be taken on the commutativity or anti-commutativity of the products; also many terms drop out because of degree considerations.

We are now in a position to prove the formulas (34) and (35). By definition we have, since $dK=0$ and $d^2v \in M_{11}^{10}$,

$$\begin{aligned}(d'-d'')y_k &= -t_k + (k-1)v\bar{v}\alpha_1^{k-2}K_1^{n-k}(-d^2v\bar{\beta}+d^2\bar{v}\beta) \\ &= -t_k + (k-1)|v|^2 v\bar{v}\alpha_1^{k-2}K_1^{n-k}(\bar{\beta}\gamma-\beta\bar{\gamma}) \\ &= -t_k - \frac{k-1}{n-k+1}|v|^2 t_{k-1} - \frac{2(k-1)^2}{n-k+1}|v|^2 y_{k-1}(d'-d'')\log|v|,\end{aligned}$$

the last step following from (46). From this we get (34) immediately.

Similarly, we find, by using (32), (38), (42), (43), (46), (47),

$$\begin{aligned}\frac{1}{|v|^2}(dt_k - 2w_k) = & -2kv\bar{v}\varphi\alpha_1^{k-1}K_1^{n-k} + 2k(n-k)v\bar{v}\gamma\bar{\gamma}\alpha_1^{k-1}K_1^{n-k-1} \\ & - k(k-1)v\bar{v}(\bar{\theta}-\theta)(\gamma\bar{\beta}+\bar{\gamma}\beta)\alpha_1^{k-2}K_1^{n-k} \\ = & \, \frac{2k}{n-k+1}w_{k-1} + \frac{2k(k-1)}{n-k+1}t_{k-1}\,d\log|v|.\end{aligned}$$

From this (35) follows.

From (39) we have

$$w_n = \alpha^n = n^2 v\bar{v}\theta\bar{\theta}\alpha_1^{n-1}.$$

Comparsion with (46) gives

$$w_n'' = 0. \tag{48}$$

Similarly we verify that

$$t_n'' = 0. \tag{48a}$$

Equations (35) and (48) permit us to express $w_0 = K^n$ as a linear combination of dt'_k. The result is

$$w_0 = -\frac{1}{2} \sum_{1 \leqq k \leqq n} (-1)^k \binom{n}{k} dt'_k. \tag{49}$$

Similarly, from (34) and (48a) we get

$$(d' - d'') \sum_{1 \leqq k \leqq n} (-1)^k \binom{n-1}{k-1} y'_k = 0, \tag{50}$$

and

$$\begin{aligned} (d' - d'') \sum_{1 \leqq k \leqq n-1} (-1)^k \binom{n-1}{k-1} \left(\frac{1}{k} + \cdots + \frac{1}{n-1} \right) y^1_k \\ = -\frac{1}{n} \sum_{1 \leqq k \leqq n-1} (-1)^k \binom{n}{k} t''_k. \end{aligned} \tag{51}$$

It follows from (33) that

$$\begin{aligned} \sum_{1 \leqq k \leqq n} (-1)^k \binom{n}{k} t'_k = & \sum_{1 \leqq k \leqq n-1} (-1)^k \binom{n}{k} t''_k \\ & -2 \sum_{1 \leqq k \leqq n} (-1)^k k \binom{n}{k} y'_k (d' - d'') \log |v| \\ = & \, (d' - d'') \sigma, \end{aligned} \tag{52}$$

where

$$\begin{aligned} \sigma = & -n \sum_{1 \leqq k \leqq n-1} (-1)^k \binom{n-1}{k-1} \left(\frac{1}{k} + \cdots + \frac{1}{n-1} \right) y'_k \\ & -2n \sum_{1 \leqq k \leqq n} (-1)^k \binom{n-1}{k-1} y'_k \log |v|. \end{aligned} \tag{53}$$

Combining (49) and (52) we get the following basic formula for complex transgression:

$$\pi^* K^n = -\tfrac{1}{2} d(d' - d'') \sigma. \tag{54}$$

We note that for $n = 1, 2$ we have respectively

$$\sigma = 2 v \bar{v} \log |v|, \tag{55}$$

$$\sigma = \frac{2}{|v|^2} v \bar{v} K + \frac{4}{|v|^4} (-v \bar{v} \alpha + |v|^2 v \bar{v} K) \log |v|. \tag{56}$$

6. Proof of the Theorem

We first observe that

$$K^n \in \tilde{\Lambda}^{nn}_{nn}, \quad \sigma \in \tilde{M}^{nn}_{n-1,n-1}.$$

Let χ be locally an element of $\tilde{\Lambda}^{n0}_{00}$, with norm equal to 1. It is defined up to a factor which is a complex number of absolute value 1, so that $\chi\bar{\chi} \in \tilde{\Lambda}^{nn}_{00}$ is globally defined in X. Since the connection is of type (1,0), we have

$$d'(\chi\bar{\chi}) = d''(\chi\bar{\chi}) = 0. \tag{57}$$

Recall that with the choice of the frame field e_i, $1 \leqq i \leqq n$, in § 4 we will have

$$\chi = (\det H)^{-\frac{1}{2}}(e_1 \wedge \cdots \wedge e_n). \tag{58}$$

From the expressions

$$K = \sum_{1 \leqq i, j \leqq n} \bar{e}_i e_j \Omega^{ij},$$

$$c_n(E) = \left(\frac{i}{2\pi}\right)^n \det(\Omega^j_i)$$

we derive immediately

$$K^n = (-1)^{\frac{n^2}{2}} n!(2\pi)^n c_n(E)\chi\bar{\chi}.$$

The theorem in § 1 follows if we set

$$\sigma = (-1)^{\frac{n^2-1}{2}} n!(2\pi)^{n-1} \rho\chi\bar{\chi}. \tag{59}$$

Remark. Our proof does not use the assumption that the dimension of the fibre be equal to the dimension of X. The theorem is therefore valid without this dimension restriction. Our formulas seem to show that the use of the spaces $\tilde{\Lambda}^{pq}_{rs}$, $\tilde{M}^{pq}_{rs}$ is very natural to the geometrical situation under consideration.

Bibliography

[1] Bott, R., and S. Chern: Hermitian vector bundles and the equidistribution of the zeroes of their holomorphic sections. Acta Math. **114**, 71–112 (1965).

[2] Chern, S.: Complex manifolds without potential theory. Princeton: Van Nostrand 1967.

[3] Flanders, H.: Development of an extended exterior differential calculus. Transactions of American Mathematical Society **75**, 311–326 (1953).

ON MINIMAL SPHERES IN THE FOUR-SPHERE

SHIING-SHEN CHERN[1] (陳省身)

To Yu-Why

1. Introduction.

We will study the immersions of the differentiable two-sphere S^2 as a minimal surface of $S^4(1)$, the four-dimensional sphere of radius 1 in the Euclidean 5-space. If the image belongs to an equator of $S^4(1)$, a theorem of Almgren-Calabi says that it must be a two-dimensional equator of $S^4(1)$, i. e., the intersection of $S^4(1)$ with a linear space of dimension 3 through the center. (Cf. Bibliography [3]). We will therefore consider only those immersions for which the image belongs to no equator of $\mathbf{S}^4(1)$. Although such minimal immersions are described in principle by Calabi [2], the only explicit example known is the Veronese surface. The latter is the only minimal immersion of S^2 in $S^4(1)$ for which the induced metric has constant Gaussian curvature < 1. We will give a general construction of minimal spheres in $S^4(1)$ from its *directrix curve*, from which follow numerous examples of minimal spheres of non-constant Gaussian curvature.

It will be an interesting problem to classify and parametrize the minimal spheres in $S^4(1)$, up to motions of the ambient space.

2. Surface theory in $S^4(1)$.

Let E be the Euclidean space of 5 dimensions, whose scalar product we denote by (x, y), x, y∈E. Later on we will extend the scalar product to complex vectors x, y, so that it will be complex linear in each of its arguments. The unit sphere $S^4(1)$ is defined by the equation

(1) $$|x|^2=(x, x)=1.$$

Received January 19, 1970

1) Work done under appointment to Miller Institute, University of California, Berkeley 1969-70.

We consider the space of orthonormal frames $x, e_1, \cdots, e_4$ of $S^4(1)$, so that

(2) $$(x, e_i)=0, \quad (e_i, e_j)=\delta_{ij}, \quad 1\leqq i, j\leqq 4.$$

Then we have

(3) $$\begin{cases} dx=\sum_i \omega_i e_i, \\ de_i=-\omega_i x+\sum_j \omega_{ij} e_j, \end{cases}$$

with

(4) $$\omega_{ij}+\omega_{ji}=0.$$

Exterior differentiation of (3) gives the structure equations

(5) $$\begin{aligned} d\omega_i &= \sum_j \omega_j \wedge \omega_{ji}, \\ d\omega_{ij} &= \sum_k \omega_{ik} \wedge \omega_{kj} - \omega_i \wedge \omega_j, \quad 1\leqq i, j, k\leqq 4. \end{aligned}$$

Let S^2 be a differentiable two-sphere and let

(6) $$x: S^2 \longrightarrow S^4(1)$$

be an immersion. We will restrict to frames such that e_1, e_2 are tangent vectors to $x(S^2)$ at x. Then we have

(7) $$\omega_3=\omega_4=0.$$

Exterior differentiation of (7) and use of (5) give

$$\begin{aligned} \omega_1\wedge\omega_{13}+\omega_2\wedge\omega_{23}=0, \\ \omega_1\wedge\omega_{14}+\omega_2\wedge\omega_{24}=0. \end{aligned}$$

By Cartan's lemma we can put

(8) $$\omega_{\alpha r}=\sum_\beta h_{r\alpha\beta}\omega_\beta, \quad \alpha, \beta=1, 2; \ r=3, 4,$$

where

(9) $$h_{r\alpha\beta}=h_{r\beta\alpha}.$$

The immersion is minimal, if and only if

(10) $$h_{r11}+h_{r22}=0,$$

in which case we will call the sphere a *minimal sphere.*

For minimal immersions of S^2 it is extremely convenient, and even essential, to introduce the complex notation. In fact, we put

$$(11)\qquad \begin{aligned} E_1&=e_1+ie_2,\\ \varphi&=\omega_1+i\omega_2. \end{aligned}$$

Then equations (3) give

$$(12)\qquad \begin{aligned} dx&=\frac{1}{2}(\bar{\varphi}E_1+\varphi\bar{E}_1),\\ dE_1&=-\varphi x-i\omega_{12}E_1+\bar{\varphi}\sum_r H_r e_r, \end{aligned}$$

where the dash denotes the complex conjugate and

$$(13)\qquad H_r=h_{r11}+ih_{r12},\qquad r=3,\ 4.$$

We introduce the complex-valued normal vector field

$$(14)\qquad V=\sum_r H_r e_r.$$

From this situation we can construct a complex-valued ordinary differential form globally defined on S^2 as follows: The vector E_1 is defined up to the transformation

$$(15)\qquad E_1\longrightarrow E_1^*=\exp(i\tau)E_1,$$

where τ is real. Under such a change, φ and V are changed according to

$$(16)\qquad \begin{aligned} \varphi&\longrightarrow\varphi^*=\exp(i\tau)\varphi,\\ V&\longrightarrow V^*=\exp(2i\tau)V. \end{aligned}$$

It follows that the complex-valued form of degree 4,

$$(17)\qquad \Phi\underset{\text{def}}{=}(\bar{V},\ \bar{V})\varphi^4=(\bar{H}_3^2+\bar{H}_4^2)\varphi^4$$

is globally defined on S^2.

Let z be an isothermal coordinate on S^2, so that we can write, *locally*,

$$(18)\qquad \varphi=\lambda dz,\qquad \lambda\neq 0.$$

Then $\Phi=f(z)dz^4$. We shall show that $f(z)$ is a holomorphic function of z.

This is to be a consequence of the structure equations. In fact, exterior differentiation of the first equation of (12) gives

$$(19)\qquad d\varphi=i\varphi\wedge\omega_{12},$$

while exterior differentiation of the second equation of (12) gives, on equating to zero the coefficients of e_3 and e_4,

$$(20)\qquad \begin{aligned} &d\bar{H}_3 - 2i\omega_{12}\bar{H}_3 - \omega_{34}\bar{H}_4 \equiv 0, \\ &d\bar{H}_4 - 2i\omega_{12}\bar{H}_4 + \omega_{34}\bar{H}_3 \equiv 0, \ \mathrm{mod}\ \varphi. \end{aligned}$$

From (19) we get, on substituting the expression of φ in (18),

$$d\lambda + i\lambda\omega_{12} \equiv 0, \ \mathrm{mod}\ \varphi.$$

It follows that

$$d\{(\bar{H}_3^2 + \bar{H}_4^2)\lambda^4\} \equiv 0, \ \mathrm{mod}\ \varphi \text{ or mod } dz.$$

This proves that the expression in the braces is a holomorphic function of z

Thus Φ is an abelian differential of degree 4 on S^2. It must be identically zero, i.e.,

$$(21)\qquad H_3^2 + H_4^2 = 0.$$

It is in this conclusion that we make essential use of the fact that the surface is a differentiable two-sphere. Condition (21) means that the vector V in (14) is "isotropic":

$$(22)\qquad (V,V) = 0.$$

If V identically zero, the 3-dimensional space spanned by x, e_1, e_2 will be fixed and the surface is totally geodesic, i. e. is an equatorial sphere. From now on we suppose V not identically zero.

We shall show that the set Σ where $V=0$ must be finite. From (21) this is also the set of zeroes of the function H_3 or H_4. Let z be an isothermal coordinate on S^2 such that the point $z=0$ belongs to Σ but is not an interior point of Σ. By choosing a frame field in a neighborhood of $z=0$, ω_{12} and ω_{34} will become linear combinations of φ, $\bar{\varphi}$ and equations (20) give an equation of the form

$$(23)\qquad \frac{\partial \bar{H}_3}{\partial \bar{z}} = a\bar{H}_3.$$

By a theorem established in §4, [4], such a function must be locally of the form

$$\bar{H}_3 = z^m \bar{H}'_3, \quad m \geqq 1,$$

with $H'_3(0)\neq 0$. Thus the point $z=0$ is an isolated point of Σ.

It follows that a point of Σ is either isolated or an interior point of Σ. Suppose $p\in\Sigma$ be an interior point of Σ. Since $\Sigma\neq S^2$, the largest open subset of Σ containing p must have a boundary point. The latter belongs to Σ, because Σ is closed. But such a point is neither isolated nor an interior point of Σ, which is impossible. Hence Σ has only isolated points and they must be finite in number. For simplicity of terminology we will refer to the points of Σ as *singular points*.

At an isolated point $z=0$ of Σ we can write

$$H_r=\bar{z}^m H'_r, \qquad r=3,4, \tag{24}$$

where $H_r'(0)\neq 0$. The non-singular points where V is not zero can be included in this representation, corresponding to $m=0$.

From (21) we get

$$H_4=\epsilon i H_3, \qquad \epsilon=\pm 1.$$

Since ϵ is a continuous function on S^2, it must take one of the two values. By changing e_4 to $-e_4$ if necessary, we can suppose $\epsilon=+1$. This choice can be called a second "orientation" of the minimal sphere.

After this choice being made we have

$$V=H_3(e_3+ie_4)=\bar{z}^m H_3'(e_3+ie_4) \tag{25}$$

and we set

$$E_2=e_3+ie_4. \tag{26}$$

E_2 is determined up to the multiplication by a complex number of absolute value 1.

Observe that our complex vectors E_1, E_2 satisfy the following relations on their scalar products:

$$\begin{gathered}(x,E_1)=(x,E_2)=(E_1,E_1)=(E_1,E_2)=(E_2,E_2)=(E_1,\bar{E}_2)=0,\\ (E_1,\bar{E}_1)=(E_2,\bar{E}_2)=2,\end{gathered} \tag{27}$$

together with relations obtained from them by complex conjugation. Expressing that their differentials are zero and using (12), we get in particular

$$(28)\quad \begin{aligned}(dE_2, x) = (dE_2, E_1) = (dE_2, E_2) = 0,\\ (dE_2, \overline{E}_1) = -2\overline{H}_3\varphi.\end{aligned}$$

On the other hand, the second equation of (3) gives

$$(29)\quad (dE_2, \overline{E}_2) = -2i\omega_{34}.$$

Now dE_2 is a linear combination of $x, E_1, E_2, \overline{E}_1, \overline{E}_2$, whose coefficients are determined by the relations (28),(29). We can therefore write

$$(30)\quad dE_2 = -\overline{H}_3\varphi E_1 - i\omega_{34}E_2.$$

The second equation of (12) can be re-written

$$(31)\quad dE_1 = -\varphi x - i\omega_{12}E_1 + H_3\overline{\varphi}E_2.$$

Equations (30),(31), together with the first equation of (12), will be called the *Frenet-Boruvka formulas* for a minimal surface on the sphere.

Exterior differentiations of (30) and (31) give, after simplification,

$$(32)\quad \begin{aligned}d\omega_{12} + i\left(\frac{1}{2} - |H_3|^2\right)\varphi\wedge\overline{\varphi} = 0,\\ d\omega_{34} + i|H_3|^2\varphi\wedge\overline{\varphi} = 0,\\ d(H_3\overline{\varphi}) + i(\omega_{12} - \omega_{34})\wedge H_3\overline{\varphi} = 0.\end{aligned}$$

The Gaussian curvature K of the induced metric is defined by

$$(33)\quad d\omega_{12} + \frac{i}{2}K\varphi\wedge\overline{\varphi} = 0.$$

Comparison with the first equation of (32) gives

$$(34)\quad 2|H_3|^2 + K = 1.$$

This implies that $K \leqq 1$. The singular points are exactly the points at wnich $K=1$.

Consider the case of minimal spheres with $K=\text{const}$ or $|H_3|^2=\text{const}$. The third equation of (32) gives, when expanded,

$$\{dH_3 + i(2\omega_{12} - \omega_{34})H_3\}\wedge\overline{\varphi} = 0,$$

or

$$(35)\quad dH_3 + i(2\omega_{12} - \omega_{34})H_3 = L\overline{\varphi}.$$

Under conjugation we have

$$d\overline{H}_3 - i(2\omega_{12} - \omega_{34})\overline{H}_3 = \overline{L}\varphi.$$

By multiplying the first equation by $\overline{H}_3$ and the second equation by H_3 and adding, we get

$$L\overline{H}_3\varphi + \overline{L}H_3\varphi = 0.$$

If $|H_3|^2 = 0$, the surface will be totally geodesic. Excluding this case, we get $L=0$. Differentiating (35), we get

$$2d\omega_{12} - d\omega_{34} = 0.$$

From the first two equations of (32) it follows that

(36) $$|H_3|^2 = \frac{1}{3} \quad \text{or } K = \frac{1}{3}.$$

It is well-known (cf. [1]) that the Veronese surface has this property and that it is the only such surface up to position in $S^4(1)$; all other minimal spheres are of non-constant Gaussian curvature.

We return to the general case. From (30) we derive

(37) $$d(\rho E_2) = -\rho\overline{H}_3\varphi E_1 + (d\rho - i\rho\omega_{34})E_2.$$

By choosing ρ to satisfy locally the condition

(38) $$d \log \rho - i\omega_{34} \equiv 0, \text{ mod } dz,$$

the components of the vector ρE_2 are holomorphic functions in z. We can interpret them as the homogeneous coordinates in the complex projective space $P_4(C)$ of dimension 4 and we get a rational curve $\triangle: S^2 \longrightarrow P_4(C)$. This will be called the *directrix curve* of the minimal sphere. If we denote by Z the homogenous coordinate vector in $P_4(C)$, then $\triangle$ lies on the hyperquadric Q:

(39) $$(Z, Z) = 0.$$

Moreover, its tangent lines, which are spanned by E_2 and E_1, also lie completely on Q.

Remark. In $P_4(C)$ we have the scalar product (Z, W) and the positive definite hermitian product $(Z, \overline{W})$ for complex vectors Z, W. Thus $P_4(C)$ is endowed with the Fubini-Study metric. We will see that the geometry of minimal spheres on $S^4(1)$ is equivalent to that of a pair of conjugate-complex totally isotropic rational curves E_2, $\overline{E}_2$ in $P_4(C)$ with the Fubini-Study metric.

(A curve is totally isotropic if all its tangent lines belong to the hyperquadric Q.)

3. Construction of the minimal sphere from the directrix curve.

At a point E_2 of the directrix curve, the tangent line is spanned by E_2, E_1, the osculating plane by E_2, E_1, x, and the osculating hyperplane by E_2, E_1, x, $\overline{E}_1$. Since $\varphi \neq 0$, the latter is fixed only when $H_3=0$, in which case the directrix curve is a point and the minimal sphere is totally geodesic. We state this result as a theorem:

Theorem. *If the directrix curve lies in a hyperplane, it must be a point and the minimal sphere is totally geodesic.*

We wish to construct the minimal sphere from its directrix curve. In view of the above theorem we can suppose that it does not lie in a hyperplane. As shown in the end of §2, a necessary condition for a directrix curve is that the curve itself, and all its tangent lines, belong to the hyperquadric (39). It suffices to give the construction locally. In a neighborhood of the point $z=0$, where z is an isothermal coordinate, let the directrix curve be defined by

$$\xi(z)=(\xi_1(z),\cdots,\xi_5(z)), \tag{40}$$

where $\xi_A(z)$, $A=1,\cdots,5$, are holomorphic functions of z and are homogeneous coordinates in $P_4(C)$, with $\xi(0)\neq 0$. The above necessary condition is expressed analytically by

$$(\xi,\xi)=(\xi,\xi')=(\xi',\xi')=0, \tag{41}$$

where the prime denotes differentiation with respect to z. Since the curve does not lie on a hyperplane, we have

$$\xi\wedge\xi'\wedge\xi''\wedge\xi'''\not\equiv 0. \tag{42}$$

We will express the vectors E_2, E_1, x in terms of ξ, the first two being determined up to multiplications by numbers of absolute value 1 and x up to sign.

We have

$$E_2=\frac{\sqrt{2}}{(\xi,\bar{\xi})^{\frac{1}{2}}}\xi \tag{43}$$

and it satisfies the relations

(44) $$(E_2, E_2)=0, \qquad (E_2, \bar{E}_2)=2.$$

To determine E_1 we first prove the lemma:

Lemma. *Suppose that* $\xi(0) \neq 0$ *and*

(45) $$\xi\wedge\xi'\equiv 0, \text{ mod } z^m, \ m\geqq 0.$$

Then there exists a constant vector $A\neq 0$ *such that*

(46) $$\xi\equiv A(1+a_1 z+\cdots+a_m z^m), \text{ mod } z^{m+1}, \ a_i\in C,$$

and

(47) $$\xi'-\frac{(\bar{\xi}, \xi')}{(\bar{\xi}, \xi)}\xi\equiv 0, \text{ mod } z^m.$$

We prove this by induction on m. For $m=0$ there is nothing to prove. For $m\geqq 1$ suppose the lemma be true for $m-1\geqq 0$. To prove it for m we write

$$\xi=A(1+a_1 z+\cdots+a_{m-1}z^{m-1})+Bz^m+\cdots,$$

so that

$$\xi'=A(a_1+2a_2 z+\cdots+(m-1)a_{m-1}z^{m-2})+mBz^{m-1}+\cdots.$$

The hypothesis gives, by the consideration of the term in z^{m-1} in $\xi\wedge\xi'$,

$$A\wedge B=0 \text{ or } B=bA.$$

This proves the first statement. There exists therefore a function ρ such that

$$\xi'-\rho\xi\equiv 0, \text{ mod } z^m$$

Taking the scalar product of this equation with $\bar{\xi}$, we can choose

$$\rho=(\bar{\xi}, \xi')/(\bar{\xi}, \xi),$$

and the lemma is proved.

The relevance of the lemma arises from the fact that the directrix curve is not necessarily an immersed curve in $P_4(C)$. The vector E_1, which is to determine a point on the tangent line to the directrix curve at E_2 and which is to satisfy the conditions

(48) $$(E_1, E_1)=(E_1, E_2)=(E_1, \bar{E}_2)=0, \ (E_1, \bar{E}_1)=2,$$

is given by

(49) $$E_1 = \frac{u}{z^m}\left(\xi' - \frac{(\bar{\xi}, \xi')}{(\bar{\xi}, \xi)}\xi\right),$$

where the expression at the right is meaningful because of the lemma. The $|u|$ is to be determined by the last condition of (48). In fact, we have

(50) $$\left(\xi' - \frac{(\bar{\xi}, \xi')}{(\bar{\xi}, \xi)}\xi,\ \bar{\xi}' - \frac{(\xi, \bar{\xi}')}{(\xi, \bar{\xi})}\bar{\xi}\right) = \{(\xi, \bar{\xi})(\xi', \bar{\xi}') - (\xi, \bar{\xi}')(\bar{\xi}, \xi')\}/(\xi, \bar{\xi})$$
$$= (\xi\wedge\xi',\ \bar{\xi}\wedge\bar{\xi}')/(\xi, \bar{\xi}).$$

We suppose that z^m is the highest power of z in $\xi\wedge\xi'$, i.e.,

(51) $$(\xi\wedge\xi')z^{-m}|_{z=0} \neq 0.$$

Then $|u|$ is given by

(52) $$|u|^2 = \frac{2(\xi, \bar{\xi})(z, \bar{z})^m}{(\xi\wedge\xi', \bar{\xi}\wedge\bar{\xi}')}$$

and $|u|(0) \neq 0$.

By calculating the coefficient of E_1 in dE_2, we have, with E_1, E_2 given by (49),(43) respectively,

(53) $$\bar{H}_3\varphi = -\frac{\sqrt{2}}{(\xi, \bar{\xi})^{\frac{1}{2}}u} z^m dz.$$

Thus the point $z=0$ is singular if and only if $m \geqq 1$.

It remains to determine x, which is the intersection of the osculating planes to the directrix curve and its complex conjugate curve at corresponding points. We first notice that by differentiation of (41) we have

(54) $$(\xi, \xi'') = (\xi', \xi'') = 0.$$

We put

(55) $$\eta = \xi'' - \frac{1}{2}(\bar{E}_1, \xi'')E_1 - \frac{1}{2}(\bar{E}_2, \xi'')E_2.$$

Then both η and $\bar{\eta}$ have scalar products zero when paired with E_1, E_2, $\bar{E}_1$, $\bar{E}_2$ respectively. Hence they are both multiples of x, say

(56) $$\eta = vx, \qquad \bar{\eta} = \bar{v}x,$$

x being real. Since $(x, x) = 1$, v is determined by the equations

(57) $$v^2 = (\eta, \eta), \qquad |v|^2 = (\eta, \bar{\eta}).$$

These equations are consistent, because (56) implies $\eta\wedge\bar{\eta}=0$ and hence

$$(\eta,\eta)(\bar{\eta},\bar{\eta})-(\eta,\bar{\eta})^2=(\eta\wedge\bar{\eta},\eta\wedge\bar{\eta})=0.$$

The vector x, which describes the surface, is determined up to sign. From (55) we find

$$(\eta,\eta)=(\xi'',\xi''), \tag{58}$$

$$(\eta,\bar{\eta})=\frac{1}{4}(\xi''\wedge E_1\wedge E_2,\ \bar{\xi}''\wedge\bar{E}_1\wedge\bar{E}_2)=\frac{|u|^2}{2(\xi,\bar{\xi})|z|^{2m}}(\xi\wedge\xi'\wedge\xi'',\ \bar{\xi}\wedge\bar{\xi}'\wedge\bar{\xi}''). \tag{59}$$

Calculating mod E_1, E_2, we get, from the coefficient of x in dE_1,

$$\varphi=-\frac{uv}{z^m}dz, \tag{60}$$

so that

$$\varphi\wedge\bar{\varphi}=\frac{|u|^2|v|^2}{|z|^{2m}}dz\wedge d\bar{z} \tag{61}$$

It follows that the points where $\varphi=0$, i.e., where the mapping $x\colon S^2\longrightarrow S^4(1)$ ceases to be an immersion, are exactly the points $z=0$ where

$$(\xi\wedge\xi'\wedge\xi'')z^{-2m}=0.$$

In the notation of the above lemma we put

$$\xi=AP_m(z)+A_1z^{m+1}+A_2z^{m+2}+\cdots, \tag{62}$$

where A, A_1, A_2 are constant vectors, $A\neq 0$, and $P_m(z)$ is a polynomial of degree m in z with constant term 1. It follows that

$$\xi\wedge\xi'\wedge\xi''=(m+1)^2(m+2)(A\wedge A_1\wedge A_2)z^{2m}+\cdots.$$

Hence the surface is an immersion at $z=0$ if and only if $A\wedge A_1\wedge A_2\neq 0$.

The vectors x, E_1, E_2 determined satisfy the equations (12),(30),(31), so that x describes a minimal sphere on $S^4(1)$.

Combining (52),(53),(57),(59),(60), we get

$$|H_3|^2=\frac{1}{2}\frac{(\xi\wedge\xi',\ \bar{\xi}\wedge\bar{\xi}')^3}{(\xi,\bar{\xi})^3(\xi\wedge\xi'\wedge\xi'',\ \bar{\xi}\wedge\bar{\xi}'\wedge\bar{\xi}'')}, \tag{63}$$

with which the Gaussian curvature is related by (34). Both $|H_3|^2$ and K are thus expressed in terms of the directrix curve.

Remark. It will be of interest to note that in the classical Weierstrass formulas for a minimal surface in Euclidean 3-space the surface is derived from its spherical image by integration, while in our case the surface is constructed from its directrix curve through differentiation. Perhaps the reason lies in the periodicity of a Laplace sequence of conjugate nets, but we will not enter into its discussion here (cf. [6]).

4. Line geometry.

The directrix curves are related to curves which have played an important role in line geometry, and we wish to explain this relationship.

Let P_4 be the four-dimensional complex projective space and Q_3 a non-degenerate hyperquadric in it. The lines of Q_3 form a 3-dimensional irreducible algebraic variety and we will be interested in the developable surfaces generated by these lines. Q_3 can be considered as the intersection of a non-degenerate hyperquadric Q_4 in the projective 5-space P_5 by a hyperplane. On the other hand, by the Plücker line coordinates, the points of Q_4 can be identified with the lines of the projective 3-space P_3. Following Felix Klein we list the correspondance in the following table:

Point of Q_4	Line in P_3
Point of Q_3	Line of a linear complex Π of P_3
Line of Q_3	Pencil of lines of Π
Intersecting lines of Q_3	Pencils of lines of Π with a common line
Developable of Q_3	Curve belonging to Π

Thus corresponding to our directrix curves are the curves in P_3 studied by S. Lie, E. Picard, etc., whose tangents belong to a linear line complex. Our curves have the additional property being rational. Among such curves are the rational cubics of P_3 and, more generally, the so-called W-curves

$$\xi_1:\xi_2:\xi_3:\xi_4=1:z^m:z^p:z^{m+p},\quad 1\leqq m<p. \tag{64}$$

For literature on the subject cf. [7].

5. Examples.

To find examples of minimal spheres in $S^4(1)$ it therefore suffices to give the directrix curves and verify by the criterion in §3 that the surfaces are immersed. As mentioned above, the directrix curve itself is not necessarily immersed in $P_4(C)$. Carrying out analytically the correspondance in §4 to the curve (64), we get the curve

$$(65)\quad \begin{aligned} &\xi_1 = im\left(1+\frac{p-m}{p+m}z^{2p}\right), \\ &\xi_2 = m\left(1-\frac{p-m}{p+m}z^{2p}\right), \\ &\xi_3 = ipz^{p-m}\left(1-\frac{p-m}{p+m}z^{2m}\right), \\ &\xi_4 = pz^{p-m}\left(1+\frac{p-m}{p+m}z^{2m}\right), \quad \xi_5 = -2i(p-m)z^p, \end{aligned}$$

where $\xi_1,\cdots,\xi_5$ are homogeneous coordinates and the parametric value $z=\infty$ is admitted. It can be verified that the vector ξ satisfies the conditions (41) for a directrix curve. Moreover, we have

$$(66)\quad \xi\wedge\xi'\wedge\xi''\neq 0,$$

except possibly at $z=0$ and $z=\infty$. At these points, by using the above criterion, we see that the surface is immersed if $m=1$.

A slight variation of (65) gives the directrix curve

$$(67)\quad \begin{aligned} &\xi_1 = 2\sqrt{3}(1+z^4), \quad \xi_2 = -12z^2, \quad \xi_3 = 2\sqrt{3}i(1-z^4), \\ &\xi_4 = \sqrt{3}i(-4z-4z^3), \quad \xi_5 = \sqrt{3}(-4z+4z^3), \end{aligned}$$

which leads to the Veronese surface. It is an interesting exercise to verify from (63) that the Gaussian curvature is 1/3. Let us note that the Veronese surface has the parametric equations

$$(68)\quad \begin{aligned} &x_1 = \frac{\sqrt{3}}{2}(t^2-u^2), \quad x_2 = \frac{1}{2}(t^2+u^2-2v^2), \\ &x_3 = \sqrt{3}\,tu, \quad x_4 = \sqrt{3}\,uv, \quad x_5 = \sqrt{3}\,vt, \end{aligned}$$

defined on the sphere $t^2+u^2+v^2=1$.

BIBLIOGRAPHY

[1] Calabi, E., *Minimal immersions of surfaces in Euclidean spheres*, Jour. Diff. Geom. 1, 111-125 (1967).

[2] Calabi, E., *Quelques applications de l'analyse complexe aux surfaces d'aire*

minima, Topics in Complex Manifolds, Univ. of Montreal (Canada), pp. 59-81 (1968).

[3] Chern, S., *Simple proofs of two theorems on minimal surfaces, L'Enseig. math.* 15, 53-61 (1969).

[4] Chern, S., *On the minimal immersions of the two-sphere in a space of constant curvature*, to appear in volume dedicated to S. Bochner.

[5] Chern, S., de Carmo, M, and Kobayashi, S., *Minimal submanifolds on the sphere with second fundamental form of constant length*, to appear in *"Functional Analysis and Related Fields"*.

[6] Darboux, G., *Théorie des surfaces*, t.2, livre 4.

[7] Rohn, K. und Berzolari, L., *Algebraische Raumkurven und abwickelbare Flächen, Enz. der math. Wiss.* III C 9, 1359-1387.

Academia Sinica
and
University of California,
Berkeley, California

SCRIPTA MATHEMATICA, Volume XXIX, No. 3–4

MEROMORPHIC VECTOR FIELDS AND CHARACTERISTIC NUMBERS

By SHIING-SHEN CHERN[1]

Dedicated to the memory of A. Adrian Albert

1. **Introduction and Statement of Results:** In several papers (cf. in particular [2], [3]) Bott, and Baum and Bott expressed the characteristic numbers of a compact complex manifold in terms of the properties of a holomorphic or meromorphic vector field at its zeros. We shall begin by stating these results. Let M be a compact complex manifold of dimension m, TM or simply T be its tangent bundle, and $c_r(T)\in H^{2r}(M;Z)$, $0\leqslant r\leqslant m$, be its rth Chern class. Let L be a holomorphic line bundle over M. A holomorphic section $\xi\in\Gamma(T\otimes L)$ is described locally by

$$\xi=\sum_i \xi^i\frac{\partial}{\partial z^i}\qquad 1\leqslant i,j\leqslant m, \tag{1}$$

where z^i form a local coordinate system in M and $\xi^i\in\Gamma(L)$ are holomorphic sections of L. At a point where $\xi=0$ the matrix of partial derivatives

$$\Xi=\left(\frac{\partial\xi^i}{\partial z^j}\right) \tag{2}$$

is defined up to the transformation

$$\Xi\to gJ\Xi J^{-1}, \tag{3}$$

where g is a non-zero holomorphic function and J is the Jacobian matrix in a change of the local coordinate system. The zero of ξ is called non-degenerate if $\det\Xi\neq 0$.

We shall consider complex-valued functions $P(A_1,\ldots,A_r)$ whose arguments A_α are complex-valued $(m\times m)$-matrices, which have the properties: 1) P is linear in each A_α; 2) P remains unchanged under a permutation of the A's; 3) for any nonsingular matrix X we have

$$P(X^{-1}A_1X,\ldots,X^{-1}A_rX)=P(A_1,\ldots,A_r). \tag{4}$$

The function

$$P(A)=P(A,\ldots,A),\qquad r\text{ arguments}, \tag{5}$$

Received by the editors, February 4, 1972.

[1]Work done under partial support of National Science Foundation grant GP29697. The author wishes to thank Douglas Dunham for many discussions.

is a polynomial of degree r in A and will be called an invariant polynomial. Conversely, $P(A_1,\ldots,A_r)$ is completely determined by $P(A)$ and is usually called the complete polarization of $P(A)$. Examples of invariant polynomials are the coefficients of the polynomial in t:

$$\det(tI+A)=t^m+c_1(A)t^{m-1}+\cdots+c_m(A). \tag{6}$$

If

$$A=(A_i^{\,j}), \tag{7}$$

we have explicitly

$$c_r(A)=\frac{1}{r!}\sum \delta^{i_1,\ldots,i_r}_{j_1,\ldots,j_r} A_{i_1}^{\,j_1}\ldots A_{i_r}^{\,j_r}, \tag{8}$$

where $i_1,\ldots,i_r$ is a permutation of $j_1,\ldots,j_r$, $\delta^{i_1,\ldots,i_r}_{j_1,\ldots,j_r}$ is the sign of the permutation, and the summation is over all the i's. By putting A in the diagonal form, it is easily seen that an invariant polynomial is a polynomial in $c_r(A)$, $1\leqslant r\leqslant m$. In particular, any invariant polynomial of degree m is a linear combination of the following:

$$c^\alpha(A)=c_1^{\alpha_1}(A)\cdots c_m^{\alpha_m}(A), \qquad \alpha=(\alpha_1,\ldots,\alpha_m), \tag{9}$$

where $\alpha_1+2\alpha_2+\cdots+m\alpha_m=m$. If P is an invariant polynomial of degree m, we see easily that at a non-degenerate zero of a holomorphic section $\xi\in\Gamma(T\otimes L)$, the ratio $P(\Xi)/\det\Xi$ is independent of the choice of local coordinates and is an invariant of ξ at a zero.

We now review some well-known facts on hermitian geometry (cf. [4]). An hermitian structure on M is given, relative to the local coordinates z^i, $1\leqslant i,j,k\leqslant m$, by the scalar products

$$H\left(\frac{\partial}{\partial z^i},\frac{\partial}{\partial z^j}\right)=h_{ij}=\bar{h}_{ji}, \tag{10}$$

where the functions at the right-hand side are complex-valued and C^∞ and the matrix

$$H=(h_{ij}) \tag{11}$$

is hermitian positive-definite. There is a uniquely defined connection of bidegree (1,0) which preserves the hermitian structure. Its connection matrix is

$$\omega=(\omega_i^{\,j})=\partial H\cdot H^{-1}, \qquad \omega_i^{\,j}=\sum_k \Gamma^j_{ik}\,dz^k. \tag{12}$$

From this we verify that

$$\partial\omega=\omega\wedge\omega, \tag{13}$$

so that the curvature matrix is

$$\Omega = (\Omega_i^{\,j}) \underset{\text{def}}{=} d\omega - \omega \wedge \omega = \bar{\partial}\,\omega, \tag{14}$$

whose entries are forms of bidegree (1,1). As a consequence we have

$$\bar{\partial}\,\Omega = 0. \tag{15}$$

The torsion form is by definition

$$\tau^j = \sum_i dz^i \wedge \omega_i^{\,j}, \tag{16}$$

from which it follows that

$$\bar{\partial}\tau^j = -\sum_i dz^i \wedge \Omega_i^{\,j}. \tag{17}$$

Let L be a holomorphic line bundle over M. This is given, relative to an open covering $\{U,V,\ldots\}$ of M, by the local coordinates ξ_U with the coordinate change described by

$$\xi_U\, g_{UV} = \xi_V, \tag{18}$$

where g_{UV} is a non-zero holomorphic function defined in $U \cap V \neq \varnothing$ An hermitian structure in L is given by a C^∞-function $a_U > 0$ in each U, such that

$$\|\xi\|^2 \underset{\text{def}}{=} a_U|\xi_U|^2 = a_V|\xi_V|^2, \quad \text{in } U \cap V \neq \varnothing. \tag{19}$$

This condition can also be written

$$a_U = a_V |g_{UV}|^2. \tag{20}$$

As in (12) the hermitian structure in L defines a connection with the connection form

$$\phi_U = \partial \log a_U. \tag{21}$$

Its curvature form is

$$\Phi = \bar{\partial}\phi_U = -\partial\bar{\partial}\log a_U, \tag{22}$$

which is independent of U. Also independent of U are the forms of bidegree (0, 1):

$$\psi_k = \bar{\partial}\,\frac{\partial \log a_U}{\partial z^k}. \tag{23}$$

They have the following properties, easily verified:

$$\bar\partial \psi_k = 0, \tag{24}$$

$$\sum_k dz^k \wedge \psi_k = -\Phi. \tag{25}$$

We now introduce the matrix

$$\tilde\Omega = (\tilde\Omega_i^j), \tag{26}$$

where

$$\tilde\Omega_i^j = \Omega_i^j - dz^j \wedge \psi_i. \tag{27}$$

The theorem of Baum-Bott can be stated as follows:

THEOREM 1. *Let M be a compact complex manifold of dimension m. Let L be a holomorphic line bundle over M. Consider a holomorphic section $\xi \in \Gamma(T \otimes L)$, which has only isolated non-degenerate zeros. If P is an invariant polynomial of degree m, we have*

$$\left(\frac{i}{2\pi}\right)^m \int_M P(\tilde\Omega) = \sum P(\Xi)/\det \Xi \tag{28}$$

where the summation is extended over all the zeros of ξ.

The left-hand side of (28) can be expressed in terms of the Chern classes of M and the Chern class of L. In fact, we put $P = c^\alpha$ as defined in (9). By substituting the curvature matrix Ω into c_r, we find $c_r(\Omega)$ to be a closed form of bidegree (r, r). Its cohomology class in the sense of de Rham's theorem, when multiplied by $(i/2\pi)^r$, i.e., the class $(i/2\pi)^r\{c_r(\Omega)\}$, is the Chern class $c_r(T)$. Similarly, the cohomology class $(i/2\pi)\{\Phi\}$ is the Chern class $c_1(L)$ of the line bundle L. We introduce the cohomology classes

$$c_r(T-L^{-1}) = c_r(T) + c_{r-1}(T)c_1(L) + \cdots + (c_1(L))^r, \qquad 1 \leqslant r \leqslant m, \tag{29}$$

and

$$c^\alpha(T-L^{-1}) = c_1^{\alpha_1}(T-L^{-1}) \cdots c_m^{\alpha_m}(T-L^{-1}). \tag{30}$$

Then we have the theorem:

THEOREM 2. *Under the hypotheses of Theorem 1, we have*

$$\int_M c^\alpha(T-L^{-1}) = \sum c^\alpha(\Xi)/\det(\Xi), \qquad \alpha_1 + 2\alpha_2 + \cdots + m\alpha_m = m, \tag{31}$$

where the summation is extended over all the zeros of ξ.

2. Local Formulas Involving a Meromorphic Vector Field: Consider a section ξ of the bundle $T\otimes L$. Locally it can be given by

$$\xi=\sum_i \xi_U^i \frac{\partial}{\partial z^i}. \tag{32}$$

It is called holomorphic if the components ξ_U^i are holomorphic functions in z^j. In the following we will consider only holomorphic sections ξ. The covariant differential of ξ is an element of $\Gamma(T\otimes T^*\otimes L)$ (for a bundle W, $\Gamma(W)$ will always denote the space of sections of W) and can be written

$$D\xi=\sum_i D\xi_U^i\otimes\frac{\partial}{\partial z^i}, \tag{33}$$

where

$$D\xi_U^i=d\xi_U^i+\sum_{j,k}\Gamma_{jk}^i\xi_U^j dz^k+\xi_U^i\phi_U. \tag{34}$$

The covariant derivatives, i.e., the coefficients of dz^k of the expression at the right-hand side, are

$$\frac{D\xi_U^i}{\partial z^k}=\frac{\partial\xi_U^i}{\partial z^k}+\sum_j\Gamma_{jk}^i\xi_U^j+\frac{\partial\log a_U}{\partial z^k}\xi_U^i. \tag{35}$$

We introduce another element of $\Gamma(T\otimes T^*\otimes L)$, given by the components

$$-E_k^i=\frac{\partial\xi_U^i}{\partial z^k}+\sum_j\Gamma_{kj}^i\xi_U^j+\frac{\partial\log a_U}{\partial z^k}\xi_U^i. \tag{36}$$

Following Bott we will make use of the operators $\bar\partial$ and $i(\xi)$, where $i(\xi)$ is the interior product with ξ. Both are anti-derivations and satisfy the relations

$$\bar\partial^2=i(\xi)^2=0, \tag{37}$$

$$\bar\partial i(\xi)+i(\xi)\bar\partial=0, \tag{38}$$

$$i(\xi)(dz^j)=\xi^j,\qquad i(\xi)(d\bar z^j)=0. \tag{39}$$

From (27) and (36) we have immediately the important relation

$$\bar\partial E_k^i=i(\xi)\tilde\Omega_k^i. \tag{40}$$

With a given invariant polynomial P we introduce the forms

$$P^{(r)}(\tilde{\Omega}) = \binom{m}{r} P(\underbrace{E,\dots,E}_{m-r},\underbrace{\tilde{\Omega},\dots,\tilde{\Omega}}_{r}), \qquad 0 \leqslant r \leqslant m, \tag{41}$$

where $E=(E^i_k)$. $P^{(r)}(\tilde{\Omega})$ is a form in M of bidegree (r,r), with value in the tensor product $L^{m-r}=L\otimes\cdots\otimes L$ ($m-r$ factors). We find immediately

$$\bar{\partial} P^{(r)}(\tilde{\Omega}) = i(\xi) P^{(r+1)}(\tilde{\Omega}). \tag{42}$$

The section ξ, when it is non-zero, defines a form of bidegree (1,0), with value in L^{-1}, which is

$$\pi = \sum_{i,k} h_{ik}\bar{\xi}^k_U dz^i \Big/ \sum_{i,k} h_{ik}\xi^i_U\bar{\xi}^k_U, \qquad \xi \neq 0. \tag{43}$$

We have

$$i(\xi)\pi = 1 \tag{44}$$

and it is easily verified that

$$i(\xi)\,\bar{\partial}\pi = 0. \tag{45}$$

Then we have

$$i(\xi)\bar{\partial}[\pi\wedge(\bar{\partial}\pi)^{m-r-1}\wedge P^{(r)}(\tilde{\Omega})]$$

$$= (\bar{\partial}\pi)^{m-r}\wedge i(\xi)P^{(r)}(\tilde{\Omega}) - (\bar{\partial}\pi)^{m-r-1}\wedge i(\xi)P^{(r+1)}(\tilde{\Omega}), \qquad 0 \leqslant r \leqslant m-1. \tag{46}$$

It follows that

$$i(\xi)\left\{ \bar{\partial} \sum_{0\leqslant r\leqslant m-1} \pi\wedge(\bar{\partial}\pi)^{m-r-1}\wedge P^{(r)}(\tilde{\Omega}) + P^{(m)}(\tilde{\Omega}) \right\} = 0. \tag{47}$$

The form between the braces is a form of bidegree (m,m) in $M-$(zero of ξ). Since $\xi\neq 0$, the above equation implies that the form itself is zero. Moreover, the expression under $\bar{\partial}$ is a form of bidegree $(m,m-1)$, so that $\bar{\partial}$ can be replaced by d. We have therefore

$$P^{(m)}(\tilde{\Omega}) = d\Pi \qquad \text{in } M-(\text{zero of } \xi), \tag{48}$$

where

$$\Pi = -\sum_{0\leqslant r\leqslant m-1} \pi\wedge(\bar{\partial}\pi)^{m-r-1}\wedge P^{(r)}(\tilde{\Omega}). \tag{49}$$

3. Proof of the Theorems: By (8) and (27) we have

$$c_r(\tilde{\Omega}) = \frac{1}{r!}\sum \delta^{i_1\ldots i_r}_{k_1\ldots k_r}\tilde{\Omega}^{k_1}_{i_1}\wedge\cdots\wedge\tilde{\Omega}^{k_r}_{i_r}$$

$$= \frac{1}{r!}\sum_{0\leqslant s\leqslant r}(-1)^{r-s}\binom{r}{s}\sum \delta^{i_1\ldots i_r}_{k_1\ldots k_r}\Omega^{k_1}_{i_1}\wedge\cdots\wedge\Omega^{k_s}_{i_s}\wedge dz^{k_{s+1}}\wedge\psi_{i_{s+1}}\wedge\cdots\wedge dz^{k_r}\wedge\psi_{i_r}.$$

In the second summation $k_1,\ldots,k_r$ are a permutation of $i_1,\ldots,i_r$ and the resulting terms are summed with respect to all the i's. Among the permutations there are two possibilities: either $k_1,\ldots,k_s$ are a permutation of $i_1,\ldots,i_s$, and then $k_{s+1},\ldots,k_r$ are a permutation of $i_{s+1},\ldots,i_r$, or otherwise. In the latter case each term contains a factor such as $\sum_j \Omega^k_j\wedge dz^j$, which by (17) is equal to $-\bar{\partial}\tau^k$. Since $\bar{\partial}\Omega^k_i=\bar{\partial}\psi_i=0$, such a term is in the image of $\bar{\partial}$. We have therefore

$$c_r(\tilde{\Omega})\equiv\frac{1}{r!}\sum_{0\leqslant s\leqslant r}(-1)^{r-s}\binom{r}{s}\Big\{\sum\delta^{i_1\ldots i_s}_{k_1\ldots k_s}\Omega^{k_1}_{i_1}\wedge\cdots\wedge\Omega^{k_s}_{i_s}$$

$$\wedge\sum\delta^{i_{s+1}\cdots i_r}_{k_{s+1}\cdots k_r}dz^{k_{s+1}}\wedge\psi_{i_{s+1}}\wedge\cdots\wedge dz^{k_r}\wedge\psi_{i_r}\Big\},$$

where the congruence means that the two sides differ by a term which lies in the image of $\bar{\partial}$. Noticing that

$$\frac{(-1)^{r-s}}{(r-s)!}\sum\delta^{i_{s+1}\ldots i_r}_{k_{s+1}\ldots k_r}dz^{k_{s+1}}\wedge\psi_{i_{s+1}}\wedge\cdots\wedge dz^{k_r}\wedge\psi_{i_r}=\Phi^{r-s},$$

we get

$$c_r(\tilde{\Omega})\equiv\sum_{0\leqslant s\leqslant r}c_s(\Omega)\wedge\Phi^{r-s},\qquad 1\leqslant r\leqslant m. \tag{50}$$

Put

$$c_r(\Omega,\Phi)=\sum_{0\leqslant s\leqslant r}c_s(\Omega)\wedge\Phi^{r-s},\qquad 1\leqslant r\leqslant m \tag{51}$$

and

$$c^{\alpha}(\Omega,\Phi)=c_1^{\alpha_1}(\Omega,\Phi)\wedge\cdots\wedge c_m^{\alpha_m}(\Omega,\Phi). \tag{52}$$

Then we have

$$\left(\frac{i}{2\pi}\right)^{|\alpha|}\{c^{\alpha}(\Omega,\Phi)\}=c^{\alpha}(T-L^{-1}), \qquad |\alpha|=\alpha_1+2\alpha_2+\cdots+m\alpha_m, \tag{53}$$

where the braces denote the cohomology class represented by the differential form inside. If P is an invariant polynomial of degree m, $P(\tilde{\Omega})$ is a linear combination of $c^{\alpha}(\tilde{\Omega})$ with $\alpha_1+2\alpha_2+\cdots+m\alpha_m=m$. By the above discussion every such $c^{\alpha}(\tilde{\Omega})$ differs from $c^{\alpha}(\Omega,\Phi)$ by a term of the form $\bar{\partial}Q$, where Q is of bidegree $(m, m-1)$. For such a Q we have $\bar{\partial}Q=dQ$. Hence

$$\int_M c^{\alpha}(\tilde{\Omega})=\int_M c^{\alpha}(\Omega,\Phi) \tag{54}$$

and the integral depends only on the Chern classes of T and L.

To prove Theorem 1 we use the fact established above that the integral at the left-hand side of (28) is independent of the hermitian structures in M and L. By Stokes theorem formula (48) reduces this integral to a sum of integrals of Π over sufficiently small spheres above the zeros of ξ, which are finite in number. Consider one such point p and let z^i, $1\leqslant i,j\leqslant m$, be a local coordinate system in M centered at p. In a neighborhood of p we choose the hermitian metric $\sum_i dz^i d\bar{z}^i$ in M and the hermitian metric in L with $a_U=1$. This is done at all the zeros of ξ and is extended, by a partition of unity, over M. If we restrict to a neighborhood of p, we have therefore $\tilde{\Omega}=0$ and

$$\Pi=(-1)^{m+1}P(\Xi)\pi\wedge(\bar{\partial}\pi)^{m-1}, \qquad \Xi=\left(\frac{\partial\xi^i}{\partial z^j}\right), \tag{55}$$

where

$$\pi=\sum_i \bar{\xi}^i dz^i \Big/ \sum_i |\xi^i|^2. \tag{56}$$

It follows that

$$\left(\sum_i |\xi^i|^2\right)^m \pi\wedge(\bar{\partial}\pi)^{m-1}$$

$$=(-1)^{\frac{1}{2}m(m-1)}(m-1)!\,dz^1\wedge\cdots\wedge dz^m\wedge\Theta, \tag{57}$$

where

$$\Theta=\sum_{1\leqslant i\leqslant m}(-1)^{i-1}\bar{\xi}^i d\bar{\xi}^1\wedge\cdots\wedge d\bar{\xi}^{i-1}\wedge d\bar{\xi}^{i+1}\wedge\cdots\wedge d\bar{\xi}^m. \tag{58}$$

Under our hypothesis that $\det \Xi \neq 0$ we can write this expression (57) as a numerical multiple of

$$\frac{1}{\det \Xi} d\xi^1 \wedge \cdots \wedge d\xi^m \wedge \Theta. \tag{59}$$

Consider the space C_m with the coordinates $\xi^1, \ldots, \xi^m$, in which we put

$$(\sum |\xi^i|^2)^m \Sigma_\xi = C d\xi^1 \wedge \cdots \wedge d\xi^m \wedge \Theta \tag{60}$$

where C is a numerical constant. It is easily verified that $d\,\Sigma_\xi = 0$ and that, when C is chosen properly, Σ_ξ restricts to the volume element with total volume 1 on the unit sphere defined by $\sum |\xi^i|^2 = 1$. Let f denote the mapping $(z^1, \ldots, z^m) \to (\xi^1(z), \ldots, \xi^m(z))$, which is one-to-one in a neighborhood of $(0, \ldots, 0)$. Let S_ϵ be the sphere $\sum |z^i|^2 = \epsilon^2$. Then the mapping f on S_ϵ followed by $(\xi^1, \ldots, \xi^m) \to (\xi^1/|\xi|, \ldots, \xi^m/|\xi|)$, $|\xi|^2 = \sum |\xi^i|^2$, has degree 1 and we have

$$\int_{S_\epsilon} f^* \Sigma_\xi = 1.$$

It follows that, up to a numerical factor, the integral $\int_{S_\epsilon} \Pi$ tends, as $\epsilon \to 0$, to the value of $P(\Xi)/\det \Xi$ at p. Thus we proved that

$$\left(\frac{i}{2\pi}\right)^m \int_M P(\tilde{\Omega}) = c' \sum P(\Xi)/\det \Xi,$$

where the summation is over the zeros of ξ and c' is a numerical constant, independent of P. By choosing $P = \det$, we find $c' = 1$. This proves Theorem 1.

Theorem 2 follows immediately from Theorem 1 and the formulas (53), (54).

REFERENCES

1. M. F. Atiyah and I. M. Singer, "The index of elliptic operators: III", *Ann. of Math.* **87** (1968), 546–604.

2. R. Bott, "Vector fields and characteristic numbers", *Mich. Math. J.* **14** (1967), 231–244.

3. P. F. Baum and R. Bott, "On the zeros of meromorphic vector fields", *Essays on Topology and Related Topics,* Springer 1970, pp. 29—47.

4. S. Chern, "Complex manifolds without potential theory", Van Nostrand Notes 1967.

University of California
Berkeley, California

Characteristic forms and geometric invariants

By SHIING-SHEN CHERN AND JAMES SIMONS*

1. Introduction

This work, originally announced in [4], grew out of an attempt to derive a purely combinatorial formula for the first Pontrjagin number of a 4-manifold. The hope was that by integrating the characteristic curvature form (with respect to some Riemannian metric) simplex by simplex, and replacing the integral over each interior by another on the boundary, one could evaluate these boundary integrals, add up over the triangulation, and have the geometry wash out, leaving the sought after combinatorial formula. This process got stuck by the emergence of a boundary term which did not yield to a simple combinatorial analysis. The boundary term seemed interesting in its own right and it and its generalization are the subject of this paper.

The Weil homomorphism is a mapping from the ring of invariant polynomials of the Lie algebra of a Lie group, G, into the real characteristic cohomology ring of the base space of a principal G-bundle, cf. [5], [7]. The map is achieved by evaluating an invariant polynomial P of degree l on the curvature form Ω of a connection, θ, on that bundle, and obtaining a closed form on the base, $P(\Omega^l)$. Because the lift of a principal bundle over itself is trivial, the forms $P(\Omega^l)$ are exact in the bundle. Moreover, in a way that is canonical up to an exact remainder one can construct a form $TP(\theta)$ on the bundle such that $dTP(\theta) = P(\Omega^l)$. Under some circumstances, e.g., $\dim P(\Omega^l) >$ dim base, $P(\Omega^l) = 0$ and $TP(\theta)$ defines a real cohomology class in the bundle. Our object here is to give some geometrical significance to these classes.

In § 2 we review standard results in connection theory. In § 3 we construct the forms $TP(\theta)$ and derive some basic properties. In particular we show that if $\deg P = n$ and the base manifold has dim $2n - 1$ that the forms $TP(\theta)$ lead to real cohomology classes in the total space, and, in the case that $P(\Omega^l)$ is universally an integral class, to R/Q characteristic numbers. Both the class above and the numbers depend on the connection.

In § 4 we restrict ourselves to the principal tangent bundle of a

* Work done under partial support of NSF Grants GP-20096 and GP-31526.

manifold and show that if θ, θ', Ω, Ω' are the connection and curvature forms of conformally related Riemannian metrics then $P(\Omega^l) = P(\Omega'^l)$. Moreover, if $P(\Omega^l) = 0$ then $TP(\theta)$ and $TP(\theta')$ determine the same cohomology class and thus define a conformal invariant of M. In § 5 we examine the question of conformal immersion of an n-dim manifold into R^{n+k}. We show that a necessary condition for such an immersion is that in the range $i > [k/2]$ the forms $P_i^{\perp}(\Omega^{2i}) = 0$, and the classes $\{(1/2)TP_i^{\perp}(\theta)\}$ be integral classes in the principal bundle. Here $P_i^{\perp}$ is the ith inverse Pontrjagin polynomial. In § 6 we apply these results to 3-manifolds.

In a subsequent paper, [3], by one of the present authors and J. Cheeger, it will be shown that the forms $TP(\theta)$ can be made to live on the manifold below in the form of "differential characters". These are homomorphisms from the group of smooth singular cycles into R/Z, subject to the restriction that on boundaries they are the mod Z reduction of the value of a differential form with integral periods evaluated on a chain whose boundary is the given one. These characters form a graded ring, and this ring structure may be exploited to perform vector bundle calculations of geometric interest.

We are very happy to thank J. Cheeger, W. Y. Hsiang, S. Kobayashi, J. Roitberg, D. Sullivan, and E. Thomas for a number of helpful suggestions.

2. Review of connection theory*

Let G be a Lie group with finitely many components and Lie algebra $\mathfrak{G}$. Let M be a C^∞ oriented manifold, and $\{E, M\}$ a *principal G-bundle* over M with projection $\pi: E \to M$. Rg: $E \to E$ will denote right action by $g \in G$. If $\{E', M'\}$ is another principal G-bundle and $\varphi: E \to E'$ is a C^∞ map commuting with right action, φ is called a *bundle map*. Such a map defines $\varphi: M \to M'$, and the use of the same symbol should lead to no confusion.

Let $\{E_G, B_G\}$ be a universal bundle and classifying space for G. B_G is not a manifold. Its key feature is that every principal G-bundle over M admits a bundle map into $\{E_G, B_G\}$, and any two such maps of the same bundle are homotopic. If Λ is any coefficient ring, $u \in H^k(B_G, \Lambda)$, and $\alpha = \{E, M\}$, then the *characteristic class*

$$u(\alpha) \in H^k(M, \Lambda)$$

is well-defined by pulling back u under any bundle map. Since G is assumed to have only finitely many components it is well-known that

$$H^{2l-1}(B_G, R) = 0 \quad \text{all } l . \tag{2.1}$$

We finally recall that E_G is *contractible.*

* This chapter summarizes material presented in detail in [7].

Let $\mathfrak{G}^l = \overbrace{\mathfrak{G} \otimes \mathfrak{G} \otimes \cdots \otimes \mathfrak{G}}^{l}$. Polynomials of degree l are defined to be symmetric, multilinear maps from $\mathfrak{G}^l \to R$. G acts on $\mathfrak{G}^l$ by inner automorphism, and polynomials invariant under this action are called *invariant polynomials* of degree l, and are denoted by $I^l(G)$. These multiply in a natural way, and if $P \in I^l(G)$, $Q \in I^{l'}(G)$ then $PQ \in I^{l+l'}(G)$. We set $I(G) = \sum \otimes I^l(G)$, a graded ring.

These polynomials give information about the real cohomology of B_G. In fact, there is a *universal Weil homomorphism*

$$W: I^l(G) \longrightarrow H^{2l}(B_G, R) \tag{2.2}$$

such that $W: I(G) \to H^{\text{even}}(B_G, R)$ is a ring homomorphism.

If $\{E, M\}$ is a principal G-bundle over M we denote by $\Lambda^{k,l}(E)$ k-forms on E taking values in $\mathfrak{G}^l$. We have the usual exterior differential $d: \Lambda^{k,l}(E) \to \Lambda^{k+1,l}(E)$. If $\varphi \in \Lambda^{k,l}(E)$ and $\psi \in \Lambda^{k',l'}(E)$ define

$$\varphi \wedge \psi \in \Lambda^{k+k',l+l'}(E)$$

$$\varphi \wedge \psi(x_1, \cdots, x_{k+k'}) = \sum_{\pi \text{ shuffle}} \sigma(\pi)\varphi(x_{\pi_1}, \cdots, x_{\pi_k}) \otimes \psi(x_{\pi_{k+1}}, \cdots, x_{\pi_{k+k'}}) .$$

If $\varphi \in \Lambda^{k,1}(E)$ and $\psi \in \Lambda^{k',1}(E)$ define

$$[\varphi, \psi] \in \Lambda^{k+k',1}(E)$$

$$[\varphi, \psi](x_1, \cdots, x_{k+k'}) = \sum_{\pi \text{ shuffle}} \sigma(\pi)[\varphi(x_{\pi_1}, \cdots, x_{\pi_k}), \psi(x_{\pi_{k+1}}, \cdots, x_{\pi_{k+k'}})] .$$

Let P be a polynomial of degree l and $\varphi \in \Lambda^{k,l}(E)$. Then $P(\varphi) = P \circ \varphi$ is a *real valued* k-form on E. The following are elementary

$$[\varphi, \psi] = (-1)^{kk'+1}[\psi, \varphi] \tag{2.3}$$

$$[[\varphi, \varphi], \varphi] = 0 \tag{2.4}$$

$$d[\varphi, \psi] = [d\varphi, \psi] + (-1)^k[\varphi, d\psi] \tag{2.5}$$

$$d(\varphi \wedge \psi) = d\varphi \wedge \psi + (-1)^k \varphi \wedge d\psi \tag{2.6}$$

$$d(P(\varphi)) = P(d\varphi) \tag{2.7}$$

$$P(\varphi \wedge \psi \wedge \rho) = (-1)^{kk'} P(\psi \wedge \varphi \wedge \rho) \tag{2.8}$$

where $\varphi \in \Lambda^{k,l}$, $\psi \in \Lambda^{k',l'}$, $\rho \in \Lambda^{k'',l''}$ and in the first three lines $l = l' = 1$. If $P \in I^l(G)$ then differentiating the invariance condition shows

$$\sum_{i=1}^{l} (-1)^{k_1+\cdots+k_i} P(\psi_1 \wedge \cdots \wedge [\psi_i, \varphi] \wedge \cdots \wedge \psi_l) = 0 \tag{2.9}$$

where $\psi_i \in \Lambda^{k_i,1}(E)$ and $\varphi \in \Lambda^{1,1}(E)$.

For $e \in E$, let $T(E)_e$ denote the tangent space of E at e and $V(E)_e = \{x \in T(E)_e \mid d\pi(x) = 0\}$. $V(E)_e$ is called the *vertical space*, and may be canonically identified with $\mathfrak{G}$. If $x \in V(E)_e$ we let $\bar{x} \in \mathfrak{G}$ denote its image under this identification. A *connection* on $\{E, M\}$ in a $\mathcal{G}$ valued 1-form, θ, on E satisfying $R_g^*(\theta) = \mathrm{ad}_g^{-1} \circ \theta$, and $\theta(v) = \bar{v}$ for vertical v. If θ is a connection, setting $H(E)_e = \{x \in T(E)_e \mid \theta(x) = 0\}$ defines a complement to $V(E)_e$ called the *horizontal space*; i.e., $T(E)_e \cong V(E)_e \oplus H(E)_e$ and $dR_g(H(E)_e) = H(E)_{R_g(e)}$. The *structural equation* states

$$d\theta = \Omega - \frac{1}{2}[\theta, \theta] \tag{2.10}$$

where Ω is the *curvature form*. $\Omega \in \Lambda^{2,1}(E)$ and is horizontal, i.e., $\Omega(x, y) = \Omega(H(x), H(y))$, $H(x)$, and $H(y)$ denoting the horizontal projections of x and y. (2.4) and (2.5) show

$$d\Omega = [\Omega, \theta]. \tag{2.11}$$

An element $\varphi \in \Lambda^{k,l}$ is called *equivariant* if $R_g^*(\varphi) = \mathrm{ad}\, g^{-1} \circ \varphi$. A connection is equivariant by definition, and so is its curvature by (2.10), as equivariance is preserved under d, wedge products, and brackets. If $\varphi \in \Lambda^{k,l}(E)$ is equivariant and $P \in I^l(G)$ then $P(\varphi)$ is a real valued *invariant* k-form on E. In particular, $\Omega^l = \overbrace{\Omega \wedge \cdots \wedge \Omega}^{l}$ is equivariant, and so $P(\Omega^l)$ is real valued, invariant and horizontal, and so uniquely defines a $2l$-form on M whose lift is $P(\Omega^l)$. We will also denote this form on M by $P(\Omega^l)$. Formulae (2.11) and (2.9) immediately show this form is *closed*.

THEOREM 2.12 (*Weil homomorphism*). *Let $\alpha = \{E, M, \theta\}$ be a principal G-bundle with connection, and let $P \in I^l(G)$. Then*

$$P(\Omega^l) \in W(P)(\alpha) \,;$$

i.e., $P(\Omega^l)$ represents the characteristic class corresponding to the universal Weil image of P.

For some of the calculations in the sections that follow it will be convenient to have classifying bundles equipped with connections. To do this we use a theorem of Narasimhan-Ramanan [10]. We introduce the category $\varepsilon(G)$. *Objects* in $\varepsilon(G)$ are triples $\alpha = \{E, M, \theta\}$ where $\{E, M\}$ is a principal G-bundle with connection θ. *Morphisms* are connection-preserving bundle maps; i.e., if $\alpha = \{E, M, \theta\}$ and $\hat{\alpha} = \{\hat{E}, \hat{M}, \hat{\theta}\}$, and $\varphi\colon \{E, M\} \to \{\hat{E}, \hat{M}\}$ is a bundle map, then $\varphi\colon \alpha \to \hat{\alpha}$ is a morphism if $\varphi^*(\hat{\theta}) = \theta$. An object $A \in \varepsilon(G)$ is called *n-classifying* if two conditions hold: First for every $\alpha \in \varepsilon(G)$ with $\dim M \leq n$ there exists a morphism $\varphi\colon \alpha \to A$. Second, any two such morphisms are homotopic through bundle maps. We do not require the homotopy to be via morphisms.

THEOREM 2.13 (*Narasimhan-Ramanan*). *For each integer n there exists an n-classifying* $A \in \varepsilon(G)$.

3. The forms $TP(\theta)$

Let $\alpha = \{E, M, \theta\} \in \varepsilon(G)$. The bundle $\{\pi^*(E), E\}$ is trivial as a principal G-bundle, and so all of its characteristic cohomology vanishes. Thus $\pi^*(P(\Omega^l)) = P(\Omega^l)$ is exact when considered as a form on E. Set $\varphi_t = t\Omega + 1/2(t^2 - t)[\theta, \theta]$, and set

$$TP(\theta) = l\int_0^1 P(\theta \wedge \varphi_t^{l-1})dt \,. \tag{3.1}$$

$P \in I^l(G)$, and $TP(\theta)$ is a real-valued invariant $(2l - 1)$-form on E. It is of course not horizontal.

PROPOSITION 3.2. $dTP(\theta) = P(\Omega^l)$.

Proof. Set $f(t) = P(\varphi_t^l)$. Then $f(0) = 0$ and $f(1) = P(\Omega^l)$. Thus

$$P(\Omega^l) = \int_0^1 f'(t)dt \,. \tag{3.3}$$

We claim

$$f'(t) = ldP(\theta \wedge \varphi_t^{l-1}) \,. \tag{3.4}$$

We first observe

$$\frac{d}{dt}(\varphi_t) = \Omega + \left(t - \frac{1}{2}\right)[\theta, \theta] \,.$$

Using (2.3)-(2.8) we have

$$\begin{aligned} f' &= lP\left(\frac{d}{dt}(\varphi_t) \wedge \varphi_t^{l-1}\right) \\ &= lP(\Omega \wedge \varphi_t^{l-1}) + l\left(t - \frac{1}{2}\right)P([\theta, \theta] \wedge \varphi_t^{l-1}) \,. \end{aligned}$$

On the other hand,

$$\begin{aligned} ldP(\theta \wedge \varphi_t^{l-1}) &= lP(d\theta \wedge \varphi_t^{l-1}) - l(l-1)P(\theta \wedge d\varphi_t \wedge \varphi_t^{l-2}) \\ &= lP(\Omega \wedge \varphi_t^{l-1}) - \frac{1}{2}lP([\theta, \theta] \wedge \varphi_t^{l-1}) - l(l-1)P(\theta \wedge d\varphi_t \wedge \varphi_t^{l-2}) \end{aligned}$$

by the structural equation (2.10). Now using (2.10), (2.11), and (2.4)

$$d\varphi_t = t[\varphi_t, \theta] \,.$$

Plugging this into the formula above and using the invariance formula (2.9) on the last piece we get

$$l dP(\theta \wedge \varphi_t^{l-1}) = lP(\Omega \wedge \varphi_t^{l-1}) - \frac{1}{2} lP([\theta, \theta] \wedge \varphi_t^{l-1}) + ltP([\theta, \theta] \wedge \varphi_t^{l-1}) = f'$$

by the computation above. This shows (3.4) and the proposition follows from (3.3).

The form $TP(\theta)$ can of course be written without the integral, and, in fact, setting

$$A_i = (-1)^i l!\,(l-1)!/2^i(l+i)!\,(l-1-i)!$$

one computes

$$TP(\theta) = \textstyle\sum_{i=0}^{l-1} A_i P(\theta \wedge [\theta, \theta]^i \wedge \Omega^{l-i-1}) . \tag{3.5}$$

The operation which associates to $\alpha \in \varepsilon(G)$ the form $TP(\theta)$ is natural; i.e., if $\varphi\colon \alpha \to \hat{\alpha}$ is a morphism, since $\varphi^*(\hat{\theta}) = \theta$ and thus $\varphi^*(\hat{\Omega}) = \Omega$, clearly $\varphi^*(TP(\hat{\theta})) = TP(\theta)$. This naturality characterizes T up to an exact remainder:

PROPOSITION 3.6. *Given $P \in I^l(G)$, let S be another functor which associates to each $\alpha \in \varepsilon(\theta)$ a $(2l-1)$-form in E, $SP(\theta)$, which satisfies $dSP(\theta) = P(\Omega^l)$. Then $TP(\theta) - SP(\theta)$ is exact.*

Proof. Let $\alpha = \{E, M, \theta\}$ with $\dim M = n$. Choose $\hat{\alpha} = \{\hat{E}, \hat{M}, \hat{\theta}\} \in \varepsilon(G)$ so that $\hat{\alpha}$ is m classifying with m sufficiently greater than n. Let $\varphi\colon \alpha \to \hat{\alpha}$ be a morphism. Now in $\hat{E}$ we have $dSP(\hat{\theta}) = dTP(\hat{\theta}) \Rightarrow SP(\hat{\theta}) - TP(\hat{\theta})$ is closed. But since $\hat{E}$ is an approximation to E_G its $2l-1$ cohomology vanishes for sufficiently large m. Thus $SP(\hat{\theta}) = TP(\hat{\theta}) +$ exact. So by the naturality assumption on S, $SP(\theta) = \varphi^* SP(\hat{\theta}) = \varphi^* TP(\hat{\theta}) + \varphi^*$ exact $= TP(\theta) +$ exact. □

PROPOSITION 3.7. *Let $P \in I^l(G)$ and $Q \in I^s(G)$.*
(1) $PQ(\Omega^{l+s}) = P(\Omega^l) \wedge Q(\Omega^s)$
(2) $TPQ(\theta) = TP(\theta) \wedge Q(\Omega^s) +$ *exact* $= TQ(\theta) \wedge P(\Omega^l) +$ *exact.*

Proof. (1) is immediate. To prove (2) we may use naturality and work in a classifying bundle. But there, $d(TP(\theta) \wedge Q(\Omega^s)) = P(\Omega^l) \wedge Q(\Omega^s) = PQ(\Omega^{l+s}) = d(TPQ(\theta))$. Similarly $d(TQ(\theta) \wedge P(\Omega^l)) = d(TPQ(\theta))$. (2) then follows by low dimensional acyclicity of the total space of the classifying bundle. □

We are interested in how the forms $TP(\theta)$ change as the connection changes.

PROPOSITION 3.8. *Let $\theta(s)$ be a smooth 1-parameter family of connections on $\{E, M\}$ with $s \in [0, 1]$. Set $\theta = \theta(0)$ and $\theta' = (d/ds)(\theta(s))|_{s=0}$. For $P \in I^l(G)$*

$$\frac{d}{ds}(TP(\theta(s)))\,|_{s=0} = lP(\theta' \wedge \Omega^{l-1}) + \mathit{exact}\,.$$

Proof. Building on the theorem of Narasimhan-Ramanan it is not difficult to show that one can find a principal G-bundle $\{\hat{E}, \hat{M}\}$ which classifies bundles over manifolds of dim $\geqq m \geqq$ dim M, and to equip this bundle with a smooth family of connections $\hat{\theta}(s)$, and to find a bundle map φ: $\{E, M\} \rightarrow \{\hat{E}, \hat{M}\}$ such that $\varphi^*(\hat{\theta}(s)) = \theta(s)$ $s \in [0, 1]$. It thus suffices to prove the theorem in $\{\hat{E}, \hat{M}\}$, and by choosing m large enough $\hat{E}$ will be acyclic in dimensions $\leqq 2l-1$. We now drop all "hats" and simply assume $H^{2l-1}(E, R) = 0$. Thus it is sufficient to prove both sides of the equation have the same differential. Now

$$d\Big(\frac{d}{ds}(TP\theta(s))\,|_{s=0}\Big) = \frac{d}{ds}(dTP(\theta(s)))\,|_{s=0}$$
$$= \frac{d}{ds}(P(\Omega(s)^l)\,|_{s=0}) = lP(\Omega' \wedge \Omega^{l-1})$$

where $\Omega' = (d/ds)(\Omega(s))\,|_{s=0}$. Also

$$\begin{aligned} d(lP(\theta' \wedge \Omega^{l-1})) &= lP(d\theta' \wedge \Omega^{l-1}) - l(l-1)P(\theta' \wedge d\Omega \wedge \Omega^{l-2}) \\ &= lP(d\theta' \wedge \Omega^{l-1}) - l(l-1)P(\theta' \wedge [\Omega, \theta] \wedge \Omega^{l-2}) \text{ by (2.11)} \\ &= lP(d\theta' \wedge \Omega^{l-1}) + lP([\theta', \theta] \wedge \Omega^{l-1}) \text{ by (2.9)}\,. \end{aligned}$$

Now $d\theta' = d((d/ds)(\theta(s))|_{s=0}) = (d/ds)(d\theta(s))|_{s=0} = (d/ds)(\Omega(s) - (1/2)[\theta(s), \theta(s)])|_{s=0} = \Omega' - [\theta', \theta]$. Putting this in the calculation above shows

$$dlP(\theta' \wedge \Omega^{l-1}) = lP(\Omega' \wedge \Omega^{l-1})$$

and this with the first calculation completes the proof. □

If $P \in I^l(G)$ and $P(\Omega^l) = 0$ then $TP(\theta)$ is closed in E and so defines a cohomology class in E. We denote this class by $\{TP(\theta)\} \in H^{2l-1}(E, R)$.

THEOREM 3.9. *Let* $\alpha = \{E, M, \theta\}$ *with* dim $M = n$. *If* $2l - 1 = n$ *then* $TP(\theta)$ *is closed and* $\{TP(\theta)\} \in H^n(E, R)$ *depends on* θ. *If* $2l - 1 > n$ *then* $TP(\theta)$ *is closed and* $\{TP(\theta)\} \in H^{2l-1}(E, R)$ *is independent of* θ.

Proof. $P(\Omega^l)$ is a horizontal $2l$-form. If $2l - 1 \geqq n$ then $2l > n$ and since the dimension of the horizontal space is exactly n, $P(\Omega^l) = 0$. Thus $TP(\theta)$ is closed, and $\{TP(\theta)\}$ is defined. We will see in a later section that when $2l - 1 = n$, $\{TP(\theta)\}$ depends on the connection. However, suppose $2l - 1 > n$. Since any two connections may be joined by a smooth 1-parameter family, it is sufficient to show, using the notation of the previous proposition that

$$\frac{d}{ds}(TP(\theta(s)))\,|_{s=0} = \text{exact}\,.$$

By that proposition it is sufficient to show $P(\theta' \wedge \Omega^{l-1}) = 0$. Since θ' is the derivative of a family of connections, all of which must agree on vertical vectors, $\theta'(v) = 0$ for v vertical. Thus $P(\theta' \wedge \Omega^{l-1})$ is a horizontal $(2l-1)$-form, and thus must vanish for $2l - 1 > n$. □

The equation in E, $dTP(\theta) = P(\Omega^l)$, implies that $TP(\theta)\,|\,E_m$ is a closed form, where E_m is the fibre over $m \in M$. Formula (3.5) shows that $TP(\theta)\,|\,E_m$ is expressed purely in terms of $\theta\,|\,E_m$, which is independent of the connection. More specifically, let ω denote the Maurer-Cartan form on G, which assigns to each tangent vector the corresponding Lie algebra element. Set

(3.10) $$TP = \frac{(-1)^{l-1}}{2^l\binom{2l-1}{l}} P(\omega \wedge [\omega, \omega]^{l-1})\,.$$

TP is a real valued, bi-invariant $(2l-1)$-form on G. It is closed and represents an element of $H^{2l-1}(G, R)$. For $m \in M$ and $e \in E_m$ let $\lambda\colon G \to E_m$ by $\lambda(g) = R_g(e)$. Then $\lambda^*(\theta) = \omega$, and (3.5) shows

(3.11) $$\lambda^*(TP(\theta)) = TP\,.$$

The class $\{TP\} \in H^{2l-1}(G, R)$ is universally transgressive in the sense of [1]. In fact, recalling Borel's definition of transgressive ([1], p. 133), a class $h \in H^k(G, \Lambda)$ is called transgressive in the fibre space $\{E, M\}$ if there exists $c \in C^k(G, \Lambda)$ so that $c\,|\,G \in h$ and δc is a lift of a cochain (and thus a cocycle) from the base. It is called universally transgressive if this happens in the classifying bundle. In this case the transgression goes from $\{TP\}$ via $TP(\theta)$ to $P(\Omega^l)$. One can do this over the integers as well as the reals, and if we set

$$I_0^l(G) = \{P \in I^l(G)\,|\,W(P) \in H^{2l-1}(B_G, Z)\}$$

one can easily show

(3.12) $$P \in I_0^l(G) \Longrightarrow \{TP\} \in H^{2l-1}(G, Z)$$

and (3.11) shows this is equivalent to

(3.13) $$P \in I_0^l(G) \Longrightarrow TP(\theta)\,|\,E_m \in H^{2l-1}(E_m, Z)$$

where in all these equations we mean the real image of the integral cohomology. The following proposition will provide a proof of this, but also will give us some extra understanding of the form $TP(\theta)$ when $P \in I_0^l(G)$.

For a real number a let $\tilde{a} \in R/Z$ denote its reduction, and similarly for any real cochain or cohomology class $\sim$ will denote its reduction mod Z. The Bockstein exact sequence

$$(3.14) \quad \longrightarrow H^i(X, Z) \xrightarrow{r} H^i(X, R) \xrightarrow{\sim} H^i(X, R/Z) \xrightarrow{b} H^{i+1}(X, Z) \longrightarrow$$

shows that a real class, U, is an integral class if and only if $\tilde{U} = 0$. For X any manifold and Λ any coefficient group we let $C(X, \Lambda)$ denote the cochain group with respect to the group of *smooth singular* chains. If φ is a differential form on X then $\varphi \in C(X, R)$, and by $\tilde{\varphi} \in C(X, R/Z)$ we mean its reduction mod Z as a real cochain.

PROPOSITION 3.15. *Let $\alpha = \{E, M, \theta\} \in \varepsilon(G)$. Then if $P \in I_0^l(G)$ there exists $u \in C^{2l-1}(M, R/Z)$ so that*

$$\widetilde{TP(\theta)} = \pi^*(u) + coboundary\ .$$

Proof. Let $\hat{\alpha} = \{\hat{E}, \hat{M}, \hat{\theta}\} \in \varepsilon(G)$ be k-classifying with k sufficiently large. Since $P \in I_0^l$ we know that $P(\hat{\Omega}^l)$ represents an integral class in $\hat{M}$. Thus the R/Z cocycle $\widetilde{P(\hat{\Omega}^l)}$ vanishes on all cycles in $\hat{M}$, and so is an R/Z coboundary; i.e., there exists $\hat{u} \in C^{2l-1}(\hat{M}, R/Z)$ such that $\delta\hat{u} = \widetilde{P(\hat{\Omega}^l)}$. Thus

$$\begin{aligned} \delta\pi^*(\hat{u}) &= \pi^*(\delta\hat{u}) = \pi^*(\widetilde{P(\hat{\Omega}^l)}) \\ &= \widetilde{\pi^*(P(\hat{\Omega}^l))} = d\widetilde{TP(\hat{\theta})} = \delta\widetilde{TP(\hat{\theta})} = \delta(\widetilde{TP(\hat{\theta})})\ . \end{aligned}$$

So $\delta\pi^*(\hat{u}) = \delta(\widetilde{TP(\hat{\theta})})$. Since we have chosen k large, $\hat{E}$ is acyclic in dim $2l-1$, and so

$$\widetilde{TP(\hat{\theta})} = \pi^*(\hat{u}) + \text{coboundary}\ .$$

The proposition then follows in general by choosing a morphism $\varphi: \alpha \to \hat{\alpha}$ and taking $u = \varphi^*(\hat{u})$. □

We note that (3.13) and hence (3.12) follow directly from this. We also note that for these special polynomials, the classes $\{TP(\theta)\}$, when they exist, have the property that their mod Z reductions are already lifts. That is

THEOREM 3.16. *Let $\alpha = \{E, M, \theta\} \in \varepsilon(G)$ and let $P \in I_0^l(G)$. Suppose $P(\Omega^l) = 0$. Then there exists $U \in H^{2l-1}(M, R/Z)$ so that*

$$\{\widetilde{TP(\theta)}\} = \pi^*(U)\ .$$

Proof. Choose $u \in C^{2l-1}(M, R/Z)$ as in Proposition 3.14. The assumption $P(\Omega^l) = 0$ implies $\pi^*(\delta u) = 0$. Since every chain in M comes from one in E this means $\delta u = 0$. Thus u is an R/Z cocycle in M, and Proposition 3.14 shows $\pi^*(u) \sim \widetilde{TP(\theta)}$. Letting $U \in H^{2l-1}(M, R/Z)$ denote the class represented by u the theorem follows.

Characteristic numbers in R/Q. An interesting special case of this theorem occurs when M is compact, oriented, and $\dim M = 2l - 1$. Then for each $P \in I_0^l(G)$ we know that $P(\Omega^l) = 0$ and $\{TP(\theta)\} \in H^{2l-1}(E, R)$ depends on the connection. On the other hand, reducing mod Z, $\{\widetilde{TP(\theta)}\} = \pi^*(U)$ for some $U \in H^{2l-1}(M, R/Z) \cong R/Z$. Thus U is determined up to an element of $\ker \pi^*$. Now, either $\ker \pi^* = H^{2l-1}(M, R/Z)$, or $\ker \pi^*$ is a finite subgroup of $H^{2l-1}(M, R/Z)$. In the second case, since all finite subgroups of R/Z lie in Q/Z, U is determined uniquely in $R/Z/Q/Z \cong R/Q$. Let μ denote the fundamental cycle of M. Define $SP(\theta) \in R/Q$ by

$$SP(\theta) = 0 \qquad \text{if } \ker \pi^* = H^{2l-1}(M, R/Z)$$
$$SP(\theta) = u(\mu)/Q \quad \text{otherwise}\,.$$

Examples in the last section will show that these numbers are nontrivial invariants.*

COROLLARY 3.17. *Suppose* $\dim M < 2l - 1$. *Then for* $P \in I_0^l(G)$

$$\{TP(\theta)\} \in H^{2l-1}(E, Z)\,.$$

Proof. Since $\dim M < 2l - 1$, $H^{2l-1}(M, R/Z) = 0$ and so $\{\widetilde{TP(\theta)}\} = 0$. Thus from (3.14) $\{TP(\theta)\}$ is the image of an integral class.

4. Conformal invariance

In this section we suppose $G = Gl(n, R)$. $\mathfrak{G}$ consists of all $n \times n$ matrices, and we define the basic invariant polynomials $Q_1, \cdots, Q_n$

$$Q_l(A_1 \otimes \cdots \otimes A_l) = \frac{1}{l!} \sum_\pi \operatorname{tr} A_{\pi_1} A_{\pi_2} \cdots A_{\pi_l}\,.$$

It is well known that the Q_i generate the ring of invariant polynomials on $\mathfrak{G}$. If $\alpha = \{E, M, \theta\}$ is a principal G bundle then $\theta = \{\theta_{ij}\}$ and $\Omega = \{\Omega_{ij}\}$, matrices of real valued 1 and 2-forms respectively. One verifies directly that for any $\varphi = \{\varphi_{ij}\} \in \Lambda^{k,1}(E)$

$$Q_l(\Omega^l) = \sum\nolimits_{i_1,\cdots,i_l=1}^{n} \Omega_{i_1,i_2} \wedge \Omega_{i_2,i_3} \wedge \cdots \wedge \Omega_{i_l,i_1} \tag{4.1}$$

$$Q_l(\varphi \wedge \Omega^{l-1}) = \sum\nolimits_{i_1,\cdots,i_l=1}^{n} \varphi_{i_1,i_2} \wedge \Omega_{i_2,i_3} \wedge \cdots \wedge \Omega_{i_l,i_1}\,. \tag{4.2}$$

These polynomials have different properties. In particular the Weil map

* This construction was made in discussion with J. Cheeger. It is an easy way of producing the mod Q reductions of R/Z invariants developed in [3].

(see (2.2)) takes the ring generated by $\{Q_{2l}\}$ isomorphically onto the real cohomology of $B_{Gl(n,R)} = B_{O(n)}$, while the kernel of the Weil map is the ideal generated by the $\{Q_{2l+1}\}$.

PROPOSITION 4.3. *Let* $\alpha = \{E, M, \theta\} \in \mathcal{E}(Gl(n, R))$. *Suppose* θ *restricts to a connection on an* $O(n)$ *subbundle of* E. *Then* $Q_{2l+1}(\Omega^{2l+1}) = 0$, *and* $TQ_{2l+1}(\theta)$ *is exact.*

Proof. The first fact is well known and is one way to prove $Q_{2l+1} \in \operatorname{Ker} W$. Our assumption on θ is that there is an $O(n)$ subbundle $F \subseteq E$ such that at each $f \in F$, $H(E)_f \subseteq T(F)_f$, or equivalently that at all x tangent to F, $\theta_{ij}(x) = -\theta_{ji}(x)$. It easily implies that at all points of F, $\Omega_{ij} = -\Omega_{ji}$ as a form on E. Now if A is a skew symmetric matrix then $\operatorname{tr}(A^{2l+1}) = 0$ and by polarization we see $Q_{2l+1}(A_1 \otimes \cdots \otimes A_{2l+1}) = 0$ when all A_i are skew symmetric. Since $Q_{2l+1}(\Omega^{2l+1})$ is invariant, we need only show it vanishes at points in F, but at these points the range of Ω^{2l+1} lies in the kernel of Q_{2l+1}. Thus $Q_{2l+1}(\Omega^{2l+1}) = 0$. The same argument shows that $TQ_{2l+1}(\theta) \mid F = 0$. (Here we mean the form restricted to the submanifold, F, and not simply as a form on E considered at points of F.) Thus $TQ_{2l+1}(\theta)$ is a closed form in E whose restriction to F is 0. Since E is contractible to F, $TQ_{2l+1}(\theta)$ can carry no cohomology on E and hence must be exact. □

Let us now specialize to the case where $E = E(M)$, the bundle of bases of the tangent bundle of M. Points in E are $(n+1)$-tuples of the form $(m; e_1, \cdots, e_n)$ where $m \in M$ and $e_1, \cdots, e_n$ is a basis of $T(M)_m$. E comes equipped with a natural set of horizontal, real valued forms $\omega_1, \cdots, \omega_n$, defined by

$$d\pi(x) = \sum_{i=1}^{n} \omega_i(x) e_i$$

where $x \in T(E)_e$, and $e = (m; e_1, \cdots, e_n)$. Now let g be a Riemannian metric on M, and let θ be the associated Riemannian connection of $E(M)$. Let $E_1, \cdots, E_n$ be horizontal vector fields which are a dual basis to $\omega_1, \cdots, \omega_n$. Let $F(M)$ denote the orthonormal frame bundle. $F(M) \subseteq E(M)$ is the $O(n)$ subbundle consisting of orthonormal bases, and since θ is the Riemannian connection, θ restricts to a connection on $F(M)$.

Let h be a C^∞ function on M, and consider the curve of conformally related metrics

$$g(s) = e^{2sh} g\,, \qquad s \in [0, 1]\,.$$

Let $\theta(s)$ denote the curve of associated Riemannian connections on $E(M)$. Let $\theta = \theta(0)$, $\theta' = (d/ds)(\theta(s))\,|_{s=0}$, and $F(M)$ the frame bundle with respect to $g = g(0)$.

LEMMA 4.4. At points in $F(M)$

$$\theta'_{ij} = \delta_{ij} d(h \circ \pi) + E_i(h \circ \pi)\omega_j - E_j(h \circ \pi)\omega_i .$$

Proof. This is a standard computation, and is perhaps most easily done by using the formula for the Riemannian connection in terms of covariant differentiation (cf. [7]). It is easily seen how the connection changes under conformal change of metric, and one then translates this result back into bundle terminology.

THEOREM 4.5. *Let g and $\hat{g}$ be conformally related Riemannian metrics on M, and let θ, Ω, $\hat{\theta}$, $\hat{\Omega}$ denote the corresponding connection and curvature forms. Then for any $P \in I^l(Gl(n, R))$*

(1) $TP(\hat{\theta}) = TP(\theta) + exact$,

(2) $P(\hat{\Omega}^l) = P(\Omega^l)$.

COROLLARY. $P(\Omega^l) = 0$ *implies that the cohomology class* $\{TP(\theta)\} \in H^{2l-1}(E(M), R)$ *is a conformal invariant.*

The corollary follows immediately from the theorem, and (2) follows immediately from (1) and Proposition 3.2. So it remains to prove (1). Since the Q_i generate $I(Gl(n, R))$ we can assume P is a monomial in the Q_i. Using Proposition 3.7, an inductive argument shows that it is sufficient to prove (1) only in the case $P = Q_l$. Proposition 4.3 shows that for any Riemannian connection $Q_{2l+1}(\Omega^{2l+1}) = 0$ and $TQ_{2l+1}(\theta)$ is exact, so we can assume l is even.

Any two conformally related metrics can be joined by a curve of such metrics, with associated connections $\theta(s)$. By integration it is sufficient to prove

(*) $$\frac{d}{ds}(TQ_{2l}(\theta(s))) = \text{exact} .$$

Since each point on the curve is the initial point of another such curve, it is enough to prove (*) at $s = 0$. By Proposition 3.8 it will suffice to prove

(**) $$Q_{2l}(\theta' \wedge \Omega^{2l-1}) = 0 .$$

We use the notation and formula of Lemma 4.4, and work at $f \in F(M)$. Set

$$\alpha = (\delta_{ij} d(f \circ \pi))$$
$$\beta = (E_i(f \circ \pi)\omega_j - E_j(f \circ \pi)\omega_i) .$$

Then $\theta' = \alpha + \beta$. Now (4.2) shows

$$Q_{2l}(\alpha \wedge \Omega^{2l-1}) = d(f \circ \pi) \wedge Q_{2l-1}(\Omega^{2l-1}) = 0$$

by Proposition 4.3. Also using (4.2),

$$Q_{2l}(\beta \wedge \Omega^{2l-1}) = \sum_{i_1,\dots,i_{2l}} \begin{array}{l} E_{i_1}(f\circ\pi)\omega_{i_2} \wedge \Omega_{i_2,i_3} \wedge \cdots \wedge \Omega_{i_{2l},i_1} \\ - E_{i_2}(f\circ\pi)\omega_{i_1} \wedge \Omega_{i_2,i_3} \wedge \cdots \wedge \Omega_{i_{2l},i} \end{array}.$$

But, since θ is a Riemannian connection, the Jacobi identity holds. This states

$$\sum_{i=1}^{n} \omega_i \wedge \Omega_{ij} = 0\ ,$$

and shows $Q_{2l}(\beta \wedge \Omega^{2l-1}) = 0$. Thus at points in $F(M)$, $Q_{2l}(\theta' \wedge \Omega^{2l-1}) = 0$, and (**) follows by invariance. □

5. Conformal immersions

Let $G = U(n)$. Let A be a skew Hermitian matrix and define the ith Chern polynomial $C_i \in I_0^i(U(n))$

(5.1) $$\det\left(\lambda I - \frac{1}{2\pi\sqrt{-1}}A\right) = \sum_{i=0}^{n} C_i(\overbrace{A\otimes\cdots\otimes A}^{i})\lambda^{n-i}$$

where C_i is extended by polarization to all tensors. Let c_i denote the ith integral Chern class in $B_{U(n)}$. Then $c_i \in H^{2i}(B_G, Z)$, and letting $r(c_i) \in H^{2i}(B_G, R)$ denote its real image,

(5.2) $$W(C_i) = r(c_i)\ .$$

We also define the inverse Chern polynomials and classes $C_i^{\perp}$ and $c_i^{\perp}$

(5.3) $$\begin{gathered}(1 + C_1^{\perp} + \cdots + C_i^{\perp} + \cdots)\ (1 + C_1 + \cdots + C_n) = 1 \\ (1 + c_1^{\perp} + \cdots + c_i^{\perp} + \cdots) \cup (1 + c_1 + \cdots + c_n) = 1\ .\end{gathered}$$

These formulae uniquely determine $C_i^{\perp}$ and $c_i^{\perp}$, and since W is a ring homomorphism

(5.4) $$W(C_i^{\perp}) = c_i^{\perp}\ .$$

The inverse classes are so named because, for vector bundles, they are the classes of an inverse bundle. That is, if V, W are complex vector bundles over M with $V \oplus W$ trivial, then using the product formula for Chern class, cf. [9], one knows

(5.5) $$c_i(W) = c_i^{\perp}(V)\ .$$

Let $G_{n,k}(c)$ denote the Grassmann manifold of complex n-planes in C^{n+k}, and let $E_{n,k}(c)$ denote the Stiefel manifold of orthonormal n-frames in C^{n+k}, with respect to the Hermitian metric. Then $\{E_{n,k}(c), G_{n,k}(c)\}$ is a principal $U(n)$ bundle. There is a natural connection in this bundle most easily visualized by constructing it in the associated canonical n-dim vector bundle over $G_{n,k}(c)$. Let $\gamma(t)$ be a curve in $G_{n,k}(c)$ and let $\rho(t)$ be a curve in the vector bundle with $\pi\circ\rho=\gamma$. So for each t, $\gamma(t)$ is an n-plane in C^{n+k}, and $\rho(t)$ is a vector in C^{n+k} with $\rho(t)\in\gamma(t)$. Then $\rho'(t)=(d/dt)(\rho(t))$ is a vector in C^{n+k}, and the covariant derivative of $\rho(t)$ along γ is obtained by orthogonally projecting $\rho'(t)$ into $\gamma(t)$. We let θ denote this connection and set

$$\alpha_{n,k}(c)=\{E_{n,k}(c),\,G_{n,k}(c),\,\theta\}\,.$$

PROPOSITION 5.6. *For* $i>k$

(1) $$C_i^\perp(\Omega^i)=0$$

(2) $$\{TC_i^\perp(\theta)\}\in H^{2i-1}(E_{n,k}(c),\,Z)\,.$$

Proof. Since the n-dim vector bundle associated to $\{E_{n,k}(c),G_{n,k}(c)\}$ has a k-dim inverse, (5.5) shows that $c_i^\perp(\alpha_{n,k}(c))=0$ for $i>k$. Thus the form $C_i^\perp(\Omega^i)$ is exact on $G_{n,k}(c)$. Now $G_{n,k}(c)$ is a compact, irreducible Riemannian symmetric space, and it is easily checked that the forms $P(\Omega^l)$ are invariant under the isometry group. Thus $C_i^\perp(\Omega^i)$ is invariant and exact, and therefore must vanish. So the class $\{TC_i^\perp(\theta)\}\in H^{2i-1}(E_{n,k}(c),R)$ is defined. Since $W(C_i^\perp)=c_i^\perp\in H^{2i}(B_{U(n)},Z)$ we see that $C_i^\perp\in I_0^i(U(n))$. Using Theorem 3.16 we see that $\{\widetilde{TC_i^\perp}(\theta)\}$ is a lift of a $2i-1$ dimensional R/Z cohomology class of $G_{n,k}(c)$. But the odd dimensional cohomology of this space is zero (for any coefficient group), and thus $\{\widetilde{TC_i^\perp}(\theta)\}=0$. The Bockstein sequence (3.14) then shows that $\{TC_i^\perp(\theta)\}\in H^{2i-1}(E_{n,k}(c),Z)$. □

Now let $G=0(n)$. Let A be a skew symmetric matrix and define for $i=1,\cdots,[n/2]$ the ith Pontrjagin polynomial $P_i\in I_0^{2i}(0(n))$

$$\det(\lambda I-(1/2\pi)A)=\sum_{i=0}^{[n/2]}P_i(\overbrace{A\otimes\cdots\otimes A}^{2i})\lambda^{n-2i}+Q(\lambda^{n-\text{odd}}) \tag{5.7}$$

where we ignore the terms involving the n-odd powers of λ. Also let $p_i\in H^{4i}(B_{0(n)},Z)$ denote the ith integral Pontrjagin class. Then

$$W(P_i)=r(p_i)\,.$$

Let $\rho\colon 0(n)\to U(n)$ be the natural map. Then ρ induces $\rho^*\colon I(U(n))\to I(0(n))$, $\rho^*\colon H^*(U(n))\to H^*(0(n))$, and $\rho\colon B_{0(n)}\to B_{U(n)}$. Using Theorem 2.12 one easily sees

$$W(\rho^*(Q))=\rho^*(W(Q))$$

for any $Q\in I^l(U(n))$. The definitions of P_i and p_i are such that

$$\rho^*(C_{2i})=(-1)^iP_i\,,\qquad \rho^*(c_{2i})=(-1)^ip_i\,. \tag{5.8}$$

We also define the inverse Pontrjagin polynomials $P_i^\perp$

$$(1+P_1+\cdots+P_{[n/2]})(1+P_1^\perp+\cdots+P_i^\perp+\cdots)=1 \tag{5.9}$$

and note that $P_i^\perp \in I_0^{2i}(0(n))$ since $\rho^*(c_i^\perp) \in H^{2i}(B_{0(n)}, Z)$, and one easily sees that

$$W(P_i^\perp) = (-1)^i r(\rho^*(c_{2i}^\perp)) .$$

Formula (5.9) shows $P_i^\perp = -P_i - P_{i-1}P_1^\perp - \cdots - P_1 P_{i-1}^\perp$. Proposition 3.7 shows that $TP_i^\perp(\theta) = -TP_i(\theta) +$ terms involving curvature. Thus for any $\alpha = \{E, M, \theta\} \in \varepsilon(0(n))$

$$TP_i^\perp(\theta) \mid E_m = -TP_i(\theta) \mid E_m . \tag{5.10}$$

We now define the real Grassmann manifold, $G_{n,k}$, the real Stiefel manifold $E_{n,k}$, and the canonical connection θ on $\{E_{n,k}, G_{n,k}\}$ exactly as in the complex case. We set $\alpha_{n,k} = \{E_{n,k}, G_{n,k}, \theta\} \in \varepsilon(0(n))$.

PROPOSITION 5.11. *For* $i > [k/2]$

(1) $P_i^\perp(\Omega^{2i}) = 0$

(2) $\{(1/2)TP_i^\perp(\theta)\} \in H^{4i-1}(E_{n,k}, Z)$.

Proof. The natural map $R^n \to C^n$ induces the commutative diagram

$$\begin{array}{ccc} E_{n,k} & \xrightarrow{\varphi} & E_{n,k}(c) \\ \downarrow & & \downarrow \\ G_{n,k} & \xrightarrow{\varphi} & G_{n,k}(c) . \end{array}$$

It is straightforward to check that

$$P_i^\perp(\Omega^{2i}) = (-1)^i \varphi^*(C_{2i}^\perp(\Omega^{2i})) ,$$
$$TP_i^\perp(\theta) = (-1)^i \varphi^*(TC_{2i}^\perp(\theta)) .$$

Since $i > [k/2] \Rightarrow 2i > k$, (1) follows from Proposition 5.6, and from (2) of that proposition we see that

$$\{TP_i^\perp(\theta)\} = (-1)^i \varphi^*\{TC_{2i}^\perp(\Omega^{2i})\} \in H^{4i-1}(E_{n,k}, Z) .$$

We will be finished when we show

LEMMA 5.12. *Let* $\gamma \in H^*(E_{n,k}(c), Z)$. *Then* $\varphi^*(\gamma)$ *is an even integral class in* $H^*(E_{n,k}, Z)$.

Proof. For any Lie group G and any coefficient group Λ we want to consider the inverse transgression map $\tau\colon H^i(B_G, \Lambda) \to H^{i-1}(G, \Lambda)$. This map is defined as follows. Let $u \in H^i(B_G, \Lambda)$ be given and choose $\gamma \in Z^i(B_G, \Lambda)$ with $\gamma \in u$ and with $\gamma \mid \{m\} = 0$ for all $m \in B_G$. Letting $\pi\colon E_G \to B_G$ be the projection map, and recalling that E_G is acyclic, we see that $\pi^*(\gamma) = \delta\beta$, where $\beta \in C^{i-1}(E_G, \Lambda)$. Since $\gamma \mid \{m\} = 0$, $\beta \mid G$ is closed, and thus defines $\tau(u) \in H^{i-1}(G, \Lambda)$. Acyclicity of E_G guarantees the map independent of choice of β, and it is easy to check it is also independent of choice of γ. Thus $\tau: H^i(B_G, \Lambda) \to H^{i-1}(G, \Lambda)$ is well-defined. τ is in fact the inverse of the transgression mapping considered in [1]. We remark that if Λ is a ring then $\tau(u \cup u) = 0$. This follows since if $\gamma \in u$ with $\pi^*(\gamma) = \delta\beta$, then $\pi^*(\gamma \cup \gamma) = \delta(\beta \cup \pi^*(\gamma))$, and $\beta \cup \pi^*(\gamma) \mid G = 0$.

We first consider the case $k = 0$, i.e., $E_{n,0} = 0(n)$, $E_{n,0}(c) = U(n)$, and $\varphi: 0(n) \to U(n)$ is the natural map. We consider the diagram

$$\begin{array}{ccc} H^*(U(n)) & \xrightarrow{\varphi^*} & H^*(0(n)) \\ \uparrow \tau & & \uparrow \tau \\ H^*(B_{U(n)}) & \xrightarrow{\rho^*} & H^*(B_{0(n)}) \end{array}$$

and note that it is commutative. The Bockstein exact sequence of cohomology corresponding to the coefficient sequence $0 \to Z \xrightarrow{2} Z \to Z_2 \to 0$ shows that an integral class is *even* if and only if its mod 2 reduction is zero. Thus it is sufficient to show that for any $u \in H^*(U(n), Z_2)$, $\varphi^*(u) = 0$. Let $\hat{c}_i \in H^{2i}(U(n), Z_2)$ denote the mod 2 reduction of c_i. Now it is well-known that

$$\rho^*(\hat{c}_i) = W_i \cup W_i$$

where W_i is the ith Stiefel-Whitney class. Thus

$$\varphi^*(\tau(\hat{c}_i)) = \tau(\rho^*(\hat{c}_i)) = \tau(W_i \cup W_i) = 0 .$$

On the other hand, $H^*(U(n), Z_2)$ is generated by the set $\{\tau(\hat{c}_i)\}$, and thus $\varphi^*(u) = 0$ for any $u \in H^*(U(n), Z_2)$.

To do the general case consider the commutative diagram

$$\begin{array}{ccc} H^*(E_{n,k}(c), Z_2) & \xrightarrow{\varphi^*} & H^*(E_{n,k}, Z_2) \\ \downarrow \pi^* & & \downarrow \pi^* \\ H^*(U(n+k), Z_2) & \xrightarrow{\varphi^*} & H^*(0(n+k), Z_2) \end{array}$$

where $\pi: U(n+k) \to U(n+1)/U(k) = E_{n,k}(c)$, and $\pi: 0(n+k) \to 0(n+k)/0(k) = E_{n,k}$ are the quotient maps. It is known, cf. [2], that the π^* on the right is injective. Thus, since the image of the lower φ^* is zero from our special case, so is that of the upper φ^*. This completes the proof of the lemma and the proposition follows. □

By restricting this proposition to the fibre and using (5.10) and (3.11) we obtain the well-known fact that

(5.13) $$\frac{1}{2}\{TP_i\} \in H^{4i-1}(0(n), Z) .$$

The polynomials P_i and $P_i^\perp$ were considered on the Lie algebra of $0(n)$, but they also live on that of $Gl(n, R)$, and pull back under $O(n) \to Gl(n, R)$. We will also denote these by P_i, $P_i^\perp \in I_0^{2i}(Gl(n, R))$.

THEOREM 5.14. *Let M^n be an n-dim Riemannian manifold. Let $\alpha(M) = \{E(M^n), M^n, \theta\}$ denote the $Gl(n, R)$ basis bundle of M equipped with the Riemannian connection θ. A necessary condition that M^n admit a global conformal immersion in R^{n+k} is that $P_i^\perp(\Omega^{2i}) = 0$ and $\{(1/2)TP_i^\perp(\theta)\} \in H^{4i-1}(E(M), Z)$ for $i > [k/2]$.*

Proof. Let $\varphi: M^n \to R^{n+k}$ be a conformal immersion. By Theorem 4.5 we may assume φ is an isometric immersion. Let $F(M^n)$ denote the orthonormal frame bundle of M^n, and consider the Gauss map Φ

$$\begin{array}{ccc} F(M^n) & \xrightarrow{\Phi} & E_{n,k} \\ \downarrow & & \downarrow \\ M^n & \xrightarrow{\Phi} & G_{n,k} \end{array}$$

which is defined as usual by mapping a point into the tangent plane at its image. Letting θ denote the canonical connection on $E_{n,k}$, it is a standard fact that $\Phi^*(\theta) = \theta$, the Riemannian connection on $F(M^n)$; i.e.,

$$\Phi: \{F(M^n), M^n, \theta\} \longrightarrow \alpha_{n,k}$$

is a morphism. Thus by naturality and the previous proposition, in $F(M^n)$, $P_i^\perp(\Omega^{2i}) = 0$ and $\{(1/2)TP_i^\perp(\theta)\} \in H^{4i-1}(F(M^n), Z)$ for $i > [k/2]$. By invariance, $P_i^\perp(\Omega^{2i}) = 0$ in all of $E(M^n)$, and since $\{(1/2)TP_i^\perp(\theta)\} \in H^{4i-1}(E(M^n), R)$ it must actually be an integral class there since its restriction to the retract $F(M^n)$ is integral. □

Remark. This theorem is probably of interest only for the codimension $k \leqq n/2$. This is because if $k > n/2$ our condition $i > [k/2]$ already implies $P_i^\perp(\Omega^{2i}) = 0$ for dimension reasons, and the corresponding class, $\{TP_i^\perp(\theta)\}$, is independent of connection (see Theorem 3.9). At the same time Corollary 3.17 already shows that $\{TP_i^\perp(\theta)\} \in H^{4i-1}(E(M), Z)$, and it seems likely that the same is true for $\{(1/2)TP_i^\perp(\theta)\}$.

6. Applications to 3-manifolds

In this section M will denote a compact, oriented, Riemannian 3-manifold, and $F(M) \xrightarrow{\pi} M$ will denote its $SO(3)$ oriented frame bundle equipped with the Riemannian connection θ and curvature tensor Ω. For A, B skew symmetric matrices, the specific formula for P_1 shows $P_1(A \otimes B) = -(1/8\pi^2)\operatorname{tr} AB$. Calculating from (3.5) shows

$$(6.1) \qquad TP_1(\theta) = \frac{1}{4\pi^2}\{\theta_{12} \wedge \theta_{13} \wedge \theta_{23} + \theta_{12} \wedge \Omega_{12} + \theta_{13} \wedge \Omega_{13} + \theta_{23} \wedge \Omega_{23}\}\,.$$

Since $\dim M = 3$, $dTP_1 = 0$. By (5.13), $(1/2)TP_1(\theta) \mid F(M)_m \in H^3(F(M)_m, Z)$. We will thus be interested in the class

$$\left\{\frac{1}{2}TP_1(\theta)\right\} \in H^3(F(M), R)\,.$$

From the general considerations at the end of § 3 this data is enough to produce an R/Q invariant of M, but since M is an oriented 3-manifold, $F(M)$ is trivial; and we define the R/Z invariant,* $\Phi(M)$, as follows: Let $\chi: M \to F(M)$ be any cross-section. Then set

$$(6.2) \qquad \Phi(M) = \widetilde{\int_\chi \frac{1}{2}TP_1(\theta)} \in R/Z\,.$$

If χ' were another such, then homologically $\chi' = \chi + nF(M)_m +$ torsion, where n is an integer. Thus since $(1/2)TP_1(\theta) \mid F(M)_m$ is integral, and forms integrated over torsion classes give 0, $\Phi(M)$ is well-defined. Recalling that $P_1^\perp = -P_1$ we immediately get the following two special cases of Theorems 4.5 and 5.14.

THEOREM 6.3. *$\Phi(M)$ is a conformal invariant of M.*

THEOREM 6.4. *A necessary condition that M admit a conformal immersion in R^4 is that $\Phi(M) = 0$.*

Example 1. Let $M = RP^3 = SO(3)$ together with the standard metric of constant curvature 1. Let E_1, E_2, E_3 be an orthonormal basis of left invariant fields on $SO(3)$, oriented positively. Then it is easily seen that $\nabla_{E_1}E_2 = E_3$, $\nabla_{E_1}E_3 = -E_2$, and $\nabla_{E_2}E_3 = E_1$. Let $\chi: M \to F(M)$ be the cross-section determined by this frame. The above equations and (6.1) show

$$\chi^*\left(\frac{1}{2}TP_1(\theta)\right) = \frac{-1}{2\pi^2}\omega$$

where ω is the volume form on $SO(3)$. Thus from (6.1)

$$\widetilde{\Phi(SO(3)) = \frac{1}{2\pi^2}\operatorname{Vol}(SO(3))} = \widetilde{\frac{1}{2}}$$

since $\operatorname{Vol}(SO(3)) = (1/2)\operatorname{Vol}(S^3) = \pi^2$. Using Theorem 6.4 we see that $SO(3)$ admits no global conformal immersion in R^4. This is interesting since, being parallelizable, it certainly admits a C^∞ immersion in R^4, and locally it is isometrically imbeddable in R^4.

---

* Atiyah has subsequently shown that $2\Phi(M)$ is the mod Z reduction of a real class. This will be discussed further in [3].

Example 2. Again let $M = SO(3)$, but this time with the left invariant metric, g_λ, with respect to which λE_1, E_2, E_3 is an orthonormal frame. Direct calculation shows

$$\Phi(SO(3), g_\lambda) = \frac{2\lambda^2 - 1}{2\lambda^4}$$

and this can take any value in R/Z.

Let M be a fixed 3-manifold and let $\mathcal{C}(M)$ denote the space of conformal structures on M. Since Φ is a conformal invariant we may regard

$$\Phi\colon \mathcal{C}(M) \longrightarrow R/Z .$$

If g_t is a C^∞ curve of conformal structures then $\Phi(g_t)$ is a C^∞ R/Z valued function (we shall see this below). We are interested in calculating the critical points of the map Φ.

Let $g = \langle\,,\rangle$ be a fixed metric on M. With respect to this we let $\nabla_X Y$ and $R_{X,Y}Z$ denote covariant differentiation and curvature; i.e.,

$$R_{X,Y}Z = \nabla_X\nabla_Y Z - \nabla_Y\nabla_X Z - \nabla_{[x,y]}Z$$

where X, Y, Z are vector fields. The operator ∇_X extends as a derivation to all tensors, and all tensors have a natural inner product induced by $\langle\,,\rangle$. We make the usual identification of Λ^2 with skew symmetric linear transformations and so for $x, y \in T(M)_m$ we often regard

$$R_{x,y} \in \Lambda^2 T(M)_m . \tag{6.5}$$

Because we are working on an oriented 3-manifold there is an identification of $T(M)_m$ with $\Lambda^2 T(M)_m$ given by the metric. We denote this by

$$* : T(M)_m \longrightarrow \Lambda^2 T(M)_m . \tag{6.6}$$

Let e_1, e_2, e_3 be an orthonormal basis of $T(M)_m$ and define

$$\begin{aligned} \delta R\colon T(M)_m &\longrightarrow \Lambda^2 T(M)_m , \\ \delta R(x) &= \textstyle\sum_{i=1}^3 \nabla_{e_i}(R)_{e_i,x} . \end{aligned} \tag{6.7}$$

This definition is independent of choice of frame. Combining (6.6) and (6.7) we define the *symmetric bilinear form* $\widehat{\delta R}$ on $T(M)_m$ by

$$\widehat{\delta R}(x, y) = \langle \delta R(x), y^* \rangle + \langle \delta R(y), x^* \rangle . \tag{6.8}$$

Now let $B = B(\,,)$ be a C^∞ field of symmetric bilinear forms on M and consider the curve of metrics

$$g_t(x, y) = \langle x, y \rangle + tB(x, y) .$$

For small t these are Riemannian.

THEOREM 6.9. *Let* $M_t = \{M, g_t\}$. *Then for small* t, $\Phi(M_t) \in C^\infty(t)$ *and*

$$\frac{d}{dt}(\Phi(M_t))\Big|_{t=0} = \frac{-1}{16\pi^2}\int_M \langle B, \widehat{\delta R}\rangle .$$

Proof. The invariant Φ was defined by choosing a cross-section in $F(M)$, but we would clearly get the same value by choosing one in $E(M)$, the full $Gl(3, R)$ basis bundle. This is more convenient. So let θ^t denote the curve of connections in $E(M)$ corresponding to the metrics g_t, and let $\theta = \theta^0$, $\Omega = \Omega^0$, and $\theta' = (d/dt)(\theta^t)|_{t=0}$. The general variation formula in Theorem 3.8 shows

$$\frac{d}{dt}(\Phi(M_t))\Big|_{t=0} = \int_M P_1(\theta' \wedge \Omega) \tag{6.10}$$

where this makes sense since the forms $P(\theta' \wedge \Omega^{l-1})$ all are horizontal and invariant. The definition of P_i given in (5.7) works as well for general matrices and one easily sees

$$P_1(A, B) = \frac{1}{8\pi^2}[\operatorname{tr} A \operatorname{tr} B - \operatorname{tr} AB] . \tag{6.11}$$

Now if we work at points in $F(M)$, range Ω is skew symmetric matrices, and so the first term vanishes, to give

$$P_1(\theta' \wedge \Omega) = \frac{1}{8\pi^2}\sum\nolimits_{i,j=1}^{3} \theta'_{ij} \wedge \Omega_{ij} \tag{6.12}$$

at points of $F(M)$.

Let $x \in T(M)_m$, Y a local vector field, and let $\nabla^t_x Y$ denote covariant differentiation with respect to the connection at time t. Differentiating we get a tensor, A, defined by

$$A_x y = \frac{d}{dt}(\nabla^t_x Y)\Big|_{t=0}(m) \tag{6.13}$$

where $y = Y(m)$. At $f = (m; f_1, f_2, f_3) \in F(M)$ the following hold

$$\theta'_{ij}(x) = -\langle A_{d\pi(x)} f_i, f_j\rangle$$
$$\Omega_{ij}(x, y) = -\langle R_{d\pi(x), d\pi(y)} f_i, f_j\rangle$$

where $x \in T(E(M))_f$ and R is the curvature of $\{M, g\}$. Now regarding $P_1(\theta' \wedge \Omega)$ as a form on M, (6.12) gives

$$\begin{aligned} P_1(\theta' \wedge \Omega)(x, y, z) &= \frac{1}{8\pi^2}[\langle A_x, R_{y,z}\rangle - \langle A_y, R_{x,z}\rangle + \langle A_z, R_{x,y}\rangle] \\ &= \frac{1}{8\pi^2}\langle A, R \circ *\rangle \omega(x, y, z) \end{aligned}$$

(ω = volume form on $\{M, g\}$). Combining this with (6.10) gives

$$\frac{d}{dt}(\Phi(M_t))\Big|_{t=0} = \frac{1}{8\pi^2}\int_M \langle A, R\circ *\rangle . \tag{6.13}$$

Since the range of $R\circ *$ is skew symmetric linear transformations, we may as well project A to have the same range; i.e., set

$$\langle \hat{A}_x y, z\rangle = \frac{1}{2}\langle A_x y, z\rangle - \frac{1}{2}\langle A_x z, y\rangle .$$

Then

$$\langle A, R\circ *\rangle = \langle \hat{A}, R\circ *\rangle . \tag{6.14}$$

Finally, using the definition of Riemannian connection in terms of covariant derivatives (cf. [7]), and setting

$$\langle DB_x y, z\rangle = \frac{1}{2}[\nabla_y(B)(z, x) - \nabla_z(B)(y, x)]$$

equation (6.13) shows

$$\hat{A} = DB$$

and thus from (6.13)

$$\frac{d}{dt}(\Phi(M_t))\Big|_{t=0} = \frac{1}{8\pi^2}\int_M \langle DB, R\circ *\rangle = -\frac{1}{16\pi^2}\int_M \langle B, \widehat{\delta R}\rangle$$

where the last equation used Stokes' theorem and integration by parts. □

We should note from the definition of $\widehat{\delta R}$ that $\operatorname{tr}\widehat{\delta R} = 0$, and this is as it should be since if $B = \lambda g$, where λ is a function on M, our metric is changing conformally, $\Phi(M_t)$ should be constant, and $\langle B, \widehat{\delta R}\rangle$ should vanish, which is implied by $\operatorname{tr}\widehat{\delta R} = 0$. More importantly, the bilinear form $\widehat{\delta R}$ is itself a conformal invariant (this can be directly checked), and it has been shown by Schouten, cf. [6], that $\widehat{\delta R} \equiv 0$ if and only if $\{M, g\}$ is *locally conformally flat*; i.e., if and only if each point of M has a neighborhood conformally equivalent to R^3. For example, S^3 is locally conformally flat. This fact is peculiar to 3-manifolds, since the integrability condition for local conformal flatness in higher dimensions involves no derivatives of curvature. We, therefore, conclude

COROLLARY 6.14. *$g \in \mathcal{C}(M)$ is a critical point of Φ if and only if $\{M, g\}$ is locally conformally flat.*

Kuiper, in [8], has shown that compact $\{M, g\}$ is locally conformally flat and simply connected if and only if $\{M, g\}$ is conformally equivalent to S^n with the usual metric. We, therefore, conclude

COROLLARY 6.15. *Suppose M is a simply connected compact oriented 3-manifold. Then either Φ has exactly one critical point and M is diffeomorphic to S^3 or Φ has no critical points and M is not diffeomorphic to S^3.*

We do not see how this helps to settle the Poincaré conjecture.

UNIVERSITY OF CALIFORNIA AT BERKELEY
UNIVERSITY OF NEW YORK AT STONY BROOK

REFERENCES

[1] A. BOREL, Sur la cohomologie des espaces fibrés principaux et des espaces homogènes de groupes de Lie compacts, Ann. of Math., **57** (1953), 115-207.
[2] ———, La cohomologie mod 2 de certains espaces homogènes, Comm. Math. Helv., **27** (1953), 165-197.
[3] J. CHEEGER and J. SIMONS, Differential characters and geometric invariants, to appear.
[4] S. CHERN and J. SIMONS, Some cohomology classes in principal fibre bundles and the application to Riemannian geometry, Proc. Nat. Acad. Sci., U.S.A., **68** (1971), 791-794.
[5] S. CHERN, Geometry of characteristic classes, Proc. of 13th Biennial Seminar, Canadian Math. Congress 1972.
[6] L. P. EISENHART, *Riemannian Geometry*, Princeton, 1949.
[7] S. KOBAYASHI and K. NOMIZU, *Foundation of Differential Geometry*, Vol. I, II, Interscience, 1969.
[8] N. H. KUIPER, Conformally flat spaces in the large, Ann. of Math., **50** (1949), 916-924.
[9] J. MILNOR, Lectures on Characteristic Classes, Princeton Lecture Notes, 1957.
[10] M. S. NARASIMHAN and S. RAMANAN, Existence of universal connections, Amer. J. Math.. **83** (1961), 563-572; **85** (1963), 223-231.

(Received June 13, 1972)
(Revised December 1, 1972)

Bibliography of the Publications of S. S. Chern

I. *Books and Monographs*

1. *Topics in Differential Geometry* (mimeographed), Institute for Advanced Study, Princeton (1951), 106 pp.
2. *Differentiable Manifolds* (mimeographed), University of Chicago, Chicago (1953), 166 pp.
3. *Complex Manifolds*
   a. University of Chicago, Chicago (1956), 195 pp.
   b. University of Recife, Recife, Brazil (1959), 181 pp.
   c. Russian translation, Moscow (1961), 239 pp.
4. *Studies in Global Geometry and Analysis* (Editor), Mathematical Association of America (1967), 200 pp.
5. *Complex Manifolds without Potential Theory*. van Nostrand (1968), 92 pp.
6. *Minimal Submanifolds in a Riemannian Manifold* (mimeographed), University of Kansas, Lawrence (1968), 55 pp.

II. *Papers*

1932

[1] Pairs of plane curves with points in one-to-one correspondence. *Science Reports Tsing Hua Univ.* **1** (1932) 145–153.

1935

[2] Triads of rectilinear congruences with generators in correspondence. *Tohoku Math. J.* **40** (1935) 179–188.

[3] Associate quadratic complexes of a rectilinear congruence. *Tohoku Math. J.* **40** (1935) 293–316.

[4] Abzählungen für Gewebe. *Abh. Math. Sem. Hamburg* **11** (1935) 163–170.

1936

[5] Eine Invariantentheorie der Dreigewebe aus n-dimensionalen Mannigfaltigkeiten in $2n$-dimensionalen Räumen. *Abh. Math. Sem. Hamburg* **11** (1936) 333–358.

1937

[6] Sur la géomètrie d'une equation difféerentielle du troisième ordre. *C. R. Acad. Sci. Paris* **204** (1937) 1227–1229.

[7] Sur la possibilité de plonger un espace à connexion projective donné dans un espace projectif. *Bull. Sci. Math.* **61** (1937) 234–243.

1938

[8] On projective normal coordinates. *Ann. of Math.* **39** (1938) 165–171.
[9] On two affine connections. *J. Univ. Yunnan* **1** (1938) 1–18.

1939

[10] Sur la géomètrie d'un systeme d'équations différentielles du second ordre. *Bull. Sci. Math.* **63** (1939) 206–212.

1940

[11] The geometry of higher path-spaces. *J. Chin. Math. Soc.* **2** (1940) 247–276.
[12] Sur les invariants intégraux en géométrie. *Science Reports Tsing Hua Univ.* **4** (1940) 85–95.
[13] The geometry of a differential equation $y''' = F(x, y, y', y'')$. *Science Reports Tsing Hua Univ.* **4** (1940) 97–111.
[14] Sur une généralisation d'une formule de Crofton. *C. R. Acad. Sci. Paris* **210** (1940) 757–758.
[15] (with Chih-ta Yen) Formula principale cinematica dello spazio ad n dimensioni. *Boll. Un. Mat. Ital.* **2** (1940) 434–437.
[16] Generalization of a formula of Crofton. *Wuhan Univ. J. Sci.* **7** (1940) 1–16.

1941

[17] Sur les invariants de contact en géométrie projective différentielle. *Acta Pontif. Acad. Sci.* **5** (1941) 123–140.

1942

[18] On integral geometry in Klein spaces. *Ann. of Math.* **43** (1942) 178–189.
[19] On the invariants of contact of curves in a projective space of n dimensions and their geometrical interpretation. *Acad. Sinica Sci. Record* **1** (1942) 11–15.
[20] The geometry of isotropic surfaces. *Ann. of Math.* **43** (1942) 545–559.
[21] On a Weyl geometry defined from an $(n-1)$-parameter family of hypersurfaces in a space of n dimensions. *Acad. Sinica Science Record* **1** (1942) 7–10.

1943

[22] On the Euclidean connections in a Finsler space. *Proc. Nat. Acad. Sci. USA*, **29** (1943) 33–37.
[23] A generalization of the projective geometry of linear spaces. *Proc. Nat. Acad. Sci. USA*, **29** (1943) 38–43.

1944

[24] Laplace transforms of a class of higher dimensional varieties in a projective space of n dimensions. *Proc. Nat. Acad. Sci. USA*, **30** (1944) 95–97.
[25] A simple intrinsic proof of the Gauss–Bonnet formula for closed Riemannian manifolds. *Ann. of Math.* **45** (1944) 747–752.

[26] Integral formulas for the characteristic classes of sphere bundles. *Proc. Nat. Acad. Sci. USA* **30** (1944) 269–273.

[27] On a theorem of algebra and its geometrical application. *J. Indian Math. Soc.* **8** (1944) 29–36.

1945

[28] On Grassmann and differential rings and their relations to the theory of multiple integrals. *Sankhya* **7** (1945) 2–8.

[29] Some new characterizations of the Euclidean sphere. *Duke Math. J.* **12** (1945) 279–290.

[30] On the curvatura integra in a Riemannian manifold. *Ann. of Math.* **46** (1945) 674–684.

[31] On Riemannian manifolds of four dimensions. *Bull. Amer. Math. Soc.* **51** (1945) 964–971.

1946

[32] Some new viewpoints in the differential geometry in the large. *Bull. Amer. Math. Soc.* **52** (1946) 1–30.

[33] Characteristic classes of Hermitian manifolds. *Ann. of Math.* **47** (1946) 85–121.

1947

[34] (with Hsien-chung Wang). Differential geometry in symplectic space. *Science Reports Tsing Hua Univ.* **4** (1947) 453–477.

[35] Sur une classe remarquable de variétés dans l'espace projectif à *n* dimensions. *Science Reports Tsing Hua Univ.* **4** (1947) 328–336.

[36] On the characteristic classes of Riemannian manifolds. *Proc. Nat. Acad. Sci. USA*, **33** (1947) 78–82.

[37] Note of affinely connected manifolds. *Bull. Amer. Math. Soc.* **53** (1947) 820–823; correction ibid **54** (1948) 985–986.

[38] On the characteristic ring of a differentiable manifold. *Acad. Sinica Sci. Record* **2** (1947) 1–5.

1948

[39] On the multiplication in the characteristic ring of a sphere bundle. *Ann. of Math.* **49** (1948) 362–372.

[40] Note on projective differential line geometry. *Acad. Sinica Sci. Record* **2** (1948) 137–139.

[41] (with Yuh-lin Jou) On the orientability of differentiable manifolds. *Science Reports Tsing Hua Univ.* **5** (1948) 13–17.

[42] Local equivalence and Euclidean connections in Finslerian spaces. *Science Reports Tsing Hua Univ.* **5** (1948) 95–121.

1949

[43] (with Ye-fon Sun). The imbedding theorem for fibre bundles. *Trans. Amer. Math. Soc.* **67** (1949) 286–303.

[44] (with Sze-tsen Hu) Parallelisability of principal fibre bundles. *Trans. Amer. Math. Soc.* **67** (1949) 304–309.

1950

[45] (with E. H. Spanier). The homology structure of fiber bundles. *Proc. Nat. Acad. Sci. USA*, **36** (1950) 248–255.

[46] Differential geometry of fiber bundles. *Proc. Int. Congr. Math.* (1950) II 397–411.

1951

[47] (with E. H. Spanier). A theorem on orientable surfaces in four-dimensional space. *Comm. Math. Helv.* **25** (1951) 205–209.

1952

[48] On the kinematic formula in the Euclidean space of *n* dimensions. *Amer. J. Math.* **74** (1952) 227–236.

[49] (with C. Chevalley). Elie Cartan and his mathematical work. *Bull. Amer. Math. Soc.* **58** (1952) 217–250.

[50] (with N. H. Kuiper) Some theorems on the isometric imbedding of compact Riemannian manifolds in Euclidean space. *Ann. of Math.* **56** (1952) 422–430.

1953

[51] On the characteristic classes of complex sphere bundles and algebraic varieties. *Amer. J. of Math.*, **75** (1953) 565–597.

[52] Some formulas in the theory of surfaces. *Boletin de la Sociedad Matematica Mexicana*, **10** (1953) 30–40.

[53] Some relations between Riemannian and Hermitian geometries. *Duke Math. J.*, **10** (1953) 575–587.

1954

[54] Pseudo-groupes continus infinis *Colloque de Geom. Diff., Strasbourg* (1954) 119–136.

[55] (with P. Hartman and A. Wintner) On isothermic coordinates. *Comm. Math. Helv.* **28** (1954) 301–309.

1955

[56] Le géométrie des sous-variétés d'un espace euclidien à plusieurs dimensions. *l'Ens. Math.*, **40** (1955) 26–46.

[57] An elementary proof of the existence of isothermal parameters on a surface. *Proc. AMS*, **6** (1955) 771–782.

[58] On special *W*-surfaces. *Proc. AMS.*, **6** (1955) 783–786.

[59] On curvature and characteristic classes of a Riemann manifold. *Abh. Math. Sem. Hamburg* **20** (1955) 117–126.

1956

[60] Topology and differential geometry of complex manifolds. *Bull. AMS*, **62** (1956) 102–117.

1957

[61] On a generalization of Kähler geometry. *Lefschetz jubilee volume*. Princeton Univ. Press (1957) 103–121.

[62] (with R. Lashof) On the total curvature of immersed manifolds. *Am. J. of Math.*, **79** (1957) 306–318.
[63] (with F. Hirzebruch and J-P. Serre) On the index of a fibered manifold. *Proc. AMS*, **8** (1957) 587–596.
[64] A proof of the uniqueness of Minkowski's problem for convex surfaces. *Am. J. of Math.*, **79** (1957) 949–950.

1958

[65] Geometry of submanifolds in complex projective space. *Symposium International de Topologia Algebraica* (1958) 87–96.
[66] (with R. Lashof) On the total curvature of immersed manifolds II. *Michigan Math. J.* 5 (1958) 5–12.
[67] Differential geometry and integral geometry. *Proc. Int. Congs. Math. Edinburgh* (1958) 441–449.

1959

[68] Integral formulas for hypersurfaces in Euclidean space and their applications to uniqueness theorems. *J. of Math. and Mech.*, **8** (1959) 947–956.

1960

[69] (with J. Hano and C. C. Hsiung) A uniqueness theorem on closed convex hypersurfaces in Euclidean space. *J. of Math. and Mech.* **9** (1960) 85–88.
[70] Complex analytic mappings of Riemann surfaces I. *Amer. J. of Math.*, **82** (1960) 323–337.
[71] The integrated form of the first main theorem for complex analytic mappings in several complex variables. *Annals of Math.* **71** (1960) 536–551.
[72] Geometrical structures on manifolds. *Amer. Math. Soc. Pub.* (1960) 1–31.
[73] La géométrie des hypersurfaces dans l'espace euclidean. *Seminaire Bourbaki*, No. **193** (1959–1960).
[74] Sur les métriques Riemanniens compatibles avec une reduction du groupe structural. *Séminaire Ehresmann*, January 1960.

1961

[75] Holomorphic mappings of complex manifolds. *L'Ens. Math.* **7** (1961) 179–187.

1962

[76] Geometry of a quadratic differential form. *J. of SIAM* **10** (1962) 751–755.

1963

[77] (with C. C. Hsiung) On the isometry of compact submanifolds in Euclidean space. *Math. Annalen* **149** (1963) 278–285.
[78] Pseudo-riemannian geometry and the Gauss–Bonnet formula. *Academa Brasileira de Ciencias* **35** (1963) 17–26.

1965

[79] Minimal surfaces in an Euclidean space of N dimensions. *Differential and Combinatorial Topology*, Princeton Univ. Press (1965) 187–198.
[80] (with R. Bott) Hermitian vector bundles and the equidistribution of the

zeroes of their holomorphic sections. *Acta. Math.* **114** (1965) 71–112.

[81] On the curvatures of a piece of hypersurface in Euclidean space. *Abh. Math. Sem. Hamburg* **29** (1965) 77–91.

[82] On the differential geometry of a piece of submanifold in Euclidean space. *Proc. of U.S. Japan Seminar in Differential Geometry* (1965) 17–21.

1966

[83] Geometry of *G*-structures. *Bulletin AMS* **72** (1966) 167–219.

[84] On the kinematic formula in integral geometry. *J. of Math. and Mech.* **16** (1966) 101–118.

[85] Geometrical structures on manifolds and submanifolds. *Some Recent Advances in Basic Sciences*, Yeshiva University Press (1966) 127–135.

1967

[86] (with Robert Osserman) Complete minimal surfaces in Euclidean *n*-space. *J. de l'Analyse Math.* **19** (1967) 15–34.

[87] Einstein hypersurfaces in a Kählerian manifold of constant holomorphic curvature. *J. Diff. Geom.* **1** (1967) 21–31.

1968

[88] On holomorphic mappings of Hermitian manifolds of the same dimension. *Pro. Sym. Pure Math.* **11**. Entire Functions and Related Parts of Analysis (1968) 157–170.

1969

[89] Simple proofs of two theorems on minimal surfaces. *L'Ens. Math.* **15** (1969), 53–61.

1970

[90] (with H. Levine and L. Nirenberg) Intrinsic norms on a complex manifold. *Global analysis*, Princeton Univ. Press (1970) 119–139.

[91] (with M. do Carmo and S. Kobayashi) Minimal submanifolds on the sphere with second fundamental form of constant length. *Functional Analysis and Related Fields*, Springer-Verlag (1970) 59–75.

[92] (with R. Bott) Some formulas related to complex transgression. *Essays on Topology and Related Topics*, Springer-Verlag, (1970) 48–57.

[93] Holomorphic curves and minimal surfaces. *Carolina Conference Proceedings* (1970) 28 pp.

[94] On minimal spheres in the four sphere, Studies and Essays Presented to Y. W. Chen, Taiwan, (1970), 137–150.

[95] Differential geometry: Its past and its future. *Actes, Congrès intern. Math.*, 1970, 1, 41–53.

[96] On the minimal immersions of the two-sphere in a space of constant curvature. *Problems in analysis*, Princeton Univ. Press, 1970, 27–40.

[97] Brief survey of minimal submanifolds. *Differentialgeometrie im Grossen, Berichte aus dem Forschungsinstitut Oberwolfach*, Edited by W. Klingenberg, (1970) 43–60.

1971

[98] (with James Simons) Some cohomology classes in principal fibre bundles and their application to Riemannian geometry. *Proc. NAS*, **68** (1971) 791–794.

1972

[99] Holomorphic curves in the plane. *Differential geometry in honor of K. Yano*, (1972), 73–94.

[100] Geometry of characteristic classes. *Proc. 13th Biennial Sem. Canadian Math. Congress*, 1–40 (1972). Also pub. in Russian translation.

1973

[101] Meromorphic vector fields and characteristic numbers. *Scripta Math.* **29** (1973) 243–251.

[102] The mathematical works of Wilhelm Blaschke. *Abh. Math. Sem. Hamburg* **39** (1973) 1–9.

1974

[103] (with James Simons) Characteristic forms and geometrical invariants. *Annals of Math.* 99 (1974) 48–69.

[104] (with M. Cowen, Al Vitter) Frenet frames along holomorphic curves. *Proc. of conf. on value distribution theory, Tulane Univ.* (1974) 191–203.

[105] (with J. Moser) Real hypersurfaces in complex manifolds. *Acta Math.* **133** (1974) 219–271.

1975

[106] (with S. I. Goldberg) On the volume-decreasing property of a class of real harmonic mappings. *Amer. J. of Math.* **97** (1975) 133–147.

[107] On the projective structure of real hypersurfaces in C_{n+1}. *Math. Scandanavia*, **36** (1975) 74–82.

1976

[108] (with J. White) Duality properties of characteristic forms. *Inv. Math.* **35** (1976) 285–297.

1977

[109] Circle bundles. *Geometry and Topology, III* Latin Amer. School of Math, Lecture Notes in Math. Springer-Verlag, no. 597 (1977), 114–131.

1978

[110] (with P. A. Griffiths) Linearization of webs of codimension one and maximum rank. *Proc. Int. Symp. on Algebraic Geometry, Kyoto 1977*, (1978),

[111] Herglotz's work on geometry. to appear in *Ges. Abh. Herglotz*.

[112] On projective connections and projective relativity. *Science of Matter*, dedicated to Ta-You Wu, 1978, 225-232.

[113] (with P.A. Griffiths) Abel's theorem and webs. *Jahresberichte deut. Math. Ver.* **80** (1978), 13-110.

[114] (with P. A. Griffiths) An inequality for the rank of a web and webs of maximum rank. To appear in *Annali Sc. Norm. Pisa.*
[115] Affine minimal hypersurfaces. Proc. *US-Japan Seminar on Minimal Submanifolds*, Tokyo 1978, 1-14.
[116] (with Chuu-Lian Terng) An analogue of Bäcklund's theorem in affine geometry. To appear in *Rocky Mountain Math. J.*
[117] From triangles to manifolds. To appear in *Amer. Math. Monthly.*

List of Ph.D. Theses Written Under the Supervision of S. S. Chern

I. *At the University of Chicago*

1. Nomizu, Katsumi, Invariant affine connections on homogeneous spaces. June 1953
2. Auslander, Louis, Contribution to the curvature theory of Finsler spaces. June 1954
3. Liao, San Dao, On the theory of obstructions of fiber bundles. March 1955
4. Spanier, Jerome, Contributions to the theory of almost complex manifolds. September 1955
5. Rodrigues, Alexandre, Characteristic classes of homogeneous spaces. March 1957
6. Hertzig, David, On simple algebraic groups. August 1957
7. Levine, Harold I., Contributions to the theory of analytic maps of complex manifolds into projective space. August 1957
8. Suzuki, Haruo, On the realization of Stiefel–Whitney characteristic classes by submanifolds. August 1957
9. Wolf, Joseph Albert, On the manifolds covered by a given compact, connected Riemannian homogeneous manifold. December 1959
10. Petridis, Nicholas C., Quasiconformal mapping and pseudo-meromorphic curves. June 1961.

II. *At the University of California at Berkeley*

1. Pohl, William Francis, Differential geometry of higher order. September 1961
2. Do Carmo, Manfredo Perdigao, The cohomology ring of certain Kählerian manifolds. January 1963
3. Amaral, Leo Huet. Hypersurfaces in non-Euclidean spaces. June 1964
4. Banchoff, Thomas Francis, Tightly embedded two-dimensional polyhedral manifolds. June 1964
5. Garland, Howard, On the cohomology of lattices in Lie groups. June 1964
6. Gardner, Robert Brown, Differential geometric methods in partial differential equations. June 1965
7. Smoke, William, Differential operators on homogeneous spaces. June 1965

8. Weinstein, Alan David, The cut locus and conjugate locus of a Riemannian manifold. March 1967
9. Shiffman, Bernard, On the removal of singularities in several complex variables. June 1968
10. Reilly, Robert, The Gauss map in the study of submanifolds of spheres. September 1968
11. Wolf, R., Some integral formulas related to the volume of tubes. September 1968
12. Eisenman, D., Intrinsic measures on complex manifolds and holomorphic mappings. June 1969
13. Jordan, Steve, Some invariants for complex manifolds. September 1970
14. Leung, Dominic, Deformations of integrals of exterior differential systems. September 1970
15. Lai, Hon-Fei, Characteristic classes of real manifolds immersed in complex manifolds. June 1971
16. Yau, Shing Tung, On the fundamental group of compact manifolds of non-positive curvature. June 1971
17. Barbosa, Lucas, On the minimal immersions of S^2 in S^{2m}. December 1972
18. Bleecker, David, Contributions to the theory of surfaces. June 1973
19. Millson, John James, Chern–Simons invariants of constant curvature manifolds. December 1973
20. Simoes, Plinio, A class of minimal cones in $\mathbb{R}^n$, $n \leqslant 8$, that minimize area. December 1973
21. Cheng, Shiu-Yuen, Spectrum of the Laplacian and its applications to differential geometry. June 1974
22. Donnelly, Harold, Chern–Simons invariants of reductive homogeneous spaces. June 1974
23. Webster, Sidney Martin, Real hypersurfaces in complex spaces. June 1975
24. Sung, C. H., Contributions to holomorphic curves in complex manifolds. December 1975
25. Dunham, Douglas, Holomorphic and meromorphic vector fields on compact hermitian symmetric spaces. 1976
26. Faran, James, Segre families and real hypersurfaces. June 1978